AF832234

Springer Series on Atomic, Optical, and Plasma Physics

Volume 120

Editor-in-Chief

Gordon W. F. Drake, Department of Physics, University of Windsor, Windsor, ON, Canada

Series Editors

James Babb, Harvard-Smithsonian Center for Astrophysics, Cambridge, MA, USA

Andre D. Bandrauk, Faculté des Sciences, Université de Sherbrooke, Sherbrooke, QC, Canada

Klaus Bartschat, Department of Physics and Astronomy, Drake University, Des Moines, IA, USA

Charles J. Joachain, Faculty of Science, Université Libre Bruxelles, Bruxelles, Belgium

Michael Keidar, School of Engineering and Applied Science, George Washington University, Washington, DC, USA

Peter Lambropoulos, FORTH, University of Crete, Iraklion, Crete, Greece

Gerd Leuchs, Institut für Theoretische Physik I, Universität Erlangen-Nürnberg, Erlangen, Germany

Alexander Velikovich, Plasma Physics Division, United States Naval Research Laboratory, Washington, DC, USA

The Springer Series on Atomic, Optical, and Plasma Physics covers in a comprehensive manner theory and experiment in the entire field of atoms and molecules and their interaction with electromagnetic radiation. Books in the series provide a rich source of new ideas and techniques with wide applications in fields such as chemistry, materials science, astrophysics, surface science, plasma technology, advanced optics, aeronomy, and engineering. Laser physics is a particular connecting theme that has provided much of the continuing impetus for new developments in the field, such as quantum computation and Bose-Einstein condensation. The purpose of the series is to cover the gap between standard undergraduate textbooks and the research literature with emphasis on the fundamental ideas, methods, techniques, and results in the field.

More information about this series at https://link.springer.com/bookseries/411

Igor Girka · Manfred Thumm

Surface Flute Waves in Plasmas

Eigenwaves, Excitation, and Applications

Second Edition

Springer

Igor Girka
Education and Research Institute
School of Physics and Technology
V. N. Karazin Kharkiv National University
Kharkiv, Ukraine

Manfred Thumm
Institute for Pulsed Power and Microwave Technology
Karlsruhe Institute of Technology
Karlsruhe, Baden-Württemberg, Germany

ISSN 1615-5653 ISSN 2197-6791 (electronic)
Springer Series on Atomic, Optical, and Plasma Physics
ISBN 978-3-030-98209-6 ISBN 978-3-030-98210-2 (eBook)
https://doi.org/10.1007/978-3-030-98210-2

1st edition: © Springer International Publishing Switzerland 2014
2nd edition: © The Editor(s) (if applicable) and The Author(s), under exclusive license to Springer Nature Switzerland AG 2022
This work is subject to copyright. All rights are solely and exclusively licensed by the Publisher, whether the whole or part of the material is concerned, specifically the rights of translation, reprinting, reuse of illustrations, recitation, broadcasting, reproduction on microfilms or in any other physical way, and transmission or information storage and retrieval, electronic adaptation, computer software, or by similar or dissimilar methodology now known or hereafter developed.
The use of general descriptive names, registered names, trademarks, service marks, etc. in this publication does not imply, even in the absence of a specific statement, that such names are exempt from the relevant protective laws and regulations and therefore free for general use.
The publisher, the authors and the editors are safe to assume that the advice and information in this book are believed to be true and accurate at the date of publication. Neither the publisher nor the authors or the editors give a warranty, expressed or implied, with respect to the material contained herein or for any errors or omissions that may have been made. The publisher remains neutral with regard to jurisdictional claims in published maps and institutional affiliations.

This Springer imprint is published by the registered company Springer Nature Switzerland AG
The registered company address is: Gewerbestrasse 11, 6330 Cham, Switzerland

Preface

A comprehensive study of the properties of electromagnetic eigenwaves propagating perpendicular to the axis of symmetry in various cylindrical waveguide-structures filled with magneto-active plasma is carried out in the present book, using the flute waves as an example. This monograph is the second, updated and significantly expanded edition of our book *Surface Flute Waves in Plasmas: Theory and Applications*, published in 2014 as Volume 79 in the "Springer Series on Atomic, Optical, and Plasma Physics". "Surface waves" means that these waves only propagate along plasma surfaces and not in uniform infinite plasmas. Their wave amplitudes decrease with going away from the plasma boundary into the plasma depth. "Flute" means that the axial wavenumbers of the waves in plasma cylinders are assumed to be zero, and the waves only propagate in azimuthal direction. In this case, the surfaces of constant density resemble fluted Greek columns.

The objective of the authors is to collect and update in a single book the materials devoted to various characteristics of these transverse surface flute waves. First of all, their dispersion properties are investigated with taking into account the impact of various design features of waveguides, like possible application of a dielectric coating on the inner wall of a cylindrical waveguide or the utilization of plasma filled coaxial waveguides. Their damping due to various mechanisms, like collisions between plasma particles, resonant damping, transformation into the coupled bulk (volume) mode, and radiation through a narrow slot antenna in the waveguide wall are also considered here. Since a real experimental device confines a non-uniform plasma, the impacts of an inhomogeneity of the plasma density and of an external static magnetic field on the frequency spectrum and the spatial field distribution of these waves are studied here as well. Change in the surface flute wave properties due to the influence of the shape of the waveguide cross-section including non-circularity of both constructive elements, namely the plasma column and the metal waveguide, and the coupling between different flute modes propagating in a current-carrying plasma waveguide are also in the scope of the book. We have even demonstrated how one can investigate the properties of surface waves, which in addition to their azimuthal wavenumber have also a small axial wavenumber, using the theory of surface flute waves as zeroth approach. It should be mentioned that all problems considered in this book are solved by the method of successive approximations. It is surprising that the theory of surface waves propagating strictly across the axis of symmetry in relatively simple models of cylindrical waveguides allows one to solve the problems of transverse wave propagation in squared and rectangular waveguides, and to develop the theory of azimuthally non-symmetric surface waves, which have a small axial wavenumber. Therefore, the material presented in this book will be useful for graduate students specializing in the field of plasma physics and radiophysics as well as for professionals who are interested in problems of specifically confined plasma structures and their applications.

This second edition of the monograph distinguishes from the first one by including new chapters based on our papers published after 2013. These chapters are devoted to the following two issues. First, the surface waves are demonstrated to rotate their energy around the axis of the waveguides. Second, excitation of these waves by a charged particle beam (in most cases these are electrons) gyrating around the axis is studied in detail.

Interest in plasma physics, observed over the last 60 years, is maintained primarily due to the prospects of energy generation by controlled thermonuclear fusion and owing to astrophysical studies. However, research for solutions in these complex fields also leads to the active development of various applied investigations in related areas of science such as plasma electronics, physics of gas discharges, collective effects of charge carriers in plasmas of solids, plasma materials processing, plasmonics, and nano-science. The plasmas, which are investigated in these fields of science in order to find some new prospective applications, have different values of characteristic parameters, and hence their properties are different. However, common for any plasma is that all internal processes are accompanied by the propagation of electromagnetic waves. Another unifying feature of plasma phenomena observed on the Earth is the limited space, which is occupied by a plasma. These two factors are common for all problems, which are discussed in this book, and which are belonging to electrodynamics of bounded magneto-active plasmas. This circumstance distinguishes our book from well-known fundamental monographs on plasma electrodynamics. Another fact, which testifies the importance of this book, is connected with the wide field of the presented existing applications of surface wave propagation and prospective ones, namely fabrication and processing of large-sized semiconductor wavers using gas discharges sustained by azimuthally non-symmetric surface waves, production of nano-crystals with outstanding mechanical and electrical properties, elaboration of new biosensors, which allow one to carry out diagnostics of animal and plant pathogens, to carry out in situ gene analysis or rapid testing of a malarial strain, water purity analysis, etc.

The theoretical study of these transverse eigenwaves at Kharkiv V. N. Karazin National University has been started by Prof. A. Kondratenko, who was the scientific advisor of Volodymyr Girka, the first author of the first edition of this monograph. Regrettably, V. Girka passed away on November 3, 2015, when the first edition was already published and the authors were invited to prepare this second edition.

One of the co-authors, Igor Girka is indebted to the German Academic Exchange Service (DAAD) that allowed him to take part in the final preparation of the text for publication as a book at the Karlsruhe Institute for Technology (KIT) through the grant No. 57552334. He is also very grateful to John Jelonnek and colleagues in the High-Power Microwave Division of the Institute for Pulsed-Power and Microwave Technology (IHM) at KIT for a lot of fruitful discussions and their kind hospitality.

Kharkiv, Ukraine Igor Girka
Karlsruhe, Germany Manfred Thumm

Contents

Chapter 1
Introduction

1.1 General Notes

Considering the possible content of this second edition of our book [1], we have again paid the main attention to the following features. First, this should be new material, which has not been completely and consistently presented in previously published books of other authors. Second, this should be a presentation of theoretical results, which have some practical applications at present time and which would be prospective for future utilizations in new branches of physics and related areas of knowledge. Finally at third, this book should be interesting and useful for young students who would like to study plasma physics and its applications.

This book is devoted to studying the properties of surface flute waves propagating in magneto-active plasma with a finite curvature radius of its boundary. The applications of surface waves are known during a long time. They were described for the first time by Lord Rayleigh. In [2], one can find a classification of different types of surface waves. For the present considerations a specific type of surface waves has been selected. The spatial distribution of their electromagnetic fields at the plasma-dielectric interface has a surface nature only in the region occupied by the plasma. In the dielectric region it is of bulk (volume) type. This is the first peculiarity of the material presented in this book. Because of the leakage of wave energy into the dielectric region, metal waveguides with plasma filling are considered here.

The second peculiarity of our book is connected with the geometry of the magneto-active waveguides, in which these waves can propagate. Most of monographs and reviews, where surface waves are considered, make their theoretical study for the case of planar geometry, or they apply either cylindrical geometry inconsistently, or they consider only the case of non-magnetized plasmas [3–6].

The next distinguishing feature of our book is connected with the fact that just flute waves are investigated. This means that these electromagnetic perturbations have no axial wavenumber. From one point of view this peculiarity allows one to separate the Maxwell equations (in the limiting case of uniformity of the considered plasma system) into two independent sets. On the other side most papers in the field

© The Author(s), under exclusive license to Springer Nature Switzerland AG 2022
I. Girka and M. Thumm, *Surface Flute Waves in Plasmas*, Springer Series on Atomic, Optical, and Plasma Physics 120, https://doi.org/10.1007/978-3-030-98210-2_1

of plasma physics are devoted to the opposite case, namely to studying azimuthally symmetric electromagnetic waves. However, in any case, the scope of our book is just restricted to properties of transverse non-symmetrical eigenwaves. Nevertheless, in some sections it is demonstrated that the developed theory can be also successfully applied for consideration of surface waves with a non-zero axial wavenumber.

The fourth distinguishing feature of the presented material is connected with the fact that here only the case of magneto-active plasmas is considered. Indicating this one should mention a physical incident, which happened with the 1962 Noble Prize winner in Physics L. Landau and his scholar, A. Akhiezer (everyone who is interested in plasma physics knows the fundamental monograph devoted to electrodynamics of plasmas, which was written by a team of physicists led by Akhiezer [7]). Lev Landau worked for a long time in Kharkiv, where he has created the kernel of his huge theoretical school. This story was told to students of Kharkiv University by old physicists from the Ukrainian Institute for Physics and Technology (present title is National Science Center "Kharkiv Institute of Physics and Technology"). Once A. Akhiezer presented his work to L. Landau. The latter read it and asked the question to his scholar: "What is the strange choice of the object of your investigation? Alexander, where have you seen any plasma immersed into a magnetic field?" Maybe it was a joke; we were not present at this dialog. But nevertheless, our book deals with just different cases, where the plasma is influenced by an external static magnetic field.

The fifth peculiarity of the book is connected with the applied theoretical method. Using the method of successive approximations, we have solved a lot of tasks devoted to the investigation of the influence of plasma density non-uniformity, different types of external magnetic field non-uniformity, spatial inhomogeneity of plasma interfaces, etc. on the frequency spectrum of the extraordinary polarized surface flute waves (X-modes). Mainly these modes are considered because the corresponding ordinary polarized waves (O-modes) in the most cases have a bulk feature, and their dispersion properties do not depend on the external static magnetic field. Thus, the book will be useful for students, who want to study such a popular analytical method like the method of successive approximations, which is widely used not only in plasma physics but also in other branches of physics.

Plasma filled waveguides are actively studied both theoretically and experimentally for the purposes of plasma physics [8], radiophysics [9], and plasma electronics [10]. Experimental investigations of various plasma filled electronic devices show prospective applications of surface wave propagation, and aim at increasing their efficiency. During recent years plasmas are widely used in various designs of antenna systems. It has been proved experimentally that plasma antennas have a lot of advantages compared with conventional metal antenna systems [11]. This is not only valid for military applications, but also for civil communication systems. Moreover, the range of their civil applications becomes more and more wide at present time [9]. Waves being excited in bounded plasma waveguide structures should be eigenwaves of such structures. Surface waves can propagate not only in the case of gaseous plasmas but also in the case of solid state plasmas, see e.g. [12, 13]. Moreover, they can propagate in multilayered planar waveguide structures where the layers can be produced from meta-materials with different electro-dynamical properties.

The possibility of surface wave propagation along a flat plasma-metal boundary was mentioned for the first time in [14] and after that their properties have been actively studied by many other authors. A review devoted to the properties of surface waves propagating along a planar plasma-metal boundary across an external static magnetic field is given in [6]. However, in experimental practice one often deals with a plasma boundary, which is characterized by a curvature radius of finite magnitude and this circumstance strongly affects the properties of surface waves propagating in such plasma-waveguide structures. The finite curvature radius causes changes of the frequency ranges where these waves can propagate, of the spatial distribution of their fields, of their polarization and so on, see e.g. [15].

The propagation of surface waves (SWs) is also widely used in various plasma technologies, which employ gas discharges for cleaning and polishing solids, for laser pumping and plasma production, elaboration of fullerenes, etc. This can be explained by the fact that these waves possess a lot of advantages over bulk waves. For instance, they can produce plasmas with spatially more uniform density profile, and the efficiency of their interaction with plasma particles is also higher [16]. Surface localized plasmons (special type of surface waves connected with perturbations of conducting electrons in metals) are actively applied for formation of nano-structures using metal plasmas. Progress in elaboration of biosensors based on the surface plasmon resonance technique indicates new prospective areas of surface-wave applications.

At present time the world community is actively searching for new sources of continuously available energy. The International Thermonuclear Experimental Reactor (ITER) is being built in Cadarache, France. Surface waves can propagate in such fusion devices as well. Their propagation has a double role. On the one hand their excitation in a magnetic confinement fusion device leads to losses of RF power intended for plasma production and heating of the fusion plasma core, to impurity generation and to other undesirable events. However, on the other hand, they could be utilized for cleaning the inner surface of the metallic reactor vessel.

1.2 Historical Notes

To our best knowledge, Robert S. Elliott from Hughes Research and Development Laboratories, California, US [17], was the first who introduced the term "azimuthal surface waves" (ASWs) to designate electromagnetic waves $\propto \exp(im\,\varphi - i\omega t)$ which propagate in cylindrical waveguide structures perpendicular to the axis of symmetry along the azimuthal angle φ. Here, m is the azimuthal mode index (a natural number), and ω is the angular frequency of the ASWs. In other words, these waves propagate with zero axial wave vector, $k_{\|} = 0$. "Surface" means, that the waves do not propagate in a homogeneous infinite medium and that the presence of a boundary between the media with different physical properties is required for their existence. The study [17] was initiated by its potential application in the field of transmission lines and antennas. The following metal-dielectric structure was under consideration (see Fig. 1.1). The internal space, $r < R_1$, was occupied by a perfect metallic conductor with inner radius

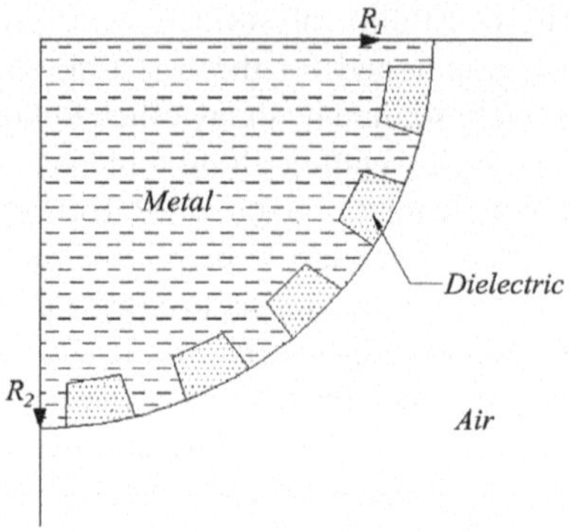

Fig. 1.1 Schematic of the problem investigated in [17]

R_1 with teeth, which extent to the outer radius R_2. The space between the teeth was assumed to be composed of a pure dielectric, which could be air. The outer space, $r > R_2$, was air. Neither a plasma nor a static magnetic field were applied. The theoretical predictions were approved by a coarse experiment.

However, the first attempt to study surface wave propagation over a coated conductor with small cylindrical curvature in the direction of propagation was reported by Kazuo Horiuchi from Laboratory of Communication Instruments, Tokyo, Japan [18]. The outline of how to study the wave fields in this case was suggested on the base of the system of plane conductors studied earlier. The schematic of the waveguide structure, perfect metal-dielectric-air, is shown in Fig. 1.2. The width of the dielectric layer was assumed to be equal to a. Again, the waveguide structure did not include neither a plasma nor a static magnetic field. The spatial distribution of the surface wave fields was derived in the approximation of small curvature in ξ direction.

James R. Wait from Central Radio Propagation Laboratory, National Bureau of Standards, Colorado, US [19, 20], was the first who studied ASWs in cylindrical plasma waveguide structures. The objective of the research was to give a possible

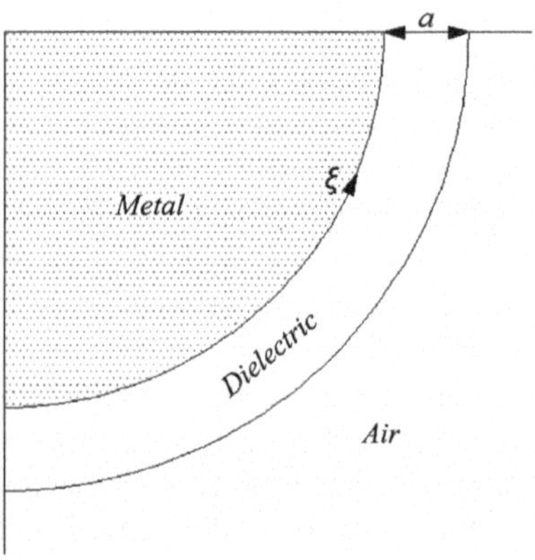

Fig. 1.2 Schematic of the problem investigated in [18]

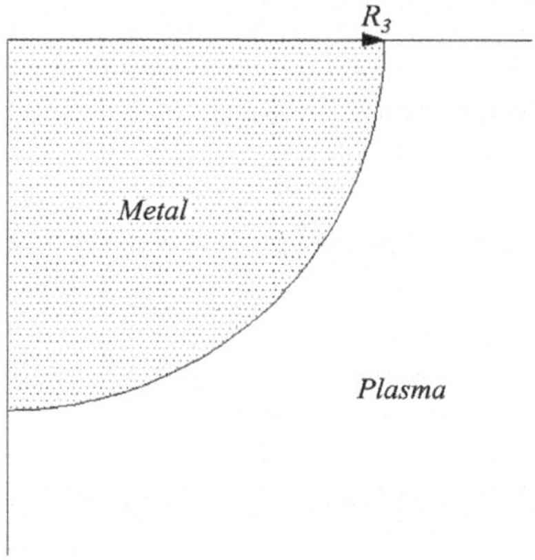

Fig. 1.3 Schematic of the problem studied in [19, 20]

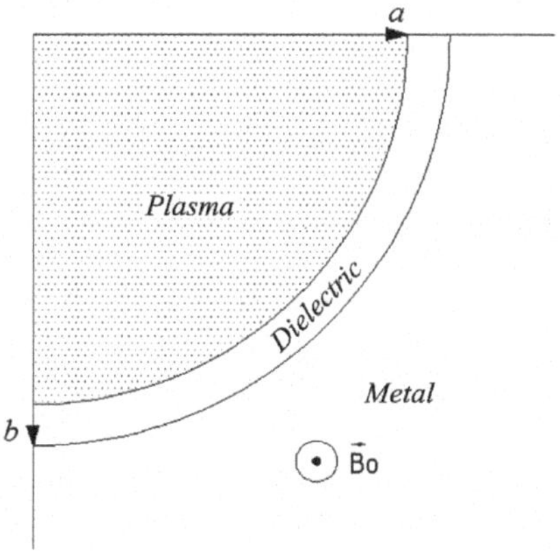

Fig. 1.4 Schematic of the problem investigated in [21, 22]

explanation for the propagation of radio waves in the spherical Earth-ionosphere system. In [19], electromagnetic waves were investigated in the following plasma-metal structure (see Fig. 1.3). An infinitely long circular metal cylinder (perfect conductor) with radius R_3 was immersed into the compressible plasma. A static magnetic field was neglected. It was concluded that in the neighborhood of the plasma interface the field amplitudes of ASWs can be comparable with, or are even much larger than those of the radiating modes.

Anatoliy M. Kondaratenko from V.N.Karazin Kharkiv National University, Kharkiv, Ukraine, with co-authors was the first in the Soviet Union to study ASWs in cylindrical plasma-dielectric-metal structures both in presence of a static axial magnetic field [21] and without any magnetic field [22]. The research was initiated by possible application in the area of plasma electronics. In these papers, the title "surface" got a new meaning to designate that the wave amplitude decreases when going from the plasma interface into the plasma core rather than oscillates as it takes place in the case of volumetric waves.

The following waveguide structure was under study (see Fig. 1.4). A homogeneous, cold, collision-less plasma column with radius a was placed inside the metal chamber with radius b. The space, $a < r < b$, was filled by a dielectric with dielectric constant ε_d. The frequency ranges were determined where electromagnetic plasma waves had surface nature. The spatial distribution of the ASW fields was studied, explicit expressions for the ASW eigenfrequencies in various limiting cases were derived, and the dependencies of the ASW eigenfrequencies on the parameters of the plasma waveguide were studied numerically.

Introduction of the plasma insert makes it possible to control the dispersion properties of the electromagnetic waves via the plasma particle density, by an external static magnetic field, and via the dimensions of both plasma column and dielectric layer. The advantages of application of ASWs for designing plasma electronic devices are as follows. The azimuthal direction of wave propagation allows to design plasma electronics devices with small axial size. The surface nature of ASWs makes it possible to reach higher efficiency of producing and maintaining gas discharges [23]. In addition, ASW excitation is possible by an electron beam gyrating around the plasma column along large Larmor orbits [24–26]. This possibility is explained by the absence of the necessity to produce a wave structure in the plasma core. The surface nature of ASWs allows also to operate in the frequency range where the plasma is non-transparent for electromagnetic waves.

If the metal waveguide is entirely filled with plasma, ASWs are unidirectional, i.e. in a definite frequency range ASWs propagate in only one direction (with only one sign of the azimuthal mode index) [27]. The plasma-metal structure considered in [27] can be easily obtained from that shown in Fig. 1.4 by removing the dielectric layer (putting $a = b$).

The surface nature of ASWs makes it possible to replace the internal space of the plasma column by some other media. Nikolay A. Azarenkov from V.N.Karazin Kharkiv National University, Kharkiv, Ukraine, with co-authors proposed to place an internal perfect conductor into the structure studied in [27]. In this case, one deals with a coaxial plasma layer bordered by the inner and outer perfect metal conductors [28, 29]. ASWs propagate in opposite directions along the opposite borders of the coaxial waveguide.

1.3 Brief Overview of the Content

Chapter 2 of this book is entitled "Surface flute waves propagating in uniform magneto-active plasma filled waveguides". Here the influence of the various waveguide parameters on the properties of surface flute waves (SWFs) is examined under the condition of uniform plasma filling. The external static magnetic field is assumed to be uniform and directed strictly along the axis of the cylindrical waveguide. Application of such a simple waveguide model allows one to study properties of SFWs in a modal approximation. The main attention is paid to the properties of X-modes, because the flute O-modes are often of bulk type.

The flute X-modes are shown to propagate along both plasma-dielectric and plasma-metal boundaries. The problems of SFW damping are also studied here. It is proved as well that application of the SFW theory allows one to describe dispersion properties even of the non-symmetrical surface eigenwaves, which have a finite magnitude of the axial wavenumber. Higher radial modes are known to propagate at higher frequencies and shorter wavelengths compared to those of the zeroth mode, a feature which is of interest for practical applications. That is why higher radial modes of SFWs are studied in the present book as well.

The theory of SFWs developed in this Chapter is applied in the following chapters for solving corresponding problems by employing the method of successive approximations.

Chapter 3 is devoted to analyzing the electromagnetic energy transfer initiated by SFWs in cylindrical plasma waveguides. In University textbooks on electrodynamics, mostly propagation of plane electromagnetic waves in an infinite uniform non-dissipative medium is studied. In this case no difference between phase and group velocities can be observed. Chapter 3 presents both a short historical overview of the problem of electromagnetic energy transfer by electromagnetic waves and application of the fundamental knowledge to the problem of energy transfer in cylindrical plasma waveguides initiated by SFWs. Four different versions of the waveguides are considered. Appropriate physical observables like angular phase velocity, angular group velocity, and angular velocity of energy transfer are introduced and applied for studying the problem. The angular velocity of energy transfer is demonstrated to decrease with increasing absolute value of azimuthal mode index.

Construction of complicated plasma devices unavoidably leads to the appearance of more complicated plasma phenomena, for example, to coupling between different electromagnetic waves. Experimentalists often deal with plasmas, which are either sustained by a flow of electric current, or when the current flowing in the confined plasma is necessary for plasma equilibrium and stability (magnetic confinement). For instance, in tokamaks the plasma is confined both by toroidal and poloidal magnetic fields, the last one just being mainly produced by the axial (toroidal) plasma current. That is why the case of SFWs propagating in current-carrying plasmas is also investigated. In tokamaks and in many other applications, the value of the axial electric current is sufficiently small to take the strength of the generated external static azimuthal magnetic field as the small parameter of the studied problem. Thus, the present research is restricted to the case of small values of this azimuthal magnetic field.

Chapter 4 considers coupled SFWs, which propagate in current-carrying cylindrical plasma waveguides. The need for studying such waveguides is explained by the following issue. To produce gaseous plasmas in laboratory devices one can utilize a DC electric current. This is why the influence of an axial electric current on the dispersion properties of SFWs propagating in waveguides with non-uniform plasma filling is considered in this Chapter. The magnitude of this electric current is assumed to be sufficiently small, so that either the azimuthal component of the external magnetic field is much smaller than its axial component or the cyclotron frequency, which appears due to the azimuthal magnetic field, is much lower than the eigenfrequency of the flute modes in the case that there is no axial static magnetic field. Four different geometries of the utilized waveguide structure are studied there. They are: plasma-filled cylindrical metal waveguide, which has a dielectric coating on its inner surface; cylindrical metal rod immersed into magnetized plasma; cylindrical metal waveguide completely filled with plasma, and cylindrical coaxial waveguide completely filled with plasma.

The results of studying the propagation of SFWs in current-carrying waveguides prove the existence of coupling between O- and X-modes, where O-modes can be either of surface or bulk type. In different plasma-filled waveguides, coupling between these flute modes leads to different phenomena: appearance of non-reciprocal frequency ranges in their frequency spectrum, change of the sign of their group velocity, damping of the X-modes and so on. The power transfer between these flute modes is characterized by the nonlinear dependence on the small parameter of the problem, which is proportional to the electric current.

The fifth Chapter of this book is devoted to investigation of dispersion properties of SFWs while they propagate across the axis of symmetry in cylindrical metal waveguides with non-uniform magneto-active plasma filling. The non-uniformity can be connected with a radial dependence of the plasma particle density or with a spatial non-uniform distribution of the applied magnetic field. This could be a radial non-uniformity of the external magnetic field, a toroidal magnetic field non-uniformity utilized in metal waveguides, which are entirely filled with plasma, or a non-uniform toroidal magnetic field produced along a metal ring, and a weakly rippled magnetic field.

There the frequencies of SFWs and the spatial distribution of their fields are examined. The damping rates, which are connected both with collisions between the plasma particles and resonant conversion of these modes into bulk modes, are calculated as well. It is interesting that from the mathematical point of view radial density non-uniformity is similar to radial non-uniformity of the external magnetic field. Thus, the influences of these non-uniformities on the SFW frequency spectrum are also similar. A weak toroidicity of the external static magnetic field leads to SFWs propagation in the form of wave packets. Corrections to the wave field of their main harmonic and to their eigenfrequency are proportional to the squared small parameter of the system in this case. Ripples of the confining magnetic field in modern thermonuclear fusion machines are characterized by a small parameter $\varepsilon_m \ll 1$. Nevertheless, these ripples can influence the conversion and absorption of SFWs. In this case SFWs propagate as well as wave packets, which have six field components. The dispersion equation obtained in this Chapter contains small summands proportional to $\varepsilon_m{}^2$, thus the frequency correction is also proportional to the same value.

Chapter 6 is entitled "Surface flute waves propagating in waveguides with non-circular cross-section". The most efficient way to excite waveguide structures including those, which are corrugated, can be realized for excitation in the range of their eigenfrequencies [30]. The shape of the cross-section of the plasma-vacuum interface in fusion devices also is well-known to be not circular [8, 31]. In the case of stellarators this peculiarity of the design is particularly well seen. Using an expansion into Fourier series one can simulate any real shape of the cross-section of any plasma device. Thus, the following problems are solved in this Chapter: propagation of SFWs in metal waveguides with non-circular cross-section, which are partially filled with plasma and propagation of SFWs in a metal waveguides with non-circular cross-section, which is partially filled with plasma. The influence of a cross-section non-circularity of the plasma column on the dispersion properties of these waves propagating in magneto-active waveguides is investigated in both resonant and non-resonant cases.

Periodical changes of the curvature of the plasma column surface and/or of the metal chamber along the azimuthal angle are shown to be the reason for SFWs propagation in the form of wave packets. The SFW eigenfrequency correction, which is caused by the deviation of the waveguide cross-section from circular shape, is found to have a small magnitude of the second order over the small parameter of the problem in the non-resonant cases. But in the resonant cases, which can be realized due to degeneration of the frequency spectrum, the magnitude of the frequency correction is proportional to the first power of the small parameter of the problem. The spectral content of SFWs propagating along such waveguides is also examined here.

After detailed analysis of the SFW dispersion properties carried out in Chaps. 2–6, the next four chapters are devoted to studying the possibility of SFW excitation by charged particle beams. These chapters contain the results which were published by the authors of the present book in the years 2011–2018. All the results describe SFW excitation by an annular charged particle beam in cylindrical plasma-filled waveguides. The beam is assumed to gyrate in the external static axial magnetic field along large Larmor radii around the plasma column.

Chapter 7 deals with excitation of SFWs in the range of the electron cyclotron frequency and provides the theoretical background for the next three chapters as well as for explaining the fundamentals of the linear theory of beam-wave interaction in this case. Most of the investigated problems in the Chaps. 7–10 consider the excitation of the zeroth radial modes. This is explained by the fact that the growth rates of higher radial modes are smaller than those of the zeroth mode.

The excitation is of resonant nature. It takes place if the SFW eigenfrequency is close to cyclotron frequency harmonics. This causes the "bell" shape of plots of the growth rate versus plasma particle density and/or plasma column radius. The resonances overlap due to two mechanisms. One consists in increasing the beam particle density in the narrow gap between plasma column and waveguide wall where the beam moves such that the growth rate becomes of the order or greater than the distance between the nearest resonance frequencies. The gap which separates the plasma column from the metal wall is sufficiently narrow, that there is no room for two or more resonant beam layers with different harmonic numbers. The other mechanism consists in widening the gap where the particle flow of low particle density rotates such that more than one resonant particle orbit appears to belong to the gap.

Several peculiarities of the plasma waveguides are taken into account. They are possible toroidal inhomogeneity of the external static magnetic field, deviation of the waveguide cross-section from the circular shape, presence of dissipation, and small axial wavenumber of the excited electromagnetic waves.

The results of studying the excitation of SFWs by electron beams gyrating along large Larmor orbits in coaxial waveguides are presented in Chap. 8. The beam can be injected into the gap between the plasma coaxial layer and either the inner or the external conductor. Chapter 8 compares these two possibilities for SFWs in two frequency regions, the electron cyclotron frequency range and the range above the upper-hybrid frequency.

The nonlinear theory of SFW excitation by gyrating electron beams is studied in Chap. 9. There, the electron cyclotron frequency range is under consideration. The conditions of both resonant beam-plasma and beam-dissipative instabilities are investigated. Modulation of the charged particle beam is known to be a powerful tool for enhancing the beam-wave interaction. This explains special interest in elaborating the nonlinear theory of SFW excitation by modulated annular electron beams in this book.

SFW excitation in the electron cyclotron frequency range is of more interest from the practical point of view. However, in some cases SFWs can be efficiently excited by a gyrating electron beam in other frequency ranges as well. Chapter 10 presents the results of research on SFW excitation above the upper-hybrid frequency in waveguides partially filled by plasma including the case of a non-circular interface of the plasma column. Excitation of SFWs in the ion cyclotron frequency range by light ion beams gyrating around a plasma column is shown to be possible in strongly magnetized plasma waveguides.

Finally, Chap. 11 contains a review of the possible applications of surface wave propagation. Both, existing plasma devices, which apply surface wave propagation, and prospective possible future applications of surface waves studied in the previous chapters are discussed. These applications concern the following areas: plasma electronics, including devices utilized for generation of super-high frequency radiation, plasma-antenna systems including designs, which utilize plasmas sustained by surface-wave propagation, plasma production, especially plasma sources based on surface-wave propagation, surface-plasmon polaritons propagating in the super-high frequency range, plasmonics devices, including biosensors, photovoltaic cells etc. (see, e.g. [32–34]), and plasma nano-science. The cases of surface wave propagation in devices for investigation of magnetically confined thermonuclear fusion plasmas are also considered and the possibility to utilize SFWs for cleaning the inner walls of large-scaled fusion reactor metal vessels is discussed. There, the external magnetic field is supplied by superconducting magnetic coils.

1.4 Basic Equations for Investigation of Surface Flute Waves

1.4.1 *Electromagnetic Wave Field Distribution in Plasmas*

In this monograph, the analytical calculations are carried out in the cgs system of units. The Maxwell equations in differential form are applied. The wave electric field $\vec{E}$ and displacement vector $\vec{D}$ are assumed to be linked by the permittivity tensor ε_{ij} [7]:

$$\vec{D} = \hat{\varepsilon}\vec{E}. \tag{1.1}$$

In most of the cases studied here, the following form of the permittivity tensor for a cold plasma placed into an axial static magnetic field is used:

$$\hat{\varepsilon} = \begin{pmatrix} \varepsilon_1 & i\varepsilon_2 & 0 \\ -i\varepsilon_2 & \varepsilon_1 & 0 \\ 0 & 0 & \varepsilon_3 \end{pmatrix}. \tag{1.2}$$

The components of the tensor read

$$\varepsilon_1 = \varepsilon_r - \sum_\alpha \frac{\Omega_\alpha^2}{\omega^2 - \omega_\alpha^2},\ \varepsilon_2 = -\sum_\alpha \frac{\Omega_\alpha^2 \omega_\alpha}{\omega\left(\omega^2 - \omega_\alpha^2\right)},\ \varepsilon_3 = 1 - \sum_\alpha \frac{\Omega_\alpha^2}{\omega^2}. \tag{1.3}$$

In (1.3), Ω_α is the angular plasma frequency of the particle of species α ($\alpha = i$ for ions and $\alpha = e$ for electrons), ω_α is corresponding angular cyclotron frequency, and ε_r is the relative dielectric constant of the gaseous plasma ($\varepsilon_r = 1$), of a plasma metamaterial or of a semiconductor crystal lattice ($\varepsilon_r > 1$).

As already mentioned earlier, flute waves are considered for which the axial wavenumber is equal to zero, $k_z = 0$. Solutions of the Maxwell equations are searched, e.g. for the axial magnetic wave field, in the form $B_z(r, \varphi, z, t) = B_z(r)\exp[i(m\varphi - \omega t)]$. Faraday's law of induction, $rot\vec{E} = \frac{i\omega}{c}\vec{B}$, for flute waves reads in cylindrical coordinates

$$\frac{m}{r}E_z(r) = \frac{\omega}{c}H_r(r), \tag{1.4}$$

$$\mathrm{i}\frac{\partial E_z(r)}{\partial r} = \frac{\omega}{c}H_\varphi(r), \tag{1.5}$$

$$\frac{\mathrm{i}}{r}\frac{\partial\left(rE_\varphi(r)\right)}{\partial r} + \frac{m}{r}E_r(r) = -\frac{\omega}{c}H_z(r). \tag{1.6}$$

Ampère-Maxwell's circuital law, $rot\vec{B} = \frac{-i\omega}{c}\vec{D}$, in cylindrical coordinates in the absence of external currents reads

$$-\frac{m}{r}H_z(r) = i\frac{\omega\varepsilon_2}{c}E_\varphi(r) + \frac{\varepsilon_1\omega}{c}E_r(r), \tag{1.7}$$

$$\frac{\partial H_z(r)}{\partial r} = \frac{i\varepsilon_1\omega}{c}E_\varphi(r) + \frac{\varepsilon_2\omega}{c}E_r(r), \tag{1.8}$$

$$\frac{\mathrm{i}}{r}\frac{\partial\left(rH_\varphi(r)\right)}{\partial r} + \frac{m}{r}H_r(r) = \frac{\varepsilon_3\omega}{c}E_z(r). \tag{1.9}$$

One does not need to take care of Gauss's law for the electric wave field, $div\vec{D} = 4\pi\rho$, because there are no external charges, $\rho = 0$, and the identity $div(\frac{ic}{\omega}rot\vec{B}) \equiv 0$ is valid. The same is true for Gauss's law for the magnetic wave field, $div\vec{B} = 0$, since $div(\frac{-ic}{\omega}rot\vec{E}) \equiv 0$.

Considering the Eqs. (1.4–1.9) one can note that the amplitudes of the wave fields H_r, H_φ, and E_z take part in the Eqs. (1.4, 1.5 and 1.9) only. These amplitudes are not included in the other three Eqs. (1.6–1.8). On the other hand, the other three amplitudes E_r, E_φ, and H_z do not take part in the Eqs. (1.4, 1.5 and 1.9). In other words, the set of Eqs. (1.4–1.9) can be separated into two independent sets of equations: (1.4, 1.5, 1.9) and (1.6–1.8). This can be explained in physical terms as follows. Ordinary polarized flute waves with the components H_r, H_φ, and E_z propagate in cylindrical waveguides independently from the extraordinary polarized flute waves with the components E_r, E_φ, and H_z.

As far as the electric field of extraordinary polarized flute waves is perpendicular to the external axial static magnetic field $\vec{B}_0$ and their dispersion properties strongly depend on the magnitude of B_0, these waves are classified as extraordinary waves (X-mode waves) in the present book according to [35–37]. Equations (1.7–1.8) make it possible to derive the radial and azimuthal components of the electric wave field in terms of the axial magnetic wave field:

$$E_r = \frac{\omega}{ck_\perp^2}\left[\mu\frac{\partial H_z}{\partial r} + \frac{m}{r}H_z\right], \quad E_\varphi = \frac{i\omega}{ck_\perp^2}\left[\frac{\partial H_z}{\partial r} + \mu\frac{m}{r}H_z\right]. \tag{1.10}$$

After substituting these expressions (1.10) into Eq. (1.6) one derives the second order differential equation which determines the radial distribution of the axial magnetic wave field:

$$\frac{1}{r}\frac{d}{dr}\left(\frac{r}{k_\perp^2}\frac{dH_z}{dr}\right) - \left[1 + \frac{\mathrm{m}^2}{\mathrm{r}^2k_\perp^2} - \frac{\mathrm{m}}{\mathrm{r}}\frac{\mathrm{d}}{\mathrm{dr}}\left(\frac{\mu}{k_\perp^2}\right)\right]\mathrm{H_z} = 0. \tag{1.11}$$

In (1.11), $k_\perp^2 = \left(\frac{\omega^2}{c^2}\right)\varepsilon_1(\mu^2 - 1)$, and $\mu = \varepsilon_2/\varepsilon_1$.

In the case of homogeneous plasmas, Eq. (1.11) turns into a Bessel type equation whose surface type solutions can be expressed in terms of modified Bessel functions $I_m(k_\perp r)$ and MacDonald functions $K_m(k_\perp r)$ [38]:

$$H_z = A_1 I_m(k_\perp r) + A_2 K_m(k_\perp r). \tag{1.12}$$

Here $A_{1,2}$ are constants of integration. The physical sense of $k_\perp$ can be explained as follows. The physical observable $k_\perp^{-1}$ has the dimension of length. It is the scale at which the wave amplitude varies significantly in radial direction. It is natural to refer to this observable as to the depth of the surface wave penetration into the plasma.

To get surface type solutions of Eq. (1.11) one needs $k_\perp^2$ to be real and positive. This condition determines the frequency range where the radial distribution of the amplitude of electromagnetic flute waves is of surface nature:

$$\omega_{LH} < \omega < |\omega_e|, \quad |\omega_e| < \omega < \omega_1, \tag{1.13}$$

$$\omega_{UH} < \omega < \omega_2. \tag{1.14}$$

where $\omega_{1,2} = \mp|\omega_e|/2 + \sqrt{\Omega_e^2 + \omega_e^2/4}$ are the cut-off frequencies for bulk modes, and $\omega_{LH} = \left[(|\omega_e|\omega_i)^{-1} + \Omega_i^{-2}\right]^{-\frac{1}{2}}$ and $\omega_{LH} = \sqrt{\Omega_e^2 + \omega_e^2}$ are the lower and upper hybrid frequencies, respectively. In this monograph, the frequency ranges (1.13) and (1.14) are entitled as the low frequency (LF) and high frequency (HF) range, respectively.

The hydrodynamic approach is applied in the present book. This approach is invalid for studying electromagnetic waves whose frequencies are close to $|\omega_e|$. In the vicinity of $|\omega_e|$, one has to apply the kinetic approach, which is the only one for studying the electron cyclotron waves [39].

1.4.2 Electromagnetic Wave Field Distribution in Dielectrics

Within the isotropic dielectric layer with dielectric constant ε_d, the electric wave field is parallel to the displacement vector, $\vec{D} = \varepsilon_d \vec{E}$. Both ordinary and extraordinary SFWs are of bulk nature within the dielectric layer. The amplitudes of the extraordinary electric wave fields are linked with that of the axial magnetic wave field H_z via the Eqs. (1.10), and the radial distribution of H_z is governed by Eq. (1.11) with the following replacements: $\mu \to 0$, and $k_\perp^2 \to -\kappa^2 = -\left(\frac{\omega^2}{c^2}\right)\varepsilon_d$. The physical sense of κ is as follows. The physical observable κ has the dimension of reversed length, and $2\pi/\kappa$ is natural to be treated as the radial wavelength. The radial distribution of the SFW axial magnetic field has the form of a linear combination of Bessel functions of the first kind $J_m(\kappa r)$ and Neumann functions $N_m(\kappa r)$ [38].

1.4.3 Electromagnetic Wave Field Distribution in Layers Where an Electron Beam Gyrates

The electrons gyrate in the static axial magnetic field along large concentric axisymmetric Larmor orbits. The electrons are assumed to have no axial momentum, $p_z = 0$, but they have all the same angular velocity $\vec{\upsilon} = \upsilon_\perp \vec{e}_\varphi$, where the azimuthal component is a function of the radial coordinate, $\upsilon_\perp = r|\omega_e|\gamma(r)$. The electron beam is assumed to have low particle density n_b as compared to that of the plasma, $n_b << n_e$, and to be neutralized with respect to electric charge and current. The electrodynamic properties of the space, where the beam gyrates, are described by the permittivity tensor ${\varepsilon_{ik}}^{(b)}$ [4]. The components of this tensor being used in the present book are:

$$\varepsilon_{11}^{(b)} = 1 + \frac{\Omega_b^2}{\omega\gamma}\sum_{s=-\infty}^{+\infty} s^2\left[\frac{\gamma\left(J_s^2(m)\right)'}{(s-y)m|\omega_e|} + \frac{kJ_s^2(m)\,r^2}{(s-y)^2cm^2}\right]; \tag{1.15}$$

$$\varepsilon_{12}^{(b)} = \frac{i\Omega_b^2}{\omega|\omega_e|}\sum_{s=-\infty}^{+\infty} s\left[\frac{\left(J_s(m)J_s'(m)\right)'}{s-y} + \frac{J_s(m)J_s'(m)}{(s-y)m} + \frac{J_s(m)J_s'(m)\,k\upsilon_\perp r}{(s-y)^2cm}\right]$$
$$= -\varepsilon_{21}^{(b)}; \tag{1.16}$$

$$\varepsilon_{22}^{(b)} = 1 + \frac{\Omega_b^2}{\omega|\omega_e|}\sum_{s=-\infty}^{+\infty}\left[\frac{2\left(J'_s(m)\right)^2}{s-y} + \frac{\left(J'_s(m)\right)^2\upsilon_\perp^2 y}{(s-y)^2c^2} + \frac{2mJ'_s(m)J''_s(m)}{s-y}\right]. \tag{1.17}$$

In Eqs. (1.15–1.17), $\Omega_b^2 = 4\pi e^2 n_b m_e^{-1}$ is the square of the plasma frequency of the electron flow, $y = \omega\gamma/|\omega_e|$, m is the azimuthal mode index, $\gamma(r)$ is the relativistic Lorentz factor, and $J_s(x)$ is the Bessel function of the first kind [38]. A prime denotes the derivative with respect to the argument.

The argument of the cylindrical functions in Eqs. (1.15–1.17) for the components of the dielectric permittivity tensor is given in [4] as the combination $z = \frac{k_{perp}\upsilon_\perp}{|\omega_e|}\gamma$. Since the electrons of the flow do not move along the cylinder axis, $\upsilon = \upsilon_\perp$. SFWs propagate along the azimuthal angle only, therefore, $k_{perp} = k_\varphi$. The Larmor radius of the beam particles is defined as: $\rho_e = \frac{\upsilon}{|\omega_e|\gamma}$, therefore $z = k_\varphi\rho_e$. The wave vector k_φ is defined by the condition of periodicity along the azimuthal angle: the mode with the azimuthal mode index m should consist just of m full perturbations along the azimuthal angle for any radial coordinate. Therefore, $2\pi r = m\lambda = 2\pi/k_\varphi$. From this relation, k_{perp} for the dielectric permittivity tensor in [4] becomes immediately as $k_{perp} = m/r$. Finally, since beams of particles which rotate just along the Larmor radius are considered, $r = \rho_e$, and thus $z = m$.

In the case of a longitudinal electron beam, $\vec{\upsilon} = \upsilon_{||}\vec{e}_z$, the beam interaction with plasmas due to Cherenkov resonance is known to dominate over the interaction due to cyclotron resonance [4]. If the electrons have also a transversal component of the velocity, $\vec{\upsilon} = \upsilon_\perp\vec{e}_\varphi + \upsilon_{||}\vec{e}_z$, the interaction due to the cyclotron resonance becomes as strong as the Cherenkov one [4]. In the case of the rotating electron beam, $\vec{\upsilon} = \upsilon_\perp \vec{e}_\varphi$, considered here, only cyclotron interaction

$$\omega = \omega_0 + \delta\omega = S|\omega_e|\gamma^{-1} + \delta\omega, \tag{1.18}$$

is possible, being represented by the second order poles in (1.15–1.17). These poles indicate that the most efficient beam-wave interaction takes place for the frequencies satisfying the condition (1.18) [4]. In (1.18), ω is the complex angular frequency of the wave; ω_0 is the angular SFW eigenfrequency in the same waveguide structure without the beam; $\delta\omega$ is the correction to the frequency caused by the beam; S is a natural number. For simplicity, in the following, ω is just called the "frequency", without "angular".

The gap, where the beam gyrates, is assumed to be narrow in the most cases considered in the present book. That is why one can neglect the weak radial dependence of the components (1.15–1.17) of the permittivity tensor $\varepsilon_{ik}{}^{(b)}$ and assume the SFW radial wave vector k_b to be a constant, where $k_b^2 = k^2\left(\varepsilon_{22}^{(b)} + \varepsilon_{12}^{(b)2}/\varepsilon_{11}^{(b)}\right)$. Therefore, if the electron beam is sufficiently intense, one can assume that all the electrons of the flow contribute to SFW excitation through the resonance (1.18). In this case the electrons can be assumed to rotate along the large orbits with the same radius which is located just in the middle of the gap. Within the gap, the radial dependence of the SFW axial magnetic field is governed by the Bessel equation (1.11) with the following replacements: $k_\perp^2 \to -k_b^2$ and $\mu \to -i\varepsilon_{12}^{(b)}/\varepsilon_{11}^{(b)}$. Its solution can be written as a linear combination of the Bessel functions of the first, $J_m(k_b r)$, and second, $N_m(k_b r)$, kind.

1.4.4 Boundary Conditions

In the present book, the metal waveguide walls are assumed to have infinite conductivity, which allows to apply non-dissipative boundary conditions. That means, that the tangential components of the electric wave field vanish at metal interfaces.

Neither free surface charges nor currents are considered here, that is why the tangential components of the wave fields are continuous at the plasma-dielectric and plasma-beam interfaces.

The wave fields should not possess any singularity within the studied space. This is specifically significant for the axis of cylindrical waveguides where the MacDonald functions $K_m(\xi)$ and Neumann functions $N_m(\xi)$ are singular.

References

1. Girka, V., Girka, I., & Thumm, M. (2014). *Surface flute waves in plasmas—Theory and applications*. Springer International Publishing.
2. Moisan, M., Shivarova, A., & Trivelpiece, A. W. (1982). Experimental investigations of the propagation of surface waves along a plasma column. *Plasma Physics, 24*(11), 1331–1400.
3. Gradov, O. M., & Stenflo, L. (1983). Linear theory of a cold bounded plasma. *Physics Reports (Rev. Sec. Phys. Lett.), 94*, 111–137.
4. Alexandrov, A. F., Bogdankevich, L. S., & Rukhadze, A. A. (1984). *Principles of plasma electrodynamics*. Springer.
5. Vukovic, V. (1986). *Surface waves in plasmas and solids*. World Scientific.
6. Azarenkov, N. A., & Ostrikov, K. N. (1991). Surface magnetoplasma waves at the interface between a plasma-like medium and a metal in a Voigt geometry. *Physics Reports, 308*, 333–428.
7. Akhiezer, A. I., Akhiezer, I. A., Polovin, R. V., Sitenko, A. G., & Stepanov, K. N. (1975). *Plasma electrodynamics*. Pergamon Press.
8. Wesson, J. (2004). *Tokamaks* (3rd ed.). Clarendon Press.
9. Anderson, T. (2011). *Plasma antennas*. Artech House.
10. Kondratenko, A. M., & Kuklin, V. M. (1988). *Basis of plasma electronics*. Energoatomizdat. (in Russian).
11. Jenn, D. C. (2003). *Plasma antennas: Survey of techniques and the current state of the art*. SPAWAR PMW 189.
12. Beletski, N. N., Bulgakov, A. A., Khankina, S. I., & Yakovenko, V. M. (1984). *Plasma instabilities and non-linear phenomena in semi-conductors*. Naukova dumka (in Russian).
13. Agranovich, V. M., & Mills, D. L. (1982). *Surface polaritons*. North-Holland.
14. Toda, M. (1964). Propagation of waves in a solid state plasma waveguide in a transverse magnetic field. *Journal of the Physical Society of Japan, 19*(7), 1126–1130.
15. Girka, V. (2006). Theory of transverse surface electro-magnetic waves propagating along a plasma–metal boundary with a finite radius of curvature in a magnetic field. *Plasma Physics Reports, 32*(5), 401–410.
16. Aliev, Yu. M., Schlüter, H., & Shivarova, A. (2000). *Guided-wave-produced-plasmas*. Springer.
17. Elliott, R. S. (1955). Azimuthal surface waves on circular cylinders. *Journal of Applied Physics, 26*(4), 368–376.
18. Horiuchi, K. (1953). Surface wave propagation over a coated conductor with small cylindrical curvature in direction of travel. *Journal of Applied Physics, 24*(7), 961–962. https://doi.org/10.1063/1.1721421
19. Wait, J. R. (1965). Waves circulating around a rigid cylindrical obstacle in a compressible plasma. *Radio Science Journal of Research NBS/USBC-URSI, 69D*(4), 567–577.
20. Wait, J. R. (1966). Transverse propagation of waveguide modes in a cylindrically stratified magnetoplasma. *Radio Science, 1*(6), 641–654.
21. Girka, V. O., Girka, I. O., Kondratenko, A. N., & Tkachenko, V. I. (1988). Azimuthal surface waves of magnetoactive plasma waveguides. *Soviet Journal of Communications Technology and Electronics, 33*(8), 37–41.
22. Girka, V. O., Girka, I. O., Kondratenko, A. N., & Tkachenko, V. I. (1989). Azimuthal surface modes of isotropic plasma waveguides. *Soviet Journal of Communications Technology and Electronics, 34*(15), 103–105.
23. Girka, V. O., Girka, I. O., & Pavlenko, I. V. (2001). Electrodynamic model of the gas discharge sustained by azimuthal surface waves. *Contributions to Plasma Physics, 41*(4), 393–406.
24. Girka, V. O., Girka, I. O., & Pavlenko, I. V. (2011). Excitation of azimuthal surface modes by relativistic flows of electrons in high-frequency range. *Plasma Physics Reports, 37*(5), 447–454.
25. Girka, V. O., Girka, I. O., Morgal, Y. I., & Pavlenko, I. V. (2011). Excitation of azimuthal surface modes by annular electron beams in the range of electron cyclotron frequency. *Physica Scripta, 84*, 025505, 7.

26. Girka, V. O., Girka, I. O., & Pavlenko, I. V. (2011). Excitation of ion azimuthal surface modes in a magnetized plasma by annular flow of light ions. *Progress in Electromagnetics Research M (PIERM), 21*, 267–278.
27. Girka, V. O., Girka, I. O., Kondratenko, A. N., & Tkachenko, V. I. (1989). Azimuthal surface waves at the boundary between a magnetoactive plasma and a metal. *Soviet Journal of Communications Technology and Electronics, 34*(4), 96–99.
28. Azarenkov, N. A., Kondratenko, A. N., & Ostrikov, K. N. (1993). Surface waves in plasma-metal structures. *Radiophysics and Quantum Electronics, 36*(5), 213–247.
29. Azarenkov, N. A., Ostrikov, K. N., & Shcherbinina, I. B. (1991). Azimuthal surface waves in a coaxial semiconductor structure with metal walls. *Soviet Journal of Communication Technology and Electronics, 36*(2), 68–72.
30. Kuzelev, V. M., & Rukhadze, A. A. (1990). *Electrodynamics of dense electron beams in plasma.* Nauka. (in Russian).
31. Volkov, E. D., Suprunenko, V. A., & Shyshkin, O. O. (1983). *Stellarators.* Naukova dumka (in Russian).
32. Maier, S. A. (2007). *Plasmonics: Fundamentals and applications.* Springer.
33. Green, R. J., Frazier, R. A., & Shakesheff, K. M. (2000). Surface plasmon resonance analysis of dynamic biological interactions with biomaterials. *Biomaterials, 21*, 1823–1835.
34. Ederra, I., Azcona, L., & Alderman, B. E. J. (2007). A 250 GHz sub-harmonic mixer design using EBG technology. *IEEE Transactions on Antennas and Propagations, 55*(11), 2974–2982.
35. Brambilla, M. (1998). *Kinetic theory of plasma waves. Homogeneous plasmas.* Clarendon Press.
36. Stix, T. H. (1984). *Waves in plasmas.* American Institute of Physics (AIP).
37. Chen, F. F. (1984). *Introduction to plasma physics and controlled fusion* Vol. 1: *Plasma physics.* Plenum Press.
38. Abramowitz, M., & Stegun, I. A. Eds. (1964). *Handbook of mathematical functions.* National Bureau of Standards, Applied Mathematics, Series 55.
39. Girka, V., Girka, I., & Thumm, M. (2019). *Surface electron cyclotron waves in plasmas.* Springer Nature Switzerland AG.

Chapter 2
Surface Flute Waves in Uniform Magneto-active Plasma Filled Waveguides

The properties of electromagnetic waves propagating perpendicularly to the axial direction in cylindrical metal waveguides filled by magneto-active plasmas are considered in this Chapter. The influence of the waveguide parameters, such as radius of the plasma column, magnitude of the external static magnetic field, constructional peculiarities etc. on the wave properties is examined here under the condition of uniform plasma filling. This assumption is valid both for the case of solid state plasmas (e.g. plasmas of semi-conductors) and for the case of gaseous plasmas with a thin transient plasma layer, whose thickness is small in comparison with the penetration depth of the studied modes. In the case of waveguides filled by magneto-active plasmas, it is assumed that the external constant magnetic field is uniform and has only an axial component.

The dependence of the electromagnetic fields of the surface flute waves (SFWs) on co-ordinates and time is assumed in the form: $f(r)\exp(im\varphi - i\omega t)$, where φ is the azimuthal angle, r is the radial co-ordinate, m and ω are the azimuthal mode index and eigenfrequency of the SFW, respectively. In the cold magneto-active plasma approximation, the Maxwell equations can be separated into two independent sets of equations, which describe extraordinary and ordinary polarized waves (or in other words: X- and O-modes, respectively). Such waves connected with the perturbations of charged plasma particles are analogous to gravitational flute modes connected with the motion of plasma density ripples across an external constant magnetic field [1]. Propagation of gravitational flute modes can be unstable under the condition of large radius of the plasma column and/or large plasma density gradient.

Since the radial distribution of O-modes has often a bulk character, and their dispersion properties do not depend on an external axial static magnetic field but strongly depend on features of the dielectric layer between the plasma column and the metal wall of the waveguide, the main attention is given in the present book just to the properties of X-modes. Thus, six of the seven sections of this Chapter are devoted to studying the dispersion properties of flute X-eigenmodes. The properties of the flute O-eigenmodes are investigated only in Sect. 2.4. Nevertheless, knowledge on O-modes is necessary for studying the case of coupled flute modes propagating in

© The Author(s), under exclusive license to Springer Nature Switzerland AG 2022

I. Girka and M. Thumm, *Surface Flute Waves in Plasmas*, Springer Series on Atomic, Optical, and Plasma Physics 120, https://doi.org/10.1007/978-3-030-98210-2_2

current-carrying waveguides. This problem is under consideration in Chap. 4 of this book.

The possibility to apply the theory of surface flute waves to the derivation of a dispersion relation for non-symmetric electromagnetic waves, which in addition to the azimuthal wavenumber, are characterized by a small but finite axial wavenumber, is also demonstrated in this Chapter. It is found that their dispersion equation has two solutions. Here the analogy between these waves and the resistive drift waves described in [1] should be mentioned once more.

2.1 Extraordinary Polarized Surface Flute Waves in Magneto-active Plasma Waveguides

The properties of SFWs, which propagate along the interface of a plasma column with radius R_1, which is separated by a dielectric layer from the inner wall of a cylindrical metal waveguide with radius $R_2 > R_1$ are considered in the present Section. An external static magnetic field is oriented along the axis of the waveguide, $\vec{B}_0 || \vec{z}$. The schematic of this waveguide is shown in Fig. 2.1. The case of a dense plasma, for which the inequality $\Omega_e^2 > \omega_e^2$ is valid, is considered.

In the region of the plasma column, the radial distribution of the SFW field is described by Eqs. (1.10) and (1.12). Application of the boundary condition at the waveguide axis, $r = 0$, causes vanishing of the constant A_2: $A_2 = 0$.

Then in the dielectric layer, $R_2 > r > R_1$, the fields of the SFW are described by a superposition of Neumann functions $N_n(z)$ and first kind Bessel functions $J_n(z)$ [2]:

$$H_z = B_1\big[N_m'(\kappa R_2)J_m(\kappa r) - J_m'(\kappa R_2)N_m(\kappa r)\big], \tag{2.1}$$

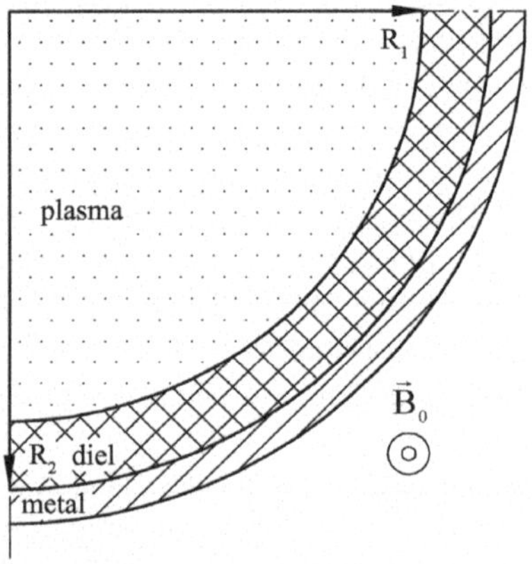

Fig. 2.1 Scheme of a metal waveguide, which is partially filled with plasma. The waveguide has an inner dielectric layer

where B_1 is a constant of integration. The other constant is chosen to satisfy the boundary condition at the metal wall, $r = R_2$, where the wave azimuthal electric field should vanish, $E_\varphi(r = R_2) = \frac{i\omega}{-c\kappa^2}\frac{\partial H_z}{\partial r} = 0$.

To determine the remaining two integration constants, one can apply the boundary conditions at the plasma-dielectric interface. First, the azimuthal electric wave field should be continuous there:

$$\begin{aligned}&\frac{i\omega}{ck_\perp^2}\left[A_1 k_\perp I'_m(k_\perp R_1) + \mu\frac{m}{R_1}A_1 I_m(k_\perp R_1)\right] \\ &= \frac{-i\omega}{c\kappa^2}\kappa B_1\left[N'_m(\kappa R_2)J'_m(\kappa R_1) - J'_m(\kappa R_2)N'_m(\kappa R_1)\right]\end{aligned} \quad (2.2)$$

Second, the SFW axial magnetic field should be also continuous at this interface:

$$A_1 I_m(k_\perp R_1) = B_1\left[N'_m(\kappa R_2)J_m(\kappa R_1) - J'_m(\kappa R_2)N_m(\kappa R_1)\right]. \quad (2.3)$$

Equations (2.2) and (2.3) form a system of two linear equations, which has a non-trivial solution in respect of the two constants A_1 and B_1 if the determinant constructed from the coefficients of them is equal to zero. (In other words, one can derive the dispersion relation by equating the ratio of the SFW fields at the interface to the dielectric layer: $E_\varphi/B_z(r = R_1 - 0) = E_\varphi/B_z(r = R_1 + 0)$.)

Thus, the following dispersion relation $D_x(\omega) = 0$ for the flute X-eigenmodes is derived:

$$D_x = \frac{\kappa I'_m(k_\perp R_1)}{k_\perp I_m(k_\perp R_1)} + \frac{\kappa\mu m}{k_\perp^2 R_1} + \frac{\left[N'_m(\kappa R_2)J'_m(\kappa R_1) - J'_m(\kappa R_2)N'_m(\kappa R_1)\right]}{\left[N'_m(\kappa R_2)J_m(\kappa R_1) - J'_m(\kappa R_2)N_m(\kappa R_1)\right]}. \quad (2.4)$$

Here and below, the prime symbol denotes the derivative of the corresponding cylindrical function over its argument.

In the following, the most interesting case is considered, when the thickness of the dielectric layer which separates the plasma column from the inner wall of the metal waveguide is sufficiently small so that the following inequality is valid: $\varDelta = 1 - R_1 R_2^{-1} \ll 1$. Then the effect of the plasma on the dispersion properties of the surface flute X-modes (SFXMs) is most pronounced. In this case the dispersion relation (2.4) can be rewritten in the simpler form:

$$\frac{\kappa I'_m(k_\perp R_1)}{k_\perp I_m(k_\perp R_1)} + \frac{\kappa\mu m}{k_\perp^2 R_1} \approx \frac{m^2 - \kappa^2 R_1^2}{\kappa R_1}\varDelta. \quad (2.5)$$

Analyzing Eq. (2.5) one can see that the azimuthal mode index m cannot be equal to zero if $\Delta \neq 0$; and only in the special case of the plasma-metal boundary when $\Delta = 0$ one can realize axially symmetric oscillations with $m = 0$ [3].

One can obtain simple expressions for the solutions of dispersion relation (2.5) in some limiting cases. They are given in the following formulas.

In the HF range (1.14), the sign of the azimuthal mode index can be only negative, and for $|m| \gg 1$ the flute X-eigenmode frequency can be written in the form:

$$\omega \approx \omega_1 \left(1 + \frac{m^2 \delta^2}{2R_1^2} - \Delta m \frac{\omega_e}{\omega_1} \right), \tag{2.6}$$

where $\delta \equiv c/\Omega_e$ is the skin-depth, and the inequality $1 \gg m^2 \delta^2 R_1^{-2} \gg m\omega_e \Delta \omega_1^{-1}$ is assumed to be valid. In the limiting case of large magnitudes of the waveguide radius ($k_\perp R_1 \gg |m|$) the flute X-mode frequency can be expressed in another form:

$$\omega \approx \omega_1 + \Omega_e \frac{m^2 \delta^2}{2R_1^2} \left(1 - 2m\Delta \frac{\omega_e}{\omega_1} \right), \tag{2.7}$$

where $R_1^2 > \delta^2 m^2 \Omega_e / |\omega_e|$, and $\Delta < \omega_1 / m\omega_e$.

For the SFXMs, whose frequency belong to the LF range (1.13), the dispersion relation (2.5) has the following solution:

$$\omega^2 \approx \frac{|m|\Delta \Omega_e^2}{1 + |m|\Delta} \left\{ 1 + \frac{\Delta R_1^2 [1 + 2\Delta(1 + |m|)]}{2(1 + |m|\Delta)^2 \delta^2 (1 + |m|)} \right\}, \tag{2.8}$$

which is derived in the limiting case $R_1 \ll \delta$, $\delta |m| < 1$. If the inequality $|m| \gg 1$ is valid, one can find the following expression for the flute X-mode frequencies:

$$\omega \approx \frac{\pm|\omega_e|}{2\psi_1} + \sqrt{\frac{\Delta |m| \Omega_e^2}{\psi_1} + \frac{\omega_e^2}{4\psi_1^2}}, \tag{2.9}$$

where $\psi_1 \approx 1 + \Delta R_1^2 \delta^{-2} |m|^{-1} + 2.7 R_1 \delta^{-1} |m|^{-1}$, the upper sign (+) relates to the case of positive m, and the lower sign (−) to the case where $m < 0$. In the limiting case, when $k_\perp R_1 \gg |m|$, the analysis of the dispersion relation (2.5) allows one to find the following solution:

$$\omega \approx \frac{-m\omega_e}{2\psi_2} + \sqrt{\frac{m^2 \Omega_e^2 \Delta}{\psi_2} + \frac{m^2 \omega_e^2}{4\psi_2^2}}, \tag{2.10}$$

where $\psi_2 \approx R_1 \delta^{-1} + R_1^2 \delta^{-2} \Delta$, and $R_1 \gg \delta |m|$. Thus, the azimuthal mode index m can be both positive and negative in the LF range unlike in the case of HF flute X-modes. Analyzing the analytical expressions derived for LF flute X-modes, one can make the following conclusions:

- their dispersion is normal;
- their frequency increases with increasing thickness of the dielectric layer.

Figure 2.2 demonstrates the dependence of the LF SFXM frequency normalized to the electron cyclotron frequency, $\omega/|\omega_e|$, versus the effective

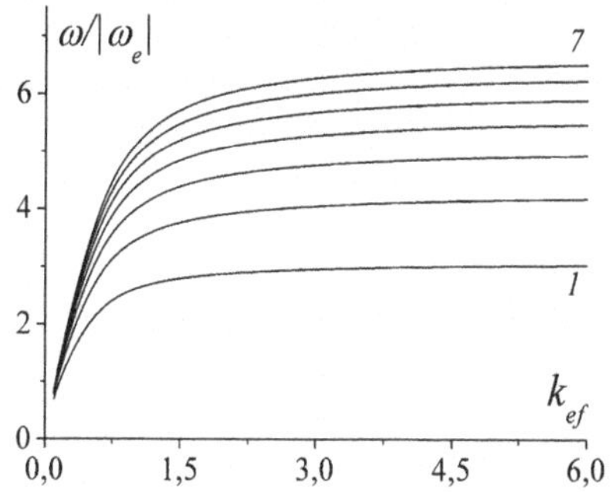

Fig. 2.2 Normalized frequency of LF SFXMs propagating in a metal waveguide, which is partially filled by magneto-active plasma versus effective wavenumber. The *numbers* indicate the azimuthal mode index m

wavenumber $k_{ef} = |m|\delta/R_1$. The following plasma waveguide parameters have been chosen for the calculations: $Z \equiv \Omega_e/|\omega_e| = 10$, $\Delta = 0.2$, $\varepsilon_d = 1$, and $m = 1$–7.

Dispersion curves of HF SFXMs are plotted in Fig. 2.3 by the solid lines. The horizontal dashed lines show the boundaries of the HF range. The dependencies of the HF SFXM frequencies normalized to the electron cyclotron frequency are shown in Fig. 2.3 as functions of the effective wavenumber k_{ef}. For the calculation of these dispersion curves, the following plasma waveguide parameters have been chosen: $Z = 6$, $\Delta = 0.2$, and $\varepsilon_d = 1$. In the limiting case of large values of the plasma radius R_1 the dispersion curves of these SFXMs with different azimuthal mode indices differ from each other only slightly; they are confined near the lower boundary of the HF range (1.14). Their dispersion curves are restricted as well by the upper boundary of the HF range.

The results of the numerical analysis of Eq. (2.4) show that increasing the radius R_1 of the plasma column, the strength of the external magnetic field B_0 or the thickness Δ of the vacuum layer leads to decreasing frequency of the HF flute X-modes. In contrast to this, the frequencies of LF flute X-modes increase with increasing Δ and with decreasing R_1 and B_0 values. Changing the sign of the azimuthal mode index makes a substantial effect on the SFXM dispersion properties. The HF branch of these X-modes vanishes for the case $m > 0$, and the frequency of their LF branch with $m > 0$ considerably grows up in comparison with the corresponding

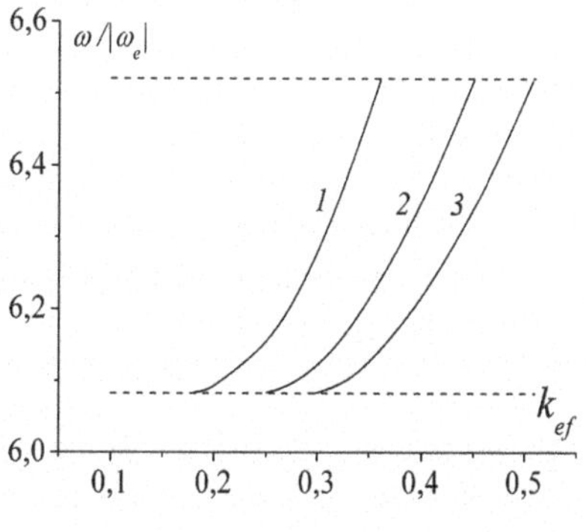

Fig. 2.3 Normalized frequency of HF SFXMs propagating in a metal waveguide, which is partially filled by magneto-active plasma versus effective wavenumber. The *numbers* indicate the absolute value of the azimuthal mode index |m|. The horizontal *dashed lines* show the boundaries of the HF range

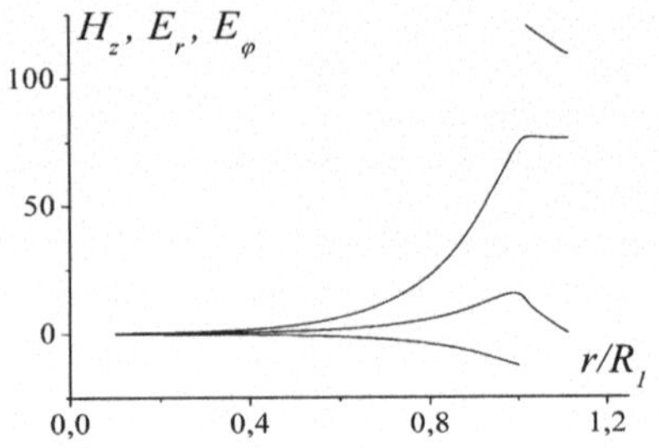

Fig. 2.4 Radial distribution of the wave fields of LF SFXMs (in arb. units) in the case of propagation along a plasma-dielectric interface

magnitudes of their frequencies in the case $m < 0$. Consequently, flute X-eigenmodes in such plasma structures are non-mutual: the frequencies of the waves which are characterized by the same absolute value of the azimuthal mode index but with opposite signs are different under the same other conditions. This is typical for SWs propagating across an external magnetic field oriented along the plasma boundary [1, 4].

Typical radial distributions of LF SFXMs fields are shown in Fig. 2.4. The axial magnetic wave field H_z is continuous and is not equal to zero on the metal wall, $r = R_2$. The radial electric wave field of LF SFXMs E_r is discontinuous at the plasma-dielectric interface; however, the azimuthal electric wave field E_φ is continuous and is equal to zero on the metal wall. The following parameters of the plasma waveguide system have been chosen for the calculations: azimuthal mode index $m = -2$, $Z = 10$, $\Delta = 0.11$, $\varepsilon_d = 1$, $k_{ef} = 0.3$, and $\omega = 1.8831|\omega_e|$.

Studying the results of the numerical analysis of Eq. (2.4) allows one to find that SFWs can propagate as well in the case of isotropic plasma. In this case, their frequency decreases with increasing radius of the plasma column and decreasing thickness of the dielectric layer. The graphic dependence of their frequency on the plasma column radius is similar to the hyperbolic one. It also proves that the SFW frequency depends strongly on the azimuthal mode index, thickness of dielectric layer and radius of the plasma column. In this case the considered modes can be used in antenna devices of plasma electronics. Radiation of their energy can be emitted through dielectric windows in the metal wall of the waveguide. Since the wavelength of the SFW radiation is determined by the relation $\lambda \approx 6 \times 10^6 n_{pl}^{-1/2}$ cm, one can obtain for the plasma density $n_{pl} \sim 10^{12}$ cm^{-3} radiation in the centimeter wavelength range, which could be gradually tuned.

2.2 Extraordinary Polarized Surface Flute Waves Propagating Along Magneto-active Plasma-Metal Interface

The properties of SFWs, which propagate along the interface between a cold magneto-active plasma and the metal wall [5, 6] of the corresponding waveguide, are considered in the present Section. Such structures can be realized for the following cases: the cylindrical metal waveguide is entirely filled by plasma or a cylindrical metal rod is immersed into the plasma. Thus, one can apply the co-ordinate system that is the same, as in the previous section, however, the radii of the metal and plasma surfaces are equal: $R_1 = R_2 = R$. The other initial assumptions, which have been chosen in Sect. 2.1, are also valid here. That is why the spatial distribution of the SFW fields in the plasma region is described in general by expression (1.12). However, there is a difference: in the case of a plasma cylinder which is surrounded by a metal wall with ideal conductivity the second constant of integration is equal to zero: $A_2 = 0$, whereas, in the case of a cylindrical metal rod, which is immersed into magneto-active plasma, the constant $A_1 = 0$.

As far as the boundary condition $E_\varphi(r = R) = 0$ is valid on the plasma-metal interface one can derive from Eq. (1.10) the following dispersion relation for SFXMs propagating along a plasma-metal interface with finite radius of curvature:

$$k_\perp R F_m'(k_\perp R) + m\mu F_m(k_\perp R) = 0. \tag{2.11}$$

In the case of a cylindrical metal waveguide, which is entirely filled by plasma (see Fig. 2.5), $F_m(\xi) = I_m(\xi)$, and for the case of a cylindrical metal rod, which is immersed into magneto-active plasma (see Fig. 2.6), $F_m(\xi) = K_m(\xi)$. In the approximation of large magnitude of the azimuthal mode index: $|m| \gg 1$, Eq. (2.11) transforms into the dispersion relation for a SW which propagates exactly perpendicularly to the external magnetic field along a planar plasma-metal boundary [4]. In this case the ratio m/R plays the role of a wave vector. The analysis of dispersion relation (2.11) proves that in the case under consideration here, the frequency range of flute X-modes diminishes as compared to the case of a three-component waveguide structure: metal-dielectric-plasma. Equation (2.11) has solutions in the HF range (1.14) and as well in the LF range (1.13), but in contrast to the previous section, the frequency cannot be higher than the electron cyclotron frequency: $\omega_{LH} < \omega < |\omega_e|$.

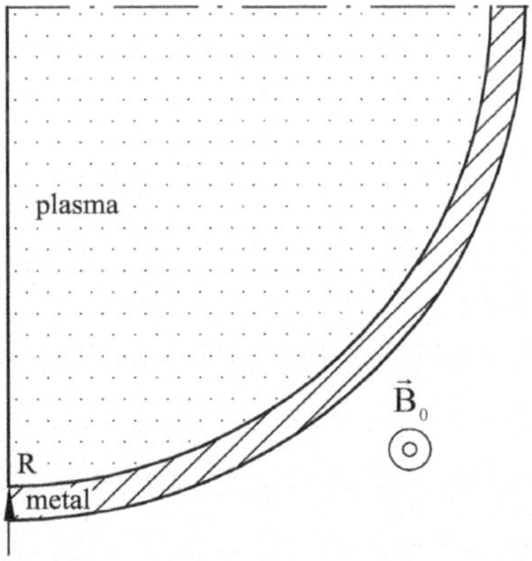

Fig. 2.5 Scheme of a metal waveguide, which is entirely filled by magneto-active plasma. Its axis is parallel to an external static magnetic field

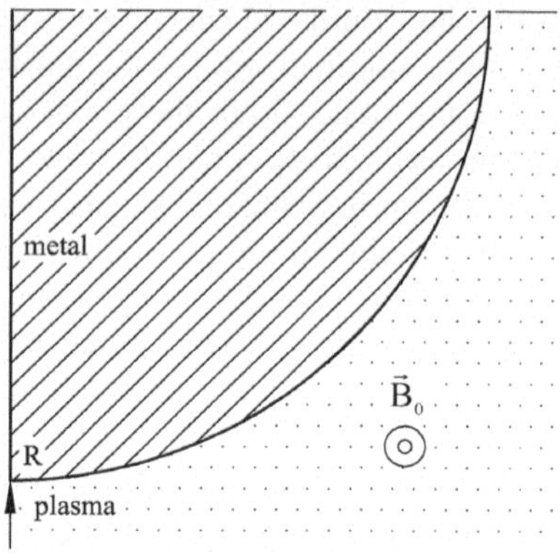

Fig. 2.6 Scheme of a cylindrical metal rod, which is immersed into infinite magneto-active plasma. Its axis is parallel to an external static axial magnetic field

Unlike the case that is investigated in Sect. 2.1, SFWs propagate along the plasma-metal boundary across the external constant axial magnetic field only in one direction in all frequency ranges and they can be only X-polarized modes in the fluid approximation. This corresponds to the general conception of SW theory in the case of their propagation along a plasma-metal interface [4] across the external magnetic field. The sign of the azimuthal mode index m correlates uniquely with the type of plasma surface curvature. Namely, in the case $F_m = I_m$, the azimuthal mode index is positive: $m > 0$, in the frequency range $\omega < |\omega_e|$, and in the range (1.14) the SFW azimuthal mode index is negative: $m < 0$. For the case of a cylindrical metal rod, which is immersed into the plasma, the situation is opposite: $m < 0$ for LF SFXMs ($\omega < |\omega_e|$) and $m > 0$ for HF SFXMs ($\omega_{UH} < \omega$).

Analytical solutions of the dispersion relation (2.11) can be derived in some limiting cases. An axial-symmetric ($m = 0$) solution of Eq. (2.11) is realized under the condition $\varepsilon_1^2 = \varepsilon_2^2$. It describes surface oscillations of the plasma cylinder at the cut-off frequencies of bulk modes $\omega = \omega_{1,2}$. In the case $|m| \geq 1$ in the LF range, the spectrum of extraordinary polarized SFWs can be represented by the following expressions:

$$\omega \approx \begin{cases} |m\omega_e|\delta R^{-1}, \ R > |m|\delta, \ \delta = c/\Omega_e, \\ |\omega_e|\left(1 + R^2\omega_1^2 m^{-2}c^{-2}\right)^{-1/2}, \ |m| \gg 1, \\ |\omega_e|\left[1 + 0.5R^2\delta^{-2}(|m|)^{-1}(|m| \pm 1)^{-1}\right]^{-1}, \ R < \delta. \end{cases} \tag{2.12}$$

Here the upper sign (+) relates to the case $F_m = I_m$, and the lower sign (−) to the case $F_m = K_m$.

In the case of the HF range, dispersion relation (2.11) has the following solution:

$$\omega \approx \sqrt{\omega_1^2 + \frac{m^2c^2}{R^2}}, \ \text{for} \ \frac{R^2|\omega_e|}{\delta^2\Omega_e} \gg m^2 \gg 1. \tag{2.13}$$

The frequencies of HF and LF SFXMs propagating along a plasma-metal boundary increase with increasing magnitude of the azimuthal mode index $|m|$ and decreasing value of the waveguide radius R. They tend to the maximum values of the LF and HF ranges with increasing B_0 value. The difference between SFXMs frequencies in the cases $F_m = K_m$ and $F_m = I_m$ is negligibly small.

Figures 2.7 and 2.8 show the results of a numerical analysis of the dispersion relation (2.11) in the LF and HF ranges, respectively, for the lowest values of the azimuthal mode index: $|m| = 1, 2, 3$. The ratio $\omega/|\omega_e|$ is chosen as ordinate axis. Comparison of these curves and those, which are obtained for the case of waveguides which are partially filled by magneto-active plasma (the last ones are plotted in Figs. 2.2 and 2.3 for the LF and HF ranges, respectively) allows one to make the following conclusions. First, the dependence of the SFXM frequencies on the absolute value of the azimuthal mode index is opposite in the cases when these waves propagate along plasma-metal or plasma-dielectric interfaces. Second, in the case of waveguides completely filled by the plasma the LF range becomes narrower, since the upper range limit becomes $|\omega_e|$ unlike in the case of waveguides only partially filled by plasma. The range of effective wavenumbers, where the dispersion relation (2.11) has solutions for the HF range, becomes wider than that one, which characterizes the dispersion relation of SFXMs propagating along a plasma-dielectric interface (compare Figs. 2.8 with 2.3).

As one can see, the SW features, in the case of the finite value of curvature radius of the plasma interface, are different from those in the case of planar plasma-metal [4] structures. One can point out the following differences. First, the flute

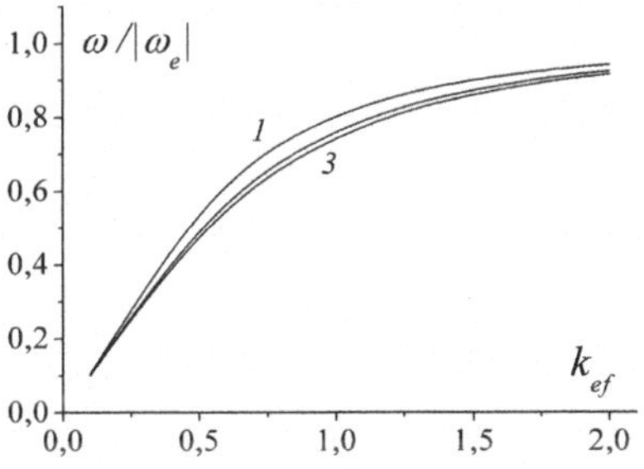

Fig. 2.7 Dispersion curves of LF SFXMs propagating along a plasma-metal interface for $\Omega_e = 10|\omega_e|$. The *numbers* indicate the absolute value of the azimuthal mode index, $|m|$

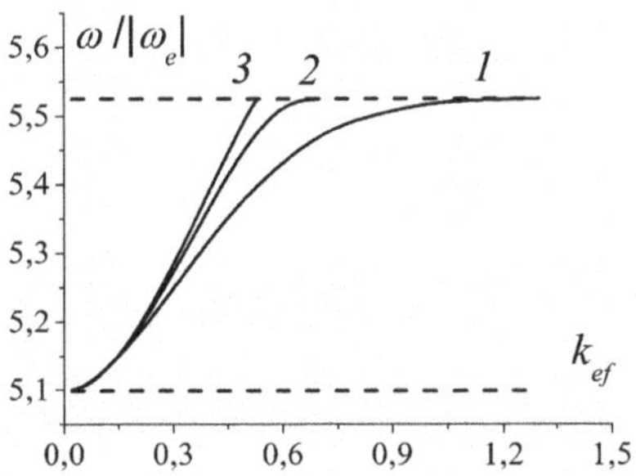

Fig. 2.8 Dispersion curves of HF SFXMs propagating along a plasma-metal interface for $\Omega_e = 5|\omega_e|$. The *numbers* indicate the absolute value of the azimuthal mode index, $|m|$. The horizontal *dashed lines* show the boundaries of the HF range

X-mode frequency has two limiting values in both HF and LF ranges. Second, a unique correspondence exits between the sign of the azimuthal mode index and the sign of the curvature radius of the plasma surface. Third, the spatial distributions of these transverse surface modes propagating along planar and curved plasma-metal interfaces are different, namely: SFWs are extraordinary polarized waves, which have two electrical components and one magnetic component, while in the case of a planar waveguide the tangential electric field of the SW vanishes in the whole plasma volume.

The analysis of Eq. (2.11) proves that SFWs cannot propagate along a plasma-metal interface in the case of isotropic uniform cold plasma unlike this type of flute waves which propagate in the structure plasma-dielectric-metal considered in the previous section. Results of a numerical analysis of the radial distribution of the axial magnetic and azimuthal electric wave fields of SFXMs propagating along a plasma-metal interface are presented in Figs. 2.9 and 2.10, respectively. The radial distribution of the E_r component is similar to the distribution of the H_z component and is therefore not shown. However, it should be mentioned that its magnitude is six times smaller than the magnitude of the axial magnetic field and its relative contribution

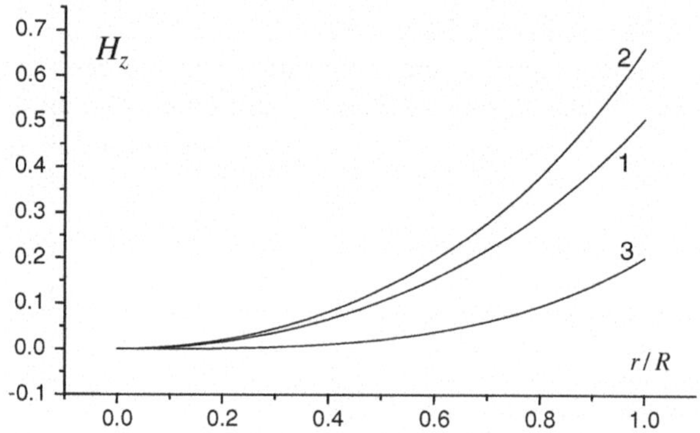

Fig. 2.9 Radial distribution of H_z for LF SFWs (in arb. units) propagating along a plasma-metal interface for the case $R = 2\delta$. The *numbers 1, 2* and *3* mark the curves obtained for 3 sets of parameters: $m = 2$ with $\Omega_e^2 = 7\omega_e^2$; $m = 2$ with $\Omega_e^2 = 50\omega_e^2$; and $m = 3$ with $\Omega_e^2 = 50\omega_e^2$, respectively

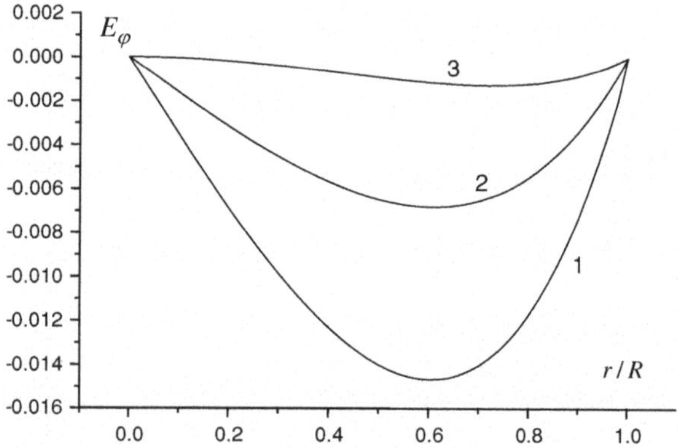

Fig. 2.10 Spatial distribution of E_φ (in arb. units) for the same parameters as in Fig. 2.9

to the complete field of SFXMs decreases with increasing plasma density unlike the contribution of the H_z component. The radial distribution of the E_φ component of SFXMs is characterized by very small amplitude values and the maximum of E_φ is located nearby the plasma-metal interface, unlike in the case of plane plasma-metal boundary [4].

2.3 Surface Flute Waves in Strongly Magnetized Plasma Waveguides

Up to now only the case of dense, $\Omega_e > |\omega_e|$, plasmas has been examined. Thus, the frequency of the considered electromagnetic waves belongs to the range, which is higher than the lower hybrid resonance. That is why the motion of plasma ions was neglected there. The above correlation between electron plasma (Langmuir) and electron cyclotron frequencies can be easily satisfied, for instance in plasmas of semiconductors [1] because they are characterized by a high concentration of free electrons. The other plasma case, where $\Omega_e{}^2 < \omega_e{}^2$, can be realized under laboratory conditions by utilizing a strong magnetic field. Thus, just this case [7], when the propagation of SFWs is connected with perturbations of the density of plasma ions in waveguides of cylindrical geometry, is considered in the present Section.

Due to the motion of plasma ions the expressions for the frequency ranges where extraordinary polarized SFWs can exist change compared to the case of application of a weak external magnetic field:

$$\omega_{LH} < \omega < \omega_1 \text{ and } \omega_{UH} < \omega < \omega_2, \tag{2.14}$$

which in this Section are called the LF and HF ranges, respectively. The frequencies $\omega_{1,2} = \pm 0.5\omega_e + \sqrt{0.25\omega_e^2 + \omega_i|\omega_e| + \Omega_e^2}$ are the cut-off frequencies of the bulk waves [8] in the case of strongly magnetized plasmas. In this case, the HF range becomes very narrow and is located nearby the electron cyclotron frequency, and therefore SFW frequencies in this range depend very weakly on the plasma parameters. Under these conditions the lower hybrid resonance is located in the range of the ion cyclotron frequency and therefore there exists the possibility for propagation of plasma ion waves perpendicularly to an external magnetic field.

Applying the boundary conditions mentioned in Sect. 1.4.4 one can derive the dispersion relation for extraordinary polarized SFWs in strongly magnetized plasmas in the form of Eq. (2.5). Analytically it can be examined in the limiting case of a narrow dielectric layer ($m\Delta \ll 1$, $(R_2 - R_1)\kappa \ll |m|$), when the influence of the plasma parameters on the dispersion properties of SFWs is the strongest. It is found that unlike in the case of SFWs propagating in waveguides, which are partially filled by a weakly magnetized dense plasma, extraordinary polarized SFWs cannot propagate in wide cylindrical metal waveguides with thin dielectric coating which are filled by a strongly magnetized plasma. Mathematically this feature can be proved by analysis of both parts of Eq. (2.5). In the case of strongly magnetized plasmas the left-hand side of Eq. (2.5) is positive and the right-hand side is negative, therefore the dispersion relation has no solutions [7].

In the following, the limiting case of a narrow waveguide, which is partially filled by a strongly magnetized plasma, is considered: $k_\perp R_1 \ll 1$. Then an analytical expression for the frequency of the flute X-mode can be written in the following form:

$$\omega \approx \omega_i \left[1 + \frac{\Omega_i^2}{\omega_i^2} \frac{\left| \frac{R_2^{2m} - R_1^{2m}}{R_2^{2m} + R_1^{2m}} \right|}{\left| \frac{R_2^{2m} - R_1^{2m}}{R_2^{2m} + R_1^{2m}} \right| + \varepsilon_d} \right]. \tag{2.15}$$

In this case, SFXMs have a unique direction of propagation, namely the azimuthal mode index can be only negative. Thus, the direction of their propagation coincides with the direction of cyclotron rotation of positively charged ions nearby the plasma surface, and the SFW frequency is approximately equal to ion cyclotron frequency.

Figure 2.11 shows the dependence of the frequency of ion SFWs normalized to the ion cyclotron frequency versus the ratio Ω_i/ω_i for three values of the azimuthal mode index $m = -2, -3, -4$. Just this frequency ratio is chosen as abscissa since the analytical solution shows that the eigenfrequency depends mostly on this parameter. The following neon plasma waveguide parameters have been chosen for this numerical analysis: $\Delta = 0.1$, $\varepsilon_d = 1$, and $|m|c/(R_1\Omega_i) = 4.8$. The curves end for small values of the abscissa Ω_i/ω_i since the frequency reaches the lower boundary of the frequency range. The numbers in Fig. 2.11 indicate the azimuthal mode index. The dispersion curve for the case $m = -1$ is not shown in the plot since these ion flute X-modes under the chosen conditions are of bulk nature.

The analysis of the spatial distribution of the ion SFXM fields proves that in this case SFWs can be considered as quasi-electric (or in other words, quasi-potential) waves; since the magnetic wave field of these SFXMs is much weaker than their electric wave fields. The amplitude of the SFXM electric wave fields with azimuthal mode index $m = -1$ remains practically independent on the radial co-ordinate even in the plasma region.

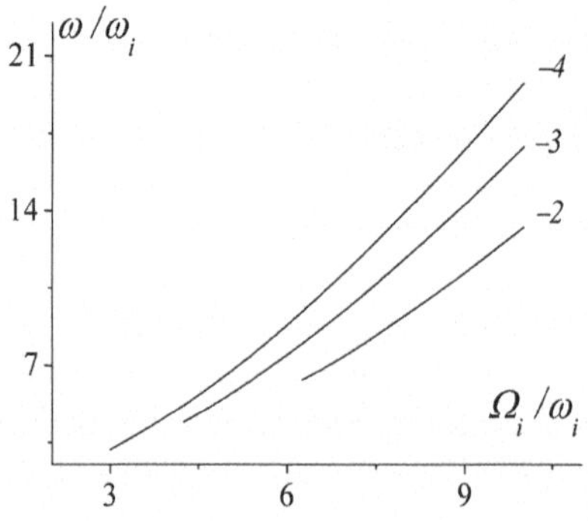

Fig. 2.11 Normalized frequency of ion SFXMs versus the ratio Ω_i/ω_i. The *numbers* indicate the azimuthal mode index m

In waveguides with a thick dielectric layer, the properties of flute X-modes depend weakly on the plasma parameters and are determined mainly by the properties of the dielectric layer that separates the plasma column from the metal wall of the waveguide. Thus, high radial harmonics of the SFW can propagate in this case, and a solution of dispersion relation (2.5) exists if the following inequality is valid:

$$(l + 0.5)\pi < (R_2 - R_1)\kappa < (l + 1)\pi; \quad l = 1, 2, 3 \ldots; \quad \kappa a \gg |m|. \tag{2.16}$$

In this case SFWs become substantially magnetic waves unlike in the previous case of strongly magnetized waveguides with thin dielectric layer. In wide waveguides, which are immersed into strong magnetic fields, SFWs propagate when the inequality $k_\perp R_1 \gg |m|$ is valid. The penetration depth of SFWs into the plasma is small and an analytical expression for the SFW frequency can be given in the form $\omega = \omega_{0l} + \delta\omega$, where $\delta\omega$ is their frequency correction. Their lth radial harmonic frequency is approximately equal to $\omega_{0l} \approx 0.5\pi lc/(R_2 - R_1)$.

Analyzing the dispersion relation one can make the following conclusions on SFW propagating in a strongly magnetized waveguide, which is partially filled by cold uniform plasma:

- the dependence of the SFW frequency on the azimuthal mode index is very weak in this limiting case;
- this dependence becomes weaker and weaker with increasing value of plasma density;
- the value of the SFW frequency correction $\delta\omega(m)$ is larger for positive mode indices m than for negative m;
- the difference between values of the SFW frequencies with mutually opposite values of azimuthal mode index is maximal nearby the upper boundary of the LF range (2.14), where the equality $\varepsilon_1 + \varepsilon_2 = 0$ is valid [7].

Results of numerical analyses prove that the frequency of the zeroth radial mode lies approximately in the range between the 20th and 40th ion cyclotron harmonics. The limiting magnitudes of the frequency range of the first radial mode are approximately equal to $60\omega_i$ and $80\omega_i$.

As an example, the eigenfrequency is calculated for the SFW with azimuthal mode index $m = -3$ which propagates in a fully ionized helium plasma under the following conditions: plasma radius $R_1 = 10$ cm, magnetic field $B_0 = 3$ kG, temperature $T = 1$ eV, plasma density $n_{pl} = 10^{10}$ cm^{-3}, gap parameter $\Delta = 0.1$. Under these conditions, the flute X-mode frequency is as follows: $\omega \approx 19\omega_i = 1.4 \times 10^8$ s^{-1}.

2.4 Ordinary Polarized Electromagnetic Flute Waves

As it was mentioned at the beginning of this Chapter, flute electromagnetic waves with ordinary polarization can also propagate in such type of cylindrical waveguides. In the limiting case of uniform plasma-filled waveguides, these O-modes are not interesting from the practical point of view as compared with the extraordinary polarized SFWs, because they cannot propagate along the plasma-metal boundary and their properties are mostly determined by the properties of the dielectric layer that separates the plasma column from the metal wall of the waveguide. However, in current-carrying plasma filled waveguides SFWs propagate as coupled X- and O-modes. Therefore, the dispersion properties of the ordinary polarized SFWs (SFOMs) propagating in waveguides filled by uniform plasmas are considered in this Section.

In the general case, the axial component of the electric field of flute O-modes is described by the following equation:

$$\frac{1}{r}\frac{d}{dr}\left(r\frac{dE_z^{(m)}}{dr}\right)+\left(k^2\varepsilon_3-\frac{m^2}{r^2}\right)E_z^{(m)}=\widehat{M}H_z^{(m)}, \tag{2.17}$$

where $k=\omega/c$. The operator $\widehat{M}H_z^{(m)}=0$ if there is no external azimuthal magnetic field: $B_{0\varphi}=0$. If an external azimuthal magnetic field is applied, $B_{0\varphi}\neq 0$, the explicit form of the operator $\widehat{M}H_z^{(m)}$ is given in Chap. 4 for each specific case. Regardless of the values of B_{0z} and $B_{0\varphi}$ the two other components of the electromagnetic field of flute O-modes are expressed through $E_z^{(m)}$ in the following way:

$$H_r^{(m)}=mE_z^{(m)}(kr)^{-1},\quad H_\varphi^{(m)}=ik^{-1}dE_z^{(m)}/dr. \tag{2.18}$$

A solution of Eq. (2.17) in a dielectric region, which satisfies the boundary condition for the tangential electric field of the waves on the metal wall of the waveguide, namely: $E_z^{(m)}(R_2)=0$, can be written in terms of Bessel and Neumann functions [2]. Application of the boundary conditions on the plasma-dielectric interface, which determine continuity of the tangential (in respect to the plasma interface) wave fields allows one to derive the dispersion relation in the form $D_o=0$, where:

$$D_o=\frac{\varphi_o}{d\varphi_o/dR_1}-\frac{1}{k}\frac{J_m(\kappa R_1)N_m(\kappa R_2)-J_m(\kappa R_2)N_m(\kappa R_1)}{J_m'(\kappa R_1)N_m(\kappa R_2)-J_m(\kappa R_2)N_m'(\kappa R_1)}. \tag{2.19}$$

In expression (2.19), $\varphi_0(r)$ is a solution of Eq. (2.17), that is finite on the waveguide axis, $r=0$. Analyzing Eqs. (2.17) and (2.18), one can make the conclusion that the properties of flute O-modes do not depend on the magnitude of an external axial static magnetic field. This is typical for all ordinary polarized modes [8].

In the case of a cylindrical metal waveguide which is partially filled by a uniform cylindrical plasma column, flute O-modes can propagate in the form of ordinary polarized SFWs for the frequencies: $\omega<\Omega_e$. In this case, the solution of Eq. (2.17) looks as follows: $\varphi_0=I_m\left(kr\sqrt{-\varepsilon_3}\right)$.

SFOMs cannot propagate with any frequency that is higher than the electron plasma (Langmuir) frequency since ordinary polarized electromagnetic flute modes acquire bulk nature in this frequency range. Therefore, the solution of Eq. (2.17) gets the following form: $\varphi_0=J_m\left(kr\sqrt{\varepsilon_3}\right)$, where the argument of the first kind Bessel function is real for $\varepsilon_3 > 0$. If the thickness of the dielectric coating (to simplify our consideration, let us assume that the dielectric constant is equal to unit) is sufficiently small, which means that the inequality $k(R_2 - R_1) \ll 1$ is valid, the dispersion relation of SFOMs can be reduced and written in the following approximate form:

$$\frac{\varphi_o}{d\varphi_o/dR_1} \approx -(R_2 - R_1). \tag{2.20}$$

It becomes clear from analysis of Eq. (2.20), that SFOMs cannot propagate in this limiting case, because the inequality $d(\ln|\varphi_0|)/dR_1 < 0$ can be valid only for some bulk flute O-modes.

In the opposite limiting case, namely, for waveguides with a thick dielectric layer: $k(R_2 - R_1) \gg |m|$, an analytical solution of dispersion relation (2.19) can be found by the method of successive approximations. Then the analytical expression for the SFOMs frequency has the form of a sum $\omega^{(0)} + \omega^{(1)}$, where $\omega^{(0)}$ is the wave frequency calculated in zeroth approximation and $\omega^{(1)}$ is a small correction to it. Since SFOMs propagate within the frequency range $\omega < \Omega_e$ (in this case $\varepsilon_3 < 0$) in the limiting case of large radius of the plasma column ($kR_1 \gg |m|$), one can obtain the following expressions for $\omega^{(0)}$ and $\omega^{(1)}$:

$$\omega^{(0)} = \frac{\pi lc}{2(R_2 - R_1)}\left[1 + \sqrt{1 + \frac{2m^2(R_2 - R_1)^2}{\pi^2 n^2 R_2 R_1}}\right], l = 1, 2, 3 \ldots \tag{2.21}$$

$$\omega^{(1)} = \frac{\omega^{(0)}\delta_o}{R_2 - R_1}\left(1 - \frac{m^2}{2k^2 R_1^2}\right)\left[1 + \frac{m^2 c^2}{2\left(\omega^{(0)}\right)^2 R_2 R_1}\right]^{-1}. \tag{2.22}$$

The influence of the plasma density on the properties of SFOMs becomes weaker with decreasing plasma column radius, i.e. if the inequality $kR_1 < 1$ is valid:

$$\omega^{(0)} = \frac{\pi lc}{R_2}\sqrt{\left(1 + \frac{|m|}{2n} - \frac{1}{4n}\right)^2 - \left(\frac{m}{\pi n}\right)^2}, \quad l = 1, 2, 3 \ldots \tag{2.23}$$

$$\omega^{(1)} = \frac{\pi|m|c}{R_2(|m|!)^2}\left(\frac{kR_1}{2}\right)^{2|m|}\left(1 - \frac{|m|\delta_o}{R_1}\right). \tag{2.24}$$

Figure 2.12 presents the results of the numerical analysis of dispersion relation (2.19). It shows the frequency of the first radial ($l = 1$) SFOMs normalized to the electron plasma (Langmuir) frequency versus skin depth normalized to the plasma radius. The numbers on the plot indicate the azimuthal mode indices $m = 1, 2, 3$. The following plasma waveguide parameters have been chosen for these calculations: $R_2 = 3R_1, \varepsilon_d = 1$. At the top, the curves are limited by the boundary of the frequency range $\omega = \Omega_e$.

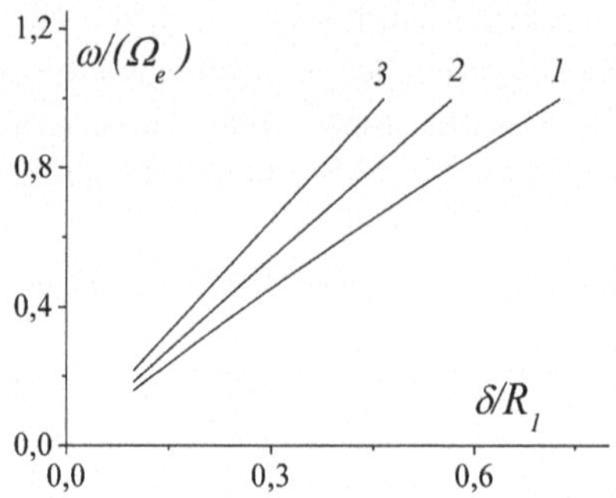

Fig. 2.12 Normalized frequency of the first radial SFOM versus skin depth normalized to the plasma radius. $R_2 = 3R_1$, $\varepsilon_d = 1$, $m = 1, 2, 3$

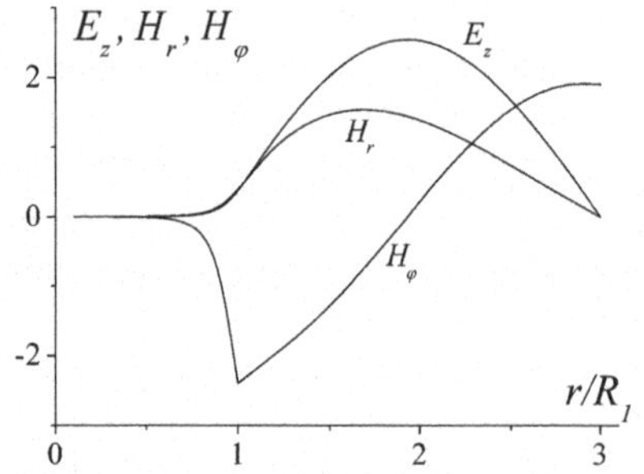

Fig. 2.13 Radial distribution of the fields of the SFOM (in arb. units) with radial mode index $l = 1$, azimuthal mode index $m = 2$, propagating in a waveguide with $\delta = 0.15R_1$, $R_2 = 3R_1$, $\varepsilon_d = 1$

Radial distributions of the fields of the SFOM which propagates with azimuthal mode index $m = 2$ in the case $\delta = 0.15R_1$ are given in Fig. 2.13. The plasma waveguide parameters, which are chosen for this calculation, are the same as for the previous figure. The axial electric wave field E_z is equal to zero on the metal as predicted by the general theory of SWs [1]. The radial magnetic field H_r of the wave is almost proportional to E_z, the weak difference is explained by the factor $1/r$ (see Eq. (2.18)). The azimuthal magnetic field H_φ of the wave differs from zero at the metal wall of the waveguide.

Thus, the dispersion properties of SFOMs in cylindrical metal waveguides, partially filled with a uniform plasma, are determined mainly by the geometry of the utilized waveguide, first of all by the thickness of the dielectric layer (within the dielectric layer they have features of bulk waves). The frequency of SFOMs increases with decreasing dielectric layer thickness and with increasing azimuthal (m) and radial (l) mode indices of these modes. The electron plasma frequency is the upper limit of the SFOM frequency range; therefore, it is $\sqrt{2}$ times larger than the maximum frequency of SWs propagating along a planar plasma interface [1]. In the frequency range $\omega > \Omega_e$ flute O-modes propagate in the form of bulk waves.

2.5 Damping of Surface Flute X-Modes

To calculate the damping rate of surface flute X-modes which is caused by collisions between the plasma particles, one can take into consideration the effective value of the collisional frequency ν in the components of the dielectric permittivity tensor ε_{jk} [8] applied in dispersion relation (2.5). This allows one to find the expression for the collisional damping rate of these modes:

$$\gamma_{col} \approx \nu/3. \tag{2.25}$$

Another reason, which can lead to damping of the surface flute X-modes is the presence of a slot in the metal wall of the utilized waveguide. Employment of such slotted waveguides allows one to radiate power of the SFXMs out of the waveguides into the free space, thus they can be applied as antenna systems.

Realization of the electromagnetic power emission out of plasma filled waveguides is an important problem for the development of generators of radio frequency (RF) and microwave radiation. The papers [9, 10] are devoted to the theory of open periodic plasma waveguides that are excited by electron beams in the regime of bulk wave generation. In the case of the excitation of bulk waves, the output coupling of microwave power is not a serious technical problem. However, in the case of surface wave excitation, one has to solve some difficult problems, e.g. to create and sustain definite profiles of plasma density or magnetic field, or to utilize a metal waveguide with special form, namely its radius can be varied according to a specific law [11].

In this Section, a waveguide is considered whose geometry differs from that one, which is studied in Sect. 2.1 by the presence of a narrow slot in the metal chamber of the waveguide (see Fig. 2.14). It is oriented along the axial direction and its angular size is small: $2\varphi_0 \ll 2\pi$. The plasma density is assumed to be relatively large: $\Omega_e^2 \gg \omega_e^2$. Determination of the spatial distribution of SFXM fields in the

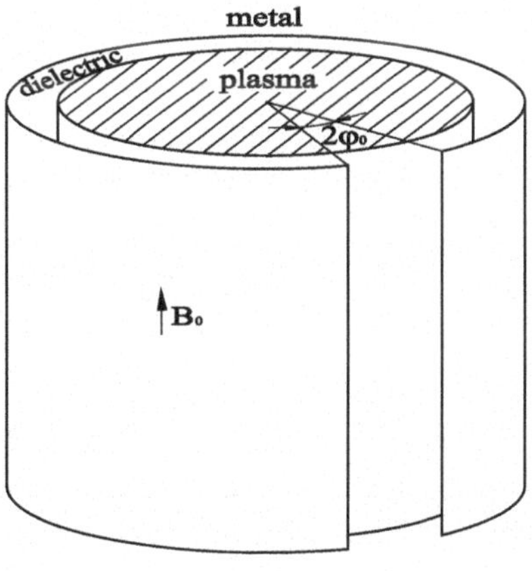

Fig. 2.14 Scheme of a plasma filled cylindrical metal waveguide with a narrow axial slot. A dielectric layer is between the plasma column and the inner waveguide wall

dielectric layer allows one to find out that the constants of integration β_1 and β_2 in the expression for the amplitude of the axial magnetic wave field

$$H_z = \beta_1 J_m(\kappa r) + \beta_2 N_m(\kappa r), \tag{2.26}$$

differ from those in (2.1). This is explained by the fact that the boundary conditions for the electromagnetic fields on the interface $r = R_2$ have in the present case another form as compared with those ones, which were applied in Sect. 2.1. These conditions should be considered in details. They can be determined as follows: first, the azimuthal component of the SFXM electric field is equal to zero on the metal wall of the waveguide:

$$\beta_2 = -\beta_1 J'_m(\kappa R_2)/N'_m(\kappa R_2) \text{ for } \varphi \in -[\pi, \varphi_0] \cup [\varphi_0, \pi]. \tag{2.27}$$

Second, there is no wave which propagates from the outer space into the inner part of the waveguide through the axial slot in its metal wall:

$$\beta_2 = i\beta_1 \text{ for } \varphi \in [-\varphi_0, \varphi_0]. \tag{2.28}$$

Since in the case when there is no axial slot ($\varphi_0 = 0$), all harmonics of the SFXMs with different values of the azimuthal mode index m propagate independently of each other, one can suppose that (in zeroth approximation) only the main harmonic of SFXMs with the index m propagates in the waveguide. Then the amplitudes of the ($m \pm 1$)th, ($m \pm 2$)th and etc. satellite harmonics have amplitudes, which are one order of magnitude smaller in comparison with the amplitude of the mth harmonic. Therefore, these small-amplitude satellite harmonics make only second order contributions to the dispersion equation of SFXMs [12].

Applying the condition that the SFXM impedance is continuous on the plasma-vacuum interface one can derive the dispersion relation for SFXMs in this case. With the accuracy up to terms of first order smallness over the parameter $\varphi_0/\pi \ll 1$ one can write this equation in the following form [12]:

$$D_x + D_1 = 0. \tag{2.29}$$

Here the expression for the D_x is defined by Eq. (2.4), and the expression for D_1 is a small quantity of the first order over the angle φ_0:

$$D_1 = \frac{\varphi_0\left[iN'_m(kR_2) + J'_m(kR_2)\right]N_m(kR_1)/\pi}{J_m(kR_1)N'_m(kR_2) - J'_m(kR_2)N_m(kR_1)}\left[\mu\frac{mk}{R_1 k_\perp^2} + \frac{kI'_m(k_\perp R_1)}{k_\perp I_m(k_\perp R_1)} + \frac{N'_m(kR_1)}{N_m(kR_1)}\right]. \tag{2.30}$$

To solve the dispersion relation (2.29) the method of successive approximations is applied, and a solution can be found as the sum: $\omega = \omega_0 + \Delta\omega - i\gamma$, where ω_0 is the solution of the dispersion equation in the zeroth ($\varphi_0 = 0$) approximation, $\Delta\omega$ is the real part of the correction to the SFXM eigenfrequency ω_0 and γ is the damping rate of the SFXMs caused by the radiation of their power through the axial slot in the metal wall of the waveguide.

In the limiting case of a narrow vacuum layer $\Delta \ll 1$, for SFXMs with large values of azimuthal mode index propagating in wide plasma waveguides $k_{ef} = m\delta R_1^{-1} \ll 1$, it is easy to derive the expression of the SFXM damping rate γ caused by radiation of power from the waveguide into the outer space:

$$\gamma \approx \frac{\Omega_e \varphi_0 (m!)^2}{4\pi^2}\left(\frac{4\delta}{m^2 R_1 \Delta}\right)^{m+0.5}. \tag{2.31}$$

The results of the numerical analysis of the dispersion relation (2.29) in the LF range are presented in Fig. 2.15. In the upper half-space one can see the frequencies of LF SFXMs (dashed curves) and damping rates caused by radiation of SFXM energy into the outer space (solid curves) normalized to the Langmuir frequency versus the effective wavenumber k_{ef}. In the bottom half-space, the dependence of the real corrections to the frequency is shown. The numbers 2, 3, 4 indicate the azimuthal mode index. The following plasma waveguide parameters are chosen for the calculation: $\Delta = 0.1$; $\Omega_e = 10|\omega_e|$; $\varphi_0 = 0.01\pi$.

The results of the numerical analysis of this phenomenon confirm that the SFW frequency ω_0 becomes lower than it was in the case of absence of an axial slot and that the additional damping described by γ is caused by the leakage of SFXM energy into the outer space. Therefore, the presence of an axial slot in the waveguide wall reduces the phase velocity of the SFXMs. In this respect one can mention the results of paper [9], where such a decrease of the frequency of bulk modes propagating in open periodic waveguides was observed in contrast to the case of closed plasma systems.

The rapid growth of the γ value with the increase of k_{ef} obtained here corresponds to the results which were obtained in paper [10]. But on the other hand, a numerical

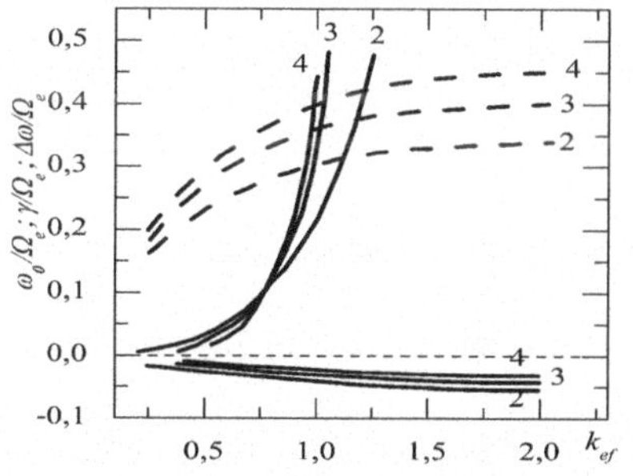

Fig. 2.15 Normalized values of eigenfrequency of LF SFXMs (*dashed curves*), damping rate (*solid curves in upper half-space*) and real part of the correction to their eigenfrequency (*solid curves in lower half-space*) caused by radiation of their power into the outer space versus effective wavenumber. The *numbers* indicate the magnitude of m

analysis testifies that the range of existence of SFXMs is located on the left-hand side of the curves $\gamma = \gamma(k_{ef})$, thus in this range of k_{ef} the damping of SFMs caused by radiation of their power through a slot antenna in the wall of the waveguide does not cancel its propagation. As one can see from (2.31) $\gamma \propto \varphi_0/\pi$ (compare with paper [9], there $\gamma \propto a/(a+b)$, where a is the size of the gap in the metal wall, b is the length of the metal sections of the open periodic waveguide examined there). However, for efficient operation of an RF generator (oscillator) based on SFW propagation it is required that the SFW damping rate has to be lower than the growth rate of their beam instability. This condition is sufficiently satisfied in the range $0.2 \le k_{ef} \le 0.6$. Numerical analysis proves that a reduction of γ can be achieved by reducing the magnitude of the magnetic field $\vec{B}_0$ and by increasing the value of the gap parameter Δ.

In the following, the case of SFXMs propagation in a metal waveguide, which is entirely filled by cold magneto-active plasma ($\Delta = 0$), is considered. The equation which describes the radiation of SFXM power through the slot in the metal wall of such a waveguide has the form of Eq. (2.29), where the term D_1 with an accuracy up to the first order terms of smallness over the parameter $\varphi_0/\pi \ll 1$ is given by:

$$D_1 = \frac{\varphi_0}{2\pi}\frac{J_m(\zeta)J'_m(\zeta) + N_m(\zeta)N'_m(\zeta)}{J_m^2(\zeta) + N_m^2(\zeta)} + \frac{i\varphi_0}{\pi^2 kR}[J_m^2(\zeta) + N_m^2(\zeta)]^{-1}, \tag{2.32}$$

and the term D_x reads as follows:

$$D_x = \frac{km\mu}{k_\perp^2 R} + \frac{kI'_m(\xi_0)}{k_\perp I_m(\xi_0)}, \tag{2.33}$$

here $\zeta = kR$, and $\xi_0 = k_\perp R$.

One can find an analytical expression for the corrections to the frequencies of SFXMs caused by radiation of their power through the axial slot in the metal wall of the waveguide, which is entirely filled with magneto-active plasma in the limiting cases of narrow ($|m| \gg \zeta$) and wide ($|m| \ll \zeta$) waveguides [6]. The real part of the frequency correction to the SFXM eigenfrequency in the first approximation over the small angular size of the slot in the metal wall of the waveguide, which is caused by radiation of the SFW power into the free space, reads:

$$\Delta\omega \approx \Omega_e\varphi_0/(\pi\varsigma), \tag{2.34}$$

where $\varsigma = 1$ in the case of a narrow waveguide and $\varsigma = 4$ in the case of wide waveguide.

The damping rates of SFXMs caused by their radiation through the axial slot antenna in the metal wall of the waveguide, which is entirely filled by magneto-active plasma, is described by the following expression:

$$\gamma \approx \varphi_0\Omega_e\varsigma_0/(2\pi), \tag{2.35}$$

where $\varsigma_0 = 1$ in the case of wide waveguide and $\varsigma_0 = \frac{4(m!)^2}{\pi|m|}(\zeta/2)^{2|m|}$ in the case of a narrow waveguide.

Figure 2.16 presents the results of numerical calculations of the damping rate of SFXMs determined by their power radiation through a narrow slot antenna in the waveguide wall into the outer space (the angular size of the axial slot is chosen as $\varphi_0 = 0.013{\cdot}2\pi$). The results of the numerical analysis of this case prove that in a wide waveguide, the value of the real corrections to the SFXM frequency increases with increasing value of azimuthal mode index, and for narrow waveguides the opposite dependence is observed. Increasing the plasma particle density and decreasing the waveguide radius lead to an increase of the magnitude of the SFXM frequency correction $\Delta\omega$. In addition, in the case of $\Delta = 0$, the real part of the correction to the eigenfrequency is positive, $\Delta\omega > 0$. This result differs from that obtained in [11] for plasma-dielectric-metal waveguide structures. The difference is explained as follows. The existence of an axial slot antenna in the metal wall of the waveguide in the present case (the metal waveguide is entirely filled by the plasma) means that the waveguide changes from a two-component plasma-metal structure into a three-component structure: plasma-insulator-metal. As it was shown in Sect. 2.2, the frequency of surface flute X-modes is larger just in the case of a three-component waveguide structure. That is why such modification of the waveguide leads to an increase of the SFW frequency.

Summarizing the results obtained for the case when the waveguide is entirely filled by a cold magneto-active plasma one can conclude that the damping rate γ of surface flute X-modes is determined by the leakage of their power through a narrow axial slot antenna. It increases with decreasing applied magnetic field, increasing radius of the waveguide and increasing angular dimension of the slot. The value of the real part of the SFXM frequency correction $\Delta\omega$ becomes positive unlike in the case of the presence of a dielectric layer between plasma column and waveguide wall

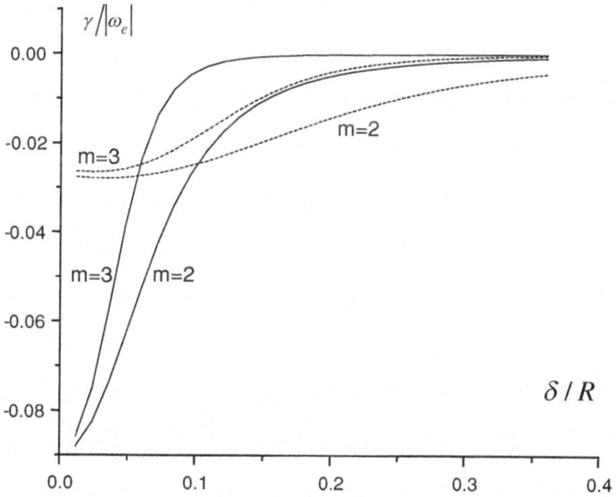

Fig. 2.16 Dependence of the normalized imaginary part of the frequency correction of the SFXMs determined by radiation of wave power from the waveguide into the free space versus skin depth normalized by the waveguide radius. *Solid* and *dotted curves* are calculated for the cases: $\Omega_e^2 = 50\omega_e^2$ and $\Omega_e^2 = 7\omega_e^2$, respectively

and $\Delta\omega$ decreases with increasing magnitude of the applied magnetic field and with increasing radius of the waveguide.

2.6 Long-Wavelength Azimuthally Non-symmetrical Surface Waves

In all the previous sections of this Chapter only those waves have been examined which propagate exactly perpendicularly to the axis of symmetry of the corresponding waveguide structures. This of course puts some limitations on the scope of application of the presented theoretical results, because electromagnetic waves often have a definite axial wavenumber with finite magnitude. That is why this Section presents the results of studying the properties of extraordinary polarized electromagnetic waves, which are characterized not only by the azimuthal mode index m, but also by the axial wavenumber k_z. Of course, these modes cannot be entitled as flute modes, but nevertheless, we shall conduct the study with the theory of surface flute waves (SFW) propagating in plasma filled waveguides. It is quite possible to do that for the case of long-wavelength azimuthally non-symmetric modes (ANSMs). The problem of studying the dispersion properties of a long-wavelength ANSM can be solved by the method of successive approximations, wherein the SFW theory is used as zeroth approximation. Here the analogy between these eigenwaves and the bulk resistive drift waves considered in [13], whose propagation is accompanied with a dissipative instability, should be mentioned. However, unlike them, long-wavelength ANSMs are stable modes and the form of their dispersion relation is similar to the form of an equation for coupled X- and O-modes.

The present results are devoted to investigation of the properties of long-wavelength ANSMs propagating in a cylindrical metal waveguide entirely filled with a cold magneto-active plasma, whose density is radially non-uniform [14]. The geometry of the waveguide and the assumptions are described in Sect. 2.2, the external magnetic field is directed along the axis of the waveguide: $\vec{B}_0 || \vec{z}$ (see Fig. 2.5). In the case of SFW propagation the Maxwell equations split into two independent subsystems, namely for X- and O-modes. However, for the long-wave ANSMs propagating in radially non-uniform plasma, if one can neglect terms of the second and higher orders of smallness of k_z, the Maxwell equations can be reduced into two second order differential equations for the axial components of electric and magnetic fields:

$$\frac{1}{r}\frac{\partial}{\partial r}\frac{r}{k_X^2}\frac{\partial H_z}{\partial r} - \left[1 + \frac{m^2}{r^2 k_X^2} - \frac{m}{r}\frac{\partial}{\partial r}\left(\frac{\mu}{k_X^2}\right)\right] H_z = \widehat{K} E_z; \tag{2.36}$$

$$\frac{1}{r}\frac{\partial}{\partial r} r \frac{\partial E_z}{\partial r} - \left(k_O^2 + \frac{m^2}{r^2}\right) E_z = \widehat{M} H_z. \tag{2.37}$$

The small value right-hand parts of the inhomogeneous differential Eqs. (2.36) and (2.37) can be written in the following forms:

$$\widehat{K}E_z = iN_z\left\{\frac{m}{r}E_z\frac{\partial}{\partial r}\left(\frac{1}{k_X^2}\right) + \frac{\mu k_O^2}{k_X^2}E_z + \frac{\partial}{\partial r}\left(\frac{\mu}{k_X^2}\right)\frac{\partial E_z}{\partial r}\right\}, \tag{2.38}$$

$$\widehat{M}H_z = iN_z\left\{\frac{1}{N_X^2}\frac{\partial\mu}{\partial r}\frac{\partial H_z}{\partial r} + \left[\frac{\omega^2}{c^2}\mu + \frac{m}{r}\frac{\partial}{\partial r}\left(\frac{1}{N_X^2}\right) - \frac{\mu m}{r}\frac{\partial}{\partial r}\left(\frac{\mu}{N_X^2}\right)\right]H_z\right\}, \tag{2.39}$$

where $N_X^2 = (\mu^2 - 1)(\varepsilon_1 - N_z^2)$, $N_z = ck_z/\omega$, and $\mu = \varepsilon_2/(\varepsilon_1 - N_z^2)$. The penetration depths of the O-mode $k_O^{-1}$and X-mode k_X^{-1}, with the field components (E_z, H_r, H_φ) and (E_r, E_φ, H_z), respectively, are determined employing the dielectric permittivity tensor ε_{ik} (see e.g. [3, 8, 15]). Their explicit forms are as follows:

$$k_O^2 = -\varepsilon_3\omega^2c^{-2}, \quad k_X^2 = N_X^2\omega^2c^{-2}. \tag{2.40}$$

The set of Eqs. (2.36) and (2.37) for the longitudinal components E_z and H_z of the electromagnetic wave field in the case of arbitrary values of k_z and radially uniform plasma immersed into an axial steady magnetic field was obtained in [16]. However, its analytical solutions for the limiting case of long-wavelength non-symmetric electromagnetic waves were found in [14] for the first time.

The solutions of Eqs. (2.36) and (2.37) are found by the method of constant variation. The azimuthal component of the ANSM electric wave field is expressed in terms of E_z according to [17]. Then, using the boundary conditions given in Sect. 1.4.4 for the tangential components of the electric wave field on the inner surface of the metal waveguide, one can derive the dispersion equation for long-wave ANSMs [14]:

$$A_{12}A_{21} = A_{11}A_{22}, \tag{2.41}$$

here

$$A_{21} = \phi(R_1), \tag{2.42}$$

$$A_{11} = \left\{\frac{ic}{\omega N_X^2}\frac{\partial\tilde{\psi}}{\partial r}\int_0^{R_1}\frac{\psi k_X^2\widehat{K}\phi dr}{W\left(\psi,\tilde{\psi}\right)} + \frac{i\mu N_\vartheta}{N_X^2}\tilde{\psi}\int_0^{R_1}\frac{\psi k_X^2\widehat{K}\phi dr}{W\left(\psi,\tilde{\psi}\right)} + \frac{N_zN_\vartheta}{N_H^2}\phi + \frac{\mu N_z c}{\omega N_H^2}\frac{\partial\phi}{\partial r}\right\}_{|r=R_1}, \tag{2.43}$$

$$A_{12} = \left\{\frac{ic}{\omega N_X^2}\frac{\partial\psi}{\partial r} + \frac{\mu N_\vartheta}{N_X^2}\psi\right\}_{|r=R_1}, \quad N_\vartheta = cm/\omega r, \tag{2.44}$$

$$A_{22} = \tilde{\phi}(R_1) \int\limits_0^{R_1} \frac{\phi \widehat{M} \psi dr}{W\left(\phi, \tilde{\phi}\right)}, \tag{2.45}$$

where ϕ and ψ are solutions of Eqs. (2.36) and (2.37), respectively, which have no singularities on the axis of the considered waveguide; $\tilde{\phi}$ and $\tilde{\psi}$ are linearly independent solutions of these equations, which have singularities on the axis of the waveguide.

It should be noted that on the one hand, the dispersion relation (2.41) is a generalization of the dispersion relation derived in [6, 17, 18] for the case of small k_z, and on the other hand it is a particular case of the general dispersion relation derived for determination of the frequency range of a coaxial plasma waveguide immersed into a finite constant magnetic field in [16]. The structure of Eq. (2.41) is typical for the case of weakly coupled electromagnetic waves. The condition $A_{12} = 0$ plays the role of the SFXM dispersion equation, whose solution ω_0 is calculated in Sect. 2.2. This means that here one can assume that its value is known. The factor A_{12} changes its sign in the frequency range found there for the SFXM. The condition $A_{21} = 0$ represents the dispersion equation, which describes the properties of bulk flute O-modes.

The right-hand side of the dispersion relation (2.41) is quadratic over the small value k_z, it plays the role of a small coupling parameter between the X- and O-modes. Therefore, the frequency correction $\Delta\omega$ to the eigenfrequency of long-wavelength ANSMs in the general case is as follows: $\Delta\omega \propto k_z^2$. However, this case needs a numerical calculation (see [16]) for studying the dispersion properties of these non-symmetrical waves. Unlike that we shall show a possibility to find analytical solutions of their dispersion relation.

In the following, analytical expressions for the solutions of Eq. (2.41) in the limiting case of electromagnetic waves with long wavelengths propagating in a plasma with uniform density are found. In zeroth approximation the eigenfrequency is the same as found in Sect. 2.2. The condition $k_X^2 > 0$ determines the frequency ranges, where ANSMs can exist, which under the condition $k_z = 0$ coincide with the frequency bands where the SFW can exist (see Sect. 2.2). Thus, ANSMs with $m > 0$ can propagate in the following frequency range, which is entitled as low-frequency (LF) range in this Section:

$$|\omega_e| \sqrt{\frac{\Omega_i^2 + c^2 k_z^2}{\Omega_e^2 + \omega_e^2}} < \omega < |\omega_e|. \tag{2.46}$$

The ANSMs with negative sign of the azimuthal mode index ($m < 0$) here are called high frequency (HF) waves. Their frequency bands are determined by the following non-equality:

$$\sqrt{\omega_e^2 + \Omega_e^2 + c^2 k_z^2} < \omega < 0.5|\omega_e| + \sqrt{0,25\omega_e^2 + \Omega_e^2 + c^2 k_z^2}. \tag{2.47}$$

Thus, the studied ANSMs are unidirectional waves, which can propagate only along a single direction in the definite frequency range. This property of SWs is typical for waves propagating across an external constant magnetic field along a plasma-metal boundary (see e.g. [4]). This can be a useful property for wave applications in different electronic devices, where one strives to avoid propagation of reflected signals. The value of k_O^2 has different signs in the ranges (2.46) and (2.47), namely it is positive, $k_O^2 > 0$, in the LF range and negative, $k_O^2 < 0$, in the HF range.

In the case of long-wavelength ANSMs propagating in a uniform plasma filled waveguide, dispersion relation (2.41) can be simplified. Its coefficients are given explicitly in [14], so they are not given here. Consideration of small values of k_z leads to the appearance of another additional solution of the dispersion equation (compared with the case of SFWs). In this Section, this solution ω_- is called as slow mode. For the LF range ω_- is almost proportional to the value of k_z, thus its value turns to the lower limit of this range at $k_z \to 0$ (see Fig. 2.17). The group velocity of the slow mode is always positive and its value exceeds the absolute value of the group velocity of the second solution of Eq. (2.41), which can be called here as fast mode ω_+. In the limiting case $k_z \to 0$ the fast mode frequency ω_+ becomes a frequency of a SFXM. The dispersion curves of the fast mode have a narrow part of k_z values, where its dispersion is normal. At this part of the ANSM dispersion curve, a small value of k_z leads to an increase of the ANSM frequency compared with the frequency of the SFW calculated for the same conditions. This allows one to calculate the frequency ω_+ of the fast mode by the method of successive approximations: $\omega_+ = \omega_0 + \Delta\omega$, using the theory of SFWs as zeroth approximation. The small frequency correction $\Delta\omega$ to the frequency of the SFW turns out to be quadratic over k_z in this case,

$$\Delta\omega \approx \left(\frac{A_{11}A_{22}}{A_{21}} - \frac{\partial A_{12}}{\partial k_z^2}_{|\omega=\omega_0} k_z^2 \right) \left[\frac{\partial A_{12}}{\partial \omega}_{|\omega=\omega_0} \right]^{-1} \propto k_z^2. \tag{2.48}$$

A simple analytical form of expression (2.48) can be obtained in two limiting cases: wide waveguide $R_1 \gg |m|\delta$, here δ is the plasma skin-depth, and narrow waveguide $R_1 \ll |m|\delta$. Explicit expressions of the LF ANSM frequencies for the

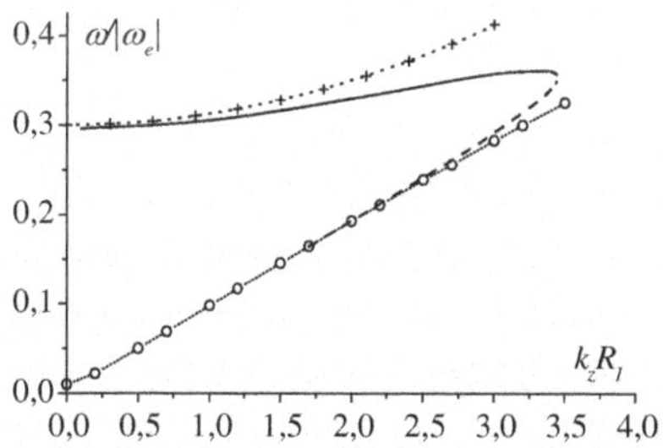

Fig. 2.17 Normalized frequency of the LF ANSM as function of normalized axial wavenumber. $m = 3$, $\Omega_e = 5|\omega_e|$, $R = 10\delta$. Fast mode (*solid curve*), slow mode (*dashed curve*), limit of the LF band (*dotted curve with hollow circles*), asymptotic solution (*dotted curve with crosses*)

both cases, respectively, can be written down in the form of a sum $\omega = \omega_0 + \Delta\omega$:

$$\omega = |\omega_e|\frac{m\delta}{R_1} + \frac{3}{8}|\omega_e|\frac{m\delta}{R_1}\left(\frac{k_z R_1}{m}\right)^2, \tag{2.49}$$

$$\omega = |\omega_e|\left(1 - \frac{R_1^2}{2m^2\delta^2}\right) + \frac{|\omega_e| R_1^2}{8\delta^2(m+1)^2}\left(1 - \frac{R_1^2}{2m^2\delta^2}\right)\left(\frac{k_z R_1}{m}\right)^2. \tag{2.50}$$

Numerical analysis proves that in practice the theory of successive approximations can be used to study the dispersion properties of the fast mode of long-wavelength ANSM in the LF range just in a fairly wide range of axial wavenumbers k_z. In the limiting case of wide waveguides this range is determined by the following inequality:

$$k_z^2 \ll 8m^2/R_1^2, \tag{2.51}$$

whereas in the limiting case of narrow waveguides it becomes even somewhat wider:

$$k_z^2 \ll 8\left(m(m+1)\delta/R_1^2\right)^2. \tag{2.52}$$

The results of the numerical analysis of dispersion relation (2.41) for this case are plotted in Fig. 2.17. In Fig. 2.17 the solid curve shows the dispersion curve of the fast mode and the dashed curve that of the slow mode for the case of LF ANSMs. The dotted line with hollow circles marks the limit of the LF ANSM frequency band (2.46). The dashed curve with crosses is drawn according to the direct calculation of the asymptotic formula (2.49). In the case of a narrow waveguide, the frequency correction is expressed very poorly, so this graph is not shown. The frequency of the fast mode of ANSMs at first increases with k_z, then it reaches a maximum, and after that it decreases so that the frequencies of the fast and slow modes become close to each other. Finally, for the axial wavenumber $k_z = k_m$ their frequencies are the same. Nearby k_m the representation of electromagnetic eigenwaves in the harmonic form, $\propto \exp\left[i(k_z z + m\varphi - \omega t)\right]$, is meaningless. Near the point in question, the dispersion curve $k_z(\omega)$ can be approximately described by the quadratic parabola, $k_z(\omega) = k_m - (\omega - \omega_m)^2/\alpha$ (where α is a constant having the dimension of acceleration, cm/s^2). In this case, an electromagnetic pulse with the field $\propto \exp\left(-t^2/\left(2\tau^2\right)\right)\cos[k_m z + m\varphi - \omega_m t]$ (where τ is the pulse duration) spreads out from the point at which it was originally formed in the axial direction over a distance of about $\alpha\tau^2/2$.

Now the case of long-wavelength HF ANSMs propagating in a uniform plasma filled waveguide is considered. In the HF range (1.14), a bulk azimuthally non-symmetric O-mode can also propagate in such waveguides under definite conditions. Thus, it can interact with the corresponding X-mode in an ordinary manner and in a resonant way as well. However, in the present book the resonant interaction is not considered (see for details [14]). The analysis of dispersion relation (2.41) shows that the HF ANSMs can propagate only in wide waveguides with radius:

$R_1/\delta \gg \Omega_e|m|/|\omega_e|$ and the bulk azimuthally non-symmetric O-mode cannot propagate if the external magnetic field and the waveguide radius are not sufficiently large, so that the following inequality is valid:

$$R_1^2\omega_e^2/c^2 < j_{|m|,s}^2 - m^2, \tag{2.53}$$

where $j_{|m|,s}$ is the sth root of the Bessel functions of the first kind $J_{|m|}\left(j_{|m|,s}\right) = 0$.

In this case, the HF ANSM frequency and its correction can be found analytically. Their approximate expressions are as follows:

$$\omega_0 \approx \sqrt{\omega_{UH}^2 + m^2c^2/R_1^2}, \tag{2.54}$$

$$\Delta\omega \approx 0.5N_z^2\omega_0 + 0.5\pi|m|N_z^2 J'_{|m|}(\varsigma^*)N_{|m|}(\varsigma^*)\omega_0, \tag{2.55}$$

where $\varsigma^* \approx \sqrt{m^2 + \omega_e^2 R_1^2 c^{-2}}$. Results of numerical analyses of Eq. (2.41) for the HF ANSM are plotted in Fig. 2.18 for the case $m = -2$, with the same notations as in Fig. 2.17. One can see that the width of this frequency range is narrower compared with the case shown in Fig. 2.17. Therefore, the slow-wave frequency does not become zero when $k_z \to 0$. It should be noted that the dispersion curve, which describes the bulk O-mode in this case, lies above the HF range (2.47) where the long-wavelength X-mode can propagate. Therefore, there is no resonant interaction between these two ANS modes in this case (see reference [14] for details of the resonant interaction).

In this Section, results of the theoretical study of the properties of non-symmetric SWs with arbitrary azimuthal mode index and small value of k_z, which propagate in a cylindrical metal waveguide completely filled with magneto-active cold plasma, are presented. The dispersion equation for these SWs has been obtained in the case of a non-uniform radial plasma density profile. However, its analysis is carried out

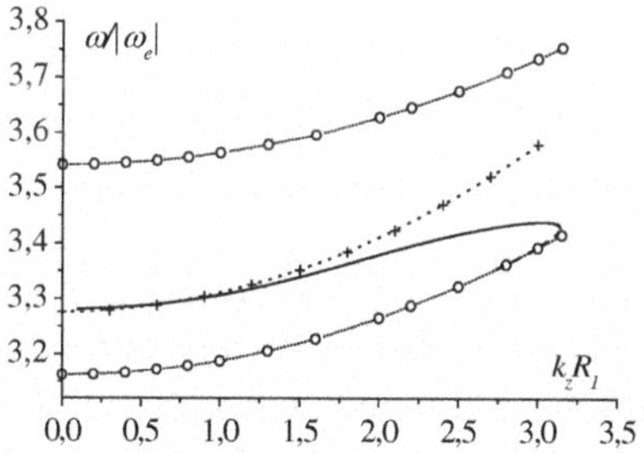

Fig. 2.18 Normalized frequency of the HF ANSM as function of normalized axial wavenumber. $m = -2$, $\Omega_e = 3|\omega_e|$, $R_1 = 7\delta$. Fast mode (*solid curve*), slow mode (*dashed curve*), limit of the LF band (*dotted curve with hollow circles*), asymptotic solution (*dotted curve with crosses*)

analytically in the limiting case of a uniform plasma. Results of the numerical study confirm the analytical ones.

2.7 Higher Radial Modes of Surface Flute Waves in Magneto-active Cylindrical Plasma Waveguides

Most results in Sects. 2.1–2.6 are obtained for the case of a narrow dielectric gap which separates the plasma column from the metal wall, $R_2 - R_1 \ll R_1$, and where the SFWs are of bulk nature. If the dielectric layer is assumed to be uniform, the electromagnetic fields of SFWs can be expressed in the gap, $R_1 < r < R_2$, as a linear combination of Bessel functions of the first and second kind, $J_m(\xi)$ and $N_m(\xi)$ [2]. In the case of the narrow gap, the radial change of the waves is small and the zeroth radial mode of SFWs has been considered. The advantage of the zeroth radial mode is as follows: the effect of plasma parameters on its dispersion properties is the most pronounced as compared to those of the higher radial modes. On the other hand, practical needs of electronics require higher frequencies and shorter wavelengths of electromagnetic radiation which can be provided by application of the higher radial modes of SFWs without any change of the waveguide design.

The dispersion properties of SFW higher radial modes in cylindrical plasma-filled metal waveguides without external static magnetic field were studied in detail in [19]. It was shown that the eigenfrequencies of higher radial modes of SFWs can reach the upper limit of the frequency range where SFWs exist in contrast to those of the zeroth radial mode. Higher radial modes of SFWs can propagate in waveguides where the radius and plasma particle density are not very small.

The present Section is devoted to studying the dispersion properties of SFW higher radial modes in magneto-active plasma waveguides. It generalizes the study of SFW dispersion properties carried out for the case of the zeroth radial mode [20] to the case of higher radial modes, and the investigation done in [19] for plasma waveguides without external static magnetic field to the case of a uniform static axial magnetic field.

In the present Section, the plasma-dielectric-metal waveguide structure shown in Fig. 2.1 is considered. The structure is assumed to be homogeneous in axial direction, $\partial/\partial z = 0$, in cylindrical coordinates. The dispersion relation of SFXWs (2.4) is derived in Sect. 2.1.

In the case of higher radial modes, one can use the asymptotic expansions for the cylindrical functions [2] and derive the following approximate expressions for the eigenfrequencies of SFXW with lth radial mode index. The asymptotic solution of the dispersion relation (2.4) in the case of LF SFXWs and small depth of the wave penetration into the plasma, $k_\perp R_1 \gg 1$ (that is $k_{ef} \ll 1$), was given in [21]:

$$\omega \approx \frac{\pi l c}{\sqrt{\varepsilon_d}(R_2 - R_1)}\left[1 - \frac{m\mu c^2}{\Omega_e^2 R_1 (R_2 - R_1)}\right]. \tag{2.56}$$

Nearby the upper limit of the LF frequency range (1.13), the depth of the wave penetration into the plasma becomes large, $k_\perp R_1 \ll 1$, and eigenfrequency of LF SFXWs with positive azimuthal mode index writes as [22]:

$$\omega \approx \frac{c}{\sqrt{\varepsilon_d}(R_2 - R_1)}\left[\pi(l - 0.5) + \frac{R_1 \Omega_e}{2c|m|\sqrt{\varepsilon_d}}\right]. \tag{2.57}$$

In the case of $m < 0$, the asymptotic expression for LF FXSWs looks like Eq. (2.57), however, with a different second term in the square brackets:

$$\omega \approx \frac{c}{\sqrt{\varepsilon_d}(R_2 - R_1)}\left[\pi(l - 0.5) + \frac{2(|m| + 1)}{\kappa R_1}\right]. \tag{2.58}$$

Nearby the upper limit of the HF frequency range (1.14), one can apply Eq. (2.57) as asymptotic expression for the eigenfrequency of HF SFXWs, which propagate with negative azimuthal mode indices, and Eq. (2.58) in the case of HF SFXWs which propagate with positive m with replacing l by $(l + 1)$.

Since $\mu \equiv \varepsilon_2/\varepsilon_1 < 0$ within the LF frequency range (1.13) under consideration, a simple analytical analysis shows that the eigenfrequencies of SFXWs with positive azimuthal mode index are higher than those of SFXWs with $m < 0$. The same relation is known for the zeroth radial mode [20]. The splitting of the spectrum is more pronounced for SFXWs with higher radial mode indices l and higher azimuthal mode indices $|m|$. Regardless of the smallness of the splitting, it is of favor for the problem of SFXW excitation with an electron beam gyrating along large Larmor orbits around the plasma column in the structure shown in Fig. 2.1 in which the dielectric layer is replaced by vacuum [23], since the direction of the electron gyration in the axial static magnetic field coincides with the direction of SFXW propagation with azimuthal mode index $m > 0$. It is superfluous to say that the expression (2.56) coincides with that derived in [19] for the case of isotropic plasma waveguides in the limit $|\omega_e| \to 0$ ($\mu \equiv \varepsilon_2/\varepsilon_1 \to 0$).

Here follows the numerical study of the dispersion relation (2.4). In Figs. 2.19, 2.20 and 2.21, the eigenfrequency normalized to the electron cyclotron frequency, $\omega/|\omega_e|$, is used as ordinate axis. Figure 2.22 shows the study of the effect of the external static magnetic field on the dispersion properties of SFXWs. In this case, their eigenfrequency normalized by the Langmuir frequency, ω/Ω_e, is used as ordinate axis. When studying the dispersion properties, one plots the curves of the dependence of eigenfrequency versus wave vector. In the case of SFXWs, the magnitude of the characteristic wave vector is $|m|/R_1$. In Figs. 2.19, 2.20, 2.21 and 2.22, this vector normalized to the skin-depth $\delta = c/\Omega_e$, $k_{ef} = |m|\delta/R_1$, is used as abscissa axis. If not mentioned specifically, the following parameters of the plasma waveguide system are used for calculating the eigenfrequencies: $Z \equiv \Omega_e/|\omega_e| = 5$, $R_2/R_1 = 2$, and $\varepsilon_d = 4$ (quartz). This set of parameters corresponds to the conditions of a moderate experimental University laboratory.

The solid curves in Fig. 2.19 show the dependence of the SFXW eigenfrequency as function of the effective wavenumber. The numbers near the curves mark the

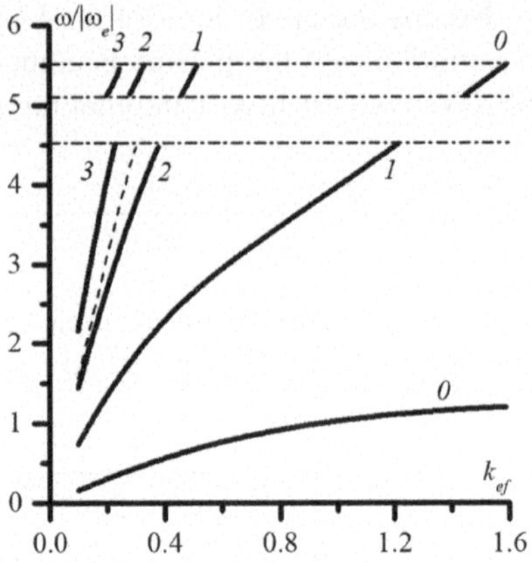

Fig. 2.19 Normalized SFXW eigenfrequency versus k_{ef} (*solid curves*). $Z \equiv \Omega_e/|\omega_e| = 5$, $R_2/R_1 = 2$, $\varepsilon_d = 4$, $m = -1$. The *numbers* mark the radial mode indices $l = 0, 1, 2, 3$. The horizontal *dash-dotted* lines show the limits of the LF and HF ranges

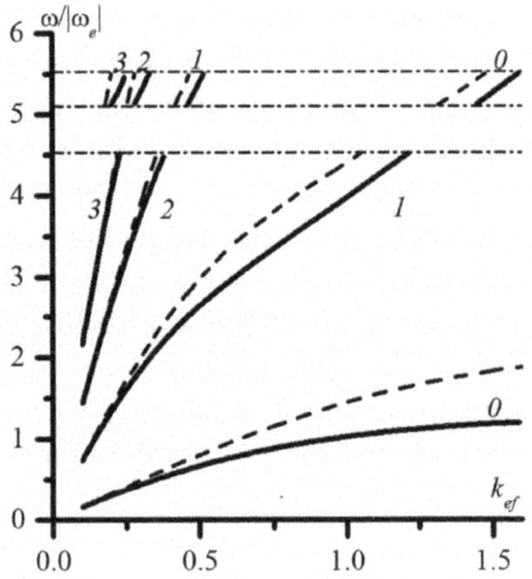

Fig. 2.20 Normalized SFXW eigenfrequency versus k_{ef}. $Z = 5$, $R_2/R_1 = 2$, $\varepsilon_d = 4$, $m = -1$ (*solid curves*), $m = +1$ (*dashed curves*). The *numbers* mark the radial mode indices $l = 0, 1, 2, 3$. The horizontal *dash-dotted* lines show the limits of the LF and HF ranges

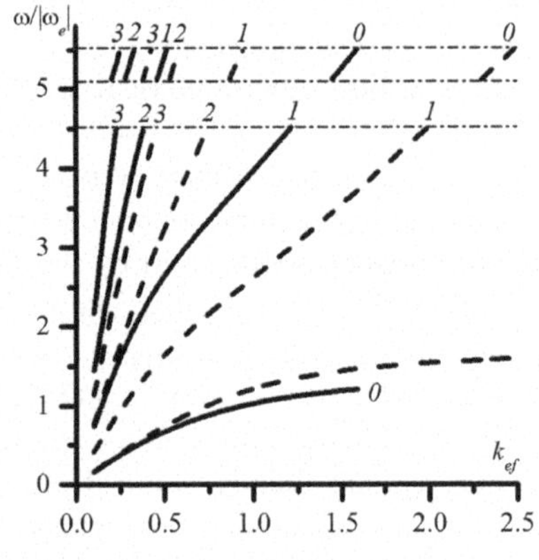

Fig. 2.21 Normalized SFXW eigenfrequency versus k_{ef}. $Z = 5$, $R_2/R_1 = 2$, $\varepsilon_d = 4$, $m = -1$ (*solid curves*), and $m = -2$ (*dashed curves*). The *numbers* mark the radial mode indices $l = 0, 1, 2, 3$. The horizontal *dash-dotted* lines show the limits of the LF and HF ranges

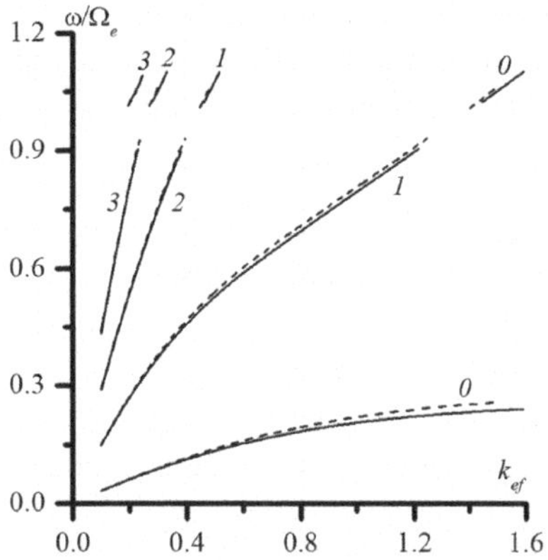

Fig. 2.22 Normalized SFXW eigenfrequency versus k_{ef}. $Z = 5$ (*solid curves*), $Z = 8$ (*dashed curves*), $b/a = 2$, $\varepsilon_d = 4$, $m = -1$. The *numbers* mark the radial mode indices $l = 0, 1, 2, 3$

radial mode index l. The horizontal dash-dotted lines correspond to the limits of the frequency ranges (1.13) and (1.14). The dashed curve shows the asymptote (2.56). It exceeds the respective curve for $l = 2$ by less than 16%, the accuracy being better for smaller k_{ef}. Although the asymptote (2.58) exceeds the magnitude of HF SFXW with $l = 1$ by less than 18%, its curve is not plotted in Fig. 2.19, since it would lie above the range (1.14).

The numerical analysis confirms the analytical result: the eigenfrequencies of higher radial modes are higher than those of the zeroth radial mode studied earlier in [20]. For example, for $k_{ef} = 0.5$, the LF SFXW with the azimuthal mode index $m = +2$ has the following eigenfrequencies: for the zeroth radial mode, $l = 0$, $\omega = 0.815|\omega_e|$; for the first radial mode, $l = 1$, $\omega = 1.844|\omega_e|$; and for the second radial mode, $l = 2$, $\omega = 3.392|\omega_e|$. The latter is higher than that of the first radial mode by the factor of 1.84 while Eq. (2.56) predicts the factor 2.

Those sections of the dispersion curves of HF SFXWs, which are described by Eqs. (2.57) and (2.58) and correspond to the frequencies nearby the upper limit of the range, are situated closely to each other. The curve with the radial mode index $(l + 1)$ is situated on the left hand of that with l. The higher is the radial mode index l, the less is the distance $\Delta k_{ef}(l + 1, l)$ between the two curves which correspond to two neighboring radial mode indices $(l + 1)$ and l, $\Delta k_{ef}(l + 1, l) \propto l^{-2}$. Increase of the radial mode index l is accompanied by the approach of the dispersion curve to the vertical disposition. The angle between the sections under the consideration and a vertical line is inversely proportional to l.

Application of the magnetic field $\vec{B}_0$ is known to remove the spectrum degeneracy in respect to the sign of the azimuthal mode index. The analytical study predicts $\omega(+|m|) > \omega(-|m|)$ for both HF and LF ranges (note, $\mu \equiv \frac{\varepsilon_2}{\varepsilon_1} < 0$ in LF range (1.13)), which is confirmed by the calculations plotted in Fig. 2.20. The difference between the curves related to LF SFXWs with $m = \pm 1$ is invisible already for $l = 3$.

The decrease of the SFXW eigenfrequency with increasing absolute value of the azimuthal mode index is demonstrated in Fig. 2.21. This behavior is in qualitative agreement with the asymptote (2.56) for LF SFXWs. Analogous qualitative dependence of the HF SFXW frequency on $|m|$ is predicted by the asymptote (2.57). In this respect SFXW higher radial modes differ from the zeroth radial modes for which the increase of the eigenfrequency with increasing m is well-known [20]. The increase in frequency versus k_{ef} for SFXWs with smaller absolute values of the azimuthal mode index is sharper than for SFXWs with larger absolute values of m. The curves which correspond to SFXWs with smaller m reach the upper boundary of the LF frequency range (1.13) for smaller values of k_{ef}. In other words, higher radial modes of SFXWs with larger $|m|$ can propagate in waveguides with smaller plasma column radius and smaller plasma particle density.

The effect of the external static axial magnetic field B_0 on the SFXW eigenfrequency is examined in Fig. 2.22. The solid curves correspond to the case of stronger B_0: $Z = 5$; the dashed curves to weaker B_0: $Z = 8$. One can draw the following conclusions on the basis of the results plotted in Fig. 2.22. Significant increase of B_0 causes a weak increase of the HF range and a weak decrease (increase) of ω for fixed k_{ef} for SFXWs with negative (positive) m. The same qualitative dependence of the SFXW eigenfrequency on B_0 is well-known (with keeping in mind the assumption $\Omega_e > |\omega_e|$ in the frame of which the present research is done) for the case of the zeroth SFXW radial mode [20] and follows from Eq. (2.56) as well.

Equation (2.56) predicts that the SFXW eigenfrequency depends on the gap parameter Δ almost inversely proportional. However, for $k_{ef} = 0.5$, the LF SFXW eigenfrequency in the case of $\Delta = 0.5$ is $\omega = 2.932|\omega_e|$, which is by the factor of 1.59 (instead of the theoretical prediction 2) larger than the SFXW eigenfrequency in the case of $\Delta = 1.0$, $\omega = 1.844|\omega_e|$, and by the factor of 2.04 (instead of the theoretical prediction 3) larger than the SFXW eigenfrequency in the case of $\Delta = 1.5$, $\omega = 1.437|\omega_e|$. Note that for the case of the zeroth LF SFXW radial mode [20], the eigenfrequency increases with increasing Δ.

Numerical analysis of the effect of the permittivity ε_d of the dielectric layer on the SFXW eigenfrequencies shows a monotonous decrease of the eigenfrequency with increasing dielectric constant. Equation (2.56) predicts that the LF SFXW frequency in the case of $\varepsilon_d = 4$ should be twice less than in the case of $\varepsilon_d = 1$. Indeed, for $k_{ef} = 0.5$, the SW eigenfrequency in the case of $\varepsilon_d = 1$ is $\omega = 3.466|\omega_e|$, and by the factor 1.88 larger than the SFXW eigenfrequency in the case of $\varepsilon_d = 4$, $\omega = 1.844|\omega_e|$. In the case of the zeroth radial mode, the decrease of the LF SFXW eigenfrequency with increasing ε_d is less pronounced.

The following conclusions can be done on the basis of the studies carried out in the present Section. A comprehensive analysis of the dispersion properties of SFXW higher radial modes is presented. The dependence of the SFXW eigenfrequency on such plasma waveguide parameters as dimensions of plasma column and dielectric layer, plasma particle density, radial and azimuthal mode indices, and dielectric permittivity is studied. The eigenfrequency ω of SFXWs is shown to increase with decreasing dielectric layer width $(R_2 - R_1)$, absolute value of azimuthal mode index $|m|$ and dielectric constant ε_d, and especially with increasing radial mode index l. An increase of the static axial magnetic field B_0 causes an insignificant increase of

LF SFXW eigenfrequency and a weak decrease in the eigenfrequency of the higher radial HF SFXW modes. The dispersion properties of SFXW higher radial modes are compared to those of the SFXW zeroth radial mode. The eigenfrequencies of SFXWs with positive azimuthal mode index, $m > 0$, are shown to be higher than those with negative m: $\omega(|m|) > \omega(-|m|)$.

Increase of the radial mode index l (decrease of the width of the dielectric and its dielectric constant) causes a shift of the HF SFXW dispersion curve towards smaller magnitudes of the effective wavenumber, which means larger magnitudes of the plasma column radius R_1 and particle density n_e. Application of higher radial modes of HF SFXWs does not give any gain for increasing the frequency compared to the zeroth mode, as it is in the case in the LF range, but expands the range of parameters of the plasma waveguide for which they can be excited towards larger values of n_e and/or R_1.

These results are of interest for the field of plasma electronics. Excitation of SFXW higher radial modes by an electron flow gyrating in a static axial magnetic field around the plasma column along large Larmor orbits is presented in Chap. 7 of this monograph. Advantages of utilization of SFXWs with extraordinary polarization excited by the rotating electron beam as compared to the bulk electromagnetic waves propagating along the axis of cylindrical waveguides excited by a longitudinal electron beam in devices of plasma electronics are discussed in Chaps. 7–10.

In the limit of a narrow depth of SFXW penetration into the plasma column, $k_\perp R_1 \gg 1$, the SFXW dispersion properties do not depend on the properties of the matter placed nearby the waveguide axis. In this case the presented results should coincide with the results of studying the dispersion properties of SFXWs in a dielectrically lined slow-wave waveguide [24]. The authors of [24] considered the increase of the number of solutions of the dispersion relation as one of the most significant results of their study. However, no explanation of the physical reasons for this increase of the number of solutions was given in [24]. The present electrodynamics of plasma waveguides proposes two possible explanations for this phenomenon. First, the increase of the number of solutions of the dispersion relation for surface type waves can be caused by introducing an additional interface, in particular, the interface "dielectric rod-plasma layer" which was done in [24]. Second, if one finds additional roots of the dispersion relation for the flute ($k_z = 0$) electromagnetic perturbations with given azimuthal mode index m, this can indicate that these roots correspond to other (higher) radial modes. Propagation of just higher radial modes is discussed in the present Section. Probably this second case took place in numerical solving the dispersion relation in [24]. It is a pity but the mistakes in deriving the expressions for the components of extraordinary polarized SFXW fields in [24] and some other papers by B. Jazi and his co-authors require revising the results obtained therein.

Studying the SFXW dispersion properties can be of interest also for nano-physics [25] and medical diagnostics [26].

2.8 Conclusions

Results of theoretical investigations on the properties of electromagnetic flute waves propagating in different cylindrical plasma filled waveguides are presented here for the case of application of an external axial static magnetic field. The investigations carried out allow one to determine limits of frequency ranges, where ordinary and extraordinary flute modes can propagate, to analyze their dispersion properties and spatial distribution of their fields, and to calculate their damping rates, which are caused by the following physical mechanisms: collisions between plasma particles and radiation of their energy through an axial slot antenna aperture in the metal wall of the waveguide. It is proved that in the presence of an external axial static magnetic field, flute eigenwaves can propagate both along plasma-dielectric and plasma-metal boundaries and that if there is no external magnetic field, SFXWs can propagate along a plasma-dielectric boundary only. Phenomena of unidirectional propagation of surface flute X-modes along a plasma-metal boundary are proved here. From the theoretical point of view such phenomena are typical for surface waves propagating along a plasma-metal interface perpendicularly to an external magnetic field. From practical point of view it is a very useful property in the case when one intends to obtain a radiofrequency device without reflected waves. As already mentioned above, SFW damping can be caused by two reasons, namely by collisions between plasma particles and by radiation of their power through a narrow axial slot antenna aperture in the metal wall of the waveguide. The magnitude of the SFXM damping rates in the both cases (presence and absence of the dielectric layer, which could separate the plasma column from the waveguides wall) is proportional to the angular size of the slot and to the electron plasma (Langmuir) frequency. In addition to the surface flute X-modes the same type of O-modes can propagate in waveguides where the plasma column is separated from the waveguide wall by a dielectric layer. The properties of these O-modes are determined by the plasma density and characteristics of the dielectric layer. It is proved as well that application of the theory of SFWs allows one to describe dispersion properties even of non-symmetrical surface eigenwaves, which have a small, but non-zero value of axial wavenumber k_z. The values of eigenfrequencies of these non-symmetrical waves are found with accuracy up to addendums proportional to the squared value of k_z. The theory of SFWs developed in this chapter is applied in Chap. 4 for studying the linear interaction between ordinary and extraordinary flute modes in current-carrying plasma waveguides.

References

1. Vukovic, V. (1986). *Surface waves in plasmas and solids*. World Scientific.
2. Abramowitz, M., & Stegun, I. A. (Eds.). (1964). *Handbook of mathematical functions. Applied Mathematics*, Series 55. National Bureau of Standards.
3. Alexandrov, A. F., Bogdankevich, L. S., & Rukhadze, A. A. (1984). *Principles of plasma electrodynamics*. Springer.

4. Azarenkov, N. A., & Ostrikov, K. N. (1991). Surface magnetoplasma waves at the interface between a plasma-like medium and a metal in a Voigt geometry. *Physics Reports, 308*, 333–428.
5. Girka, V., Girka, I., Kondratenko, A., et al. (1989). Azimuthal surface waves at the boundary between a magneto-active plasma and a metal. *Soviet Journal of Communications Technology and Electronics, 34*, 96–99.
6. Girka, V. (2006). Theory of transverse surface electro-magnetic waves propagating along a plasma–metal boundary with a finite radius of curvature in a magnetic field. *Plasma Physics Reports, 32*(5), 401–410.
7. Girka, I., & Kovtun, P. (1998). Azimuthal surface waves in a magnetized plasma. *Technical Physics, 43*(12), 1424–1427.
8. Krall, N. A., & Trivelpiece, A. W. (1973). *Principles of plasma physics*. McGraw-Hill.
9. Gestrina, G. N. (1988). Radiation of SHF oscillations of plasma from opened periodic structures. *Soviet Journal Radiophysics and Quantum Electronics, 31*, 1502–1505. (in Russian).
10. Gestrina, G. N. (1989). *On spectrum of eigen frequencies of opened periodic plasma waveguides*. Reports of Ukrainian Academy of Science, *A–10*, pp. 74–76 (in Russian).
11. Shestopalov, V. P. (1976). *Diffraction electronics*. Vyscha Shkola (in Russian).
12. Girka, V. O., & Girka, I. O. (1992). Emission of azimuthal surface waves from narrow waveguide slot. *Soviet Journal of Communications Technology and Electronics, 37*(9), 32–35.
13. Chen, F. F. (1974). *Introduction to plasma physics*. Plenum Press.
14. Girka, V., & Girka, I. (2002). Asymmetric long-wavelength surface waves in magnetized plasma waveguides entirely filled with plasma. *Plasma Physics Reports, 28*(11), 916–924.
15. Akhiezer, A. I., Akhiezer, I. A., Polovin, R. V., Sitenko, A. G., & Stepanov, K. N. (1975). *Plasma electrodynamics*. Pergamon Press.
16. Kuzelev, M. V., Romanov, R. V., Selivanov, I. A., et al. (1994). Spectrums of eigen oscillations of coaxial plasma waveguide immersed into finite value magnetic field. *Proceedings of Institute for General Physics RAS, 45*, 17–38. (in Russian).
17. Girka, O. I., Girka, I. O., Girka, V. O., et al. (2008). Propagation of azimuthal waves along the surface of a metal current-carrying cylinder immersed into a magnetized plasma. *Radiophysics and Quantum Electronics, 51*(2), 110–122.
18. Girka, V., Girka, I., Pavlenko, I., et al. (2012). Coupled azimuthal modes propagating in current-carrying plasma waveguides. *Journal of Plasma Physics, 78*(2), 105–123.
19. Girka, I. O., Omelchenko, I. V., & Sydora, R. (2017). Higher radial modes of azimuthal surface waves in cylindrical waveguides without external magnetic field. *Progress in Electromagnetics Research, M (PIERM), 54*, 1–7.
20. Girka, V. A., Girka, I. A., Kondratenko, A. N., et al. (1988). Azimuthal surface waves of magnetoactive plasma waveguides. *Soviet Journal of Communications Technology and Electronics, 33*(8), 37–41.
21. Girka, I. O., Kondratenko, V. M., & Thumm, M. (2018). Higher radial modes of azimuthal surface waves in magnetoactive cylindrical plasma waveguides. *Journal of Plasma Physics, 84*(6), 905840603.
22. Girka, I., & Kondratenko, V. (2020). Higher radial modes of azimuthal surface waves above the upper-hybrid frequency in cylindrical waveguides partially filled by plasma. *Problems of atomic science and technology* (No. 6 (130)). *Plasma Physics* ((26), pp. 22–25).
23. Girka, I., Pavlenko, I., & Thumm, M. (2018). Excitation of higher radial modes of azimuthal surface waves in the electron cyclotron frequency range by rotating relativistic flow of electrons in cylindrical waveguides partially filled by plasmas. *Physics of Plasmas, 25*, 052109.
24. Jazi, B., & Mehdian, H. (2004). Dispersion relation of azimuthal electromagnetic surface waves on a magnetized plasma column in a dielectric lined slow-wave waveguide. *Plasma Physics and Controlled Fusion, 46*(3), 507–518.
25. Ostrikov, K., Neyts, E. C., & Meyyappan, M. (2013). Plasma nanoscience: From nano-solids in plasmas to nano-plasmas in solids. *Advances in Physics, 62*(2), 113–224.
26. Baieva, S., Ihalainen, J. A., & Toppari, J. J. (2013). Strong coupling between surface plasmon polaritons and β-carotene in nanolayered system. *Journal of Chemical Physics, 138*, 044707.

Chapter 3
Electromagnetic Energy Rotation Caused by Surface Flute Waves

The difference between "velocity of transmission of phase" υ_{ph} and "velocity of propagation of vibratory motion" υ_g was established for the first time in [1] as one of the conclusions of researches in the field of theory of light. Thirty-four years later, in studies of elastic waves in continuous media, the average wave energy flux density was found to be proportional to the average wave energy density [2]. The brief and complete overview of the history of implementation of the term "group velocity" in [3] makes it possible not to repeat it herein. However, the significance of the group velocity as a tool for studying a wide scope of physical problems keeps it in the focus of many investigations up to now.

One can find three different definitions of the group velocity, e.g. in [4]. The first one is the group velocity $\upsilon_g = d\omega/dk$. In [4], Rayleigh was noted as the person who was the first to introduce this definition while studying sound waves. However, as it is already mentioned above, Hamilton had operated with the group velocity already in 1839 [1] while studying the theory of light. The second definition is the signal velocity υ_s of Sommerfeld [4], who introduced this value while investigating in a one-dimensional approach how a well-defined signal propagates in a material medium. The third one is the velocity of energy transfer υ_{en}, which is the proportionality coefficient between the energy flux surface density averaged over the time period (in electrodynamics, this is the well-known Poynting vector) and the averaged wave energy volume density. For instance, a rigorous identity between the group velocity υ_g and the velocity of energy transfer υ_{en} in non-uniform non-dissipative media with or without anomalous dispersion was demonstrated under very general conditions in [5, 6]. The precise physical meaning of the complex group velocity in absorbing and active media is a matter of extensive research (see, e.g., [7] and references therein). These three velocities were shown to be identical for non-absorbing media and differ considerably in an absorption band.

The group velocity plays an important role also in the theory of linear particle accelerators [8]. In particular, the group velocity was shown to be a significant characteristic observable in studies of power flow in waveguides [9].

© The Author(s), under exclusive license to Springer Nature Switzerland AG 2022

I. Girka and M. Thumm, *Surface Flute Waves in Plasmas*, Springer Series on Atomic, Optical, and Plasma Physics 120, https://doi.org/10.1007/978-3-030-98210-2_3

Group velocity uses to be introduced in classical textbooks on electrodynamics, including plasma electrodynamics. It is demonstrated on pages 246–248 in [10] that in vacuum waveguides, the axial energy flux density is in general case proportional to the energy density with the group velocity $\upsilon_{gz} \equiv \partial\omega/\partial k_z$ as coefficient of proportionality.

The term of group velocity is widely used also in plasma electrodynamics. The group velocity $\upsilon_g = d\omega/dk$ was introduced, for example, on page 33 in the textbook [11] as the quantity which "characterizes the velocity of amplitude (and with that energy) displacement" without any reasoning. However, the authors of this textbook almost did not apply the term in presenting selected chapters of plasma electrodynamics. This explains the fact that this mistake in the physical sense of the group velocity did not hinder the authors of textbook [11] to obtain correct physical results in the field of plasma electrodynamics and avoiding mistakes in physical issues.

The other popular comprehensive textbook in plasma electrodynamics [12] does not discuss the relation between the group velocity $\upsilon_g = d\omega/dk$ and velocity of energy transfer υ_{en}. In studies on plane electromagnetic waves in infinite homogeneous plasmas, the group velocity is applied for explaining the phenomenon of conical refraction [13] in magneto-hydrodynamics, which is analogous to that in crystal optics [14].

The energy transfer by electromagnetic waves is usually considered in non-bounded media. Even the waveguide theory considers usually the energy transfer along the waveguide axis only (along the boundary). In this case, the axial wave vector k_z is obtained from the dispersion relation. Then both phase and group wave velocities are defined in a usual way but using the axial component of the wave vector instead of the complete wave vector in the medium. However, it is clear that the electromagnetic energy in waveguides with arbitrary cross section can rotate around the axis, in addition to propagation along the axis.

3.1 Electromagnetic Energy Rotation Along Plasma-Metal Interface in Cylindrical Waveguides

The objective of the present Section is to introduce an angular velocity of energy transfer ω_{en} for description of electromagnetic energy rotation around the axis of cylindrical plasma waveguides, as well as an angular group velocity ω_g, and angular phase velocity ω_{ph}. Propagation of surface flute waves (SFWs) with extraordinary polarization (SFXWs) along the plasma-metal interface in cylindrical waveguides is considered as an example. Here, SFXW propagation along the plasma-metal interface inside a cylindrical waveguide (interior problem) is under the consideration. The cylindrical metallic waveguide with radius R is entirely filled by a plasma (see Fig. 2.5).

The radial distribution of the axial magnetic wave field H_z is determined in Sects. 1.4.1 and 2.2. (see Eq. (1.12)). The boundary condition of finite magnitude of the wave fields allows to determine the constant of integration A_2.This integration constant is set to $A_2 = 0$ to eliminate the pole of $K_m(\xi)$, which is known to diverge at the axis $(r = 0)$, $lim_{\xi \to 0} K_m(\xi) = \infty$. The radial and azimuthal electric wave fields E_r and $E\varphi$ are given in Sect 1.4.1 by Eq. (1.10). SFXWs can propagate in the HF range (1.14) and LF range $\omega_{LH} < \omega < |\omega_e|$.

As it is found in Sect. 2.2, SFXWs are unidirectional waves [15, 16], which means that in specific frequency ranges, their dispersion relation allows solutions with only one sign of the azimuthal mode index. In other words, SFXWs propagate in specific frequency ranges in a definite direction without reflection. This feature can be of interest for radio frequency (RF) engineering. The azimuthal mode index of SFXWs can be only positive, $m>0$ in the LF range. In this range SFXWs propagate in the same direction in which electrons gyrate in an external static axial magnetic field. In the HF range (1.14), the azimuthal mode index of SFXWs is always negative, $m<0$.

Since the waveguide metal walls are assumed to have infinite electric conductivity, application of a non-dissipative boundary condition at the plasma-metal interface, $r = R$, allows to derive the dispersion relation of SFXWs in the form (2.11) with $F_m(\xi) = I_m(\xi)$.

The dispersion relation (2.11) describes also the dispersion properties of extraordinary polarized electromagnetic bulk flute waves (BFXWs), whose real parts of the frequency lie outside of the HF range (1.14) and LF range $\omega_{LH} < \omega < |\omega_e|$, and which are not the subject of detailed analysis in the present monograph. In this case $F_m(\xi) = J_m(\xi)$, where $J_m(\xi)$ is a Bessel function of the first kind [17].

To study electromagnetic energy rotation carried by SFXWs, one has to use the Poynting vector [2, 10, 18]

$$\vec{S} \equiv \frac{c}{4\pi} \vec{E} \times \vec{H}. \quad (3.1)$$

Since flute waves (FWs, both bulk and surface) do not propagate in axial direction, $k_z = 0$, they naturally do not transfer any energy along the $\vec{z}$ axis. This is confirmed via direct calculation of the axial component of the Poynting vector

$$S_z \equiv \frac{c}{4\pi} \left(E_r H_\varphi - E_\varphi H_r \right). \quad (3.2)$$

For ordinary polarized FWs, the electric wave field components E_r and E_φ are equal to zero, and the Poynting vector component S_z is hence also equal to zero. Extraordinary polarized FWs do not contain neither radial H_r nor azimuthal H_φ magnetic fields, therefore their axial energy flux is also zero.

The absence of the axial component of the energy flux of azimuthal waves is also clear in terms of group velocity:

$$\upsilon_{gz} \equiv \frac{\partial \omega}{\partial k_z} = -\frac{\partial D(\omega, m)/\partial k_z}{\partial D(\omega, m)/\partial \omega}. \quad (3.3)$$

Indeed, since e.g. the SFXW dispersion relation (2.11) is independent of the axial wavenumber k_z, $D(\omega,m)$ is independent of k_z, and hence $\upsilon_{gz} = 0$.

This result for the axial group velocity is expected in advance, since the axial power flux in the waveguide of arbitrary cross-section is proportional to the axial wavenumber [19]. Generally speaking, the dispersion relation of electromagnetic waves in cylindrical waveguides with axial magnetic field can depend on the axial wavenumber k_z . However, if the wave group velocity in axial direction is zero, one can conclude that the dispersion relation of electromagnetic waves is an even function of k_z. In this case, the first derivative of the dispersion relation in respect of k_z is an odd function of k_z. As a consequence, the electromagnetic wave power does not propagate axially if $k_z = 0$ as it should be from the physical point of view. In other words, the spectra of electromagnetic waves in cylindrical waveguides with axial magnetic field are degenerate in respect of the sign of k_z. This is well-known, but it is very pleasant to see that the present theory is self-coordinated and internally consistent.

Since FWs (both bulk and surface) propagate in cylindrical waveguides, their energy flux in radial direction is also zero. This is obvious, since eigenwaves of cylindrical waveguides are standing waves in radial direction. This can also be checked by calculation of the Poynting vector. Its radial component is:

$$S_r \equiv \frac{c}{4\pi} E_\varphi H_z. \tag{3.4}$$

In non-dissipative approach, $\varepsilon_{1,2}$ are real values, and the phase shift of E_φ compared to H_z is $\pi/2$ (see Eq. (1.10)). This means that averaging of S_r over the time period T gives zero:

$$\langle S_r \rangle_T \equiv \frac{c}{4\pi} E_\varphi(r) H_z(r) \frac{\int_0^T (-\cos(m\varphi - \omega\, t + \varphi_0)\sin(m\varphi - \omega t + \varphi_0))dt}{T} = 0. \tag{3.5}$$

Finally, the energy flux of FWs (both bulk and surface) in azimuthal direction is non-zero. This is expected from the point of view of general physics, but the calculation of the Poynting vector directly confirms this statement. Its azimuthal component is

$$S_\varphi \equiv \frac{-c}{4\pi} E_r H_z. \tag{3.6}$$

In non-dissipative approach, the phase shift of E_r compared to H_z is 0 (see Eq. (1.10)). This means that averaging of S_φ over the period T does not give zero:

$$\langle S_\varphi \rangle_T \equiv \frac{-c}{4\pi} E_r(r) H_z(r) \frac{\int_0^T \cos^2(m\varphi - \omega\, t + \varphi_0)dt}{T} = \frac{-cE_r(r)H_z(r)}{8\pi} \neq 0. \tag{3.7}$$

The linear phase velocity $v_{ph} = \omega/k_\varphi$ cannot be considered as an appropriate tool for describing the "velocity of phase transmission" of FWs since one has to choose an azimuthal wave vector k_φ in some way. The simplest solution would be $k_\varphi = m/r$. However, since it depends on the radial coordinate r, it does not fit. One has to introduce the angular phase velocity to describe the wave propagation in azimuthal direction. It can be easily derived from differentiating the phase $\zeta = (\zeta_0 + m\varphi - \omega t) = \text{Const}$, where: $d\zeta = (md\varphi - \omega dt) = 0$, that is why

$$\omega_{ph} = \omega / m. \tag{3.8}$$

In the case of the cylindrical plasma waveguide structure the angular group velocity

$$\omega_g \equiv \frac{\partial \omega}{\partial m} = -\frac{\partial D(\omega, m)/\partial m}{\partial D(\omega, m)/\partial \omega} \tag{3.9}$$

can be introduced.

In contrast to the angular phase velocity ω_{ph} given in Eq. (3.8) and the angular group velocity ω_g of Eq. (3.9), which are integral characteristics of the azimuthal energy transfer, the angular velocity of energy transfer $\omega_{en}(r)$ can be introduced as the radial dependence of the ratio of the time-averaged azimuthal Poynting flux

$$\langle S_\varphi(r) \rangle_T = \frac{-c}{8\pi} E_r(r) H_z(r) \equiv \frac{-c}{8\pi} E_r(R) H_z(R) \beta(r), \tag{3.10}$$

to the time-average angular electromagnetic energy density

$$\begin{aligned} \frac{d^3 W}{drdzd\varphi} &\equiv \frac{r}{16\,\pi} \\ &\times \left\{ |H_z(r)|^2 + \frac{\partial(\omega\varepsilon_1)}{\partial\omega}\left[|E_r(r)|^2 + \left|E_\varphi(r)\right|^2 \right] - 2\frac{\partial(\omega\varepsilon_2)}{\partial\omega}\mathrm{Im}\left[E_r(r)E_\varphi(r)\right] \right\} \\ &= \frac{R}{16\,\pi}\left\{ |H_z(R)|^2 + \frac{\partial(\omega\varepsilon_1)}{\partial\omega}|E_r(R)|^2 \right\}\chi(r). \end{aligned} \tag{3.11}$$

In (3.10) and (3.11), $\beta(r)$ is the normalized azimuthal component of the time-averaged Poynting vector, and $\chi(r)$ is the normalized angular electromagnetic energy density. Finally, the local angular velocity of energy transfer $\omega_{en}(r)$ reads as.

$$\omega_{en}(r) \equiv \frac{-2c}{\mathrm{R}} \frac{\beta(r)}{\chi(r)} \frac{E_r(R) H_z(R)}{|H_z(R)|^2 + \frac{\partial(\omega\varepsilon_1)}{\partial\omega}|E_r(R)|^2}. \tag{3.12}$$

Since, as mentioned above, in the simplest cases of Cartesian coordinates the velocity of energy transfer is equal to the group velocity, it is convenient to introduce the integral azimuthal velocity of energy transfer ω_{en}. First, one has to calculate the time averaged energy flux (not energy flux density, which is the Poynting vector)

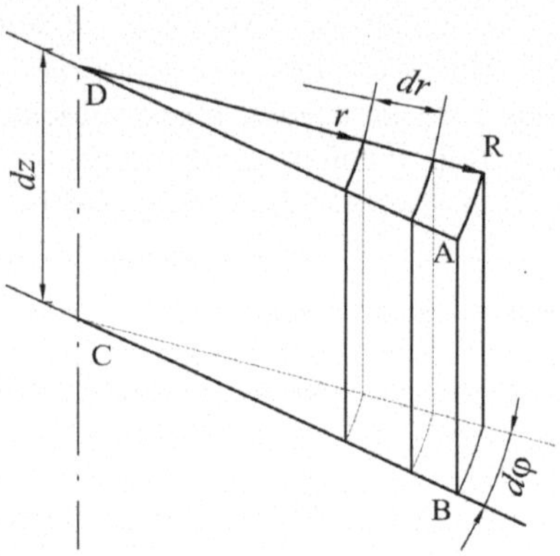

Fig. 3.1 Schematic for definition of the angular velocity of energy transfer

which is carried by the FWs through the rectangle ABCD (see Fig. 3.1). One side is a segment on the axis of the waveguide, its length is arbitrary, let it be DC = dz. The opposite side is the axial segment of the same length AB = dz at the metallic wall of the waveguide. The other two sides are the radii DA and CB, which connect the ends of the first and second sides. This time averaged energy flux is:

$$dz \int_0^R \langle S_\varphi \rangle_T dr \, . \tag{3.13}$$

This flux can be calculated also as the product of the angular velocity of energy transfer ω_{en} and the time averaged energy of the electromagnetic wave, which is contained inside the elementary volume drawn in Fig. 3.1.

Note that this elementary volume is not (!) a rectangular parallelepiped even in the limit $d\varphi << 1$. Like a piece of cake, the volume is restricted by the rectangle ABCD and the other one, which can be obtained by rotating ABCD by the elementary angle $d\varphi$ (see Fig. 3.1). The time averaged electromagnetic energy dW is given by:

$$dW = dz\, d\varphi \int_0^R r \langle u \rangle_T dr, \tag{3.14}$$

where the time averaged electromagnetic energy density $\langle u \rangle_T$ reads as [6]:

$$\langle u \rangle_T = \frac{1}{16\pi} \left\{ |H_z(r)|^2 + \frac{\partial(\omega\varepsilon_1)}{\partial\omega} \left[|E_r(r)|^2 + \left|E_\varphi(r)\right|^2 \right] \right.$$
$$\left. -2\frac{\partial(\omega\varepsilon_2)}{\partial\omega} \mathrm{Im}\left[E_r(r) E_\varphi(r)\right] \right\} . \tag{3.15}$$

Then the definition of the integral angular velocity of energy transfer is given by:

$$\omega_{en} = \int_0^R \langle S_\varphi \rangle_T dr / \int_0^R r\langle u\rangle_T dr. \tag{3.16}$$

In general, monochromatic electromagnetic FWs (both bulk and surface) propagate in cylindrical waveguides with the azimuthal phase velocity (3.8) and carry on their energy with the angular velocity of energy transfer (3.16). FWs waves do not carry their energy along the $\vec{z}$ axis, $v_{enz} = 0$.

The radial dependencies of normalized azimuthal Poynting flux $\beta(r)$, normalized angular electromagnetic energy density $\chi(r)$, normalized angular velocity of energy rotation $\omega_{en}(r)/|\omega_e|$, and linear velocity of energy transfer $r\omega_{en}(r)/c$ are plotted in Figs. 3.2 and 3.3 for $Z \equiv \Omega_e/|\omega_e| = 5$ and $m = 2$. The electromagnetic energy is concentrated mainly near the plasma-metal interface, where the electromagnetic energy flux also is the largest. However, the angular velocity of the energy transfer has its maximum near the waveguide axis. The latter means that the efficiency of energy rotation is largest just there: the full energy of the elementary cylindrical volume dr makes the full circle faster than in the other elementary cylindrical layers. Figure 3.3 presents the linear velocity of the energy rotation to confirm the fact that it does not exceed the velocity of light. The solid and dotted curves in Figs. 3.2 and 3.3 correspond to the case of small wave penetration depth into the plasma: $k_{ef} = 0.649$, $k_\perp R = 3.0$, $\omega/|\omega_e| = 0.594$, and the dashed and dash-dotted curves describe the case of larger wave penetration depth: $k_{ef} = 5.77$, $k_\perp R = 0.333$, $\omega/|\omega_e| = 0.99$. To make the dashed curve visible in Fig. 3.3 it is multiplied by 10.

The angular phase velocity ω_{ph} of SFXWs given in Eq. (3.8) is a very simple function of the SFXW eigenfrequency ω. The dependence of ω on the plasma waveguide parameters was analyzed in detail in [15, 16]. That is why not too much attention is paid in the present Section to ω_{ph}. It is only used to be compared with the angular velocity of energy transfer ω_{en} in Figs. 3.4 and 3.5. The dependence of ω_{en} on the plasma waveguide parameters is studied in detail in Figs. 3.4, 3.5, 3.6 and 3.7. Thereafter, the dependence of the angular group velocity ω_g is analyzed and compared with ω_{en} in Figs. 3.8, 3.9, 3.10 and 3.11. The numbers nearby the curves mark the azimuthal mode index m.

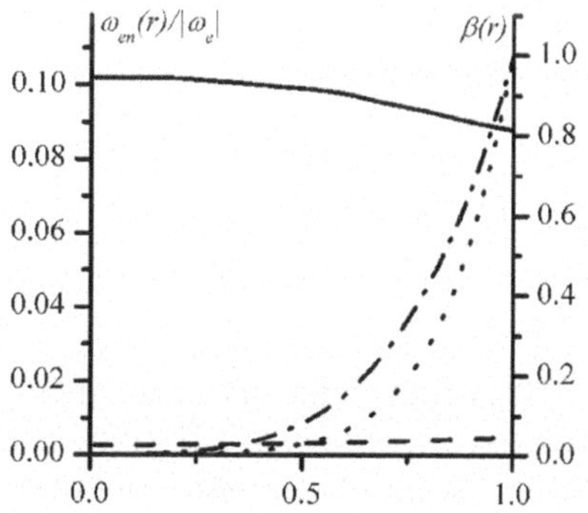

Fig. 3.2 Radial profile of normalized azimuthal energy flux density $\beta(r)$—right ordinate axis (*dash-dotted and dotted curves*), and that of angular velocity of energy rotation, $\omega_{en}(r)/|\omega_e|$—left ordinate axis (*dashed and solid curves*). $Z = 5$ and $m = 2$

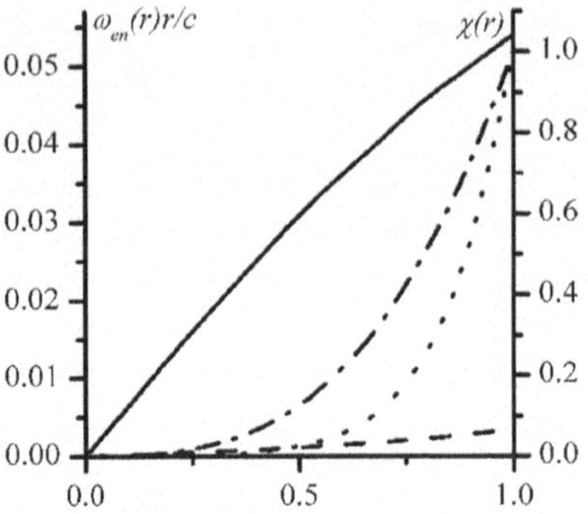

Fig. 3.3 Radial profile of normalized angular electromagnetic energy density $\chi(r)$—right ordinate axis (*dash-dotted and dotted curves*), and that of local linear velocity of energy transfer, $r\omega_{en}(r)/c$—left ordinate axis (*dashed and solid curves*). $Z = 5$ and $m = 2$

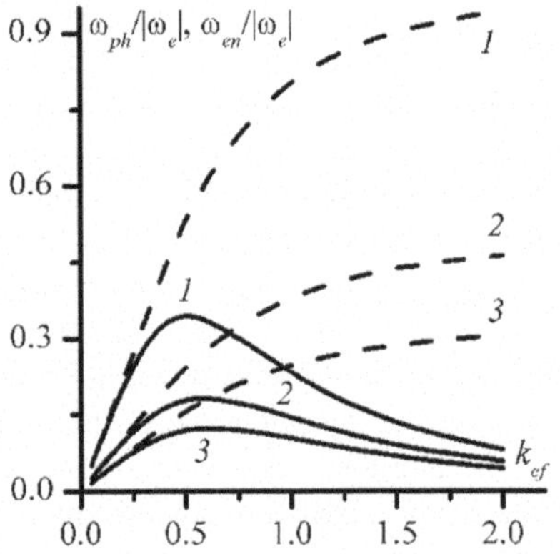

Fig. 3.4 Normalized LF SFXW angular phase velocity (*dashed lines*) and angular velocity of energy transfer (*solid lines*) for $Z = 5$. The *numbers* indicate the azimuthal mode index m

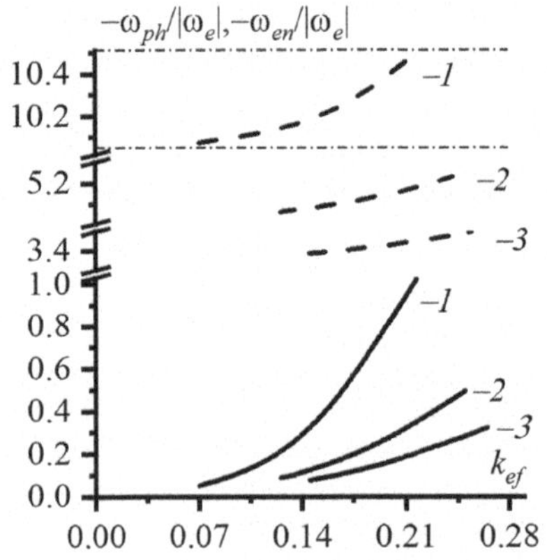

Fig. 3.5 Normalized HF SFXW angular phase velocity (*dashed lines*) and angular velocity of energy transfer (*solid lines*) for $Z = 10$. The *numbers* indicate the azimuthal mode index. The horizontal *dash-dotted lines* are the boundaries of the HF range

They present the dispersion dependencies as a function of eigenfrequency versus wave vector, $\omega = \omega\,(k)$. The ratio m/R can be seen as a characteristic azimuthal wave vector in the present problem. That is why the effective wavenumber $k_{ef} = |m|\delta/R$ is chosen as abscissa in the Figs. 3.4, 3.5, 3.6, 3.7, 3.8, 3.9, 3.10 and 3.11, where

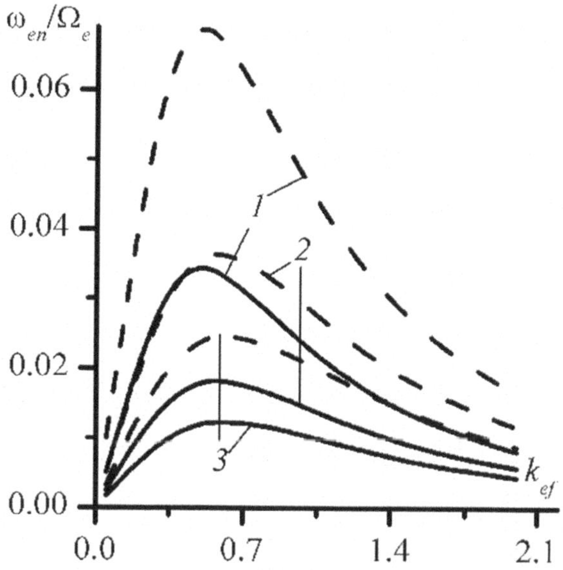

Fig. 3.6 Normalized LF SFXW angular velocity of energy transfer for two different strengths of the external static axial magnetic field B_0: Z = 5 (*dashed lines*) and Z = 10 (*solid lines*). The *numbers* indicate the azimuthal mode index *m*

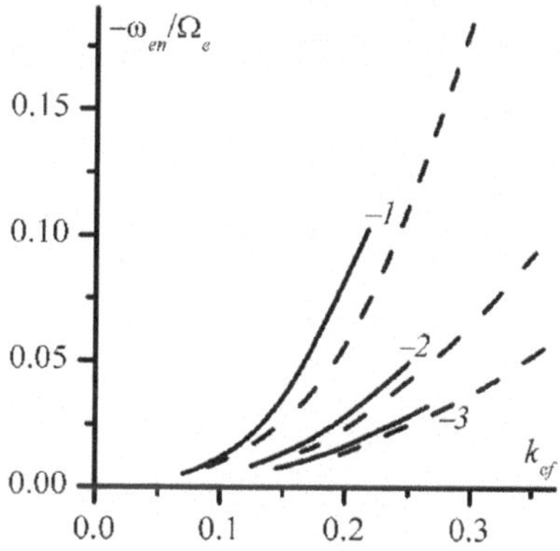

Fig. 3.7 Normalized HF SFXW angular velocity of energy transfer for two different strengths of the external static axial magnetic field B_0: Z = 5 (*dashed lines*) and Z = 10 (*solid lines*). The *numbers* indicate the azimuthal mode index *m*

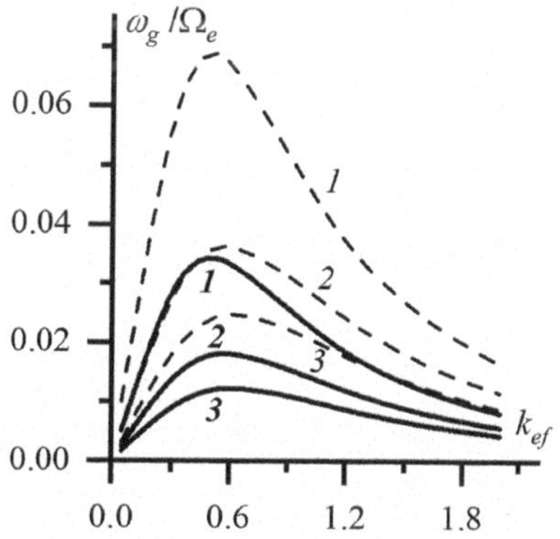

Fig. 3.8 Normalized angular group velocity of LF SFXWs versus effective wavenumber for Z = 5 (*dashed lines*) and Z = 10 (*solid lines*). The *numbers* indicate the azimuthal mode index *m*

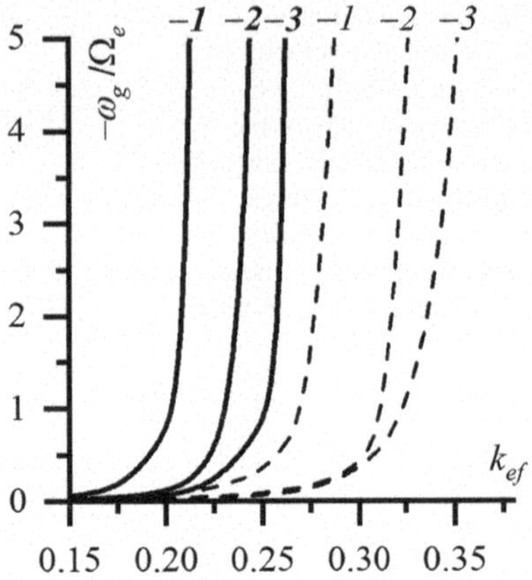

Fig. 3.9 Normalized angular group velocity of HF SFXWs versus effective wavenumber for Z = 5 (*dashed lines*) and Z = 10 (*solid lines*). The *numbers* indicate the azimuthal mode index *m*

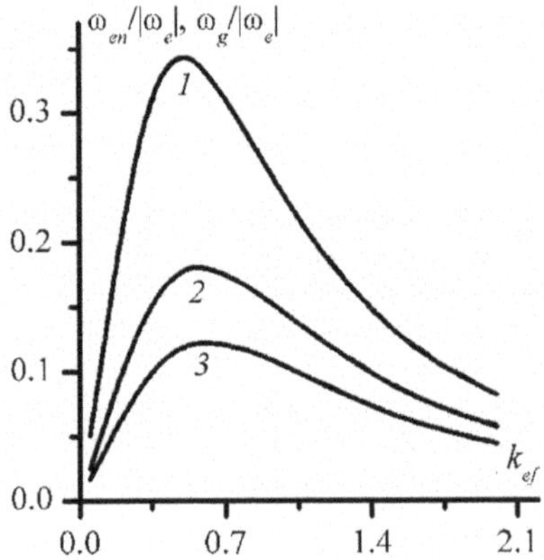

Fig. 3.10 Normalized angular group velocity (*solid curves*) and angular velocity of energy transfer (*dashed curves*) of LF SFXWs versus effective wavenumber for Z = 10. The difference between them is almost invisible in this scale. The *numbers* indicate the azimuthal index *m*

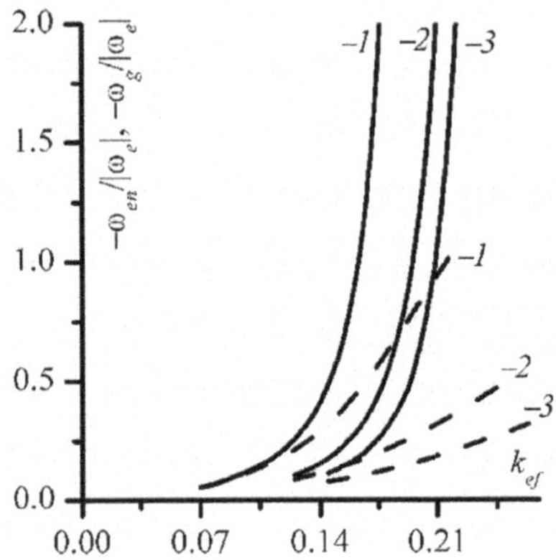

Fig. 3.11 Normalized angular group velocity (*solid curves*) and angular velocity of energy transfer (*dashed curves*) of HF SFXWs versus effective wavenumber for Z = 10. The *numbers* indicate the azimuthal mode index *m*

$\delta = c/\Omega_e$ is the skin-depth. For LF SFXWs, the value of k_{ef} is closely related to the magnitude of the argument $k_\perp R$ of the Bessel functions in the dispersion relation (2.13): $k_{ef} \approx |m|/(k_\perp R)$. According to this definition increasing k_{ef} corresponds to decreasing waveguide radius R and decreasing plasma particle density. However, the effective wavenumber k_{ef} is not a proper abscissa if one studies the dependencies on azimuthal mode index m, since k_{ef} is proportional to $|m|$. To investigate these dependencies, separate numerical runs were carried out, whose results are given in the below text but not presented in the form of figures.

The ranges of k_{ef} for which numerical analyses were performed are chosen on the base of the following reasons. For LF SFXWs the calculations are stopped for those values of k_{ef} for which the eigenfrequency becomes close to $|\omega_e|$, which is the upper limit of the LF range. That is why $0 < k_{ef} < 2$ for LF SFXWs.

In Figs. 3.5, 3.7, 3.9 and 3.11, in which the dependencies of the HF SFXW angular phase velocity, angular velocity of energy transfer and angular group velocity on k_{ef} are plotted, the curves are calculated for those ranges of k_{ef} where these waves can propagate. These ranges are known to be limited by zero on the left side and a moderate value of k_{ef} on the right side [15, 16]. For HF SFXWs, small values of k_{ef} also correspond to plasma waveguides with thin penetration depths ($k_\perp R \gg 1$), and the values of k_{ef} close to the right limit of the range correspond to plasmas wherein HF SFXWs penetrate deeply ($k_\perp R \ll 1$).

In Figs. 3.4 and 3.5, the absolute values of angular phase velocity and angular velocity of energy transfer normalized by the electron cyclotron frequency, $|\omega_{en,ph}/\omega_e|$, are chosen as ordinate axis. The dashed curves correspond to the angular phase velocities, and the solid curves to angular velocities of energy transfer.

In Fig. 3.4 one can see that the absolute value of the angular velocity of energy transfer of LF SFXWs (solid curves) is smaller than that of corresponding angular phase velocity (dashed curves). The absolute value of the angular phase velocity is known [15, 16] to approach to $\omega_{LH}/|m| \ll |\omega_e|$ with decreasing effective wavenumber, $k_{ef} \rightarrow 0$. The dispersion of LF SFXWs is normal: the signs of their angular velocities of energy transfer and phase velocities are the same. The dependence of the angular velocity of energy transfer ω_{en} of LF SFXWs on k_{ef} is not monotonous. The maximum of the dependence is reached for $k_{ef} \approx 0.5$ ($m = 1$), and $k_{ef} \approx 0.6$ ($m = 2,3$). The dependence of $|\omega_{en}|$ on the absolute value $|m|$ of the azimuthal mode index is not clear from the Fig. 3.4, since the effective wavenumber k_{ef} also depends on $|m|$. An additional numerical analysis proves that ω_{en} of LF SFXWs decreases with increasing m. For small values of k_{ef}, both $|\omega_{en}|$ and $|\omega_{ph}|$ are approximately proportional to k_{ef}.

Figure 3.5 shows that the angular phase velocities (dashed curves) of HF SFXWs are larger than their velocities of energy transfer (solid curves) in absolute value. The dash-dotted lines indicate the boundaries of the frequency range (1.14) for HF SFXWs. With decreasing effective wavenumber ($k_{ef} \rightarrow 0$) the absolute values of the phase velocities approach to the finite values $\omega_{UH}/|m|$, and $|\omega_{en}|$ of HF SFXWs decreases to a value which is much lower than $|\omega_e|$. The absolute values of ω_{ph} approach the finite values $\omega_2/|m|$ for moderate values of $\mathrm{Max}\{k_{ef}\} < 1$ [15, 16], and HF SFXWs do not propagate in waveguides with parameters corresponding to $k_{ef} > \mathrm{Max}\{k_{ef}\}$. For $0 < k_{ef} < \mathrm{Max}\{k_{ef}\}$, $|\omega_{en}|$ monotonously increases. The absolute values of both angular phase velocity and angular velocity of energy transfer of HF SFXWs decrease with increasing $|m|$. The dispersion of HF SFXWs also appears to be normal.

The dispersion properties of SFXWs are known to strongly depend on the external static magnetic field B_0 [15, 16]. Figures 3.6 and 3.7 present the results of studying the influence of an external static magnetic field on the angular velocity of energy transfer. That is why the absolute value of ω_{en} normalized by the Langmuir frequency is used in these figures as ordinate axis. In this case, both abscissa and ordinate contain the Langmuir frequency in the denominator. In Figs. 3.6 and 3.7, the dashed curves are used for the dependencies plotted for $Z \equiv \Omega_e/|\omega_e| = 5$, and the solid curves for weaker external static magnetic field, $Z = 10$. In Fig. 3.6 one can see that $|\omega_{en}|$ of LF SFXWs increases with increasing B_0. The opposite behavior is shown in Fig. 3.7 for HF SFXWs, where $|\omega_{en}|$ decreases with increasing B_0. However, increasing B_0 is accompanied by broadening the range of k_{ef}, for which HF SFXWs can propagate, towards larger magnitudes of k_{ef} (smaller magnitudes of the plasma particle density and/or radius of the plasma column) where $|\omega_{en}|$ is larger than in the case of smaller k_{ef}.

Results of numerical studies on the effect of the external static magnetic field B_0 on the dependence of the angular group velocity of SFXWs versus effective wavenumber are plotted in Figs. 3.8 and 3.9. By definition, larger Z corresponds to smaller B_0. The angular group velocity of LF SFXWs is plotted in Fig. 3.8. It increases with increasing external static magnetic field, unlike $|\omega_g|$ for HF SFXWs, which is smaller for larger B_0 as shown in Fig. 3.9. This characteristic of the dependence of $|\omega_g|$ on B_0 is in qualitative agreement with the dependence of the angular velocity of energy transfer $|\omega_{en}|$ (see Figs. 3.6 and 3.7). The angular group velocity of LF SFXWs demonstrates non-monotonous behavior, unlike $|\omega_g|$ for HF SFXWs, which infinitely increases with increasing effective wavenumber.

In Figs. 3.10 and 3.11 the dependence of the magnitude of the angular group velocity $|\omega_g|$ of SFXWs on k_{ef} is compared with that of $|\omega_{en}|$. The absolute value $|\omega_g|$ of HF SFXWs infinitely increases when k_{ef} approaches to the right limit of the k_{ef} range where HF SFXWs can propagate. For sufficiently small values of k_{ef} the absolute value of the angular group velocity is smaller than that of the angular velocity of energy transfer, $|\omega_g|<|\omega_{en}|$. The value of k_{ef} for which $|\omega_g| = |\omega_{en}|$ is larger for SFXWs with larger absolute values of the azimuthal mode index m. The angular group velocity of LF SFXWs almost completely coincides with ω_{en}. The absolute value of the angular group velocity of both LF and HF SFXWs decreases with increasing absolute value of the azimuthal mode index.

One can make the following conclusions from the research carried out in the present Section [20]. When non-axisymmetric electromagnetic waves (whose axial wavenumber and azimuthal mode index are non-zero, $k_z \neq 0$, $m \neq 0$) propagate in cylindrical waveguides, their energy is transferred in two directions. One part is transferred along the waveguide axis. This is described and studied elsewhere. The linear velocity of energy transfer, which is the proportionality coefficient between the Poynting vector given in Eq. (3.1) and the spatial electromagnetic energy density Eq. (3.15), is the appropriate tool for description of this motion. At the same time, the electromagnetic energy is rotating around the waveguide axis. The linear velocity of energy transfer cannot be used for the description of such energy motion.

The radial dependence of the angular velocity of the energy rotation shows that the electromagnetic power rotates strongest near the waveguide axis although the electromagnetic energy density and electromagnetic energy flux are largest near the plasma-metal interface for SFXWs [20]. Since the radial dependence of the angular velocity of the energy transfer is not monotonous the integral value of the angular velocity of the energy rotation ω_{en} is introduced.

The angular phase velocity ω_{ph}, angular group velocity ω_g and angular velocity of energy transfer ω_{en} should be applied to study the electromagnetic energy rotation in cylindrical waveguides. The definitions of these physical values are introduced for the case of flute electromagnetic waves ($k_z = 0$). These angular velocities are studied in the present Section in their dependence on the plasma waveguide parameters for a cylindrical metal chamber entirely filled by plasma in two frequency ranges, low frequencies (LF), $\omega_{LH} < \omega < |\omega_e|$, and high frequencies (HF) defined by (1.14).

The angular phase velocity ω_{ph} given by Eq. (3.8) is defined as the angular velocity with which the half-plane of the fixed wave phase ($m\varphi - \omega t = \text{Const}$) rotates around the waveguide axis. The angular velocity of energy transfer ω_{en} described by Eq. (3.16) is defined as the proportionality coefficient between the averaged electromagnetic energy flux and the averaged electromagnetic energy contained in an angular unit volume (see Fig. 3.1, in which the angle $d\varphi$ should be put equal to unit). The angular group velocity ω_g given by Eq. (3.9) is defined as the angular velocity with which the wave packet consisting of modes with different azimuthal index m rotates around the waveguide axis without change in shape.

In the present Section it is shown that the dispersion of SFXWs is normal: the directions of the angular velocity of energy transfer ω_{en} and the angular phase velocity ω_{ph} are the same [20]. For both LF and HF SFXWs, the absolute value of the angular velocity of energy transfer is shown to be smaller than that of the angular phase velocity ($|\omega_{en}|<|\omega_{ph}|$).

Increasing external static magnetic field results in increasing absolute value of the angular velocity of LF SFXW energy transfer and decreasing of the $|\omega_{en}|$ for HF SFXWs. Both angular velocity of energy transfer and angular phase velocity increase with increasing effective wavenumber for HF SFXWs as well as angular phase velocity for LF SFXWs. The maximum values of the angular velocity of energy transfer for LF SFXWs are observed for moderate magnitudes of the effective wavenumber.

The dependence of the angular group velocity ω_g of SFXWs on plasma waveguide parameters is derived analytically and studied numerically. The difference between ω_g and the angular velocity of energy transfer ω_{en} is observed for those plasma waveguide parameters for which the SFXW eigenfrequency is close to the upper limits of these wave frequency ranges: electron cyclotron frequency $|\omega_e|$ and cut-off frequency for bulk modes ω_2. This result is in agreement with the conclusion of [4], where it is noticed that for electromagnetic waves, the group velocity can be considered as the velocity of energy transfer only if they propagate in media whose permittivity is a slowly varying function of the wave frequency.

The presented analysis of electromagnetic energy transfer in cylindrical plasma-filled waveguides can be of interest, especially for plasma electronics [16], plasma production in gas discharges [21–23], development of new nano-dimensional photonic elements and surface enhanced Raman spectroscopy (see, e.g. [24] and references therein).

3.2 Electromagnetic Energy Rotation Around a Cylindrical Metallic Rod in Magneto-active Plasmas

The dispersion properties of extraordinary surface flute waves (SFXWs) propagating around cylindrical metallic rods placed into infinite cold collisionless plasma, parallel to an external static uniform magnetic field were studied in detail in Sect. 2.2. [15]. There, SFXWs were shown to be unidirectional waves. This means that LF SFXWs propagate along a metal-plasma interface around a cylindrical metallic rod with negative azimuthal mode index, $m < 0$, whereas HF SFXWs propagate with positive azimuthal mode index, $m > 0$.

The present Section complements the results obtained in Sect. 2.2 by the analysis of SFXW electromagnetic energy transfer with respect to their angular phase velocity and angular velocity of energy transfer [25]. These physical observables were introduced in Sect. 3.1 for studying the electromagnetic energy rotation along the plasma-metal interface inside cylindrical waveguides entirely filled by cold collisionless plasma [20].

Electromagnetic energy rotation around a cylindrical metallic rod placed into infinite cold collisionless plasma in parallel to an external static magnetic field is studied here. This rotation is compared with the case of SFXW propagation inside cylindrical metallic waveguides entirely filled by plasma.

The following metal-plasma waveguide structure is under consideration. A metal rod with radius R is assumed to be an ideal conductor (see Fig. 2.6). It is placed into a plasma which is assumed to be cold and collisionless. The rod is oriented parallel to an external static magnetic field $\vec{B}_0 \| \vec{z}$. In mathematical physics, such position of media and boundary is called as exterior problem unlike the interior problem studied in the previous Sect. 3.1 [20], where SFXWs propagate inside a cylindrical metallic waveguide entirely filled by plasma. The two-component structures being under consideration in the interior [20] and the present exterior problems can be obtained from each other by geometrical inversion.

The present exterior problem assumes an infinite plasma which encircles the metallic rod. This model of infinite plasma can be applied in practice if the plasma radial dimension R_p is much larger than the depth of SFXW field penetration into the plasma, which means $k_\perp R_p >> 1$.

The expressions (1.12) for the amplitude of the SFXW magnetic field with $A_1 = 0$ and (1.10) for the amplitudes of the components of the SFXW electric field can be calculated using the Maxwell equations.

The SFXW dispersion relation can be derived as the result of application of the linear dissipative-less boundary condition at the metal-plasma interface $r = R$ which requires that the azimuthal (tangential) component of the electric wave field is equal to zero. The dispersion relation of SFXWs in the structure considered in the present Section was studied in detail in [15]. It has the form (2.11) with $F_m(\xi) = K_m(\xi)$. This relation is known to transform into the dispersion relation for surface-type waves propagating perpendicularly to the external static magnetic field nearby a planar plasma-metal boundary in the limit of large azimuthal mode index , $|m| \gg 1$, [26].

Some conclusions about electromagnetic energy transfer initiated by SFXW propagation can be done on the base of analytical treatment of the problem. First, the electromagnetic energy of SFXWs principally does not propagate in axial direction. This can be seen from the expression (3.2) for the axial component of the electromagnetic energy flux known as Poynting vector [10, 18]. Since SFXWs do not have neither radial nor azimuthal components of the magnetic wave field, $H_r = H_\varphi = 0$, the axial component of the Poynting vector is equal to zero, $S_z = 0$. Consequently, the axial component of the velocity of energy transfer initiated by SFXWs is equal to zero, $\upsilon_{en\,z} = 0$.

Second, the electromagnetic energy of SFXWs oscillates in radial direction with twice the eigenfrequency (2ω). Being averaged over the temporal wave period $T = 2\pi/\omega$, the radial component of the Poynting vector $S_r = (c/4\pi)\ E_\varphi\ H_z$ is equal to zero, (see Eq. (3.5)). Consequently, the radial component of the velocity of energy transfer initiated by SFXWs is also equal to zero, $\upsilon_{en\,r} = 0$.

Third, to describe the rotation of electromagnetic energy around the metallic rod one has to introduce the angular velocity of energy transfer.

The local angular velocity of energy transfer $\omega_{en}(r)$ is the coefficient of proportionality between the following two physical values. One is the time-averaged electromagnetic energy flow through the elementary surface $drdz$ per unit time (1 s). The other is the time-averaged electromagnetic energy inside the elementary volume $rd\varphi\ drdz$ with the unit angle $d\varphi = 1$ (see Fig. 3.12).

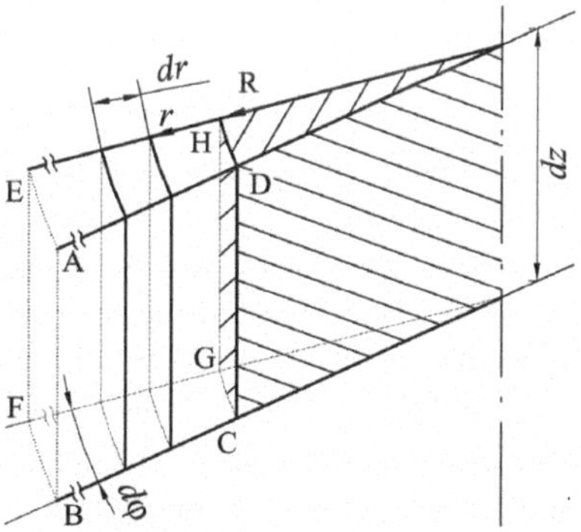

Fig. 3.12 Schematic of definition of angular velocity of energy transfer

The time-averaged azimuthal component of the Poynting vector is given by Eq. (3.10). The time-averaged angular electromagnetic energy density is calculated according to Eq. (3.11). Then, the local angular velocity of energy transfer $\omega_{en}(r)$, given by Eq. (3.12), characterizes at each radial coordinate the time-averaged angular velocity of the electromagnetic wave energy rotation around the rod. The local angular velocity of energy transfer $\omega_{en}(r)$ is used below in carrying out numerical analyses to validate the correctness of the calculations via comparison of the product $r\omega_{en}(r)$ with the speed of light. This is especially significant in the present exterior problem in which the radial coordinate can have an infinitely large magnitude.

To get the physical value which integrally characterizes the electromagnetic energy rotation around the rod one has to calculate the time- and spatially averaged angular velocity of energy transfer ω_{en}:

$$\omega_{en} = \int_R^\infty \langle S_\varphi \rangle_T dr / \int_R^\infty r \langle u \rangle_T dr. \tag{3.17}$$

The definition (3.17) naturally differs from that in the case of the interior problem (3.16). The physical meaning of the integral angular velocity of energy transfer ω_{en} is clear. First, one has to calculate the time-averaged energy flux through the surface ABCD (see Fig. 3.12) per unit time as.

$$dz \int_R^\infty \langle S_\varphi \rangle_T dr. \tag{3.18}$$

The surface ABCD is infinite in radial direction. However, it is elementary in axial direction. Second, it is necessary to calculate the time-averaged electromagnetic energy inside the volume ABCDEFGH (see Fig. 3.12) with the unit angle $d\varphi = 1$:

$$dW = dz \int_R^\infty r \langle u_T \rangle dr . \tag{3.19}$$

The volume ABCDEFGH is infinite in radial direction, finite in azimuthal direction and elementary in axial direction. Then, the integral angular velocity of energy transfer ω_{en} (Eq. (3.17)) is the ratio of the expressions (3.18) and (3.19).

Figures 3.13, 3.14 and 3.15 present the radial distribution of the normalized azimuthal component of the time-averaged Poynting vector $\beta(r)$ (in Fig. 3.13 from Eq. (3.10), the normalized angular electromagnetic energy density $\chi(r)$ (in Fig. 3.14 from Eq. (3.11)), the local angular velocity of energy transfer normalized by the electron cyclotron frequency $\omega_{en}(r)/|\omega_e|$ (in Figs. 3.13 and 3.15 from Eq. (3.13), and the local linear velocity of energy transfer normalized by the speed of light $r\omega_{en}(r)/c$ (in Figs. 3.14 and 3.15). The radial coordinate normalized by the rod radius R is chosen as abscissa axis, r/R. The calculations are carried out for $|m| = 2$, and $Z \equiv \Omega_e/|\omega_e|$

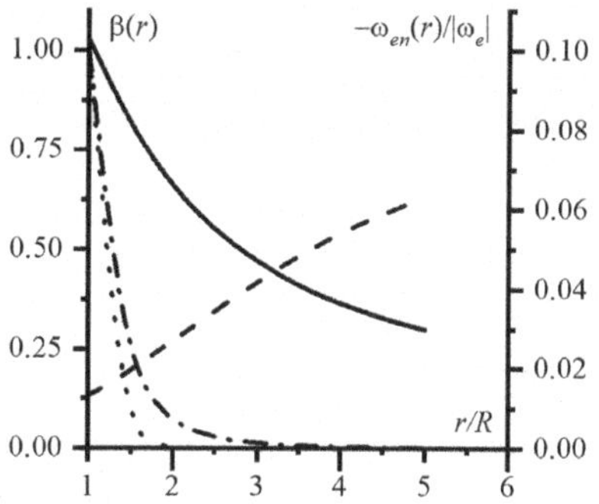

Fig. 3.13 Normalized azimuthal component of the time-averaged Poynting vector $\beta(r)$ (*dash-dotted and dotted curves*), and local angular velocity of energy transfer (*dashed and solid curves*) of LF SFXWs for two different penetration depths (see the text)

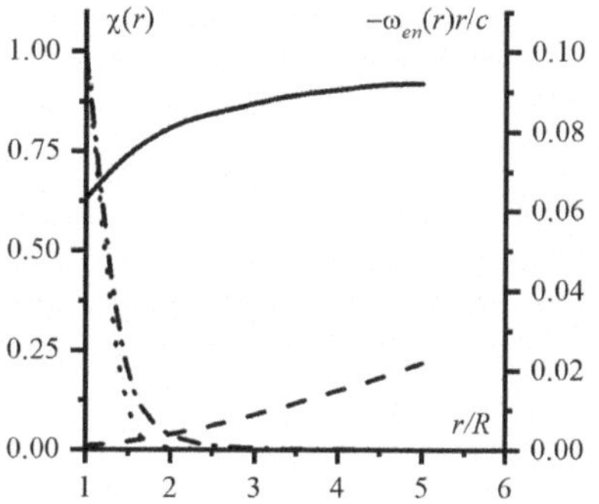

Fig. 3.14 Normalized angular electromagnetic energy density $\chi(r)$ (*dash-dotted and dotted curves*), and local linear velocity of energy transfer (*dashed and solid curves*) of LF SFXWs for two different penetration depths (see the text)

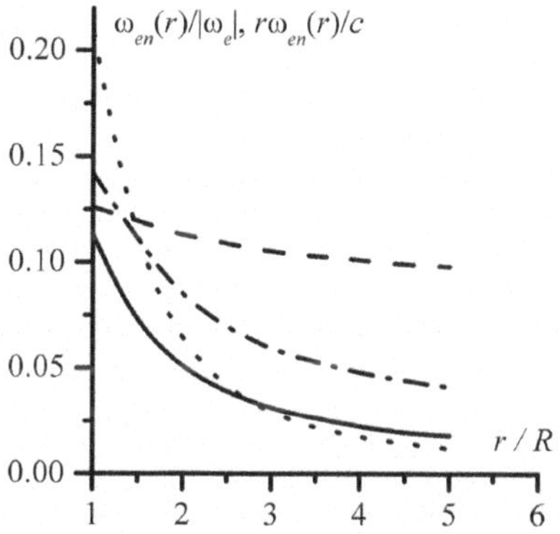

Fig. 3.15 Normalized local angular velocity of energy transfer (*dashed and solid curves*) and local linear velocity of energy transfer (*dash-dotted and dotted curves*) of HF SFXWs for two different penetration depths (see the text)

= 5. The solid and dotted curves are calculated for the case of small depth of wave field penetration into the plasma, $k_{\perp}R = 3.0$. For LF SFXWs, this takes place for the effective azimuthal wavenumber $k_{ef} \equiv |m|c/(R\Omega_e) = 0.649$. In this case the SFXW eigenfrequency is $\omega = 0.49|\omega_e|$. The dashed and dash-dotted curves are calculated for the case of large depth of wave field penetration into the plasma, $k_{\perp}R = 0.333$. For LF SFXWs (see Figs. 3.13 and 3.14), this takes place for $k_{ef} = 5.765$ with $\omega = 0.973|\omega_e|$. For HF SFXWs (Fig. 3.13) these opposite correlations between the metal rod radius R and the depth of the wave penetration into the plasma $k_\perp^{-1}$ ($k_\perp R = 3.0$ and $k_\perp R = 0.333$) are realized for the effective wavenumbers which are naturally different from the case of LF SFXWs. Namely, the metal rod can be considered as a thin one, $k_\perp R = 0.333$, for HF SFXWs with azimuthal mode index $m = 2$ and $Z = 5$ if $k_{ef} = 0.586$ and $\omega = 5.515|\omega_e|$. If $k_{ef} = 0.357$ and $\omega = 5.326|\omega_e|$ then the metal rod is thick, $k_\perp R = 3.0$, for HF SFXWs in the case of $m = 2$ and $Z = 5$.

The dash-dotted curve in Fig. 3.13 shows a strong decrease of the normalized azimuthal component of the time-averaged Poynting vector with distance r, $\beta(r) \propto r^{-5}$. The dotted curve in Fig. 3.13 also demonstrates the surface nature of SFXWs. It decreases even more rapidly ($\beta(r) \propto \exp(-2k_\perp r)$) when going away from the metal-plasma interface ($r/R = 1$) than the dash-dotted curve. This is in qualitative agreement with nine times faster decreasing wave field penetration into the plasma as compared to the case of the dash-dotted curve. The electromagnetic energy rotates in the direction opposite to that of electron gyration in the external static magnetic field. LF SFXWs in the present exterior problem are characterized by normal dispersion. The directions of angular phase velocity and that of energy transfer are the same. In the case of small depth of the wave field penetration into the plasma (solid curve), electromagnetic energy rotates faster nearby the waveguide metal wall. The absolute value of the local angular velocity of energy transfer decreases with going away from the wall. If the wave field penetrates into the plasma deeply (dashed curve), $|\omega_{en}|$ has its maximum, $|\omega_{en}| \approx 0.072|\omega_e|$, for $r \approx 8.5R$ and decreases with further increase of the radial coordinate.

The localization of the SFXW electromagnetic energy density nearby the metal-plasma interface ($r/R = 1$) is more pronounced in the case of small depth of wave penetration into the plasma (dotted curve) than in the case of large penetration depth (dash-dotted curve). This is convincingly demonstrated in Fig. 3.14. The local linear velocity of energy transfer is much slower than the speed of light. Its dependence on the radial coordinate does not show any clear decrease with increasing r/R. That is why one has to evaluate the upper limit of $r\omega_{en}(r)$ for infinitely large radii. The latter can be estimated for LF SFXWs as $\lim_{r\to\infty}|r\omega_{en}(r)| < 2c/Z$.

In the HF range, SFXWs propagate with positive azimuthal mode index ($m = +2$ in Fig. 3.15). The radial distribution of both normalized azimuthal component of the Poynting vector $\beta(r)$ and normalized angular electromagnetic energy density $\chi(r)$ are not presented for HF SFXWs because the corresponding curves are absolutely the same as for LF SFXWs if the relation between the rod radius and the depth of wave penetration into the plasma (the product $k_\perp R$) is fixed. However, both local angular and linear velocities of energy transfer of HF SFXWs in Fig. 3.15 significantly differ from those of LF SFXWs in Figs. 3.13 and 3.14. This difference is explained by the difference between the magnitudes of effective wavenumber and eigenfrequency for which the correlations $k_\perp R = 3.0$ and $k_\perp R = 0.333$ are realized for LF and HF SFXWs, as mentioned above. Both local angular and linear velocities of energy transfer decrease with going away from the metal-plasma interface into the plasma. The monotonous decrease of the local linear velocities of energy transfer $r\omega_{en}(r)$ of HF SFXWs with increasing radial coordinate (unlike in the case of LF SFXWs) does not leave any doubt that it is much smaller than the speed of light.

After this analysis of the radial dependence of the SFXW energetic characteristics, in the following, the dependence of the integral angular velocity of energy transfer ω_{en} (Eq. (3.17) on the plasma waveguide parameters is studied and compared with other characteristic frequencies in Figs. 3.16, 3.17, 3.18 and 3.19. The numbers at the curves mark the azimuthal mode index m.

The effective wavenumber $k_{ef} = |m|\delta/R$ is chosen as dimensionless abscissa in the Figs. 3.16, 3.17, 3.18 and 3.19 ($\delta = c/\Omega_e$ is the skin-depth). The following reasons provide the basis for choosing the ranges of k_{ef} in carrying out the numerical analysis. For LF SFXWs (see Figs. 3.16 and 3.18), the upper limit of k_{ef} is such that the SFXW eigenfrequency comes close to the upper limit of the LF frequency range which is the electron cyclotron frequency $|\omega_e|$. Therefore, for the present exterior problem the effective wavenumber lies in the range $0 < k_{ef} < 5$.

In Figs. 3.17 and 3.19, the dependencies of the HF SFXW angular phase velocity and angular velocity of energy transfer on k_{ef} are plotted. The upper limit of k_{ef}, for which the curves in these figures are calculated, is also determined by the condition that HF SFXWs can propagate there [15, 16]. The small values of k_{ef} for HF SFXWs

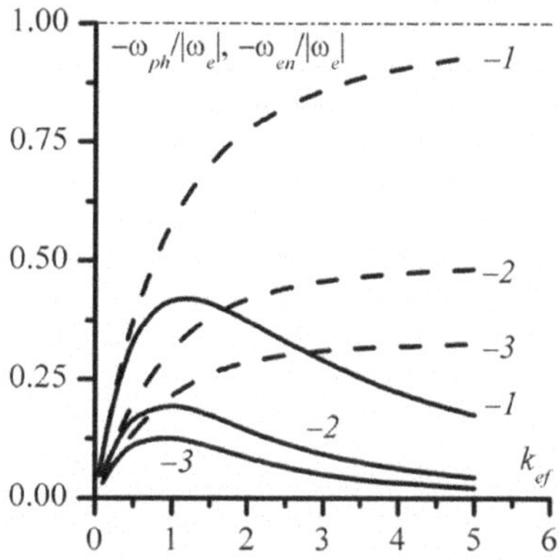

Fig. 3.16 Normalized LF SFXW angular phase velocity (*dashed lines*) and angular velocity of energy transfer (*solid lines*) for Z = 5. The *numbers* indicate the azimuthal mode index *m*. The horizontal *dash-dotted* line shows the upper limit of the LF range

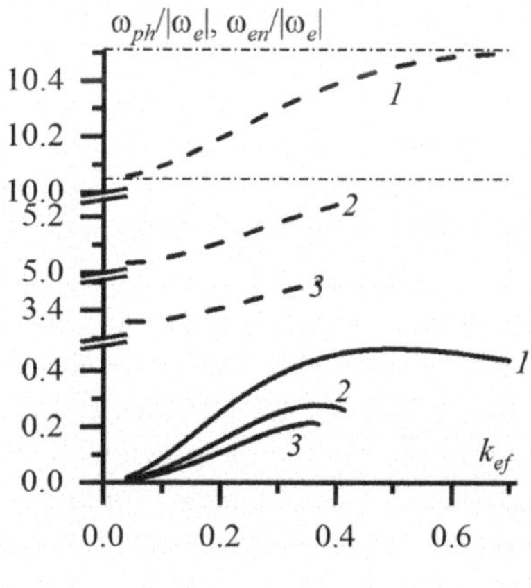

Fig. 3.17 Normalized HF SFXW angular phase velocity (*dashed lines*) and angular velocity of energy transfer (*solid lines*) for Z = 10. The *numbers* indicate the azimuthal mode index *m*. The horizontal *dash-dotted* lines show the limits of the HF range

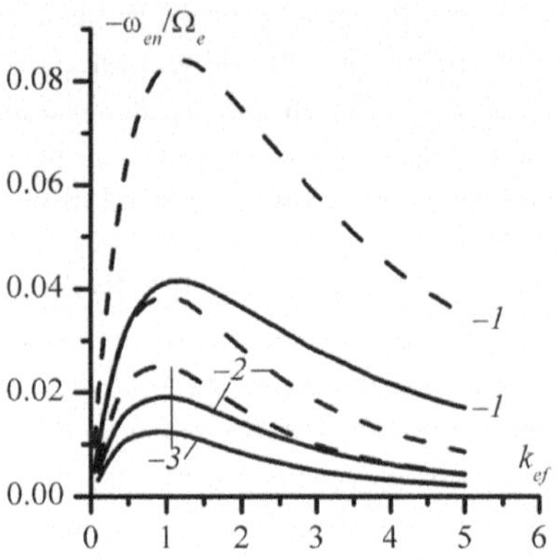

Fig. 3.18 Normalized LF SFXW angular velocity of energy transfer for two different strengths of external static axial magnetic field B_0: Z = 5 (*dashed lines*) and Z = 10 (*solid lines*). The *numbers* indicate the azimuthal mode index *m*

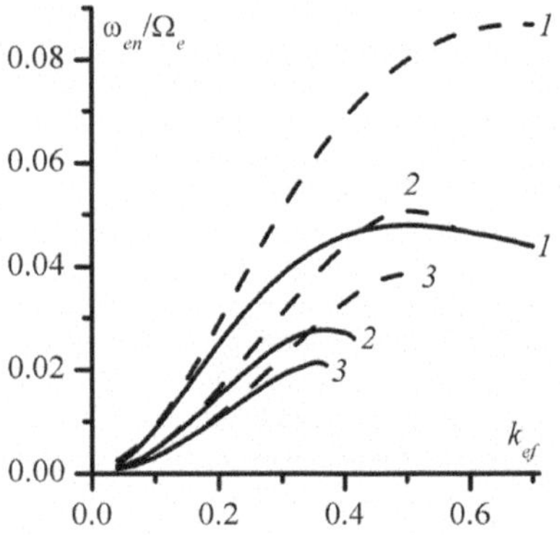

Fig. 3.19 Normalized HF SFXW angular velocity of energy transfer for two different strengths of external static axial magnetic field B_0: Z = 5 (*dashed lines*) and Z = 10 (*solid lines*). The *numbers* indicate the azimuthal mode index *m*

also correspond to the case of small penetration depths into the plasma ($k_\perp R >> 1$). The moderate values of the right limit of the k_{ef} range correspond to large depths of HF SFXWs penetration ($k_\perp R << 1$).

In Fig. 3.16, absolute values of the angular phase velocity and angular velocity of energy transfer normalized by the electron cyclotron frequency, $-\omega_{en,ph}/|\omega_e|$, are chosen as ordinate axis, $Z = 5$. The dashed curves correspond to the angular phase velocities, and the solid curves to the angular velocities of energy transfer. One can see that $|\omega_{en}/\omega_e|$ is smaller than $|\omega_{ph}/\omega_e|$ as it was for the interior problem studied in the previous Sect. 3.1. The horizontal dash-dotted line relates to the upper limit of the LF range. The absolute value of the angular phase velocity approaches to $\omega_{LH}/|m| << |\omega_e|$ with decreasing effective wavenumber, $k_{ef} \to 0$ [15, 16]. The dispersion of LF SFXWs is normal: both phase velocities and angular velocities of energy transfer are negative. The dependences of the absolute value of the angular velocity of energy transfer $|\omega_{en}|$ of LF SFXWs on k_{ef} are qualitatively similar to those studied for interior problem in the previous Sect. 3.1. The dependences are characterized by the maxima which are reached for larger values of k_{ef} as compared to the case of interior problem. The position of the maximum decreases from $k_{ef} \approx 1.2$ for LF SFXWs with

$m = -1$ to $k_{ef} \approx 1.0$ for LF SFXWs with $m = -3$. The magnitude of $|\omega_{en}|$ decreases with increasing absolute value of the azimuthal mode index $|m|$. For small magnitudes of k_{ef}, both $|\omega_{en}|$ and $|\omega_{ph}|$ are approximately proportional to k_{ef}. The absolute values of the phase velocities of LF SFXWs in the present exterior problem are approximately equal to those of the interior problem. The absolute values of the velocities of energy transfer in the case of the interior problem are smaller than those for the exterior problem with the same plasma-waveguide parameters.

The angular phase velocities (dashed curves) of HF SFXWs are larger than their velocities of energy transfer (solid curves) as it is shown in Fig. 3.17. The dash-dotted lines correspond to the boundaries of the HF range (1.14). The magnitudes of the phase velocities approach to the finite values $\omega_{UH}/|m|$ with decreasing effective wavenumber ($k_{ef} \to 0$), and $|\omega_{en}|$ of HF SFXWs decreases to a value, which is much lower than $|\omega_e|$. The magnitudes of ω_{ph} approach finite values $\omega_2/|m|$ for the upper limit of $k_{ef} < 1$ [15, 16]. The angular velocities of energy transfer demonstrate non-monotonous behavior unlike in the case of the interior problem in which $|\omega_{en}|$ monotonously increases with increasing k_{ef}. The maximum values of the angular velocity of the HF SFXW energy transfer in the present exterior problem are smaller than those for the interior problem with the parameters which correspond to the right limit of the k_{ef} range. Both angular phase velocity and angular velocity of energy transfer of HF SFXWs decrease with increasing $|m|$. The dispersion of HF SFXWs also appears to be normal.

The dispersion properties of SFXWs are known to strongly depend on the external static magnetic field B_0 [15, 16]. The results of studying the influence of B_0 on the angular velocity of energy transfer are plotted in Figs. 3.18 and 3.19. The normalization by the electron cyclotron frequency is inappropriate to study the dependence on the external static magnetic field. That is why the absolute value of ω_{en} normalized by the Langmuir frequency is used as ordinate axis in Figs. 3.18 and 3.19. In this case, both abscissa and ordinate contain the same physical value (the Langmuir frequency) in their denominators. In Figs. 3.18 and 3.19, the dashed curves are used for stronger external static magnetic field, $Z \equiv \Omega_e/|\omega_e| = 5$, and the solid curves for weaker magnetic field, $Z = 10$. One can see in Figure 3.18 that $|\omega_{en}|$ of LF SFXWs increases with increasing B_0 as it was also in studying the interior problem in Sect. 3.1. The same dependence is observed in Fig. 3.19: ω_{en} of HF SFXWs increases with increasing B_0, however, this is in contrast to the case of the interior problem.

The following conclusions can be done on the base of the research carried out in the present Section. Electromagnetic wave propagation in a cylindrical waveguide structure can be characterized by the angular phase velocity. The latter can be introduced as the angular velocity with which a half-plane of constant wave phase ($m\varphi - \omega t$ = Const) rotates around the axis of the metallic rod. It is a very simple function of the eigenfrequency ω: $\omega_{ph} = \omega/m$. The dependence of ω on the plasma waveguide parameters was analyzed in detail in Sect. 2.2 [15, 16]. That is why not too much attention is paid in the present Section to ω_{ph}.

Electromagnetic energy rotation around the metallic rod is adequately described in the present exterior problem via local (Eq. (3.12) and integral (Eq. (3.17) angular velocities of energy transfer. These two velocities are studied in their dependence on the plasma waveguide parameters: the rod radius, plasma particle density, external static axial magnetic field, and azimuthal mode index. The analysis is carried out for both the LF and HF ranges. These studies on the exterior problem in the present Section complement the research devoted to electromagnetic energy rotation in case of the interior problem carried out in the previous Sect. 3.1.

The dispersion of SFXWs is shown to be normal like for the interior problem. This means that the directions of angular phase velocity ω_{ph} and angular velocity of energy transfer ω_{en} coincide.

The local angular velocity of energy transfer as a function of radial coordinate does not exhibit monotonous behavior.

The absolute value of the integral angular velocity $|\omega_{en}|$ is shown to be smaller than that of the angular phase velocity in the interior problem. For both LF and HF SFXWs, the angular velocity $|\omega_{en}|$ shows non-monotonous behavior with increasing radius R of the metallic rod and plasma particle density. It decreases with increasing $|m|$. For LF SFXWs, $|\omega_{en}|$ is smaller in the interior problem than in the present exterior problem with the same plasma-waveguide parameters. The maxima of $|\omega_{en}|$ for HF SFXWs in the present exterior problem are smaller than those in the interior problem.

The dependence of $|\omega_{en}|$ on the external static axial magnetic field B_0 in the exterior problem is qualitatively the same as in the interior problem. The velocity $|\omega_{en}|$ of SFXWs increases with increasing B_0.

The analysis of electromagnetic energy rotation in plasmas around a metallic rod oriented along the external static magnetic field carried out in the present Section can be applied in the following fields of plasma physics: plasma electronics [16], physics of gas discharges [21–23], and development of surface enhanced Raman spectroscopy (see, e.g. [24, 27]).

3.3 Rotation of Electromagnetic Energy in Coaxial Metal Waveguides Entirely Filled by Plasmas

Coaxial lines are widely used for designing plasma sources based on the propagation of electromagnetic surface waves [28–30]. A dielectrically-loaded coaxial waveguide in the presence of an annular plasma column placed into a uniform axial magnetic field was studied theoretically in [31]. The presence of the dielectric layer on the conducting surfaces was shown to result in a considerable rise in the frequencies of the waveguide modes. The wave energy balance equation obtained from the Maxwell equations was used in [32] for optimization of the plasma parameters of a low-pressure coaxial discharge. The optimization made it possible to choose the plasma parameters to be the most appropriate for technological applications like large area waver treatments, coating of foils or deposition of diamond-like carbon. The effect of a background plasma on the electromagnetic properties of coaxial cavities with smooth and corrugated inner conductors used in very high-power gyrotrons was analyzed for the case of a 170 GHz, 2 MW $TE_{34,19}$-mode tube [33]. The effect was shown to cause a modification of the distribution of the electromagnetic energy density in such a way, that the Ohmic skin losses decrease both on the inner and outer conductors and that the resonance frequency is shifted.

Results of numerical simulations of a coaxial ubitron based on a helical electron beam efficiently interacting with the azimuthal electric field of the excited wave were presented in [34]. The linear theory of the excitation of a plasma-filled coaxial metal waveguide with a rippled wall by a longitudinal electron beam was developed in [35]. Space charge is well-known to limit the electric current which can be carried in a waveguide. The theory of the space-charge limited current in coaxial drift tube was developed in [36, 37]. A concept of gyro-devices in the form of a coaxial-waveguide cyclotron autoresonance maser (CARM) was proposed in [38], which was shown to be well suited for operation at both high power and ultrahigh gain in the millimeter-wave and far-infrared regions. Breakdown of a gas discharge by microwave radiation with a stochastically jumping phase (MWRSJP), the dispersion properties of axially symmetric ($m = 0$) electromagnetic waves in coaxial waveguides, as well as the collisionless electron heating by MWRSJP, and gas ionization were studied in [39].

The dynamics of a multipactor discharge within a coaxial line was analyzed in [40] using the mono-energetic electron emission model taking into account the space charge effects. A Monte Carlo model was applied in [41] to investigate the multipactor discharge in a coaxial waveguide structure loaded by a ceramic window, which was applied to a high-power coupler. The application of an external magnetic field was theoretically analyzed and simulated with the intention to increase the transmitted power through the high-power microwave window. Regardless of the fact that the papers [40, 41] do not present any study of eigenwaves and directed energy fluxes in coaxial lines, these and other practical applications explain the on-going interest to carry out further extensive studies on the physics of coaxial plasma waveguides.

SFXWs are the eigenmodes of coaxial metal waveguides entirely filled by plasma. SFWs with ordinary polarization, with the components H_r, H_φ, E_z, do not propagate in these waveguides [16, 42]. As far as the electric field of SFXWs is perpendicular to an external axial static magnetic field $\vec{B}_0$ and their dispersion properties strongly depend on the magnitude of B_0, these waves are classified as extraordinary waves in [15, 16, 20, 25, 26, 43] according to [6, 44, 45]. Their dispersion properties were studied for the first time in [42]. The waves were shown to propagate in two frequency ranges if the plasma particle density is sufficiently high and the external static axial magnetic field is sufficiently small, so that $\omega_e^2 < \Omega_e^2$. The ranges are the HF range (1.14) and the LF range $\omega_{LH} < \omega < |\omega_e|$.

SFXWs are unidirectional waves. The sign of the azimuthal mode index m correlates uniquely with the type of plasma surface curvature. Namely, within the LF range, SFXWs with positive azimuthal mode index, $m > 0$, propagate along the outer wall of a coaxial waveguide, while those with negative azimuthal mode index, $m < 0$, propagate along the inner wall. This is vice versa within the HF range, where SFXWs propagate along the outer wall with $m < 0$, as well as along the inner wall with $m > 0$. The presence of the second metal wall influences the dispersion properties of SFXW propagation along both inner and outer walls, which was wrongly called in [26] as mode coupling. There is no wave coupling, and the SFXWs with different/opposite azimuthal mode index propagate independently from each other.

In [43], the linear interaction between azimuthal modes (extraordinary surface waves and ordinary bulk waves) of a coaxial metal waveguide entirely filled with a current-carrying plasma was studied. The presence of the azimuthal component of the external static magnetic field, caused by the axial electric current in the plasma, was shown to convert azimuthal electromagnetic wave modes into coupled-wave perturbations, leading to the appearance of the axial electric wave field (as well as the radial and azimuthal components of the magnetic wave field), and to result in a slight decrease of the axial magnetic wave field.

The limiting case of SFXW propagation along the outer plasma-metal interface of a coaxial waveguide with finite radius of curvature, in which the radius of the inner conductor is sufficiently smaller compared with that of the outer wall, can be considered as a cylindrical metal waveguide entirely filled by plasma. In this case, studying the SFXW propagation along the inner plasma-metal interface becomes meaningless. The other limiting case of SFXW propagation along the inner plasma-metal interface of the coaxial waveguide with finite radius of curvature, in which the radius of the outer wall is sufficiently larger compared with that of the inner conductor, can be considered as a metal rod placed into infinite plasma along an external static magnetic field. In this case, studying the SFXW propagation along the outer plasma-metal interface also has no sense. Dispersion properties of SFXWs in these limiting cases were studied in detail in Sect. 2.2 [15], and electromagnetic energy rotation initiated by SFXWs in Sects. 3.1 and 3.2 [20, 25]. One limiting case more, which is a thin coaxial waveguide, is investigated in the present Section by the asymptotic method of a thin layer [46].

The goal of this Section is to complement the previous study of SFXW dispersion properties in a coaxial metal waveguide entirely filled by cold plasma placed in an external static axial magnetic field [42] by the analysis of electromagnetic energy transfer. Although the influence of inhomogeneous plasma particle density and finite plasma pressure on the dispersion properties of SFXWs were reported previously [26, 47, 48], the present Section is intended to outline just the power transfer in the coaxial geometry. Sometimes the approach of a homogeneous plasma is valid for the plasma waveguides [26, 28, 33–35, 39]. However, account for the plasma inhomogeneity and plasma finite temperature on the power transfer requires additional detailed research. The other objective of the present Section is to merge the investigation of electromagnetic energy rotation along the plasma-metal interface in cylindrical waveguides initiated by SFXWs [20, 25] with the consideration of the coaxial waveguide for which the waveguides studied earlier in this Chapter play the role of two limiting cases.

The coaxial metal waveguide with inner radius a and outer radius b is assumed to be entirely filled by cold homogeneous plasma, and placed into an external static axial magnetic field $\vec{B}_0$ (Fig. 3.20). The metal walls are supposed to have infinite electrical conductivity (to be loss free).

The radial distribution of the SFXW axial magnetic field can be derived from the Maxwell equations as a linear combination of modified Bessel functions $I_m(\xi)$ and MacDonald functions $K_m(\xi)$ [17] in the form (1.12) with non-zero constants

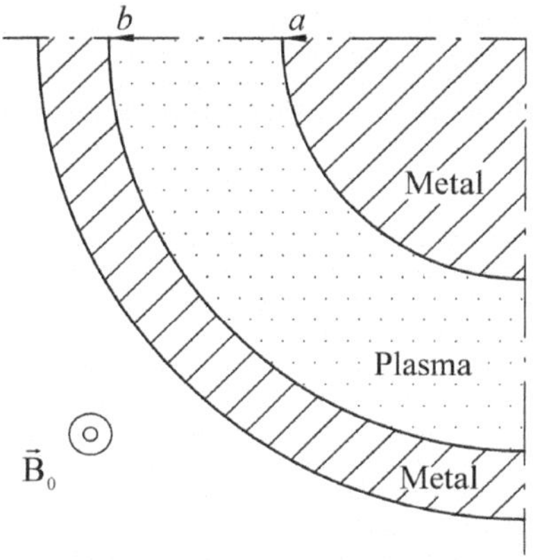

Fig. 3.20 Schematic of the coaxial plasma waveguide placed into an external static axial magnetic field

A_1 and A_2. Application of the boundary conditions of zero tangential (azimuthal) components of the electric wave field at the plasma-metal interfaces results in the following dispersion relation [42]:

$$\begin{aligned}&[I'_m(\xi_a)\xi_a + \mu m I_m(\xi_a)][K'_m(\xi_b)\xi_b + \mu m K_m(\xi_b)]\\&\quad -[I'_m(\xi_b)\xi_b + \mu m I_m(\xi_b)][K'_m(\xi_a)\xi_a + \mu m K_m(\xi_a)] = 0.\end{aligned} \tag{3.20}$$

In (3.20), $\xi_a = k_\perp a$, and $\xi_b = k_\perp b$.

The axial energy flow caused by SFXWs was shown in Sect. 3.1 to be identically equal to zero [20]. The radial component of the electromagnetic energy flow of SFXWs oscillates with the frequency 2ω, and the time averaged radial energy flow is also equal to zero. The only non-zero component of the time averaged energy flow is the azimuthal one. The averaged (over temporal period and radial cross-section) angular velocity of electromagnetic energy transfer [49] reads as follows:

$$\omega_{en} = \int_a^b \langle S_\varphi \rangle_T dr \Big/ \int_a^b r\langle u\rangle_T dr. \tag{3.21}$$

In (3.21), the azimuthal component of the Poynting vector $\langle S_\varphi\rangle_T$ and the electromagnetic energy density $\langle u\rangle_T$ are defined in Sect. 3.1. by Eqs. (3.7) and (3.15). It is difficult to treat analytically the expression (3.21) for arbitrary relation between the inner and outer radii of the coaxial metal waveguide. Therefore, the angular velocity of the energy transfer is analyzed numerically.

The dependencies of ω_{en} of SFXWs in coaxial waveguides on most of the plasma waveguide parameters: plasma particle density, external static axial magnetic field B_0, radii of inner and outer metal wall, as well as azimuthal mode index m, are qualitatively the same as described in Sects. 3.1 and 3.2 [20, 21]. The dependence of ω_{en} on the radii of the outer and inner metal walls is presented in Figs. 3.21,

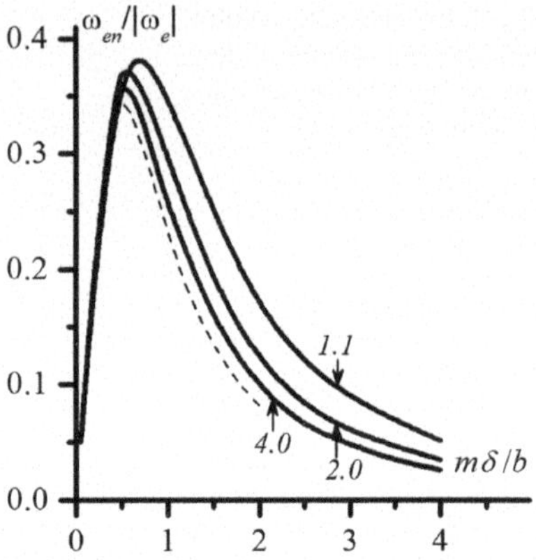

Fig. 3.21 Normalized angular velocity of energy transfer versus effective wavenumber for LF SFXWs propagating along the outer plasma-metal interface at different ratios $b/a = 1.1$, 2.0, 4.0 (*solid curves*), and $b/a = \infty$ (*dashed curve*), for $m = +1$, and $Z = 5$

3.22, 3.23 and 3.24 for SFXWs with azimuthal mode index $|m| = 1$ for $Z \equiv \Omega_e/|\omega_e| = 5$. The larger $|m|$ and Z are, the smaller the angular velocity of energy transfer

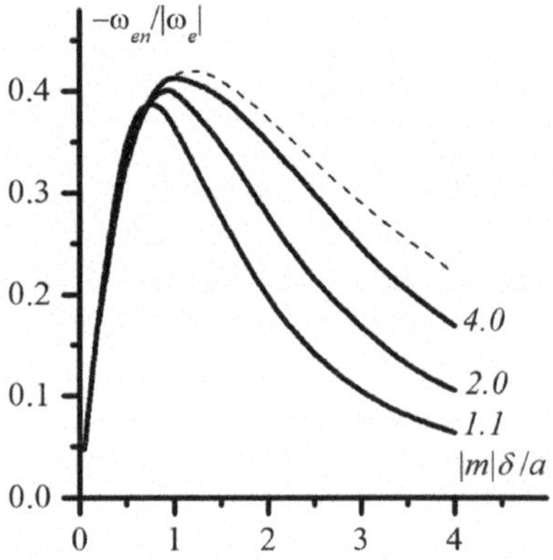

Fig. 3.22 Normalized angular velocity of energy transfer versus effective wavenumber for LF SFXWs propagating along the inner plasma-metal interface at different ratios b/a = 1.1,2.0,4.0 (*solid curves*), and $b/a = \infty$ (*dashed curve*), for $m = -1$, and $Z = 5$

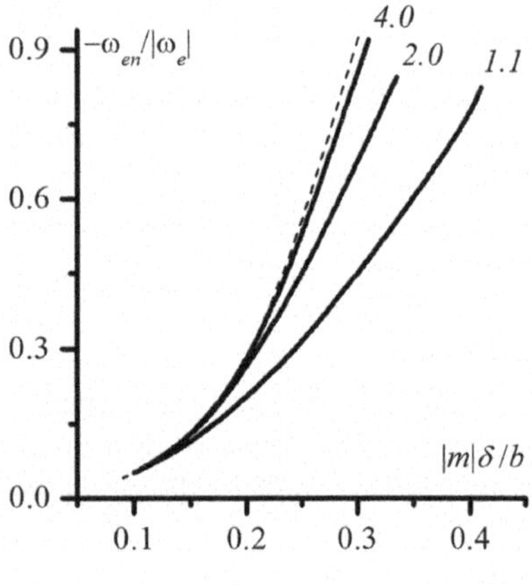

Fig. 3.23 Normalized angular velocity of energy transfer versus effective wavenumber for HF SFXWs propagating along the outer plasma-metal interface at different ratios b/a denoted by *numbers*: b/a = 1.1,2.0,4.0 (*solid curves*), and $b/a = \infty$ (*dashed curve*), for $m = -1$, and $Z = 5$

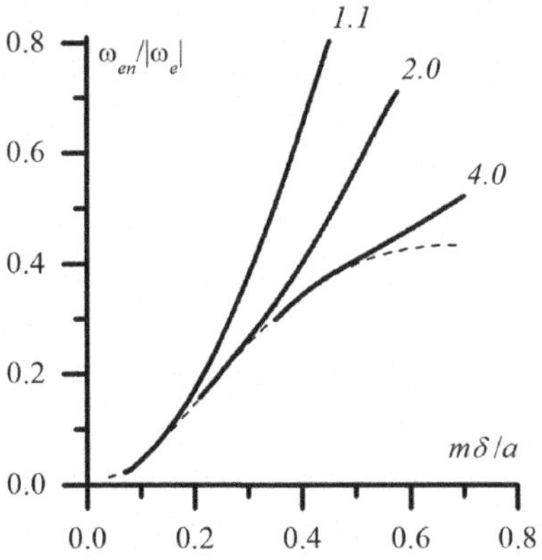

Fig. 3.24 Normalized angular velocity of energy transfer versus effective wavenumber for HF SFXWs propagating along the inner plasma-metal interface at different ratios *b/a* denoted by *numbers*: b/a = 1.1,2.0,4.0 (*solid curves*), and $b/a = \infty$ (*dashed curve*), for $m = +1$, and $Z = 5$

ω_{en} is. Note that Z is assumed to be larger than one in the present Section, which means sufficiently large plasma particle density and sufficiently small external static magnetic field.

The angular velocity of energy transfer normalized to the electron cyclotron frequency, $|\omega_{en}/\omega_e|$, is chosen as ordinate axis in the Figs. 3.21, 3.22, 3.23 and 3.24. The numbers near the curves show the value of the ratio *b/a*. The dashed curves correspond to the limiting cases $b >> a$. The effective wavenumber k_{ef} is chosen as abscissa axis on these figures as it was done in Sects. 3.1 and 3.2. In the Figs. 3.21 and 3.23, which describe ω_{en} of SFXWs propagating along the outer plasma-metal interface, $r = b$, the abscissa is $k_{ef} = |m|c/(\Omega_e b)$. In the Figs. 3.22 and 3.24, ω_{en} of SFXWs propagating along the inner plasma-metal interface, $r = a$, is shown. That is why the effective wavenumber is chosen there as follows: $k_{ef} = |m|c/(\Omega_e a)$. Such a choice of the abscissa allows the analysis of the dependence of ω_{en} on the inner and outer radii a and b of the coaxial waveguide, and the plasma particle density n_e, since $\Omega_e \propto n_e^{1/2}$.

In the Figs. 3.21 and 3.22, where ω_{en} of SFXWs propagating in the LF range is studied, the calculations are stopped after the curves demonstrate gradual decrease. In the Figs. 3.23 and 3.24, ω_{en} of HF SFXWs is plotted. HF SFXWs are known not to propagate under the conditions which correspond to large values of k_{ef} [16, 42]. That is why the calculations in Figs. 3.23 and 3.24 are stopped when the frequencies of HF SFXWs reach the upper limit of the HF range (1.14).

In the case of a thin coaxial waveguide, $(b-a)/a \ll 1$, the spatial distribution of the axial magnetic field of SFXWs can be found asymptotically from the Maxwell equations in the form of series in $(r-a)$:

$$H_z(r) \approx H_z(a)\left[1 + \frac{k_\perp^2 a^2 + m^2}{2}\left(\frac{r-a}{a}\right)^2\right]$$
$$+ a\frac{dH_z(r)}{dr}_{|r=a}\left(\frac{r-a}{a}\right)\left[1 - \frac{r-a}{2a} + \frac{k_\perp^2 a^2 + m^2}{6}\left(\frac{r-a}{a}\right)^2\right]. \quad (3.22)$$

Expression (3.22) is obtained as result of integrating the second order differential equation of Bessel type. Its two linearly independent solutions are $H_z(a)\left[1 + \sum_{l=1}^{\infty} \alpha_l \left(\frac{r-a}{a}\right)^{2l}\right]$ and $a\frac{dH_z(r)}{dr}_{|r=a} \ln\left(\frac{r}{a}\right) \cdot \left[1 + \sum_{l=1}^{\infty} \beta_l \left(\frac{r-a}{a}\right)^{2l}\right]$ with α_l and β_l being definite constants. In the square brackets of (3.22), the terms of the order higher than the second one in (r–a) are neglected. This is sufficient to derive SFXW electric fields with an accuracy up to terms linear in (r–a).

Substituting the expression (3.22) into the definition (3.21) makes it possible to derive the asymptotic expression for the angular velocity of electromagnetic energy transfer ω_{en} in a thin coaxial waveguide as follows:

$$\omega_{en} = \frac{2\left[m - \frac{b-a}{2a}\left(\frac{\omega^2}{c^2}\varepsilon_2 a^2 + m + m^2\mu\right)\right]c^2}{\varepsilon_1 \omega a^2\left[1 + \frac{\partial}{\partial\omega}(\omega\varepsilon_1)\frac{c^2 m^2}{\omega^2 \varepsilon_1^2 a^2} + \frac{b-a}{2a}\left(1 - 2m\mu - \frac{\partial}{\partial\omega}(\omega\varepsilon_1)\frac{m}{\varepsilon_1^2}\left(\frac{mc^2}{a^2\omega^2} + 2\varepsilon_2\right)\right)\right]}. \quad (3.23)$$

In course of deriving Eq. (3.23), the following link between the constants $H_z(a)$ and $\frac{dH_z(r)}{dr}_{|r=a}$ known from the boundary condition at the inner metal wall ($E_\varphi(a) = 0$) is used,

$$\frac{dH_z(r)}{dr}_{|r=a} = -\frac{\mu m}{a} H_z(a). \quad (3.24)$$

In Figs. 3.25 and 3.26, the influence of the azimuthal mode index and the external static axial magnetic field B_0 on the dependencies of the normalized ω_{en} versus

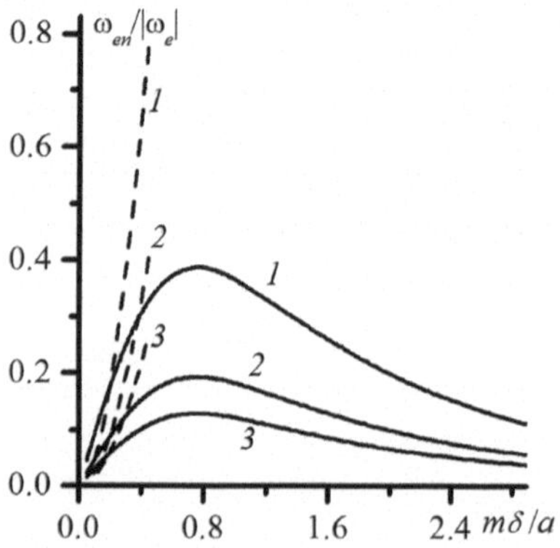

Fig. 3.25 Normalized angular velocity of energy transfer of LF SFXWs (*solid curves*) and HF SFXWs (*dashed curves*) versus effective wavenumber for $Z = 5$, $b/a = 1.1$. The *numbers* denote $|m|$

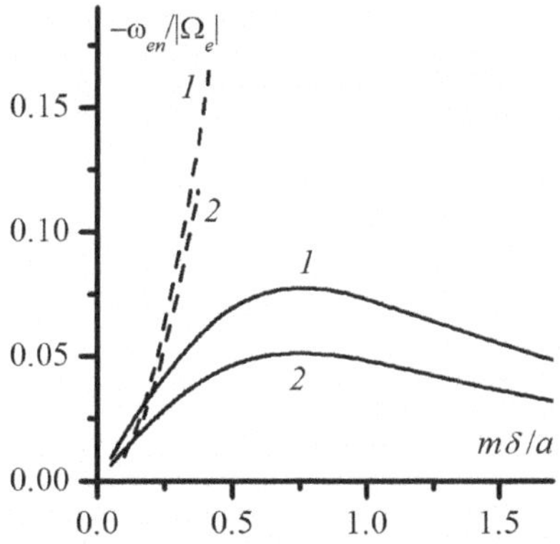

Fig. 3.26 Normalized angular velocity of energy transfer of LF SFXWs (*solid curves*) and HF SFXWs (*dashed curves*) versus effective wavenumber for $m = -1$, $b/a = 1.1$ and two different strengths of static magnetic field. The *numbers* denote the magnitude of the ratio $Z = \Omega_e / |\omega_{en}|$, where "1" corresponds to Z = 5, and "2" to Z = 7.5

effective wavenumber is demonstrated for the case of a thin plasma waveguide with $b/a = 1.1$. The effective wavenumber $k_{ef} = |m|c/(\Omega_e \alpha)$ is chosen as abscissa as it was done in Sect. 3.2 [25].

In Fig. 3.25, ω_{en} is normalized to the absolute value of the electron cyclotron frequency. The numbers at the curves denote the azimuthal mode index *m*. The dashed curves correspond to the case of HF SFXWs, which propagate with positive azimuthal mode index, and the solid curves to the LF SFXWs, which also propagate with $m > 0$. The following plasma parameters are chosen for the calculations: $Z = 5$. Figure 3.25 convincingly demonstrates a decrease of $|\omega_{en}|$ which is approximately inversely proportional to $|m|$ for both LF and HF SFXWs. The dependencies of $|\omega_{en}|$ for SFXWs, which propagate with negative azimuthal mode index in a thin waveguide, are not shown here since the magnitudes of $|\omega_{en}|$ for SFXWs with $m < 0$ exceed those for SFXWs with $m > 0$ only by less than 5%.

The influence of B_0 on the dependence of $|\omega_{en}|$ on the effective wavenumber is shown in Fig. 3.26 for SFXWs propagating with $m = -1$. The electron cyclotron frequency cannot be applied for normalization of $|\omega_{en}|$ in this case. Here the Langmuir frequency Ω_e is used instead of $|\omega_e|$, which results in a similar dependence of both abscissa and ordinate on the plasma particle density, $\propto n_e^{-1/2}$, independent of the magnitude of B_0. The dashed and solid curves correspond to the case of HF and LF SFXWs, respectively. The same case of a thin waveguide as for calculations on Fig. 3.25 is chosen, $b/a = 1.1$. The numbers at the curves denote the ratio Z, where "1" corresponds to $Z = 5$, and "2" to a lower external static axial magnetic field with $Z = 7.5$. The behavior of the curves in Fig. 3.26 is in a good agreement with those obtained in the previous Sects. 3.1 and 3.2: $|\omega_{en}|$ is approximately inversely proportional to Z. The dependencies of ω_{en} for SFXWs, which propagate with positive azimuthal mode index in a thin waveguide, are not shown in Fig. 3.26, since these curves almost coincide with those for negative $m = -1$.

The accuracy of the dependencies of the curves calculated for the Figs. 3.25 and 3.26 from the expression (3.23) is better than 5% as compared to those calculated from the Eq. (3.21). This accuracy is in agreement with expectations for the case of $(b-a)/a = 0.1$, which assumes an accuracy of Eq. (3.23) of about 10%.

The effect of dissipations on the angular velocity of electromagnetic energy transfer ω_{en} can be studied by incorporating the effective collision frequency ν into the expressions for the components of the plasma permittivity tensor (compare with those in the collisionless case, Eq. (1.3)):

$$\varepsilon_1 = 1 - \sum_\alpha \frac{\Omega_\alpha^2(\omega + i\nu)}{\left[(\omega + i\nu)^2 - \omega_\alpha^2\right]\omega}, \quad \varepsilon_2 = \sum_\alpha \frac{-\omega_\alpha \Omega_\alpha^2}{\omega\left[(\omega + i\nu)^2 - \omega_\alpha^2\right]}. \tag{3.25}$$

Since $|\omega_{en}|$ is shown above (see Figs. 3.21, 3.22, 3.23 and 3.24) to be larger for SFXWs with negative azimuthal mode index, such SFXWs with $m < 0$ are analyzed further. For example, SFXWs which propagate with $m = -1$ in a coaxial waveguide with $Z = 5$, and $b/a = 4$ are chosen in the calculations below.

For LF SFXWs, which propagate with negative azimuthal mode index along the inner plasma-metal interface, a small collision frequency ν results in insignificant increase of the $|\omega_{en}|$. This increase is more pronounced for small magnitudes of $k_{ef} = |m|c/(\Omega_e a)$ than for $k_{ef} \gtrsim 1$. This is explained by the fact that the eigenfrequency of LF SFXWs increases almost linearly for $k_{ef} << 1$ and saturates for $k_{ef} \gtrsim 1$. For instance, an increase of ν from zero to 0.2 results in the increase of $|\omega_{en}|$ by 4% (from $|\omega_{en}| = 0.326|\omega_e|$ to $|\omega_{en}| = 0.341|\omega_e|$) for $k_{ef} = 0.5$, and less than by 2% (from $|\omega_{en}| = 0.416|\omega_e|$ to $|\omega_{en}| = 0.422|\omega_e|$) for $k_{ef} = 1.0$.

HF SFXWs propagate with negative azimuthal mode index along the outer plasma-metal interface, and dissipations cause a weaker influence on $|\omega_{en}|$ of HF SFXWs than that of LF SFXWs. For instance, an increase of ν from zero to 0.8 results in the increase of $|\omega_{en}|$ by less than 1% (from $|\omega_{en}| = 0.852|\omega_e|$ to $|\omega_{en}| = 0.857|\omega_e|$) for $k_{ef} = 0.3$, and in the decrease of $|\omega_{en}|$ by less than 2% (from $|\omega_{en}| = 0.108|\omega_e|$ to $|\omega_{en}| = 0.106|\omega_e|$) for $k_{ef} = 0.14$. Since dissipations cause a decrease of the HF SFXW eigenfrequencies, the dissipations result in an expansion of the $k_{ef} = |m|c/(\Omega_e b)$ range, where HF SFXWs propagate. The right limit of this range moves to the right into the region of smaller radii of the outer metal wall and smaller plasma particle densities.

The angular group velocity ω_g of electromagnetic energy rotation was introduced in [20] by analogy with the linear group velocity $\upsilon_g = \partial\omega/\partial k$ as follows: $\omega_g = \partial\omega/\partial m$ (3.9). It can be calculated precisely from the known expression for the dispersion relation $D(\omega,m) = 0$ (3.20). Plots of the dependence of ω_g on the coaxial plasma waveguide parameters are not shown here, since they precisely coincide with those plotted in Figs. 3.21, 3.22, 3.23, 3.24, 3.25 and 3.26.

One can conclude the following from the analysis presented above. First, SFXWs have normal dispersion in coaxial metal waveguides entirely filled by cold plasma: they propagate and transfer their electromagnetic energy in the same direction, as it is also in the cases considered in Sects. 3.1 and 3.2: a cylindrical metal waveguide entirely filled by plasma and a metallic rod placed into infinite plasma along an external static magnetic field. Second, no difference between the magnitudes of the angular group velocity ω_g of SFXWs and their angular velocities of energy rotation ω_{en} is found. Third, the electromagnetic energy of SFXWs rotates with almost the same angular velocity both in the LF and HF frequency ranges along both inner and outer plasma-metal interfaces for different magnitudes of the ratio *b/a* if the depth of the wave penetration into the plasma is small. This condition can be explained as follows. The width (*b–a*) of the coaxial waveguide should be larger than the depth $k_{\perp}{}^{-1}$ of the wave penetration into the plasma. In other words, the wave propagation along one metal-plasma interface is weakly influenced by the presence of the other metal-plasma interface. This latter condition can be realized for arbitrary fixed ratio *b/a* in the limit $a \to \infty$, which causes $k_{ef} \to 0$. The curves in Figs. 3.21, 3.32, 3.33 and 3.34 go closer to each other for the smaller values of the effective wavenumber. Fourth, LF SFXWs rotate their electromagnetic energy most fast for $k_{ef} \sim 1$. The angular velocity of HF SFXW energy transfer ω_{en} is the largest at the right limit of the k_{ef} range where HF SFXWs can propagate.

Comparison of the curves in Fig. 3.21 with those in Fig. 3.20 as well as the curves in Fig. 3.23 with those in Fig. 3.24 gives rise to the following conclusions. First, SFXWs transfer electromagnetic energy along the opposite walls of a thin coaxial plasma waveguide ($b/a \gtrsim 1$) in opposite directions with almost the same angular velocities. Second, SFXWs with positive azimuthal mode index transfer their electromagnetic energy slower, the larger the ratio *b/a* is (see Figs. 3.21 and 3.24). Third, the angular velocity of energy transfer $|\omega_{en}|$ initiated by SFXWs with negative azimuthal mode index increases with increasing ratio *b/a* (see Figs. 3.22 and 3.23). One can summarize these three conclusions in the following way. SFXWs rotate their energy most quickly in the direction opposite to that of electron gyration in the external static magnetic field. LF SFXWs do this along the inner plasma-metal interface (exterior problem), and HF SFXWs along the outer one (interior problem).

The asymptotic expression (3.23) for ω_{en} is shown to be a handy tool to calculate the angular velocity of electromagnetic energy rotation in a thin coaxial plasma-filled waveguide.

3.4 Rotation of Electromagnetic Energy in Cylindrical Metal Waveguides Partially Filled by Magneto-active Plasmas

In the present Section, the same structure is considered as in Sect. 2.1: a cold collisionless plasma column with radius R_1 is placed into the cylindrical metal waveguide with inner radius R_2 (Fig. 3.27). The external static axial magnetic field is parallel to the waveguide axis, $\vec{B}_0 \| \vec{z}$. The plasma column is separated from the metal wall by a dielectric layer with relative permittivity ε_d. The present study differs from that in Sect. 3.1 by the introduction of this dielectric protective layer, which is often present in the case of real experimental or technological devices. The present consideration generalizes also the research carried out in [50] by including an external static axial magnetic field. On one hand, the presence of $\vec{B}_0$ increases the design complexity of the device, as well as the costs of operation and manufacturing. On the other

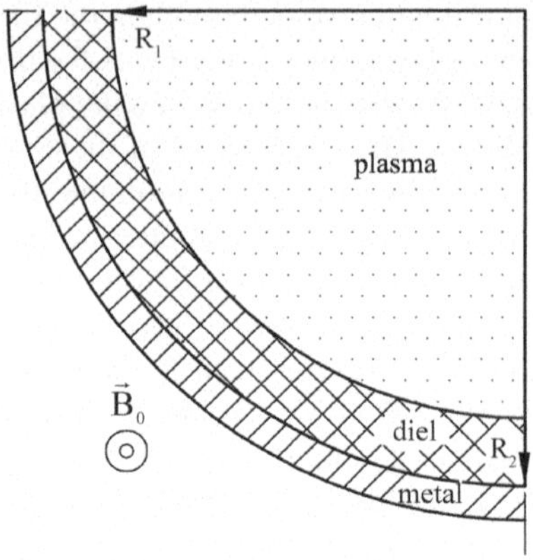

Fig. 3.27 Schematic of the plasma-waveguide structure placed into an external static axial magnetic field

hand, it provides an enrichment of the wave spectrum, and thus an additional tool for experimentalists to control the technological processes.

The flute character of the studied waves results in the absence of axial electromagnetic energy flow: the axial component of the Poynting vector is identically equal to zero. The presence of the cylindrical metal wall of the waveguide causes that also the radial component of the Poynting vector at $r = R_2$ is exactly equal to zero. However, inside the waveguide, $r < R_2$, the radial component of the electromagnetic energy flow oscillates with double the SFXW frequency (2ω). As a result, SFXWs provide electromagnetic energy transfer in just azimuthal direction. In other words, SFXWs initiate electromagnetic energy rotation, which is the objective of the present study.

This configuration, in which the waves propagate across the external static magnetic field, which in turn is parallel to the interface, is known to be called as Voigt geometry [26]. The theory of surface electromagnetic wave propagation in Voigt geometry along the plasma-metal interface in both plane and cylindrical waveguides was reviewed in [26]. On the one hand, the review did not regard the energy transfer, and the cylindrical structures considered therein can be obtained only in the limit that there is no dielectric layer (which significantly influences the spatial wave field distribution). On the other hand, the review considered physical mechanisms of SFXW parametric excitation, the influence of finite plasma temperature, and three- and four-wave interactions, as well as the relevance of SFXW theory to some experimental results.

The studied waves are of surface nature in the plasma column: their amplitude $H_z(r)$ decreases with going into the plasma depth from the plasma-dielectric interface $r = R_1$. This means, the waves do not propagate in infinite plasma. The amplitude $H_z(r)$ is governed by a second order differential equation of Bessel type. That is why it can be written in terms of modified Bessel functions of the first kind $I_m(x)$ (see. Equation (1.12) with the constant $A_2 = 0$, since the MacDonald functions $K_m(\xi)$ are known to have a singularity at the axis, $r = 0$). The constant A_2 is determined to be equal to zero to meet the boundary condition at the waveguide axis, where the wave fields should be of finite magnitude.

However, SFXWs are of volumetric nature in the dielectric layer. The dielectric is transparent for SFXWs. The radial dependence of the axial component of the magnetic wave field can be written in terms of Bessel functions of the first, $J_m(\xi)$, and second, $N_m(\xi)$, kinds (see Eq. (2.1)).

The case of sufficiently high plasma particle density and sufficiently weak external static axial magnetic field is considered. Then SFXWs are known to propagate in the waveguide structure plasma-dielectric-metal in two frequency ranges: (1.13) and (1.14).

The present study deals with the zeroth radial mode of SFXWs. Just in this case, the plasma parameters cause the strongest impact on the wave properties. The dispersion properties of higher radial SFXW modes were studied in detail in [51, 52]. Application of the first radial mode of SFXWs makes it possible to get the advantage of higher frequency during excitation by a gyrating electron beam as compared with the case of the zeroth radial mode. However, this is accompanied by a moderate decrease of the SFXW growth rate [53].

The temporally and radially averaged angular velocity of energy transfer ω_{en} is determined as the ratio of the time-averaged electromagnetic energy flow

$$0.5 \int_0^{R_2} S_\varphi(r)dr, \tag{3.26}$$

and the time-averaged electromagnetic energy in a unit angular sector

$$\int_0^{R_2} r\langle u\rangle_T dr\ , \tag{3.27}$$

where $S_\varphi(r)$ (3.6) is the azimuthal component of the Poynting vector, and $\langle u\rangle_T$ (3.15) is the time-averaged electromagnetic energy density.

The angular velocity of electromagnetic energy rotation

$$\omega_{en} = \frac{-c}{8\pi} \frac{\int_0^{R_2} E_r(r)H_z(r)dr}{\int_0^{R_2} r\langle u\rangle_T dr}, \tag{3.28}$$

is the subject of the detailed numerical study in the present Section.

The dependence of the angular velocity ω_{en} (3.28) on the effective wavenumber $k_{ef} = |m|c/(\Omega_e R_1)$ is plotted in Figs. 3.28, 3.29, 3.30, 3.31, 3.32, 3.33, 3.34, 3.35, 3.36 and 3.37. According to the definition, an increase of k_{ef} is equivalent to a decrease of the plasma column radius R_1 and plasma particle density. LF SFXW energy rotation is studied in Figs. 3.28, 3.29, 3.30, 3.31 and 3.32. The curves in Figs. 3.28, 3.29, 3.30, 3.31 and 3.32 are calculated for the k_{ef} range, for which the

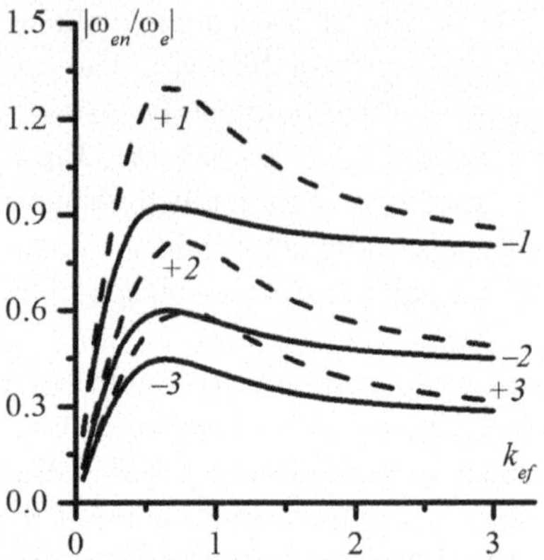

Fig. 3.28 Absolute value of normalized angular velocity of energy transfer $|\omega_{en}/\omega_e|$ of LF SFXWs versus effective wavenumber for $Z = \Omega_e/|\omega_e| = 5$, $\varepsilon_d = 1$, $b/a = 1.2$, $\nu = 0$. *Dashed curves* correspond to SFXWs with positive azimuthal mode index, *solid curves* relate to SFXWs with $m < 0$. The *numbers* denote the azimuthal wave indices

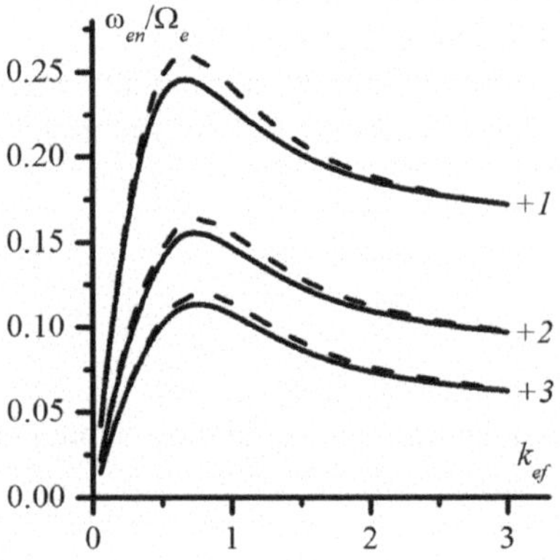

Fig. 3.29 Normalized angular velocity of energy transfer ω_{en}/Ω_e of LF SFXWs versus effective wavenumber for $\varepsilon_d = 1$, $b/a = 1.2$, $\nu = 0$. *Dashed curves* correspond to the case of stronger external axial static magnetic field, $Z = 5$, and *solid curves* relate to the case of weaker magnetic field, $Z = 7.5$. The *numbers* denote the azimuthal wave indices

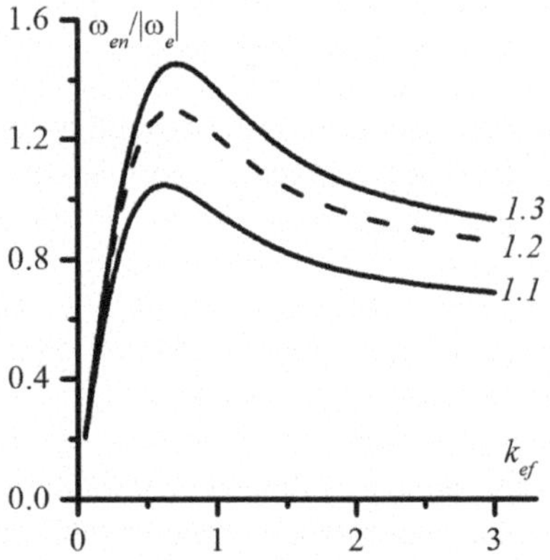

Fig. 3.30 Normalized angular velocity of energy transfer $\omega_{en}/|\omega_e|$ of LF SFXWs versus effective wavenumber for $m = 1$, $Z = 5$, $\varepsilon_d = 1$, $\nu = 0$. The *numbers* denote the relative width of the dielectric layer R_2/R_1. The *dashed curve* is the same as in Figs. 3.28, 3.31 and 3.32 for comparison

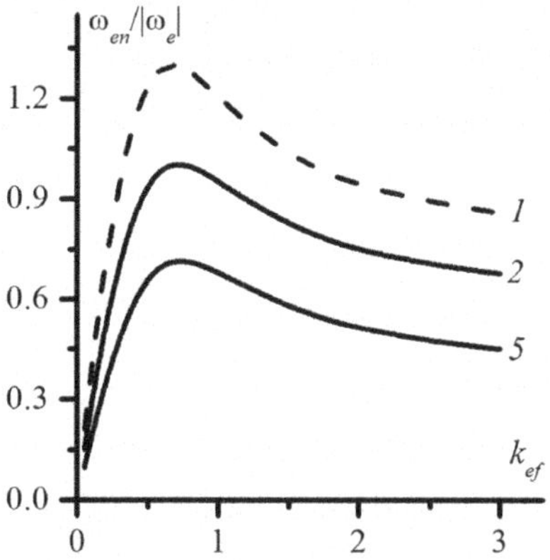

Fig. 3.31 Normalized angular velocity of energy transfer $\omega_{en}/|\omega_e|$ of LF SFXWs versus effective wavenumber for $m = 1$, $Z = 5$, $b/a = 1.2$, $\nu = 0$. The *numbers* denote the magnitude of the dielectric permittivity ε_d. The *dashed curve* is the same as in Figs. 3.28, 3.30 and 3.32 for comparison

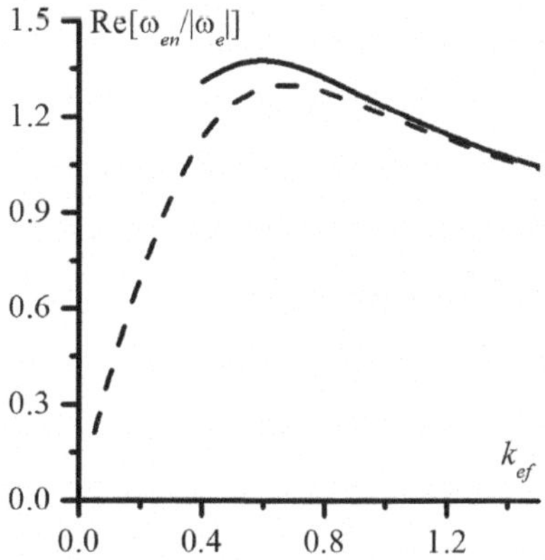

Fig. 3.32 Real part of normalized angular velocity of energy transfer $\omega_{en}/|\omega_e|$ of LF SFXWs versus effective wavenumber for $m = 1$, $Z = 5$, $R_2/R_1 = 1.2$, $\varepsilon_d = 1$. The *dashed curve* corresponds to the case of no dissipation, $\nu = 0$, and the *solid curve* relates to the case of $\nu = 0.4|\omega_e|$

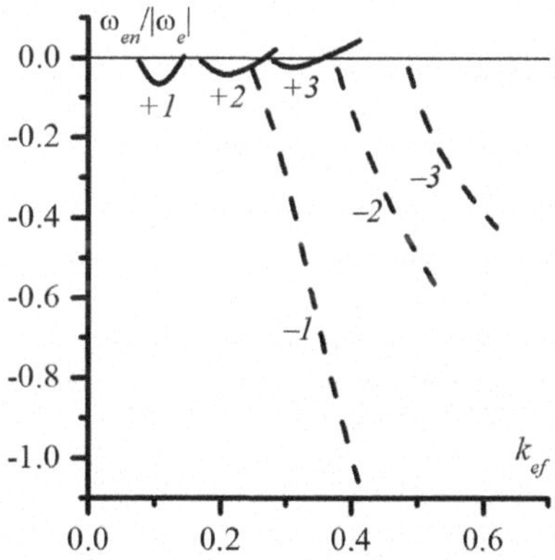

Fig. 3.33 Absolute value of normalized angular velocity of energy transfer $|\omega_{en}/\omega_e|$ of HF SFXWs versus effective wavenumber for $Z = 5$, $\varepsilon_d = 1$, $R_2/R_1 = 1.2$, $\nu = 0$. The *dashed curves* correspond to SFXWs with negative azimuthal mode index, *solid curves* relate to SFXWs with $m > 0$. The *numbers* denote the azimuthal wave indices

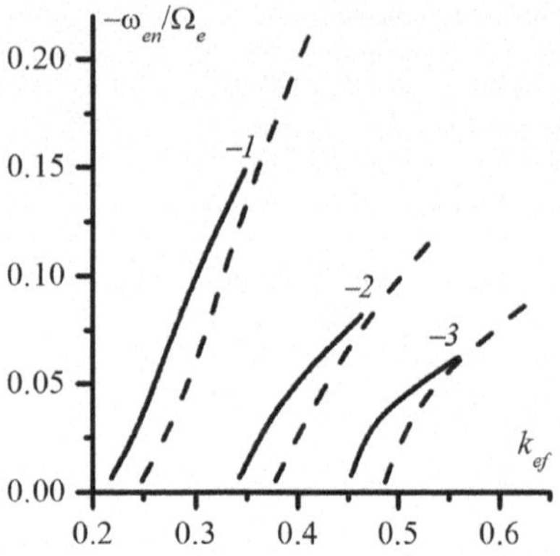

Fig. 3.34 Absolute value of normalized angular velocity of energy transfer ($-\omega_{en}/\Omega_e$) of HF SFXWs versus effective wavenumber for $\varepsilon_d = 1$, $R_2/R_1 = 1.2$, $\nu = 0$. The *dashed curves* correspond to the case of stronger external axial static magnetic field, Z = 5.0, and *solid curves* relate to the case of weaker magnetic field, Z = 7.5. The *numbers* denote the azimuthal wave indices

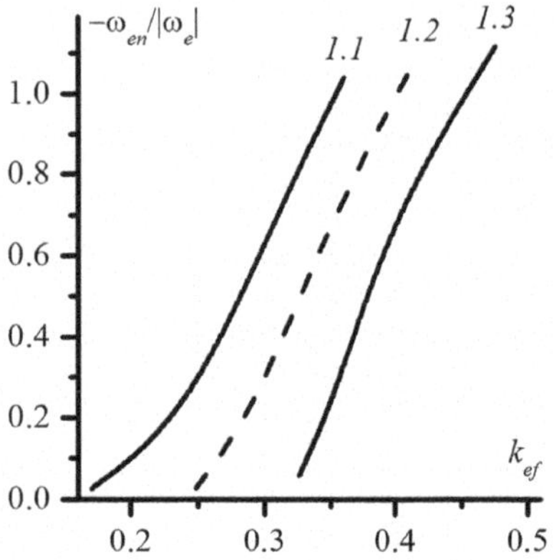

Fig. 3.35 Absolute value of normalized angular velocity of energy transfer ($-\omega_{en}/|\omega_e|$) of HF SFXWs versus effective wavenumber for $m = -1$, $Z = 5$, $\varepsilon_d = 1$, $\nu = 0$. The *numbers* denote the ratio of the dielectric layer radii R_2/R_1. The *dashed curve* is for comparison with Fig. 3.34

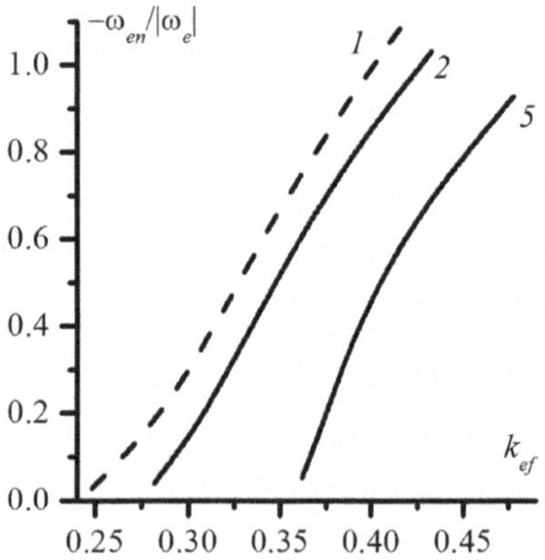

Fig. 3.36 Absolute value of normalized angular velocity of energy transfer ($-\omega_{en}/|\omega_e|$) of HF SFXWs versus effective wavenumber for $m = -1$, $Z = 5$, $R_2/R_1 = 1.2$, $\nu = 0$. The *numbers* denote the magnitude of the dielectric permittivity ε_d. The *dashed curve* is for comparison with Figs. 3.35 and 3.37

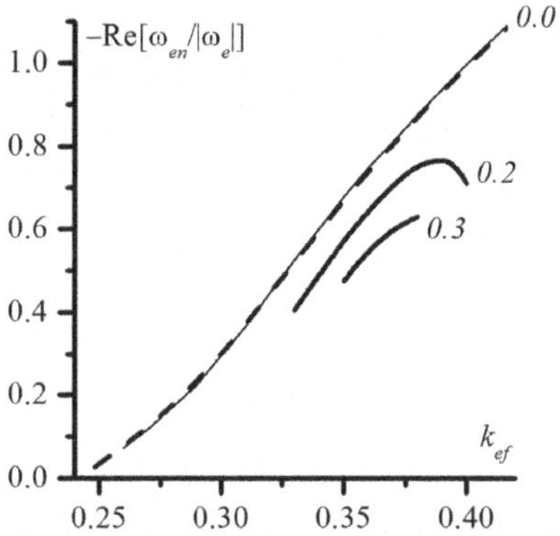

Fig. 3.37 Absolute value of normalized angular velocity of energy transfer ($-\omega_{en}/|\omega_e|$) of HF SFXWs versus effective wavenumber for $m = -1$, $Z = 5$, $R_2/R_1 = 1.2$, $\varepsilon_d = 1$. The *numbers* denote the magnitude of the effective collision frequency $\nu/|\omega_e|$. The *thin line* shows the angular group velocity, $-\mathrm{Re}[\omega_g/|\omega_e|]$, for the case $\nu = 0.3|\omega_e|$. The *dashed curve* is for comparison with Figs. 3.35 and 3.36

right limit of the magnitude of the angular velocity saturates. The absolute value of the angular velocity normalized to the absolute value of the electron cyclotron frequency, $|\omega_{en}/|\omega_e||$, is chosen as ordinate in Figs. 3.28, 3.30, 3.31, 3.32, 3.33, and 3.35, 3.36, 3.37. Figures 3.29 and 3.30 demonstrate the influence of the external magnetic field on the angular velocity. That is why ω_{en} is normalized to the electron plasma frequency therein.

LF SFXWs with positive azimuthal mode index rotate faster than those with $m < 0$ (see Fig. 3.28). That is why the angular velocity $\omega_{en}$for SFXWs with $m > 0$ is presented in Figs. 3.29, 3.30, 3.31 and 3.32. Increasing absolute value of the azimuthal mode index causes a decrease of the absolute value of the angular velocity ω_{en}. That is why $\omega_{en}$for SFXWs with $m = +1$ is shown in Figs. 3.30, 3.31 and 3.32. It is appropriate to mention here that just LF SFXWs with $m > 0$ propagate in the cylindrical metal plasma-filled waveguides in absence of the dielectric layer, $R_1 = R_2$. The decrease of $|\omega_{en}|$ with increasing $|m|$ was also noted in the case of no magnetic field B_0 in [50], as well as in the case $R_1 = R_2$ in Sect. 3.1.

The external static axial magnetic field B_0 surely influences the dispersion properties of SFXWs. B_0 enters into the calculations through the parameter $Z = \Omega_e/|\omega_e|$. The influence of B_0 is most pronounced for that k_{ef} range where ω_{en} reaches its maximum. Decrease of B_0 by a factor of 1.5 (which is equivalent to an increase of Z from $Z = 5$ to $Z = 7.5$ in Fig. 3.29) causes a reduction of the angular velocity ω_{en} by 6% for $k_{ef} = 0.7$ and $m = +1$ in Fig. 3.29. This result is in qualitative agreement with that obtained in Sect. 3.1. Note, that the reduction of the LF SFXW eigenfrequency is also about 6% in this case.

The width of the dielectric layer (R_2–R_1) is assumed to be small in this Section to provide the calculations in the framework of zeroth radial mode of SFXWs. This circumstance is checked in calculations for all the Figs. 3.28, 3.29, 3.30, 3.31, 3.32,

3.33, 3.34, 3.35, 3.36 and 3.37 via monitoring the radial change of the wave phase: $\kappa(R_2 - R_1) \lesssim \pi/2$. Within this restriction, one can conclude from Fig. 3.30 that an increase of the width (R_2–R_1) is accompanied by a reduction of the angular velocity ω_{en}. The opposite effect is caused by an increase of the permittivity of the dielectric layer, as shown in Fig. 3.31. These dependencies of ω_{en} versus (R_2–R_1) and ε_d are in qualitative agreement with those obtained in [50].

The influence of dissipation on the angular velocity is demonstrated in Fig. 3.32. Taking the effective collision frequency $\nu = 0.4|\omega_e|$ (see solid curve) into account, causes an increase of ω_{en} compared to the case $\nu = 0$. The upper curve starts from $k_{ef} = 0.4$ since the imaginary part of ω_{en} becomes of the order of or even larger than $\mathrm{Re}[\omega_{en}]$ in the vicinity of $k_{ef} = 0.3$, where the LF SFXW eigenfrequency is close to the electron cyclotron frequency, $\omega \approx |\omega_e|$. The dependence of ω_{en} on k_{ef} in the case $\nu = 0$ is shown by the dashed curve. The following plasma waveguide parameters are applied in the calculations for Fig. 3.32: $m = 1$, $Z = 5$, $R_2/R_1 = 1.2$, and $\varepsilon_d = 1$.

The results of the numerical analysis plotted in Fig. 3.31 are applied in the following to calculate the maximum magnitude of the angular velocity of energy transfer ω_{en} of LF SFXWs with azimuthal mode index $m = 1$ in a modest University sized device. In a plasma waveguide with the metal chamber of radius $R_2 = 6$ cm, the radius of the plasma column should be $R_1 = 5$ cm to meet the conditions of Fig. 3.31, which shows that the maximum magnitude of ω_{en} is reached for the case of vacuum (with dielectric permittivity $\varepsilon_d = 1$), used to separate the plasma column from the metal wall.

Maximizing the angular velocity of energy transfer is desirable for surface-wave sustained discharges inside the plasma-dielectric-metal structures placed into an external magnetic field to reach the best conditions for the energy conversion from the microwave generator to the plasma column [54]. The dashed curve in Fig. 3.31 shows that the maximum value of the angular velocity of energy transfer ω_{en} to be equal to $1.301|\omega_e| = 2.23 \times 10^9$ s^{-1} for $k_{ef} = 0.7$. The condition $k_{ef} = 0.7$ implies the plasma frequency to be $\Omega_e \approx 8.57 \times 10^9$ s^{-1}, which corresponds to the plasma particle density $n_e = 2.3 \times 10^{10}$ cm^{-3}. To meet the condition $Z = 5$, the electron cyclotron frequency should be $|\omega_e| \approx 1.71 \times 10^9$ s^{-1} which implies the magnetic field strength $B_0 \approx 97.5$ G. In this case, the SFXW propagates with the angular frequency $\omega \approx 3.1 \times 10^9$ s^{-1}.

HF SFXWs are known to propagate within the range of k_{ef}, which is limited for larger values by a moderate magnitude for which the frequencies reach the upper boundary ω_2. This circumstance determines the choice of the range of k_{ef} for which the curves in Figs. 3.33, 3.34, 3.35, 3.36 and 3.37 are calculated.

The energy of HF SFXWs with negative azimuthal mode index rotates faster than that of HF SFXWs with $m > 0$ (see Fig. 3.33). This can be explained by the fact that HF SFXWs do not propagate with $m > 0$ in cylindrical metal waveguides entirely filled by plasma, which means $R_1 = R_2$. That is why the angular velocity is calculated in Figs. 3.34, 3.35, 3.36 and 3.37 for HF SFXWs with negative azimuthal mode index. An increase of the absolute value of m influences the angular velocity ω_{en} for HF SFXWs with different signs of m, differently. It is necessary to mention here that just HF SFXWs with $m < 0$ propagate in cylindrical metal plasma-filled waveguides in absence of the dielectric layer ($R_1 = R_2$). For HF SFXWs with both $m < 0$ and $m > 0$, increasing $|m|$ causes a decrease of the maximum magnitude of $|\omega_{en}|$ and an extension of the k_{ef} range where HF SFXWs propagate. For HF SFXWs with $m > 0$, approach to the right limit of the k_{ef} range, where HF SFXWs propagate, first causes a decrease of $|\omega_{en}|$ and then can result in a change of the direction of electromagnetic energy transfer. HF SFXWs with positive azimuthal mode index are the only ones, which show non-normal dispersion: the directions of wave propagation and energy rotation are opposite in this case. This is explained by the following reasons. For SFXWs with $m < 0$, the electromagnetic energy rotates in the same direction in both plasma column and dielectric layer. For SFXWs with $m > 0$, the energy rotates in the plasma column and the dielectric layer in opposite directions. In this case, the azimuthal energy flow in the dielectric layer prevails over that in the plasma column for smaller k_{ef} and vice versa in waveguides with parameters, which correspond to the right limit of the k_{ef} range.

The influence of the external static axial magnetic field B_0 on the angular velocity $|\omega_{en}|$ of HF SFXWs is more pronounced than in the case of LF SFXWs. Decrease of B_0 by a factor of 1.5 is accompanied by the decrease of Max$\{|\omega_{en}|\}$ by $\approx$1.5 times for HF SFXW with $m = -1$ in Fig. 3.34. This is in good agreement with the case with no dielectric layer ($R_2 = R_1$), studied in [20].

The influence of the width of the dielectric layer on the angular velocity of energy transfer is shown in Fig. 3.35. Increasing width causes increasing Max$\{|\omega_{en}|\}$ from Max$\{|\omega_{en}|\} = 1.04|\omega_e|$ for the case of $R_2/R_1 = 1.1$ to Max$\{|\omega_{en}|\} = 1.115|\omega_e|$ for $R_2/R_1 = 1.3$.

An increase of the dielectric layer permittivity ε_d has the opposite impact (see Fig. 3.36). Increasing ε_d is accompanied by decreasing Max$\{|\omega_{en}|\}$ from Max$\{|\omega_{en}|\}$ $= 1.085|\omega_e|$ for the case of $\varepsilon_d = 1.0$ to Max$\{|\omega_{en}|\} = 0.928|\omega_e|$ for $\varepsilon_d = 5.0$. This result is in qualitative agreement with that obtained in [50] for the case of zero external static magnetic field.

Account for dissipation decreases the angular velocity $|\omega_{en}|$ and shortens the range of k_{ef} where $|\mathrm{Re}[\omega_{en}]| > |\mathrm{Im}[\omega_{en}]|$ (see Fig. 3.37). This character of influence of dissipation on $|\omega_{en}|$ is qualitatively similar to that obtained in the case of isotropic waveguides in [50].

The angular group velocity $\omega_g = \partial\omega/\partial m$ (see Eq. (3.9)) was introduced in Sect. 3.1 in analogy with the linear group velocity $\upsilon_g = \partial\omega/\partial k$. It can be calculated for known dispersion relation $D(\omega, m) = 0$. The dispersion relation of SFXWs considered in the present cylindrical waveguide structure—plasma-dielectric-metal (2.4) was derived in Sect. 2.1.

The dependencies of ω_g on effective wavenumber k_{ef} and plasma waveguide parameters for the waves studied here are not presented in the present Section because they almost coincide with those for ω_{en}. The most valuable difference between ω_g and ω_{en} is observed, investigating the case of non-zero dissipation. The influence of dissipation on ω_g is much weaker than that on the ω_{en}. For example, in the case considered in Fig. 3.37, account for $\nu = 0.3|\omega_e|$ causes a rise of Im(ω_g) along with a decrease of $|\mathrm{Re}(\omega_g)|$ only by less than 0.3% as compared to the case of $\nu = 0$ (no dissipation). The dependence of $|\omega_g|$ on the effective wavenumber for this case

($\nu = 0.3|\omega_e|$) is plotted in Fig. 3.37 by the thin curve, which almost coincides with the dashed line. Difference between ω_g and ω_{en} in presence of dissipation is well-known for a long time (see, e.g., [4]).

The angular velocity of electromagnetic energy rotation ω_{en} (3.28) as well as the angular group velocity ω_g (3.9) initiated by extraordinary polarized azimuthal surface waves are applied in the present Section for studying the electromagnetic energy transfer to complement the research of SFXW dispersion properties carried out in Sect. 2.1. In the present Section, the research is accomplished for plasma-dielectric-metal cylindrical waveguide structures placed into an axial external static magnetic field B_0. The study merges two limiting cases studied earlier. The first one is the cylindrical metal waveguide entirely filled by cold plasma, placed into B_0 (see Sect. 3.1). The second one is a cold plasma column separated from the cylindrical metal wall by a dielectric layer in absence of any external static magnetic field [50].

No difference has been found between ω_g and ω_{en} in most cases. Dissipation is confirmed to be the only physical mechanism which distinguishes the obtained dependencies of $\mathrm{Re}[\omega_g]$ and $\mathrm{Re}[\omega_{en}]$ on the plasma waveguide parameters.

The obtained dependencies of ω_g and ω_{en} on the plasma waveguide parameters are qualitatively similar to those obtained in the Sect. 3.1 and [50]. Specifically, $|\omega_g|$ and $|\omega_{en}|$ increase with decreasing absolute value of the azimuthal mode index $|m|$ and dielectric constant ε_d of the liner, as well as with increasing B_0 and thickness of the dielectric layer (R_2–R_1).

In the LF range (1.13), the dependencies of $|\omega_g|$ and $|\omega_{en}|$ on the effective wavenumber are not monotonous and have their maxima for $k_{ef} \sim 1$. In the HF range (1.14), the dependencies are monotonous and reach their maxima at the right limit of the k_{ef} range wherein SFXWs propagate.

The maximum values of angular velocities $|\omega_g|$ and $|\omega_{en}|$ are larger for positive azimuthal mode index, $m > 0$, in LF range. Vice versa, they are larger for $m < 0$ in the HF range. These signs of m are those for which SFXWs propagate in absence of the dielectric layer.

SFXWs are demonstrated to have normal dispersion in most cases. This means that the directions of their angular phase velocity, ω/m, and group velocity, ω_g, coincide. The only exclusions are HF SFXWs with positive azimuthal mode index. They do not propagate in absence of the dielectric layer and in absence of an external magnetic field. That is why this phenomenon was not observed in previous studies [20, 50]. This effect is explained by the fact that SFXWs carry the electromagnetic energy in opposite directions in the plasma column and in the dielectric layer. The presence of non-normal dispersion is less pronounced in waveguides with wider dielectric layers and for SFXWs with larger $|m|$.

The results obtained in the present Section can be of great interest in the field of plasma electronics as well as for designing technological surface-wave sustained plasma applicators [54–56].

3.5 Conclusions

To our best knowledge, the terms of a*ngular phase velocity, angular group velocity and angular velocity of energy transfer* were introduced in [20] for the first time. However, the terms which correspond to angular rotation are well-known in electrodynamics since a long time. For example, orbital angular momentum (OAM) carried by light beams (vortex beams) was discovered in [57] and has been widely employed in many technological applications like optical tweezers, optical drives of micro-machines, atom trapping, and optical communications.

The well-defined OAM of rotating cavity modes in high-power gyro-devices (electron cyclotron masers (ECM)) operating very close to the cutoff frequency ($k_z \approx 0$) excited by gyrating electrons was derived by photonic and electromagnetic wave approaches in [58]. While the OAM describes the phase twisting, the so-called spin angular momentum (SAM) is used for studying of twisting of the electric wave field (circular or elliptical polarization) [59, 60]. The total angular momentum is given by m/ω.

Higher-order $TE_{m,n}$-mode gyrotrons were shown to be natural sources of high-power OAM millimeter wave beams [59]. Very pure high-power OAM pulses at 140 GHz and 165 GHz with OAM mode number (topological charge $l = m{-}1$) $l = 9$ and $l = 30$, where generated by 500 kW $TE_{10,4}$- and $TE_{31,7}$-mode gyrotrons, respectively [59].

High-precision quasi-optical mode generators produce high-order rotating OAM modes for precise low-power (cold) measurements. Examples are the modes $TE_{6,2}$ with $l = 5$ and $TE_{10,1}$ with $l = 9$ at approximately 95 GHz [58] and $TE_{28,8}$ with $l = 27$ at 140 GHz [60]. These high-power gyrotron and cold-test mode generator measurements on higher-order OAM mode generation promise the realization towards wireless long-range communications using high-power gyro-devices.

In studying a capability of SFXWs to sustain a gas discharge in cylindrical waveguides, the angular discharge "length" was used in [21] rather than the linear discharge length commonly used for the case of planar discharge chambers.

The research performed in the present Chapter complements that presented in the previous chapter by comprehensive analysis of electromagnetic energy transfer initiated by SFXWs. Flute electromagnetic waves naturally do not transfer any energy in axial direction. Since cylindrical metal waveguides are under consideration, no energy transfer takes place through the metal interface. However, the electromagnetic energy oscillates in radial direction with the double eigenfrequency (2ω), and there is no time-averaged radial energy flow.

Electromagnetic energy rotation initiated by SFXWs is studied in the present Chapter in four different waveguide structures. The first one is a cylindrical metal waveguide entirely filled by magnetoactive plasma. The second structure is a metal rod placed into infinite plasma along an external static magnetic field. The third one is a coaxial metal waveguide entirely filled by plasma. Finally, the fourth waveguide structure is a cylindrical metal waveguide partially filled by magneto-active plasma. According to general theory [4], no difference is found between the angular group velocity and angular velocity of energy transfer in most cases. Deviation is found in the cases foreseen by the theory [4], specifically, if SFXWs propagate with the frequency close to the frequency range limits, were the plasma electrodynamic properties strongly depend on the wave frequency.

References

1. Hamilton, W. R. (1839). First series of researches respecting vibration, connected with the theory of light. *Proceedings of the Royal Irish Academy, 1836–1869*(1), 341–349.
2. Umov, N. A. (1874). Ableitung der Bewegungsgleichungen der Energie in continuirlichen Körpern. *Zeitschrift für Mathematik und Physik, 19*, 418–431.
3. Oughstun, K. E., & Cartwright, N. A. (2005). Physical significance of the group velocity in dispersive, ultrashort gaussian pulse dynamics. *Journal of Modern Optics, 52*(8), 1089–1104.
4. Brillouin, L. (1960). *Wave propagation and group velocity*. Academic Press Inc.
5. Biot, M. A. (1957). General theorems on the equivalence of group velocity and energy transport. *Physical Review, 105*(4), 1129–1137.
6. Brambilla, M. (1988). *Kinetic theory of plasma waves. Homogeneous plasmas*. Clarendon Press.
7. Gerasik, V., & Stastna, M. (2010). Complex group velocity and energy transport in absorbing media. *Physical Review E, 81*(5), 056602.
8. Wiedemann, H. (1999). *Particle accelerator physics I. Basic principles and linear beam dynamics*. Springer.
9. Slater, J. C. (1947). In *The design of linear accelerators*. Technical Report No. 47, Research Laboratory of Electronics. MIT.
10. Jackson, J. D. (1998). *Classical electrodynamics* (3rd ed.). John Wiley & Sons.
11. Alexandrov, A. F., Bogdankevich, L. S., & Rukhadze, A. A. (1984). *Principles of plasma electrodynamics*. Springer.
12. Akhiezer, A. I., Akhiezer, I. A., Polovin, R. V., Sitenko, A. G., & Stepanov, K. N. (1975). *Plasma electrodynamics* (p. 54). Pergamon Press.
13. Sivukhin, D. V. (1966). The conical refraction of magnetosonic waves. *Magnetohydrodynamics, 2*(1), 18–22.
14. Lloyd, H. (1833). On the phenomena presented by light in its passage along the axes of biaxial crystals. *Philosophical Magazine, 2*, 112.
15. Girka, V. A., Girka, I. A., Kondratenko, A. N., et al. (1989). Azimuthal surface waves at the boundary between a magnetoactive plasma and a metal. *Soviet Journal of Communications Technology and Electronics, 34*(4), 96–99.
16. Girka, V., Girka, I., & Thumm, M. (2014). In *Surface flute waves in plasmas. Theory and applications*. Springer.
17. Abramowitz, M., & Stegun, I. (Eds.). (1972). *Handbook of mathematical functions with formulas, graphs, and mathematical tables*. National Bureau of Standards, Applied Mathematics Series.
18. Poynting, J. H. (1884). On the transfer of energy in the electromagnetic field. *Philosophical Transactions of the Royal Society of London, 175*, 343–361.
19. Landau, L. D., Lifshitz, E. M., & Pitaevskii, L.P. (1984). In *Electrodynamics of continuous media* (2nd ed.). Butterworth-Heinemann.
20. Girka, I.O., Pavlenko, I.V., & Thumm, M. (2019). Electromagnetic energy rotation along plasma-metal interface in cylindrical waveguides initiated by azimuthal surface waves. *Physics of Plasmas, 26(2), 022113.*
21. Girka, V. O., Girka, I. O., & Pavlenko, I. V. (2001). Electrodynamic model of the gas discharge sustained by azimuthal surface waves. *Contributions to Plasma Physics, 41*(4), 393–406.

22. Azarenkov, N. A., Denisenko, I. B., & Ostrikov, K. N. (1995). A model of a large-area planar plasma producer based on surface wave propagation in a plasma-metal structure with a dielectric sheath. *Journal of Physics D: Applied Physics, 28*(12), 2465–2469.
23. Moisan, M., & Nowakowska, H. (2018). Contribution of surface-wave (SW) sustained plasma columns to the modeling of RF and microwave discharges with new insight into some of their features. A survey of other types of SW discharges. *Plasma Sources Science and Technology, 27(7), 073001.*
24. Baieva, S., Hakamaa, O., & Groenhof, G., et al. (2017). Dynamics of strongly coupled modes between surface plasmon polaritons and photoactive molecules: the effect of the Stokes shift. *ACS Photonics, 4*(1), 28–37.
25. Girka, I.O., Pavlenko, I.V., & Thumm, M. (2019). Electromagnetic energy rotation along plasma-metal interface around cylindrical metallic rod placed into infinite plasma in parallel to external static uniform magnetic field initiated by azimuthal surface waves. *Physics of Plasmas, 26*(5), 052103.
26. Azarenkov, N. A., & Ostrikov, K. (1999). Surface magnetoplasma waves at the interface between a plasma-like medium and a metal in a Voigt geometry. *Physics Reports, 308*, 333–428.
27. Stiles, P. L., Dieringer, J. A., Shah, N. C., et al. (2008). Surface-enhanced Raman spectroscopy. *Annual Review of Analytical Chemistry, 1*, 601–626.
28. Moisan, M., & Zakrzewski, Z. (1991). Plasma sources based on the propagation of electromagnetic surface waves. *Journal of Physics D. Applied Physics, 24*(7), 1025–1048.
29. Kemaneci, E., Mitschker, F., & Rudolph, M., et al. (2017). A global model of cylindrical and coaxial surface- wave discharges. *Journal of Physics D: Applied Physics, 50*(24), 245203.
30. Chen, L., Zhao, Y., & Wu, K., et al. (2016). On the generation of magnetic field enhanced microwave plasma line. *Physics of Plasmas, 23*(12), 123509.
31. Aghamir, F. M., & Zandieh, M. N. (2006). Theoretical investigation of electromagnetic waves in a dielectric loaded coaxial plasma waveguide. *Physics of Plasmas, 13*(6), 063107.
32. Neichev, Z., & Benova, E. (2004). Optimizing the plasma parameters of a low-pressure coaxial discharge. *Vacuum, 76*, 393–396.
33. Moskvitina, Yu. K., Zaginaylov, G. I., & Tkachenko, V. I. (2014). Effect of background plasma on electromagnetic properties of coaxial gyrotron cavity. *Technical Physics, 59*(7), 1065–1071.
34. Ramazanov, R. Z., Sotnikov, G. V., & Tkach, Yu. V. (2005). High-power coaxial microwave ubitron: Simulation by the particle-in-cell method. *Technical Physics, 50*(6), 747–753.
35. Onishchenko, I. N., Sidorenko, D. Y., & Sotnikov, G. V. (1995). Electron beam interaction with a plasma-filled rippled coaxial waveguide. *Plasma Physics Reports, 21*(8), 670–676.
36. Yatsenko, T., Ilyenko, K., & Sotnikov, G. V. (2012). Limiting current of axisymmetric relativistic charged-particle beam propagating in strong axial magnetic field in coaxial drift tube. *Physics of Plasmas, 19*(6), 063107.
37. Yatsenko, T., Sotnikov, G. V., & Portillo, S., et al. (2019). Relativistic charged-particle beam space-charge limited current in finite length coaxial drift tube. *IEEE Transactions on Plasma Science, 47*(5), 2602–2608.
38. Zhang, S.-C. (2004). Linear and nonlinear investigation of a coaxial-waveguide cyclotron autoresonance maser amplifier. *Physics of Plasmas, 11*(8), 3969.
39. Zagrebelny, I.A., Markov, P.I., & Podobinsky, V.O. (2008). About breakdown in a coaxial waveguide of atomic gas of low pressure by a microwave radiation with a stochastic jumping phase. *Problems of Atomic Science and Technology. Series: Plasma Electronics and New Methods Acceleration, 4*(6), 195–198.
40. Sorolla, E., Sounas, A., & Mattes, M. (2015). Space charge effects for multipactor in coaxial lines. *Physics of Plasmas, 22*(3), 033512.
41. Yao, L., Wang, Y., & Zhang, R., et al. (2019). Analysis of multiplication characteristics of coaxial waveguide loaded ceramic under external magnetic field. *Physics of Plasmas, 26*(8), 082104.
42. Azarenkov, N. A., Ostrikov, K. N., & Shcherbinina, I. B. (1991). Azimuthal surface waves in a coaxial semiconductor structure with metal walls. *Soviet Journal of Communications Technology, 36*, 68–72.

43. Girka, O. I., Girka, V. O., Girka, I. O., et al. (2008). Coupled transverse modes of coaxial metal waveguides completely filled with magnetoactive current-carrying plasma. *Plasma Physics Reports, 34*(11), 901–910.
44. Stix, T. H. (1992). *Waves in plasmas.* AIP.
45. Chen, F. F. (1984). In *Introduction to plasma physics and controlled fusion, 1: Plasma physics.* Plenum Press.
46. Dorodnitsyn, A. A. (1952). Asymptotic laws of distribution of the characteristic values for certain special forms of differential equations of the second order. *Uspehi Mat. Nauk (N.S.), 7, 6*(52), 3–96. (in Russian).
47. Girka, V. A., & Girka, I. A. (1990). Influence of plasma inhomogeneity on the spectra of azimuthal surface waves. *Radiophysics and Quantum Electronics, 33*(4), 516–517.
48. Girka, V. A., & Girka, I. A. (1991). Azimuthal surface waves in a nonuniform plasma cylinder. *Radiophysics and Quantum Electronics, 34*(4), 324–328.
49. Girka, I.O., Pavlenko, I.V., & Thumm, M. (2020). Rotation of electromagnetic energy initiated by azimuthal surface waves in coaxial metal waveguides entirely filled by plasma. *Physics of Plasmas, 27*(2), 032104.
50. Girka, I. O., Girka, O. I., & Thumm, M. (2019). Electromagnetic energy rotation along plasma-dielectric interface in isotropic cylindrical metallic waveguide caused by azimuthal surface waves. *Physics of Plasmas, 26*(12), 122104.
51. Girka, I. O., Kondratenko, V. M., & Thumm, M. (2018). Higher radial modes of azimuthal surface waves in magnetoactive cylindrical plasma waveguides. *Journal of Plasma Physics, 84*(6), 905840603.
52. Girka, I., & Kondratenko, V. (2020). Higher radial modes of azimuthal surface waves above the upper-hybrid frequency in cylindrical waveguides partially filled by plasma. *Problems of Atomic Science and Technology, 6*(130). Series: Plasma Physics (26), 22–25.
53. Girka, I. O., Pavlenko, I. V., & Thumm, M. (2018). Excitation of higher radial modes of azimuthal surface waves in the electron cyclotron frequency range by rotating relativistic flow of electrons in cylindrical waveguides partially filled by plasmas. *Physics of Plasmas, 25*(5), 052109.
54. Georgieva, V., Berthelot, A., & Silva, T., et al. (2017). Understanding microwave surface-wave sustained plasmas at intermediate pressure by 2D modeling and experiments. *Plasma Processes and Polymers, 14,* 1600185.
55. Naito, T., & Sakai, O. (2019). Analytical formulation for radiation characteristics of a surface wave sustained plasma antenna. *Physics of Plasmas, 26*(7), 073506.
56. Durocher-Jean, A., Desjardins, E., & Stafford, L. (2019). Characterization of a microwave argon plasma column at atmospheric pressure by optical emission and absorption spectroscopy coupled with collisional-radiative modeling. *Physics of Plasmas, 26*(6), 063516.
57. Allen, L., Beijersbergen, M. W., & Spreeuw, R. J. C., et al. (1992). Orbital angular momentum of light and the transformation of Laguerre-Gaussian laser modes. *Physical Review A, 45,* 8185–8189.
58. Sawant, A., Choe, M. S., & Thumm, M., et al. (2017). Orbital angular momentum (OAM) of rotating modes driven by electrons in electron cyclotron masers. *Scientific Reports-Nature, 7,* 3372.
59. Thumm, M., Sawant, A., & Choe, M. S., et al. (2017). The gyrotron—A natural source of high-power orbital angular momentum millimeter-wave beams. In *10th international workshop 2017 strong microwaves and terahertz waves: Sources and Aapplications (EPJ Web of conferences* (Vol. 149, pp. 04014).
60. Thumm, M. (2020). Gyro-devices—Natural sources of high-power high-order angular momentum millimeter-wave beams. *Terahertz Science and Technology, 13*(1), 1–21.

Chapter 4
Coupled Surface Flute Waves in Current-Carrying Plasma Waveguides

To produce and sustain gaseous plasmas in various laboratory devices one can often apply DC electrical currents, for example, in magnetically confined fusion plasmas in tokamaks or for plasma production in various technological applicators. Possible flowing of charged plasma particles has not been taken into account in the previous chapters, which are devoted to the study of the basic features of transverse eigenmodes propagating in simple models of cylindrical and coaxial waveguides with plasma filling. That is why in the present Chapter the influence of an axial electric current on the dispersion properties of flute eigenmodes propagating in various cylindrical waveguide structures with plasma filling is under consideration. The plasma is assumed to be sufficiently dense so that the electron cyclotron frequency is lower than the electron plasma frequency (Langmuir frequency). However unlike in Chaps. 2 and 3, here the plasma is affected by an azimuthal magnetic field, which is produced by the axial electric current. The value of this axial current is supposed to be sufficiently small so that either the azimuthal component of the external magnetic field is much weaker than its axial component or the cyclotron frequency due to the azimuthal magnetic field is much lower than the eigenfrequency of the flute modes in the case if there is no axial magnetic field. The dispersion equation of flute X- and O-modes becomes a coupled form leading to new wave phenomena as mode coupling, forbidden bands in the frequency spectrum of the coupled flute modes, and mode damping or radiation of power out of the waveguide.

4.1 Coupled Surface Flute Waves in Azimuthal Static Magnetic Fields

In the present Section, the following case is considered. Surface flute waves (SFWs) propagate in a waveguide, which is partially filled by cold plasma and immersed into an external azimuthal magnetic field $\vec{B}_0 = B_{0\varphi}\vec{e}_\varphi$ (there is no axial component of the magnetic field: $B_{0z} = 0$). In this case the SFWs propagate with a frequency below

© The Author(s), under exclusive license to Springer Nature Switzerland AG 2022
I. Girka and M. Thumm, *Surface Flute Waves in Plasmas*, Springer Series on Atomic, Optical, and Plasma Physics 120, https://doi.org/10.1007/978-3-030-98210-2_4

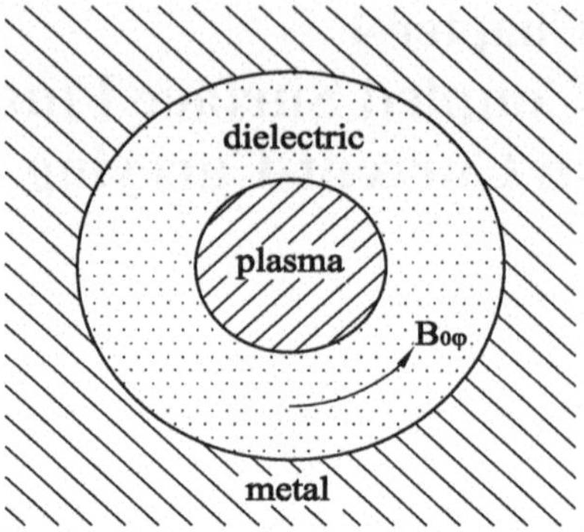

Fig. 4.1 Scheme of a metal waveguide with inner radius R_2, which is partially filled with current-carrying plasma with outer radius R_1 in the case of applying an external azimuthal magnetic field and no external axial magnetic field

the electron plasma frequency, $\omega<\Omega_e$. The geometry of the problem is illustrated in Fig. 4.1. The magnitude of the azimuthal magnetic field is assumed to be sufficiently small, so that the inequality $\beta = |\omega_e|/\omega \ll 1$ is valid. In this case the set of the Maxwell equations cannot be represented in the form of two independent sets of equations for two different modes, namely X- and O-modes. However, ignoring second and higher order terms related to the small parameter β, one can derive from the Maxwell equations a set of differential equations for the SFW fields, which consists of two equations. The first one is (2.17) where the operator, which is located at its right-hand side, has the following form:

$$\widehat{M} = im\varepsilon_2 k/(r\varepsilon_1). \tag{4.1}$$

The second equation is that for the axial component of the SFW magnetic field $H_z^{(m)}$:

$$\frac{1}{r}\frac{d}{dr}\left(\frac{r}{\varepsilon_1}\frac{dH_z^{(m)}}{dr}\right) + \left(k^2 - \frac{m^2}{r^2\varepsilon_1}\right)H_z^{(m)} = -\widehat{M}\,E_z^{(m)}. \tag{4.2}$$

The other four components of the SFW field are expressed through the axial components $E_z^{(m)}$ and $H_z^{(m)}$. In the present case ($B_{0z} = 0$), the components of the dielectric permittivity tensor can be approximately written in the following form: $\varepsilon_1 \approx \varepsilon_3$, $\varepsilon_2 \approx |\omega_e|\Omega_e^2\omega^{-3}$ [1, 2]. These approximate equalities are satisfied with an accuracy of up to terms of the second order of the small parameter β. Taking into account expression (4.1), the set of Eqs. (2.17) and (4.2) can be solved using the theory of successive approximations.

Application of the boundary conditions (see Sect. 1.4.4) allows one to derive the dispersion relation, which has the typical form for the case of coupled modes:

$$D_o D_x = D(\beta), \tag{4.3}$$

where D_o is determined by Eq. (2.19) obtained in Sect. 2.4 for the case of ordinary flute modes, $D_x = 0$ is the dispersion relation of the surface flute X-modes (SFXMs) in the case of $B_{0z} = 0$ (see Eq. (2.4) and comments at the end of Sect. 2.1), $D(\beta)$ is the coefficient of coupling between SFXMs and surface flute ordinary polarized modes (SFOMs), whose magnitude is of the second order of smallness over the parameter β. In the case of non-uniform plasma filling, the coefficient of the coupling $D(\beta)$ has the following integral form [1, 2]:

$$D(\beta) = \frac{\varepsilon_1}{\tilde{\varphi}_o^2 R_1^2 (d\varphi_o/dR_1)^2} \left[\int_0^{R_1} r\tilde{\varphi}_o \widehat{M} \varphi_o dr \right]^2, \tag{4.4}$$

here φ_o and $\tilde{\varphi}_o$ are two linearly independent solutions of the homogeneous Eq. (4.2), i.e. for $\widehat{M} = 0$, where φ_o has no singularities on the axis of the waveguide unlike $\tilde{\varphi}_o$.

A solution of the dispersion relation (4.3) can be found in the following form of a sum $\omega = \omega_{x,o} + \delta\omega_{x,o}$, where the subscripts "$x$" and "$o$" relate to the cases of X- and O-modes. Therefore, here $\omega_{x,o}$ is the angular SFW frequency calculated in the case of zero axial stationary magnetic field and $\delta\omega_{x,o}$ is the frequency correction caused by the coupling between these two modes. Assuming that the magnitudes of the SFW eigenfrequencies are known (see Chap. 2), one can find the frequency corrections for SFXMs and/or SFOMs, whose magnitudes are caused by application of a weak external azimuthal magnetic field, in the following form:

$$\delta\omega_{x,o} = \frac{D(\beta)}{D_{o,x}} \left(\frac{\partial D_{x,o}}{\partial\omega} \right)^{-1}_{|\omega=\omega_{x,o}}. \tag{4.5}$$

In most cases, the dispersion curves of SFXMs and SFOMs, which propagate in a uniform plasma along the external azimuthal magnetic field, do not intersect with each other. But under specific conditions such crossing is possible. Then the influence of the external field $B_{0\varphi}$ gets stronger and the ratio $\delta\omega_{x,o}/\omega_{x,o}$ becomes a term of the first order of smallness over the small parameter β according to the following formula:

$$(\delta\omega_{x,o})^2 = D(\beta) \left(\frac{\partial D_o}{\partial\omega} \right)^{-1}_{|\omega=\omega_o} \left(\frac{\partial D_x}{\partial\omega} \right)^{-1}_{|\omega=\omega_x}. \tag{4.6}$$

Analyzing the explicit expressions for the frequency corrections, which can be derived from formula (4.5), it is possible to examine the influence of the field $B_{0\varphi}$ on the dispersion properties of SFXMs, which propagate along the surface of a uniform plasma cylinder separated by a narrow vacuum layer, $k(R_2 - R_1) \ll 1$, from the metal wall of the cylindrical waveguide. In this limiting case, the expression for the eigenfrequency of SFXMs is the same as in the case of an isotropic plasma waveguide (one can apply expressions (2.8–2.10), putting there the value of the electron cyclotron frequency $\omega_e = 0$ (see as well the comments at the end of Sect. 2.1)). In the limiting case of uniform plasma filling, it is found that their frequency correction caused by the field $B_{0\varphi}$ has the following complicated dependence on the inner radius R_2 of the waveguide [1, 2]:

$$\delta\omega_x \approx \beta^2\omega_x k^2\phi/(8R_1\Delta), \tag{4.7}$$

where $\delta = c/\Omega_e$ is the skin-depth, and $\phi = \delta^4(R_2 - R_1 + \delta)^{-1}$ if for the plasma radius R_1 the inequality $R_1 \gg |m|\delta$ is valid (limiting case of a wide plasma waveguide).

In the opposite limiting case of a narrow waveguide $R_1 << \delta$, the factor ϕ has the following form: $\phi = {R_1}^4(R_2 - R_1 + R_1/|m|)^{-1}(m^2 + |m|)^{-2}$. Taking into account the dispersion properties of SFXMs propagating in a plasma filled waveguide under the condition $B_{0z} = 0$, one can conclude from the analysis of expression (4.7) that the influence of the external azimuthal magnetic field becomes weaker with increasing thickness of the vacuum layer Δ and increasing azimuthal mode index m.

In the case of non-uniform plasma density, a region of rarefied plasma can be located nearby the plasma surface. Then the equality $Re(\varepsilon_1(r_0)) = 0$ can be satisfied there. Such condition leads to an additional resonant damping of SFXMs. In this case during calculating of the integral, which is represented in expression (4.4) as squared multiplier, one has to take into account the small imaginary part of ε_1 which is caused by collisions between the plasma particles. One can approximately estimate the damping rate γ_r of SFXMs under the condition of this resonant damping associated with the existence of the external static azimuthal magnetic field:

$$\gamma_r \approx \frac{\pi\beta^2k^2\omega_x m^2}{2R_1^2k_*^5(R_2 - R_1)(1 + k_*(R_2 - R_1))}, \tag{4.8}$$

where $k_*^3 = k^2 d\varepsilon_1/dR_1 > 0$. It should be underlined that SFXM resonant damping, caused by their conversion into ordinary polarized flute modes in the region of rarefied peripheral plasma, where $Re(\varepsilon_1(r_0)) = 0$, becomes stronger under the condition of application of an external static azimuthal magnetic field $B_{0\varphi}$ than under the condition $B_{0\varphi} = 0$.

4.2 Coupled Surface Flute Waves in the Range of Electron Cyclotron Frequency in Current-Carrying Waveguides Partially Filled with Plasmas

In the present Section, the influence of the azimuthal magnetic field on the SFW frequency is considered under the condition of their propagation in a magneto-active plasma waveguide, assuming that the following ratio $\kappa = B_{0\varphi}/B_{0z} \ll 1$ is the small parameter of the problem. For this case the waveguide geometry is shown in Fig. 4.2.

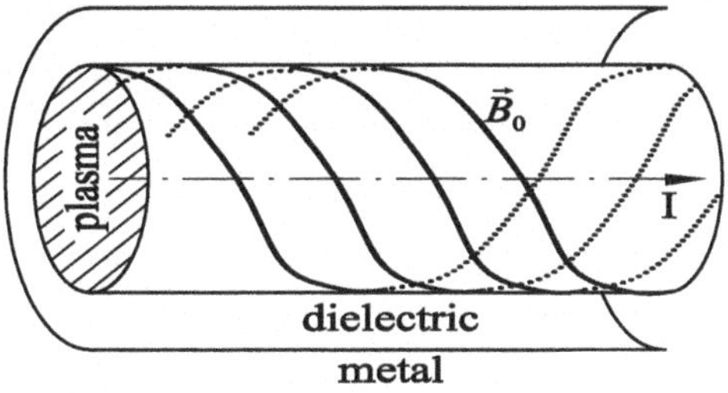

Fig. 4.2 Scheme of cylindrical metal waveguide, which is partially filled with current-carrying plasma, that produces an azimuthal static magnetic field in addition to an external axial static magnetic field

Taking into account terms of the first order of the smallness parameter κ, one can derive from the Maxwell equations a set of differential equations for the axial SFW fields $E_z^{(m)}$ and $H_z^{(m)}$. This set consists of Eq. (2.17) for the axial component of the SFW electric field, where the operator $\widehat{M}$ now has the following form:

$$\widehat{M} = -ik\kappa\frac{d}{dr} - \frac{k^2\kappa\varepsilon_3}{\varepsilon_1^2 - \varepsilon_2^2}\left(\frac{\varepsilon_1}{ik}\frac{d}{dr} + \frac{m\varepsilon_2}{ikr}\right), \tag{4.9}$$

and of the following equation for their axial magnetic field [1, 2]:

$$\widehat{L}\,H_z^{(m)} = \widehat{K}\,E_z^{(m)} \tag{4.10}$$

where

$$\widehat{L} = \frac{d^2}{dr^2} + \left(\frac{1}{r} + \frac{1}{\varepsilon_1}\frac{d\varepsilon_1}{dr} - \frac{d\ln\left|\varepsilon_1^2 - \varepsilon_2^2\right|}{dr}\right)\frac{d}{dr}$$
$$+\frac{m}{\varepsilon_1 r}\frac{d\varepsilon_2}{dr} - \frac{m\varepsilon_2}{r\varepsilon_1}\frac{d\ln\left|\varepsilon_1^2 - \varepsilon_2^2\right|}{dr} + k^2\frac{\varepsilon_1^2 - \varepsilon_2^2}{\varepsilon_1} - \frac{m^2}{r^2}, \tag{4.11}$$

$$\widehat{K} = \frac{ik}{\varepsilon_1}\left[\frac{-\kappa}{r}\left(\varepsilon_1^2 - \varepsilon_1\varepsilon_3 - \varepsilon_2^2\right) - \kappa\left(\varepsilon_1^2 - \varepsilon_1\varepsilon_3 - \varepsilon_2^2\right)\frac{d}{dr}\right.$$
$$\left. - \frac{\kappa m\varepsilon_2\varepsilon_3}{r} - \frac{d\left(\kappa\varepsilon_1^2 - \kappa\varepsilon_1\varepsilon_3 - \kappa\varepsilon_2^2\right)}{dr} + \kappa\left(\varepsilon_1^2 - \varepsilon_1\varepsilon_2 - \varepsilon_2^2\right)\frac{d\ln\left|\varepsilon_1^2 - \varepsilon_2^2\right|}{dr}\right]. \tag{4.12}$$

One can express the magnetic wave field components $H_r^{(m)}$ and $H_\varphi^{(m)}$ (which are the components of SFOMs in the case $B_{0\varphi} = 0$) through the axial electric component $E_z^{(m)}$ using Eqs. (2.18). The other electric components $E_r^{(m)}$ and $E_\varphi^{(m)}$ of the wave fields are connected with both axial components $E_z^{(m)}$ and $H_z^{(m)}$. With the aid of the theory of successive approximations, taking into account the small right-hand sides of Eqs. (2.17) and (4.10), one can find their solutions.

After that, application of the boundary conditions for the SFW fields on the plasma-vacuum interface allows one to derive the dispersion relation for coupled

SFOMs and SFXMs in the form of Eq. (4.3), in which one has to change the expression for D_x by the following one [1, 2]:

$$D_x = \frac{\varepsilon_1 d\psi_x/dR_1 + \varepsilon_2 m\psi_x/R_1}{(\varepsilon_1^2 - \varepsilon_2^2)\psi_x} - k\frac{J_m'(kR_2)N_m'(kR_1) - J_m'(kR_1)N_m'(kR_2)}{J_m'(kR_2)N_m(kR_1) - J_m(kR_1)N_m'(kR_2)}, \tag{4.13}$$

and the coefficient $D(\beta)$ should be replaced there by the following expression as well:

$$D(\kappa) = \frac{(d\varphi_o/dR_1)^{-2}}{R_1}\int_0^{R_1} r\varphi_o \widehat{M}\,\psi_x dr\cdot \left[\frac{k^2 W(\psi_x,\tilde{\psi}_x)}{k_\perp^2\psi_x^2}\int_0^{R_1}\frac{\psi_x \widehat{K}\varphi_o}{W(\psi_x,\tilde{\psi}_x)}dr - ik\kappa\left(1+\frac{k_\varphi^2}{k_\perp^2}\right)\frac{\varphi_o}{\psi_x}\right], \tag{4.14}$$

where ψ_x and $\tilde{\psi}_x$ are two linearly independent solutions of the homogeneous Eq. (3.10) (i.e. for $\widehat{K} = 0$), thus the first of them (ψ_x) has no singularities on the axis of the waveguide unlike the second one ($\tilde{\psi}_x$). The Wronskian is $W\left(\psi_x, \tilde{\psi}_x\right) = -k_\perp^2/r$. All the terms represented in expressions (4.13) and (4.14), which depend on the radial co-ordinate and are not integrands, should be calculated just on the plasma surface $r = R_1$.

One can derive simple analytical expressions for the frequency corrections of SFXMs, which propagate along the interface of a uniform plasma cylinder under the condition of a rather narrow vacuum layer $\Delta << 1$ and large plasma density $\Omega_e^2 >> \omega_e^2$, applying the Eqs. (4.3, 4.5, 4.13 and 4.14). If the plasma cylinder is narrow, which means that $\delta_x^2 = k_\perp^{-2} \approx \delta^2 >> R_1^2$, the expression for the SFXM frequency correction is as follows:

$$\delta\omega_x = \frac{\kappa^2\omega_e^2 R_1^2}{4(|m|+1)^2c^2}\left(R_2 - R_1 + \frac{R_1}{|m|}\right)^{-1}\left[\frac{m\delta^2|\omega_e|}{c^2R_1} + \frac{2(R_2-R_1)m^2}{R_1^2\omega_x}\right]^{-1}, \tag{4.15}$$

where the value of the azimuthal component of the magnetic field $B_{0\varphi}$, which determines the magnitude of the parameter κ should be calculated at the radius $r = R_1$. In this case the angular SFXW frequency ω_x calculated in the zeroth approximation ($\kappa = 0$) is presented by Eq. (2.8).

If the radius of the plasma cylinder is sufficiently large (this limiting case can be entitled as "a wide waveguide" because of the validity of the inequality $R_1 \gg \delta$), the SFXM frequency correction is equal to:

$$\delta\omega_x \approx \frac{\kappa^2\delta_x^4\omega_e^2 m^2}{4R_1^2c^2(R_2 - R_1 + \delta_0)}\left(1 - \frac{\omega|\omega_e|R_1\delta_x}{mc^2}\right)^2\left[\frac{m\delta_x^2|\omega_e|}{c^2R} + \frac{2(R_2 - R_1)m^2}{\omega_x R_1^2}\right]^{-1}. \tag{4.16}$$

It is appropriate to remind that the SFXW frequency ω_x in a wide waveguide in the zeroth approximation is given by Eq. (2.10).

While carrying out the calculations of (4.15) and (4.16), the field $B_{0\varphi}$ is considered as a linear function of the radial co-ordinate. It should be outlined as well that participation of electrons in axial motion does not lead to a Doppler shift of the frequency of these modes because of the assumption of axial homogeneity of the considered waveguide.

Comparing the expressions (4.15) and (4.16), one can make the following conclusions. First, the influence of the azimuthal magnetic field on the SFXM frequency is stronger in a wide plasma cylinder, than in the case of a narrow one. Second, the frequency correction $\delta\omega_x$ is larger in the case of a plasma-metal interface $R_2 = R_1$, than in the case of utilization of a vacuum gap for separation of the plasma column from the metal wall, $R_2 \neq R_1$. Third, in the case that all other parameters are the same, the SFXM eigenfrequency is larger under the condition of an applied external azimuthal magnetic field than in the case of its absence ($B_{0\varphi} = 0$) [1, 2].

The eigenfrequency of the lth radial mode of SFXMs, which propagates in a plasma waveguide with a wide vacuum gap ($k(R_2 - R_1) \gg |m|$), in zeroth approximation is determined by the characteristics of the vacuum layer and its value is described by the formula (2.21). The frequency correction describes the influence of the characteristics of the magnetized plasma column on the SFW properties and is determined by the following formula:

$$\omega_x^{(1)} \approx \frac{-\omega_x\delta}{R_2 - R_1}\left(1 - \frac{m\delta|\omega_e|}{\omega_x R_1}\right)\left(1 - \frac{m^2}{2k^2R_1^2}\right)^{-1}\left(1 + \frac{m^2}{2k^2R_1R_2}\right)^{-1}_{|\omega=\omega^{(0)}}. \tag{4.17}$$

If the frequencies of SFOMs and SFXMs are the same, it is possible to find the condition for crossing of their dispersion curves (specifically, by equating the expressions (2.22) and (4.17)):

$$\frac{R_2 - R_1}{R_1} = \Delta = \frac{\pi\, l|\omega_e|}{m\Omega_e}\left(1 - \frac{\omega_e^2 R_1}{2\Omega_e^2 R_2}\right)^{-1}. \tag{4.18}$$

Equation (4.18) can be satisfied only for SFXMs propagating along the direction of the gyration of the plasma electrons on Larmor orbits. To calculate the frequency corrections of flute electromagnetic waves in this case one can apply Eq. (4.6) under the indicated condition (4.18):

$$(\delta\omega)_x^2 \approx \frac{\kappa^2 m^4 c^4 \Omega_e^{-6}\omega_e^4}{16\pi^2 l^2 R_1^2 R_2^2}\left(1 - \frac{m^2}{k^2R_1R_2}\right). \tag{4.19}$$

In this case the influence of the field $B_{0\varphi}$ on the SFW frequency enhances with increasing azimuthal mode index m, and also with decreasing radial mode index l, plasma density and waveguide radius.

From comparison of the results obtained in the present Section with those in Sect. 4.1, one can conclude that the transversal component (in respect to the direction of SFW propagation) of the external magnetic field B_{0z} affects the dispersion properties of SFXMs stronger than the azimuthal $B_{0\varphi}$ component of the applied external static magnetic field. Such dependence of the eigenfrequency on the direction of the external magnetic field coincides with that one, that was obtained in [3] for surface magneto-plasma waves, which propagate along a planar plasma-metal boundary.

4.3 Coupled Surface Flute Waves Propagating Above the Upper-Hybrid Frequency Around a Current-Carrying Cylindrical Metal Rod in Magnetized Plasmas

In the present Section, the following antenna system is considered. It is modeled by a metal cylinder with radius R_1, which is placed into a cold magnetized plasma with a uniform particle density profile, where the density is sufficiently large so that the inequality $\Omega_e^2 \gg \omega_e^2$ is valid. The external static magnetic field has two components: $\vec{B}_0 = \vec{B}_{0\varphi} + \vec{B}_{oz}$. The present consideration is restricted to the case of small magnitude of the external static azimuthal magnetic field, thus the small parameter of the problem is: $\kappa = B_{0\varphi}/B_{0z} \ll 1$. The schematic of the problem is shown in Fig. 4.3.

Ignoring the terms, whose values are higher than of the first order of the smallness parameter κ, one can derive the set of differential equations for the axial components of the electric and magnetic fields of SFWs [4] from the Maxwell equations. The equation for the axial electric wave field looks like (2.17). In the considered case, the operator $\widehat{M}$ located in the right-hand side of this equation is given by Eq. (4.9). For the axial magnetic field of the coupled SFW one can obtain Eq. (2.36), where the operator $\widehat{K}$ looks as follows:

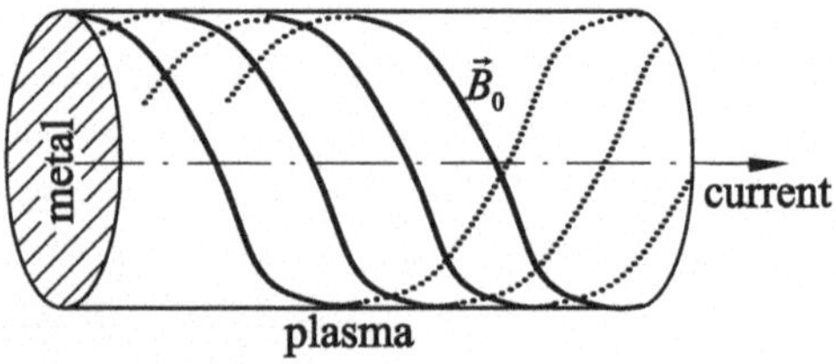

Fig. 4.3 Model of cylindrical metal rod antenna with radius R_1, surrounded with cold plasma in the case $\vec{B}_0 = \vec{B}_{0z} + \vec{B}_{0\varphi}$

$$\widehat{K} = \frac{ik}{\varepsilon_1}(\varepsilon_2^2 + \varepsilon_1\varepsilon_3 - \varepsilon_1^2)\left[\frac{\kappa}{r} + \kappa\frac{d}{dr} + \frac{\kappa\varepsilon_2\varepsilon_3 m r^{-1}}{\varepsilon_1^2 - \varepsilon_1\varepsilon_3 - \varepsilon_2^2} + \frac{d\kappa}{dr}\right]. \tag{4.20}$$

As it is shown in Sect. 2.2, SFXMs with negative azimuthal mode index m propagate in the LF range (1.13) and those with $m > 0$ in the HF range (1.14). An analysis of the physical observable $k_o^2 = k^2\varepsilon_3$ which is present in Eq. (2.17) proves that in the LF range it is negative, and in the HF range it is positive. That is why the parameter k_o can be referred to be the radial wavenumber of the bulk component of these coupled flute waves in the HF range. Consequently, the field E_z describes the bulk component of these coupled flute waves. In zeroth approximation, Eq. (2.36) can be solved in terms of modified cylindrical functions [5]. That is why the spatial distribution of the flute wave field H_z has a surface-type structure. Therefore, the parameter $k_\perp^{-1}$ determines the penetration depth of the surface component of these coupled flute modes into the plasma.

The upper edge of the HF range is determined by the equation $\varepsilon_1 = \varepsilon_2$, that is why nearby the upper limit the penetration depth of the SFXM into the plasma is considerably larger than the radius of the utilized metal cylinder: $k_\perp R_1 \ll 1$ (limiting case of a narrow waveguide). The low limit of the HF range is determined by $\varepsilon_1 = 0$, that is why nearby this lower limit the opposite inequality is realized: $k_\perp R_1 \gg 1$ (limiting case of a wide waveguide). Results of numerical analysis of the dependence of $k_\perp^{-1}$ on the value of the radius of the metal cylinder testify that increasing metal cylinder radius leads to a decrease of the penetration depth of these flute waves into the plasma. Reduction of the cylinder radius leads to sharp growing of $k_\perp^{-1}$, this growth begins sooner for those waves, which have larger azimuthal mode index m: $k_\perp^{-1} \approx 2\delta;\ 4\delta;\ 6\delta$ for the $m = 1;\ 2;\ 3$, respectively.

To determine the spatial distribution of the SFWs fields in this case one can apply the method of variation of a constant for solving the set of Eqs. (2.17) and (2.36). This provides the expressions for the axial components of the SFW electric and magnetic fields [4]. Application of the boundary conditions given in Sect. 1.4.4 allows one to make the conclusion that SFOMs cannot propagate along the interface between magneto-active plasma and metal. Instead of them, the bulk ordinary polarized flute modes (BFOMs) can propagate in such plasma-metal antenna structures under the condition of application of an external static azimuthal magnetic field. BFOM electromagnetic fields are described by cylindrical functions of the first kind [5], their amplitudes are of the first order of the small parameter κ as compared with the SFXM amplitudes.

Using the boundary condition that an axial electric field on the interface of the considered plasma-metal antenna system is equal to zero one can derive the following equation:

$$A_1 N_m(k_o R_1) + A_2 J_m(k_o R_1) = 0. \tag{4.21}$$

By using absorbing boundary conditions at infinity (see as well Sect. 2.5) one can derive the following relation between the constants of integration $A_{1,2}$ and B_1:

$$A_1 + B_1 \frac{\pi}{2} \int_{R_1}^{\infty} r J_m(k_o r) \widehat{M} K_m(k_\perp r) dr = i[A_2 - B_1 \frac{\pi}{2} \int_{R_1}^{\infty} r N_m(k_o r) \widehat{M} K_m(k_\perp r) dr]. \tag{4.22}$$

Application of both Eqs. (4.21) and (4.22) makes it possible to express the integration constants $A_{1,2}$ through the constant B_1. After that one can apply the boundary condition on the interface of the metal antenna. There the azimuthal component of the electric field of the coupled SFWs must be zero. This finally leads to the derivation of the following dispersion relation, which describes coupled flute modes propagating along a cylindrical metal rod located in a cold magnetized plasma [4]:

$$D_x^{(0)} + D^{(2)} = 0. \tag{4.23}$$

Here $D_x^{(0)} = 0$ is the dispersion relation of the SFXMs obtained in the zeroth approximation (see Eq. (2.11), where one can apply $F_m(x) = K_m(x)$). The summand $D^{(2)}$ is a correction to the SFXM dispersion relation, which is caused by application of an external static azimuthal magnetic field, its value is of the second order of the small parameter κ:

$$D^{(2)} = \frac{\left[k_\perp R_1 K_m'(k_\perp R_1) + m\mu K_m(k_\perp R_1)\right]}{J_m(k_o R_1) + i N_m(k_o R_1)}$$

$$\frac{\int_{\infty}^{R_1} r K_m(k_\perp r)\Big[N_m(k_o R_1)\, \widehat{K}\, J_m(k_o r) - J_m(k_o R_1)\, \widehat{K}\, N_m(k_o r)\Big] dr}{\frac{\pi}{2} \int_{R_1}^{\infty} r[J_m(k_o r) + i N_m(k_o r)]\, \widehat{M}\, K_m(k_\perp r) dr.} \tag{4.24}$$

It is suitable to search for the solution of Eq. (4.24) in the form: $\omega = \omega^{(0)} + \Delta\omega$, here $\omega^{(0)}$ is the eigenfrequency of the SFXMs under the condition $B_{0\varphi} = 0$, and $|\Delta\omega| \ll \omega^{(0)}$ is a small correction to their eigenfrequency. Then $\Delta\omega$ can be found via solving this dispersion relation, where $D_x^{(0)}$ can be presented as a Taylor series over the parameter $\Delta\omega$. Its imaginary part is negative, since propagation of SFXMs in this case is characterized by losses of electromagnetic energy caused by their conversion into radiated bulk waves. The damping rate $\gamma_{rad} = \mathrm{Im}(\Delta\omega)$ of the SFXM, which is caused by this process, is proportional to the second power of external static azimuthal magnetic field $B_{0\varphi}$: $\gamma_{rad} \propto D^{(2)}$.

In the further consideration within this Section, the external static azimuthal magnetic field is assumed to depend on the radial co-ordinate in the following way: $B_{0\varphi}(r) = R_1 B_{0\varphi}(R_1)/r$. This assumption relates to the case when the axial electric current is flowing only along the surface of the metal antenna, which is immersed into cold plasma, and there is no electric current in the plasma volume.

As it was proved in Sect. 2.2, SFXMs can propagate around a cylindrical metal rod immersed into magneto-active cold plasma in the HF range (1.14) under the condition that there is no external static azimuthal magnetic field, if the inequality $R_1 > m\delta$ is valid. Then the eigenfrequency of HF SFXMs can be approximately written in the following form:

$$\omega^{(0)} \approx \sqrt{\Omega_e^2 + \omega_e^2 + m^2c^2/R_1^2}. \tag{4.25}$$

In this limiting case of a wide waveguide, the asymptotical expression for the damping rate of these coupled flute waves caused by generation of electromagnetic radiation can be represented in the following form:

$$\gamma_{rad} \approx \kappa(R_1)^2 m^3 \delta^3 R_1^{-3} \omega^{(0)}. \tag{4.26}$$

Note that the damping rate (4.26) in this case is much less than the correction $\mathrm{Re}(\Delta\omega)$ to the wave frequency:

$$\mathrm{Re}(\Delta\omega) = \gamma_{rad} R_1/\delta. \tag{4.27}$$

As it was found in the Chap. 2 the collisional damping rate of these surface waves is proportional to the effective value of the collision frequency ν (see expression (2.25). Comparing this value with (4.27), it is possible to determine the conditions, when the collisional damping rate is larger than the damping rate caused by radiation of the coupled flute waves electromagnetic energy:

$$\nu/\omega^{(0)} >> \kappa^2(m\delta/R_1)^3. \tag{4.28}$$

One can estimate the conditions, under which the collisional damping rate is smaller than the damping rate caused by radiation of the coupled flute mode electromagnetic energy, on the base of inequality (4.28), e.g. for the case of an antenna with a radius, which is twice as large as the product $m\delta$. In the case, when ν is five orders less than the eigenfrequency of the coupled flute modes, for κ =0.01 the collisional damping rate is smaller than the damping rate caused by SFW power radiation due to bulk O-mode emission. Remind that if the plasma density is $n_e \approx 10^{10}$ cm^{-3}, the frequency of collisions between cold plasma particles is about $\nu \approx 10^4$ s^{-1} [6, 7].

The results of the numerical analysis of dispersion relation (4.23) allow one to make the following conclusions.

1. The dispersion curves of SFXMs with different azimuthal mode index only weakly differ from each other in the case of metal antennas with large radius.
2. Increasing of the azimuthal mode index leads to narrowing of the range of waveguide radius, where SFXM propagation is possible.

The following four diagrams present the results of the analysis. In Fig. 4.4 the results of numerical calculation of the magnitude of the damping rate of the coupled flute modes caused by conversion of SFXMs into BFOMs are plotted. It is shown that increasing azimuthal mode index m leads to increased damping of the coupled

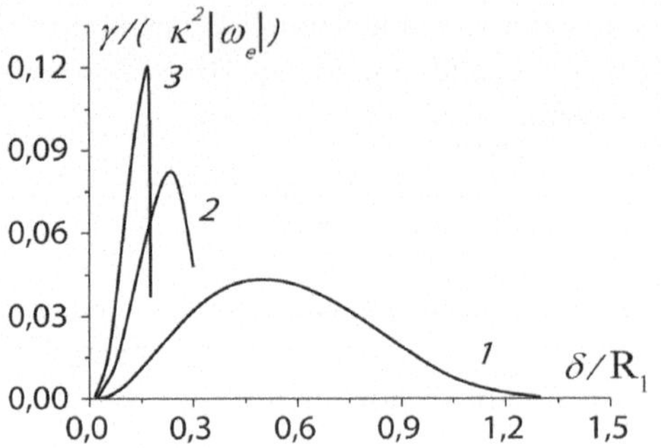

Fig. 4.4 Normalized damping rates of coupled flute modes versus normalized reversed radius of a metal rod antenna immersed into plasma; the *numbers* indicate the magnitudes of the azimuthal mode index m

flute X- and O-modes. This feature is connected with the presence of an azimuthal component of the external magnetic field. The maximal value of this damping rate grows linearly with increasing value of m.

The results shown in Fig. 4.5 prove that the larger the value of m is, the stronger is the influence caused by the application of the field $B_{0\varphi}$ on the dispersion properties of coupled flute modes. The maximal absolute value of the real part of the frequency correction $\Delta\omega$ is larger than the maximal absolute value of the corresponding damping rate γ_{rad}.

The numerical analysis shows that the radial distribution of the real part of the correction to the azimuthal magnetic wave field of the coupled flute modes, which is determined by application of the field $B_{0\varphi}$, is approximately similar to the graph of the first kind Bessel function $J_n(x)$ [5]. Its absolute value is proportional to the parameter κ, so this correction is of the first order of smallness. The axial electric field of the coupled flute waves is also approximately similar to the graph of the function $J_n(x)$.

The results of the numerical analysis of the spatial distribution of the coupled flute modes fields are depicted in Figs. 4.6 and 4.7. The radial distribution of the azimuthal

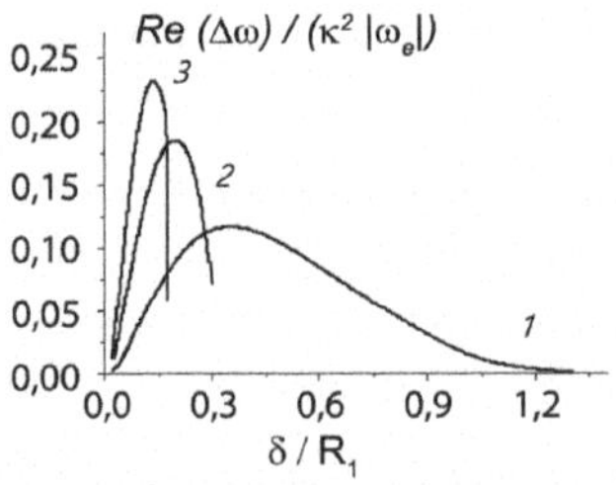

Fig. 4.5 Normalized real parts of the correction to the angular frequency of coupled flute modes propagating around a metal rod antenna immersed into plasma as a function of normalized reversed antenna radius; the *numbers* indicate the values of the azimuthal mode index m

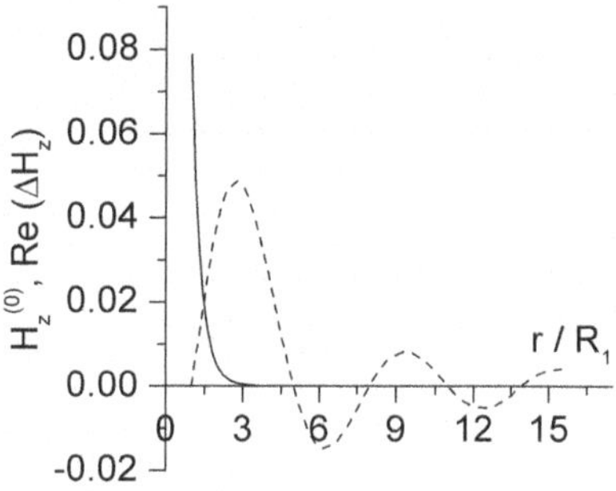

Fig. 4.6 Radial distribution of $H_z^{(0)}$ (*solid curve*) and $Re(\Delta H_z)$ (*dashed curve*) of coupled flute modes (in arb. units) excited by a metal rod antenna immersed into plasma for $\kappa(R_1) = 1$

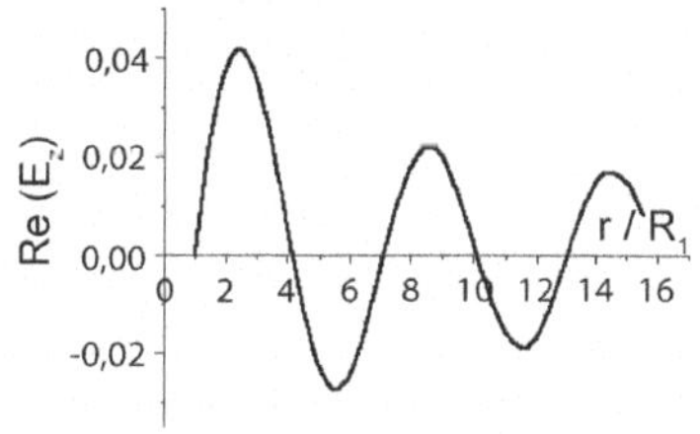

Fig. 4.7 Radial distribution of $Re(E_z)$ of coupled flute modes (in arb. units) excited by a metal rod antenna immersed into plasma for $\kappa(R_1) = 1$

magnetic field of the coupled SFWs in zeroth approximation is shown by the solid curve in Fig. 4.6. The dashed curve indicates the radial distribution of the real part of the first order (over the small parameter κ) correction to the azimuthal magnetic field of these waves, which is caused by application of the azimuthal field $B_{0\varphi}$. As one can see from Fig. 4.7, the radial wavelength of the BFOMs $\lambda \approx 6.13 R_1$ relates to the magnitude of the product $k_o\lambda \approx 2\pi$ and the absolute value of the axial electric field of the coupled SFW is approximately of the same order as the correction to the axial magnetic wave field ΔH_z nearby the antenna up to a distances of $r \approx 10 R_1$.

To make a conclusion on the possibility to observe the coupled flute X- and O-modes in a practical laboratory experiment, the following estimations are made. For instance, if a gaseous plasma is sustained in the experimental device under the following conditions: $n_e \approx 10^{10}\,\text{cm}^{-3}$ and $B_{0z} \approx 0.064$ kG, then for the coupled flute modes with $m = 1$ one can obtain $\Omega_e = 5|\omega_e|$, which is a little bit lower than the eigenfrequency $\omega^{(0)} \approx 6 \times 10^9\,\text{s}^{-1}$. Under the condition of application of an external axial current $I \approx 24\ A$, which produces an azimuthal magnetic field $B_{0\varphi} \approx 45 \cdot 10^{-5}$ kG, the small parameter of the problem is $\kappa \approx 7 \cdot 10^{-3}$, and one can find the frequency correction $Re(\Delta\omega) \approx 5700\,\text{s}^{-1}$. In this case the damping rate caused by emission of the BFOM is $\gamma_{rad} \approx 2400\,\text{s}^{-1}$. This means that the value of γ_{rad} is maximum for $R_1 \approx 2\delta \approx 10.6$ cm. To estimate the frequency of ion–electron collisions the formulas obtained in [7] are applied: $\nu \approx 25 n_e T_e^{-3/2}$ (here the density of the plasma particles is calculated in cm^{-3} and the electron temperature T_e in K), so for the temperature $T_e \approx 5$ eV one can obtain $\gamma_{col} \approx 9000 s^{-1}$, which is larger than both magnitudes of γ_{rad} and frequency correction $Re(\Delta\omega)$. The quality factor Q of such oscillations is sufficiently large: $Q = \omega^{(0)}/(2\gamma_{col}) \approx 7 \cdot 10^5$.

However, increasing the plasma temperature up to $T_e \approx 50$eV leads to decreasing value of the collisional frequency. It becomes approximately thirty times less as compared with the magnitude mentioned above. Then the damping rate γ_{rad} is about $10\gamma_{col}$ and the value of the Q-factor also increases by a factor of 3.5. Therefore, experimental investigations into the coupling between SFXMs and BFOMs can be carried out without utilization of a somewhat complicated apparatus or extreme experimental conditions.

4.4 Coupled Surface Flute Waves Propagating Above the Upper-Hybrid Frequency in Waveguides Entirely Filled with Current-Carrying Plasmas

In this Section, a cylindrical, ideally conducting metal waveguide with inner radius R_1, which is entirely filled with cold plasma, is considered. Its plasma particle density is assumed to be uniform along axial and azimuthal directions, but its radial profile could be non-uniform, in a general case. Thus, the dispersion relation for coupled flute waves is obtained for arbitrary radial profile in the present Section. However, analytical solutions are obtained for the limiting case of a plasma with uniform particle density. The waveguide is assumed to be immersed into an external steady magnetic field with two components, $\vec{B}_0 = \vec{B}_{0z} + \vec{B}_{0\varphi}$. As it is supposed in the present Chapter, here the case of small external static azimuthal magnetic field is under consideration, so that the inequality $\kappa = B_{0\varphi}/B_{0z} << 1$ is valid. The absence of a dielectric layer between the metal wall of the waveguide and the plasma column makes propagation of SFOMs impossible and ordinary polarized flute modes propagate as bulk modes. The geometry of this case is presented in Fig. 4.8.

To find the spatial distribution of the flute electromagnetic wave fields the set of Maxwell equations is applied. The presence of an azimuthal component $B_{0\varphi}(r)$ of the external static magnetic field leads to the situation, that in cylindrical co-ordinates $(\vec{e}_r, \vec{e}_\varphi, \vec{e}_z)$ just all nine elements of the dielectric permittivity tensor ε_{ik} of a cold magneto-active plasma are non-zero. That is why solving the Maxwell equations becomes in this case a rather complicated problem as compared with the cases that are studied in Chap. 2.

To simplify the problem, one can choose a special local co-ordinate system. It is proper to apply an ortho-normal basis of the co-ordinate vectors $(\vec{e}_1, \vec{e}_2, \vec{e}_3)$, which

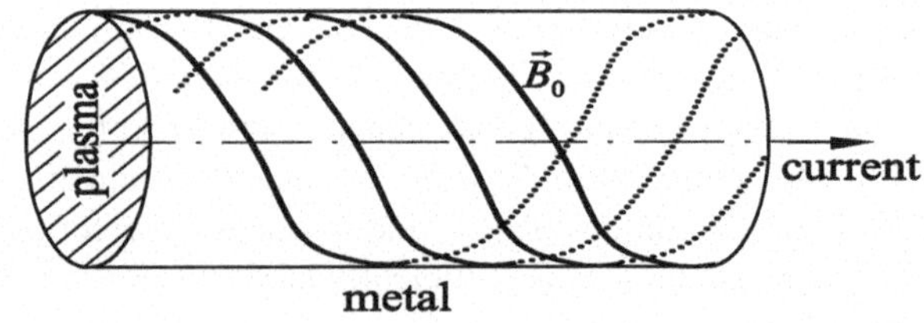

Fig. 4.8 Scheme of cylindrical metal waveguide, which is entirely filled with non-uniform current-carrying plasma

are connected with the force lines of the external static magnetic field $\vec{B}_0$. The first vector of this ortho-normal basis coincides with the radial basis vector in cylindrical co-ordinates, $\vec{e}_1 = \vec{e}_r$; the third vector is parallel to the magnetic force lines, $\vec{e}_3 = \vec{B}_0/\left|\vec{B}_0\right|$; and the second vector adds these two vectors to a right-handed coordinate system: so the following expression $\vec{e}_2 = \vec{e}_3 \times \vec{e}_1$ is valid. Within the accuracy up to terms of the second order of the small parameter $\kappa \ll 1$, which characterizes the problem, one can obtain the following approximate relations between the proposed new co-ordinate vectors and the basis co-ordinate vectors of a cylindrical system: $\vec{e}_1 = \vec{e}_r$, $\vec{e}_2 \approx \vec{e}_\varphi - \kappa\vec{e}_z$, $\vec{e}_3 \approx \vec{e}_z + \kappa\vec{e}_\varphi$. Using these local co-ordinates, one can write down the set of Maxwell equations, which incorporates the following simple relation between the components of the displacement vector $\vec{D}$ and the electric field $\vec{E}$ of the wave:

$$\vec{D} = \varepsilon_1(E_1\vec{e}_1 + E_2\vec{e}_2) + \varepsilon_3 E_3\vec{e}_3 - i\varepsilon_2\vec{e}_3 \times \vec{E}. \tag{4.29}$$

In this case the dielectric permittivity tensor of a cold magneto-active plasma with five non-zero components (1.2) can be applied [8–10]. The value of the cyclotron frequency in the expressions for these components of the tensor ε_{ik} is determined by the complete external static magnetic field $\vec{B}_0$, whose value is equal to the axial magnetic field $B_0 \approx B_{0z}$ with an accuracy up to small terms of the second order of κ. In the applied right-handed coordinate system $(\vec{e}_1, \vec{e}_2, \vec{e}_3)$, components of the rotor-operator within accuracy up to terms of the second order of smallness $\sim \kappa^2$ can be expressed through the components of this operator calculated in cylindrical co-ordinates in such way:

$$rot_1 = rot_r, \quad rot_2 \approx rot_\varphi - \kappa rot_z, \quad rot_3 \approx \kappa rot_\varphi - rot_z. \tag{4.30}$$

Taking into account all observations mentioned above, one can obtain the following three scalar equations from the set of Maxwell equations, which in zero approximation describe the ordinary polarized electromagnetic waves:

$$B_r = (m/(kr))E_z, \tag{4.31}$$

$$B_\varphi = \frac{i}{k}\frac{dE_z}{dr}, \tag{4.32}$$

$$\frac{-i\kappa}{k}\frac{dH_z}{dr} + \frac{i}{kr}\frac{d(rH_\vartheta)}{dr} + \frac{m}{kr}H_r = \varepsilon_3\left(E_z + \kappa E_\varphi\right). \tag{4.33}$$

Solving the set of Eqs. (4.31–4.33) for the axial electric wave field, one can derive a non-uniform differential equation of the second order, which has the form of Eq. (2.17) with small right-hand part. The latter is given by Eq. (4.9). The other three Maxwell equations in zeroth approximation describe extraordinary polarized surface waves. They also can be reduced to a non-uniform differential equation of the second order like Eq. (2.36), where the influence of the constant azimuthal magnetic field shows up by the appearance of a small right-hand term like expression (4.21).

As far as these waves are described by equations which are typical for weakly coupled modes it is suitable to apply the method of successive approximations, using the theory of SFXMs as the zeroth approximation for their study. Solutions of Eqs. (2.17) and (2.36) with their right-hand sides indicated above can be found by the method of constants variation. The fields of the investigated waves should satisfy the boundary conditions, which are indicated in Sect. 1.44. Thus, their application leads to the derivation of a dispersion relation for the coupled flute modes like Eq. (4.3). But the factor D_o which is written there is determined in the present case by formula (2.19), where one should set $R_1 = R_2$. Putting the factor D_x from Eq. (4.3) equal to zero, $D_x = 0$, one gets the dispersion relation of the SFXMs in absence of an external constant azimuthal magnetic field. Here, an explicit expression for D_x can be written as follows [11]:

$$D_x = -\frac{k^2}{k_\perp^2}\left(\frac{1}{\psi_x}\frac{d\psi_x}{dr} + \frac{m\mu}{R_1}\right). \tag{4.34}$$

Under the present conditions, the summand $D(\beta)$ in Eq. (4.3) should be replaced by the coupling coefficient which describes the interaction between BFOMs and SFXMs, $D(\beta) \to D(\kappa)$. Its value is quadratic over the small parameter κ:

$$D(\kappa) = \left(\frac{d\varphi_0}{dr}\right)^{-2} \int\limits_0^{R_1} \frac{r}{R_1}\varphi_o \widehat{M}\, \psi_x dr \left[\frac{k^2 W\left(\psi_x, \tilde{\psi}_x\right)}{k_\perp^2 \psi_x^2} \int\limits_0^{R_1} \frac{\psi_x \widehat{K}\, \varphi_o dr}{k_\perp^2 W\left(\psi_x, \tilde{\psi}_x\right)} - ik\left(1 + \frac{k_o^2}{k_\perp^2}\right)\frac{\varphi_o}{\psi_x}\right]. \tag{4.35}$$

The functions φ_o and ψ_x, which are utilized in formulas (4.34) and (4.35), are the solutions of Eqs. (2.36) and (2.37) with zero right-hand sides. Both functions should be of limited values within all the volume of the waveguide, in particular, in the center of the plasma column ($r = 0$). The parameters, which depend on the radial co-ordinate but are not integrand in the expression (4.35), should be calculated at $r = R_1$.

Solutions of the dispersion relation, which describe coupled flute X- and O-modes propagating in the waveguide entirely filled by cold magneto-active plasma can be found in the form $\omega = \omega_{O,X} + \delta\omega_{O,X}$, where $\omega_{O,X}$ are the eigenfrequencies of BFOMs and SFXMs, respectively, calculated in the case of absence of an external azimuthal magnetic field. The corrections $\delta\omega_{O,X}$ to the eigenfrequencies are determined by formula (4.5) in this case, $\left|\delta\omega_{O,X}\right| \ll \omega_{O,X}$.

Simple analytical solutions of the dispersion relation for coupled BFOMs and SFXMs can be obtained in the limiting case of plasmas with uniform particle density. In this case Eqs. (2.36) and (2.37) for the electromagnetic wave fields H_z and E_z are non-uniform Bessel equations. Solutions of the corresponding uniform equations, which are limited in the center of the plasma column, can be expressed in terms of

Bessel functions of the first kind [5] with the argument $k_o r$ and modified Bessel functions [5] with the argument $k_\perp r$, respectively. The analysis of k_o^2, which determines the features of the flute O-modes, testifies that it has different signs in the HF and the LF ranges. Namely, in the LF range (1.13) the inequality $k_o^2 < 0$ is valid, whereas in the HF range (1.14) it becomes positive: $k_o^2 > 0$. Thus, linear interaction between BFOMs and SFXMs is possible only in the HF range (1.14) for waveguides, which are entirely filled with cold magneto-active plasma [11].

In the case of uniform electron temperature, the external static azimuthal magnetic field depends linearly on the radial co-ordinate, $B_{0\varphi}(r) = r B_{0\varphi}(R_1)/R_1$. Then the integrals, which are included in the expression (4.35) for the coupling parameter of X- and O-modes $D(\kappa)$, can be calculated in an explicit form [11]. An analysis of the SFXM dispersion relation proves that in narrow waveguides HF SFXMs cannot propagate. In wide waveguides ($R_1 >> \delta|m|$) the eigenfrequency ω_X of HF SFXMs is approximately determined by the expression (4.25) given in the previous section.

If the values of the applied magnetic field and/or the waveguide radius are sufficiently small, so that the following inequality is valid:

$$R_1^2 \omega_e^2 / c^2 < g_{|m|,s}^2 - m^2, \tag{4.36}$$

BFOMs cannot propagate in the HF range (1.14). Here $g_{|m|,s}$ is the sth root of the first kind Bessel function of the order $|m|$ [5]. With increasing radius of the utilized waveguide and/or increasing strength of the external static magnetic field, inequality (4.36) becomes invalid. Then one can obtain a corresponding equality instead of inequality (4.36). The validity of this equality means that the frequencies of SFXMs and BFOMs are equal to each other under the corresponding conditions. Therefore, in this case, the correction to the SFXM eigenfrequency has the following form:

$$\delta\omega_X \approx \omega c^2 R_1^{-2} \sqrt{m^3 \omega_e \Omega_e^{-3}} \approx \kappa |m|^{3/2} Z^{-5/2} \Omega_e / g_{|m|,s}^2, \tag{4.37}$$

here $Z = \Omega_e/|\omega_e| > 1$. From the analysis of expression (4.37), one can conclude that the influence of $B_{0\varphi}$ on the frequency of SFXMs increases with increasing axial static magnetic field, and also with decreasing plasma particle density, absolute value of azimuthal mode index m and radius of the waveguide [12].

To investigate the dispersion properties of the coupled flute waves for arbitrary values of the parameters of the waveguide system one can apply a numerical method. It allows one to calculate, for instance the crossing-point of SFXMs with azimuthal mode index $m = -1$ and the first three radial BFOM modes. In Fig. 4.9 the dispersion curve of SFXMs (depicted by the solid line) and also the dispersion curves, which illustrate the first, second and third radial modes of BFOMs (dashed curves marked by numbers 1, 2 and 3, respectively), which propagate with azimuthal mode index $m = -1$ under the condition of zero field $B_{0\varphi}$, are presented as functions of the effective wavenumber $k_{ef} = |m|c/(R_1\Omega_e) = |m|\delta/R_1$. For the following values of the plasma waveguide parameters: κ =0.1, $Z = 3$ the crossing-points can be approximately found from expression (4.36): $k_{ef} \approx |m|/(Z g_{|m|,S}) \approx 0.087$, 0.048,

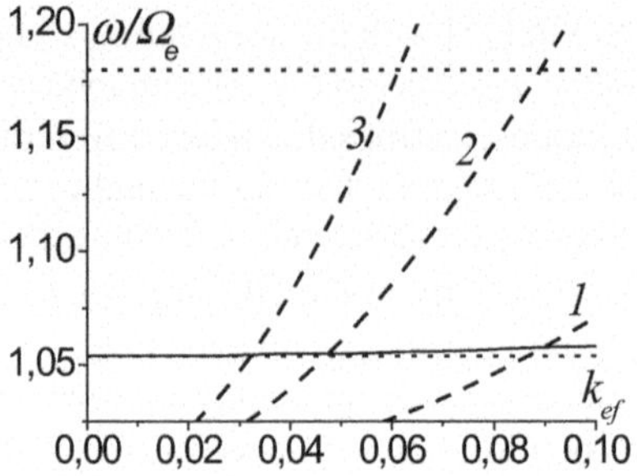

Fig. 4.9 Dependence of the normalized angular frequency of flute modes on the effective wavenumber for the case of a cylindrical waveguide entirely filled with plasma without azimuthal static magnetic field. The horizontal *dotted lines* show the limits of the HF range. The *solid curve* corresponds to the SFXMs and the *dashed curves* to BFOMs, the *numbers* indicate their radial mode index

0.033... To see the fine structure of the frequency spectrum, which illustrates the interaction between SFXMs and BFOMs one can apply the dimensionless variable $y = \sqrt[4]{\omega^2 \omega_e^{-2} - 1 - Z^2}$, where $y = 0$ is the lower limit of the HF range (1.14).

Below, the variation of the fine structure of the spectral interaction between the SFXMs and the BFOMs with decreasing effective wavenumber k_{ef} is explained. One can see from Fig. 4.10 that the dispersion curve for the SFXM (curve *1*) comes from the range of maximal effective wavenumbers and the function $y(k_{ef})$ undergoes a discontinuity near $k_{ef} \approx 0.21$. On the left side of this discontinuity, the dispersion curve for the SFXM emerges from the low boundary of the frequency range (Fig. 4.10, right-hand branch of curve 2). Near the point $k_{ef} \approx 0.087$, the SFXM transforms into the BFOM and this dispersion curve then abruptly falls to the boundary of the frequency range (left-hand branch of curve 2). The dispersion curve, which is drawn for the first radial bulk flute O-mode (right-hand branch of curve 3) comes from the range of relatively high effective wavenumbers. Near the point $k_{ef} \approx 0.087$, the BFOM transforms into the SFXM and then this curve runs almost horizontally to the left and rises rapidly when it approaches to $k_{ef} = 0.064$ (left-hand branch of curve 3) and so on in the same manner. One can see in Fig. 4.10 the similarity in

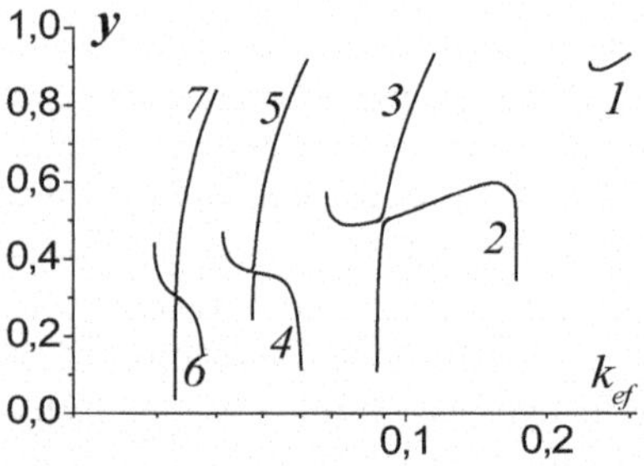

Fig. 4.10 Fine structure of interaction between SFXMs and three radial modes of BFOMs propagating in cylindrical waveguide entirely filled with plasma. The branches of the curves marked by the *numbers* $1 \div 7$ are explained in the text (see Sect. 4.4)

behavior of the curves marked by 3, 5 and 7; the same can be said on the behavior of the curves marked by numbers 2, 4 and 6, respectively.

Therefore, one can make the following conclusions concerning the interaction between the surface flute X-modes and bulk flute O-modes, if the radial mode index of the BFOM becomes larger.

1. The difference between the dispersion curves of BFOMs and SFXMs becomes less and less pronounced.
2. The range of the k_{ef} values between two neighboring mode conversion crossing points of the dispersion curves for SFWs becomes narrower.
3. Reduction of the k_{ef} value leads to a decrease of the SFXM frequency towards the lower edge of the HF range.

Thus, interaction between SFXMs and BFOMs results in the appearance of forbidden bands in the flute wave frequency spectrum, which are typical for waves propagating in media with periodic spatial non-uniformity (see, for example [13]). The appearance of the forbidden bands in the frequency spectrum, which is shown in Fig. 4.10, is equivalent to the existence of ranges of forbidden k_{ef} values. This is explained by the fact that for these values of the plasma cylinder radius, the BFOMs stop to be the eigenmodes and approaching to these forbidden ranges of k_{ef} leads to a sharp increase of the correction term to the frequency of BFOMs, which does not allow further application of the method of successive approximations.

One can also see the appearance of k_{ef} ranges, where SFXMs are characterized by reverse dispersion ($d\omega/dk_{ef} < 0$), which is caused by the interaction between SFXMs and BFOMs. This means that in waveguides with the corresponding values of parameters, flute wave energy is transported in the direction which is opposite to the direction of the SFW phase velocity (backward waves).

It should be underlined that the points $k_{ef} \approx 0.260, 0.160, 0.075, \ldots$, on the dispersion curve shown in Fig. 4.10, where direct (or in other words, normal) dispersion changes to reverse (abnormal) dispersion (or vice versa) and where $d\omega/dk_{ef} = 0$, are of particular interest. Nearby these points the dependence of the frequency of flute waves $\omega = \omega(k_{ef})$ can be represented in the following form:

$$\omega \approx \omega_0\left(1 + \alpha^2 k_{ef}^2/2\right). \tag{4.38}$$

Therefore, if at the initial time, the starting value of an electromagnetic pulse of the studied SFW has the form of a modulated sinusoidal function with Gaussian envelope:

$$H_z(\varphi, 0) = H^{(0)} \cos(m\varphi) \exp[-\varphi^2/(2\beta^2)], \tag{4.39}$$

then at any next moment of time this electromagnetic pulse will still have a modulated sine-wave form. However, the pulse broadens and the effective width β_t of its envelope depends on time in the following way:

$$\beta_t^2 = \beta^2 + \omega_0\alpha^2\delta^2 t/R_1^2. \tag{4.40}$$

Thus, the form of the electromagnetic pulse coincides with its initial form only during the following specific short interval of time: $t << R_1^2\beta^2\alpha^{-2}\delta^{-2}$ [14].

One can estimate the possibility of an experimental realization of the phenomenon described above in the following manner. For a laboratory gaseous plasma with the particle density $n_e \approx 10^{11}$ cm^{-3} it is necessary to apply the external axial magnetic field $B_{0z} \approx 0.48$ kG (then $Z = 3$). Coming from condition (4.36), one can find that the minimum value of the plasma radius, which is necessary for achieving the resonant interaction between SFXMs and BFOMs with azimuthal mode index $m = -1$, is $R_1 \approx 12$ cm. To satisfy the condition $B_{0\varphi} = 0.1B_{0z}$ one has to apply the axial electric current $I \approx 3$ kA. Then the difference between the angular frequencies nearby the crossing point of the dispersion curves of a SFXM and the first radial mode of BFOMs is equal to 1.4×10^7 s^{-1}.

Summarizing the results of numerical analysis of coupled flute modes, one can see that if the condition of linear interaction between BFOMs and SFXMs is not satisfied, the frequency correction to the SFXM eigenfrequency is proportional to the second power of the magnitude of the external static azimuthal magnetic field, while the corresponding correction determined by a constant axial magnetic field B_{0z} is directly proportional to its magnitude (i.e. $\delta\omega_x \propto B_{0z}$). Linear resonant interaction between SFXMs and BFOMs is realized nearby the crossing points of their dispersion curves. It is accompanied by the appearance of ranges of forbidden bands of their frequencies. In this case, the frequency correction becomes larger: it is directly proportional to the external static azimuthal magnetic field.

It is found, however, that a constant azimuthal magnetic field $B_{0\varphi}$ has the greatest effect on the dispersion properties of SFXMs by no means at the points of intersection between the dispersion curves of these surface X- and bulk O-modes. It should be noted that the waveguide radius accommodates an almost integer number of BFOM half-wavelengths near these points on the presented plot. Therefore, as in the case of linear SFXM–BFOM interaction, the boundary conditions are satisfied owing to a small variation of the frequency, because the field $B_{0\varphi}$ has a strong effect on the spatial distributions of both modes. Since the BFOM field distribution is of bulk type, one can select sets of waveguide parameters such that in zeroth approximation the plasma-waveguide radius can accommodate an odd number of quarters of the wavelength of the induced BFOM. Such conditions are observed almost in the middle between the points where the dispersion curves for the SFXM intersect with corresponding radial BFOM harmonics (see Fig. 4.10). In this situation, the plot of the dispersion curves for the coupled SFXMs and BFOMs is characterized by the appearance of forbidden wavenumber bands. Then, to satisfy the boundary conditions, the frequency correction must be so high that the corresponding increase (or decrease) of the wavelength removes (adds) an excessive (missing) quarter-wavelength. This also results in a considerable change in the SFXM dispersion in wavenumber ranges adjacent to the forbidden k_{ef} bands. In these ranges, the SFXM dispersion becomes of reverse type; i.e., the energy of these flute waves is transmitted in the direction opposite to their phase velocity vector (backward waves). It illustrates the qualitative difference between SFXM–BFOM linear interaction and interaction of SFXMs with the surface flute ordinarily polarized modes.

It is a matter of interest to compare the present results with those obtained in the previous Sect. 4.3. The values of SFXM frequencies, which propagate in metal waveguides entirely filled with plasma, are very close to the frequencies of SFXMs, which propagate along the boundary between a cold magneto-active plasma and a metal rod immersed into it (see Sect. 2.2). But the influence of a weak external static azimuthal magnetic field on their dispersion properties appears to be different in these two cases. In the case of a metal waveguide, which is entirely filled with plasma [11], resonant linear interaction between SFXMs and BFOMs can be realized. This phenomenon is not observed in the case of the coupled flute waves, which propagate around a metal rod immersed into plasma (just this case is studied in Sect. 4.3). On the other hand, linear interaction between SFXMs and BFOMs in a metal waveguide that is entirely filled with a cold magneto-active plasma is not accompanied by damping of the coupled flute modes. While their interaction under the condition of their propagation along the boundary of the metal bar immersed into a cold magneto-active plasma results in flute waves damping that is caused by conversion of SFXMs into BFOMs, which are radiated out of this antenna-plasma system. Thus, the amplitude of the waves, which are radiated from the plasma coated antenna, linearly depends on the strength of the electric current flowing along the antenna, whereas the damping rate of the flute waves is quadratic over the strength of electric current.

4.5 Coupled High Frequency Surface Flute Waves in Coaxial Waveguides Entirely Filled with Current-Carrying Plasmas

To increase the efficiency of operation of high-frequency electronic devices just coaxial waveguide structures are utilized more and more frequently because their characteristics are found to be better than the best ones of cylindrical structures. Therefore, the present Section is devoted to theoretical examining of a coaxial metal waveguide with inner and outer radii a and b, respectively. The space between the metal walls of the coaxial waveguide is entirely filled with magneto-active plasma that has a uniform particle density profile. The geometry of this coaxial waveguide is presented in Fig. 4.11. The case of weak azimuthal static magnetic field, $\kappa = B_{0\varphi}/B_{0z} \ll 1$, is considered. This allows one to carry out not only numerical but also analytical studies into the properties of coupled flute waves, which can propagate in such waveguides [15].

Ignoring the summands, whose order of smallness is higher than the κ value, one can derive the set of differential equations for the axial components of the electric field E_z of BFOMs and magnetic field H_z of SFXMs from the Maxwell equations. The tangential components of the electric field of the considered electromagnetic waves

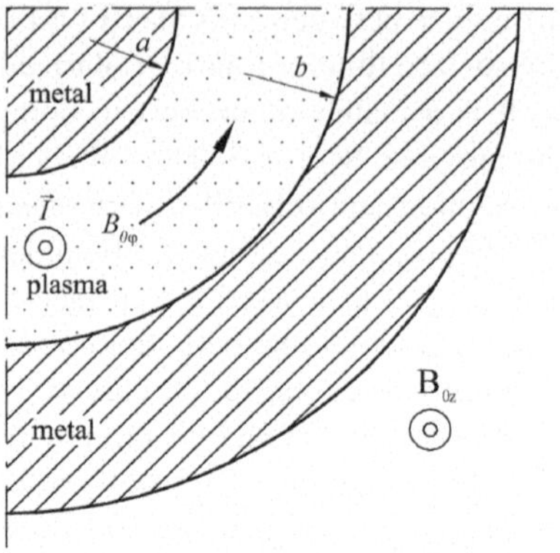

Fig. 4.11 Scheme of coaxial metal waveguide, which is entirely filled with magneto-active current-carrying plasma

are assumed to be equal to zero on the metal surfaces of the coaxial line. The frequency ranges for the existence of these flute waves are given in Sect. 1.4.1, and the signs of their azimuthal mode indices are determined in Sect. 2.2. To study the dispersion properties of such coupled flute electromagnetic waves one can use the theory of successive approximations, applying the theory of SFXMs (see Sect. 2.2) as the zeroth approximation. In this case one can find solutions of the Maxwell equations for the axial electric and magnetic fields of flute electromagnetic oscillations by the method of variation of constants [15]. All the other components of the electromagnetic field of the coupled flute modes can be expressed through the E_z and the H_z components from the Maxwell equations.

Application of the boundary conditions of Sect. 1.4.4 leads to the dispersion relation in the form (4.3), where $D_o = J_m(k_o a)N_m(k_o b) - N_m(k_o a)J_m(k_o b) = 0$ is the dispersion equation of BFOMs and $D_x = 0$ is the dispersion relation of SFXMs (both these equations are obtained for zero azimuthal external magnetic field, $B_{0\varphi} = 0$). In the present case, the SFXM dispersion relation looks as follows:

$$D_x = a_{23}a_{44} - a_{24}a_{43}. \tag{4.41}$$

In the dispersion relation obtained in [15] for the case of coaxial plasma filled waveguides, the parameter $D(\kappa)$ plays the role of the coupling coefficient between BFOMs and SFXMs that describes the influence of the azimuthal component of an external static magnetic field on the propagation of these modes:

$$
\begin{aligned}
D(\kappa) = \frac{\pi}{2}\Bigg\{\Bigg[& a_{23}\int_a^b r I_m(k_\perp r)\,\widehat{K}\,N_m(k_o r)dr \\
& - a_{24}\int_a^b r K_m(k_\perp r)\,\widehat{K}\,N_m(k_o r)dr\Bigg] J_m(k_o r) \\
& - N_m(k_o r)\Bigg[a_{23}\int_a^b r I_m(k_\perp r)\,\widehat{K}\,J_m(k_o r)dr - a_{24}\int_a^b r K_m(k_\perp r)\,\widehat{K}\,J_m(k_o r)dr\Bigg]\Bigg\} \\
& \times \{[N_m(k_o r)\int_a^b r J_m(k_o r)\,\widehat{M}\,K_m(k_\perp r)dr \\
& - J_m(k_o r)\int_a^b r J_m(k_o r)\,\widehat{M}\,K_m(k_\perp r)dr]a_{44} \\
& - a_{43}[N_m(k_o b)\int_a^b r J_m(k_o r)\,\widehat{M}\,I_m(k_\perp r)dr \\
& - J_m(k_o b)\int_a^b r N_m(k_o r)\,\widehat{M}\,I_m(k_\perp r)dr]\},
\end{aligned}
\tag{4.42}
$$

here the factors, which determine both expressions for D_x and $D(\kappa)$, have the following forms:

$$
a_{23} = \left[k_\perp K_m'(k_\perp a) + \frac{m\mu}{a}K_m(k_\perp a)\right],\quad a_{24} = \left[k_\perp I_m'(k_\perp a) + \frac{m\mu}{a}I_m(k_\perp a)\right],
\tag{4.43}
$$

$$
a_{43} = \left[k_\perp K_m'(k_\perp b) + \frac{m\mu}{b}K_m(k_\perp b)\right],\quad a_{44} = \left[k_\perp I_m'(k_\perp b) + \frac{m\mu}{b}I_m(k_\perp b)\right].
\tag{4.44}
$$

As one can see from the analysis of the expression (4.42) the coupling coefficient $D(\kappa) \propto \widehat{M}\,\widehat{K} \propto \kappa^2$ has a value of the second order of smallness.

Now, the structure of the SFXM dispersion relation (see expression (4.41)) in the case of no external static azimuthal magnetic field should be discussed in details. The condition $a_{44} = 0$ coincides with the dispersion relation for SFXMs propagating in a cylindrical metal waveguide with radius b, which is entirely filled with plasma (see Sect. 2.2). The condition $a_{23} = 0$ is the dispersion relation of SFXMs propagating along the interface of a metal rod with radius a, which is immersed into a uniform plasma and is oriented just along the external static axial magnetic field.

From the formal (mathematical) point of view, the structure of equation $D_x = 0$ is similar to the structure of the dispersion relation of coupled waves with linear conversion. Thus, one can make the following conclusions.

1. $a_{44} = 0$ plays the role of the dispersion relation of that wave, which propagates along the external metal surface of the considered coaxial waveguide.
2. $a_{23} = 0$ plays the role of the dispersion relation of the other wave, which propagates along the internal metal surface of the waveguide.
3. The product $a_{43}a_{24}$ plays the role of the weak coupling parameter between them.

However, it should be underlined here, that the investigated eigenwaves cannot interact linearly, because they cannot propagate with the same magnitude of azimuthal mode index m simultaneously. In other words, for a fixed sign of m, a solution exists either only for the equation $a_{44} = 0$ or only for the equation $a_{23} = 0$. This is the specific property of HF SFXMs: they propagate along a convex metal surface (inner rod) in wide coaxial waveguides only with positive azimuthal mode index, $m > 0$, and along a concave metal surface (external metal surface) they propagate only with negative mode index, $m < 0$.

The eigenfrequency of HF SFXMs, which propagate in a coaxial waveguide entirely filled with cold magneto-active plasma with $m > 0$, can be approximately calculated from Eq. (4.41):

$$\omega_x \approx \sqrt{\Omega_e^2 + \omega_e^2 + \frac{m^2c^2}{a^2} + 4\frac{m^2c^2}{a^2}\frac{b-a}{b+a}\exp[-2k_\perp(b-a)]}. \tag{4.45}$$

The presence of an external metal surface with a finite radius leads to some increase of the HF SFXM frequency with azimuthal mode index $m > 0$, propagating in the considered coaxial waveguide, as compared with the frequency of HF SFXMs, propagating around a metal rod immersed into an infinite uniform magneto-active plasma, under the same conditions.

Along the external metal surface, the HF SFXMs propagate in wide coaxial waveguides only with negative values of the azimuthal mode index $m < 0$. In this case, the value of the SFXM frequency is less than the frequency of HF SFXMs, which propagate in a cylindrical metal waveguide entirely filled with magneto-active plasma for the same other parameters of the waveguide structure. This is evident from comparison of the expressions for the frequencies of HF SFXMs with opposite signs of the azimuthal mode indices. As it is proved in Chap. 2, for a fixed absolute value of azimuthal mode index $|m|$, the frequency of the flute wave propagating along an internal surface is higher than the frequency of the flute wave propagating along an external surface with radius $r = b > a$ in the opposite direction.

For the case $m < 0$, the expression for the HF SFXM frequency is similar to expression (4.45). There are only two differences, in this case:

- (1) The fourth summand under the square root is negative;
- (2) In this summand, a^2 should be replaced by b^2.

Thus, in a coaxial waveguide, SFXMs have a different character of propagation and the dependence of their frequency on the effective wavenumber $k_{ef} = |m|\delta/r$ (where $r = a$ or $r = b$, respectively) is different for positive and negative signs of the azimuthal mode index m [15]. As far as HF SFXMs propagate only in wide plasma waveguides (this means that the plasma skin-depth is much less than the plasma radius) the presence of the second waveguide wall only weakly influences their dispersion properties. That is why the difference between frequencies of the HF SFXMs propagating in similar non-coaxial structures (see Chap. 2) is also weak.

As far as practical applications of the studied coupled flute modes are concerned, one can see that the considered coaxial waveguides can be used for designing waveguide phase shifters. Changes in the phases of electromagnetic signals take place on the different interfaces of the coaxial plasma filled waveguide in the different directions: with increasing azimuthal angle the signal phase decreases on the external surface $r = b$, and it increases on the internal surface $r = a$. One can tune the frequency of the phase shifter to the frequency of the input signal just by changing the external axial magnetic field. Application of such type of plasma filled waveguides allows one to get a signal with any necessary phase and to take-off simultaneously a set of signals with phase changes assigned in advance.

HF BFOMs propagate in a wide coaxial waveguide with the following frequency:

$$\omega_o \approx \sqrt{\Omega_e^2 + \pi^2 c^2 j^2 (b - a)^{-2}}, \quad \text{where } j = 1, 2, 3. \tag{4.46}$$

If the external axial magnetic field and the sizes of the waveguide are large enough, so that the following equality is valid:

$$\frac{\omega_e^2}{c^2} \approx \frac{\pi^2 j^2}{(b-a)^2} - \frac{m^2}{b^2}, \tag{4.47}$$

the dispersion curves of SFXMs and BFOMs can intersect each other. In this case, their linear conversion becomes possible under the condition of application of an external static magnetic field, which has a large axial and a small azimuthal component. Therefore, frequency corrections to the eigenvalues of the frequencies of ordinary and extraordinary flute modes can be calculated with formula (4.5) and their values are linear over the small parameter κ.

More information can be obtained by numerical studies of the properties of coupled flute modes propagating in coaxial waveguides. In Fig. 4.12, one can see the dependence of the SFW eigenfrequency on the effective wavenumber $k_{ef} = |m|\delta/b$ for the case of azimuthal mode index $m = +1$. The minimum size of the coaxial line, wherein linear interaction between SFXMs and BFOMs can be realized, is just related to the modes with the minimum value of azimuthal mode index. Increasing the Z parameter leads to narrowing (in units of electron cyclotron frequency) of the HF range wherein SFXMs can exist. It should be mentioned that this numerical analysis is carried out for the case of uniform radial profile of the electron temperature (linear radial dependence of the external static azimuthal magnetic field: $B_{0\varphi} \propto r$).

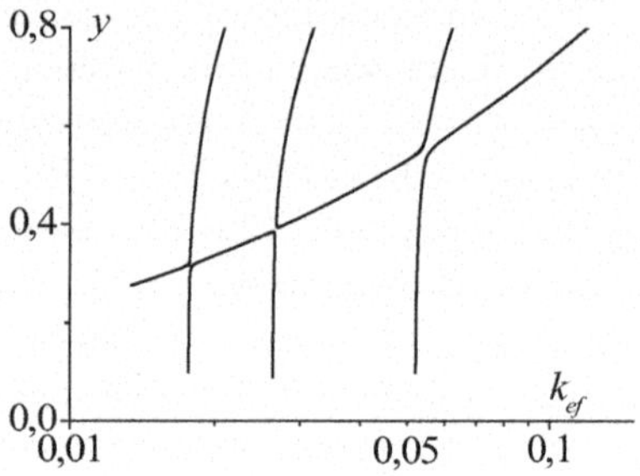

Fig. 4.12 Interaction between SFXMs and three first radial harmonics of BFOMs propagating in a plasma filled coaxial waveguide for the case: $m = +1$; $\kappa = 0.3$; $Z = 3$; $b = 2a$

The parameter $y = \sqrt[4]{\omega^2\omega_e^{-2} - 1 - Z^2}$ here has the same meaning as that one, which is applied in Fig. 4.10 (see Sect. 4.3).

In Fig. 4.13, the radial distribution of the axial magnetic and electric fields of SFWs propagating with $m = -1$ in a coaxial waveguide, whose inner radius is four times smaller than its outer radius, is plotted. The following values of the waveguide parameters were applied for the numerical analysis: $Z = 3$, $k_{ef} = 0.1$, and $y = 0.537$. The dashed curve depicts the radial dependence of the axial magnetic field of SFWs in zeroth approximation. It has the typical form for the case of surface-wave propagation. This means that the wave amplitude is localized mainly nearby the outer wall of the considered coaxial waveguide. The solid curve shows the radial distribution of the axial electric field of the SFW (it is equal to zero if there is no external azimuthal magnetic field); this dependence is normalized by the small parameter of the problem in order to be clearly visible for the readers. One can see that the shown radial distribution of the axial electric field $E_z^{(1)}$ is the first radial mode of the BFOMs. So, in this case the phase difference between the internal and external surfaces of the coaxial waveguide is $k_o(b - a) \approx 2.602$. The dashed-dotted curve gives the correction of the radial dependence of the SFW axial magnetic field caused by the presence of an external static azimuthal magnetic field.

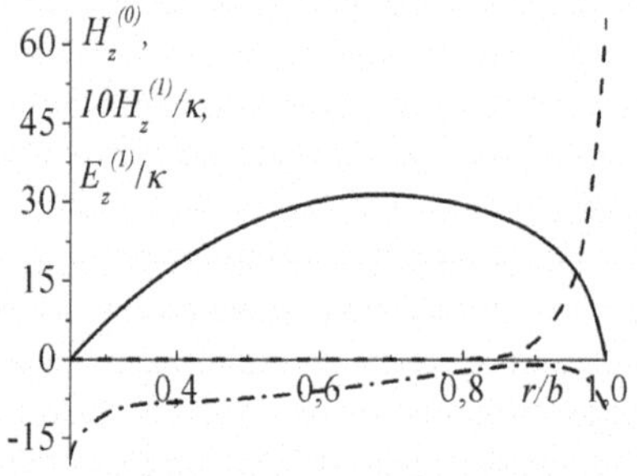

Fig. 4.13 Radial distribution of the magnetic (*dashed curve*) and normalized axial electric (*solid curve*) fields (arb. units) for coupled flute modes propagating in a plasma filled coaxial waveguide for the case $m = -1$; $\kappa = 0.3$; $a = b/4$. The *dashed-dotted* curve shows the normalized correction of the radial dependence of the axial magnetic SWF field ($\times 10$) caused by the azimuthal static magnetic field

In the following, some common remarks on the results of the numerical analysis carried out in this Section are made. The frequency spectrum of coupled flute modes propagating in coaxial waveguides has some peculiarities compared to the spectra of coupled flute modes studied in the previous sections. First, increasing radial mode index of the bulk flute modes results in the fact that the splitting of the frequencies of these flute modes becomes less pronounced nearby the point of their conversion. Second, the range of the k_{ef} values between two neighboring points of their conversion becomes narrower. Third, diminishing of the k_{ef} value leads to a decrease of the SFXM frequency, which means that their frequency approaches to the lower limit of the frequency range (dimensionless parameter $y \to 0$). Fourth, for the SFXMs with $m = +1$, which propagate along the internal metal surface, areas of non-transparency appear (this happens after the second intersection of the dispersion curves plotted in Fig. 4.12, starting from the range of large values of k_{ef}). Fifth, the frequencies of BFOMs diminish with increasing distance between the walls of the coaxial waveguide, and intersections of their dispersion curves move toward greater values of k_{ef}. Sixth, the minimum size of the coaxial waveguide (wherein linear interaction between SFXMs and BFOMs can be realized) is related just to the flute waves with the minimum value of azimuthal mode index. Seventh, increasing Z parameter (decreasing external axial static magnetic field) leads to narrowing (in units of electron cyclotron frequency) of the HF range where SFXMs can exist at all [15].

To estimate the possibility of an experimental observation of this phenomenon one can assume the following parameters: plasma density $n_e = 10^{13}$ cm^{-3}, and external magnetic field $B_{0z} \approx 4.831$ kG. For realization of linear interaction between the SFXMs and the first radial mode of the BFOMs one has to utilize a coaxial waveguide with $b = 3.12$ cm and $a = 1.56$ cm. In this case [15], the splitting of the frequencies of the SFXM and the BFOM is equal to $2\Delta\omega_X \approx 3 \cdot 10^8\ \text{s}^{-1}$ for an applied axial electric current of 90 kA.

4.6 Conclusions

Results of theoretical investigations on the properties of electromagnetic flute modes propagating in different plasma filled waveguides are presented here for the case that an axial electric current is flowing along the plasma column. The theory of SFWs developed in the Chap. 2 is applied for studying the linear interaction between the ordinary and extraordinary flute modes in current-carrying plasma waveguides.

The results of studying the propagation of SFWs in current-carrying waveguides with plasma filling seem to be interesting for various practical applications in plasma electronics and technologies. Under different conditions, flowing of an axial electric current leads to coupling between ordinary and extraordinary modes where the forced ordinary polarized modes can be either of surface or bulk types. In different plasma filled waveguides, coupling between ordinary and extraordinary polarized modes can lead to different phenomena: appearance of non-reciprocal frequency ranges in the frequency spectrum of the coupled flute modes; change of the sign of their group velocity; damping of the extraordinary flute modes caused by their conversion into radiated forced ordinary bulk modes and so on.

The phenomenon of coupling between flute X- and O-modes studied here is linear in respect to temporal evolution of the mutual power transfer of these modes. However, power transfer between these modes is characterized by the nonlinear dependence on the small parameter of the problem: $\kappa = B_{0\vartheta}/B_{0z} << 1$.

References

1. Girka, V., Girka, I., & Pavlenko, I., et al. (2012). Coupled azimuthal modes propagating in current-carrying plasma waveguides. *Journal of Plasma Physics, 78*(2), 105–123.
2. Girka, V. O., & Girka, I. O. (1992). Coupled azimuthal surface waves in a nonuniform current—carrying plasma cylinder. *Soviet Journal of Communications Technology and Electronics, 37*(4), 23–29.
3. Azarenkov, N. A., & Ostrikov, K. N. (1991). Surface magnetoplasma waves at the interface between a plasma-like medium and a metal in a Voigt geometry. *Physics Reports, 308*, 333–428.
4. Girka, O. I., Girka, I. O., & Girka, V. O., et al. (2008). Propagation of azimuthal waves along the surface of a metal current-carrying cylinder immersed into a magnetized plasma. *Radiophysics and Quantum Electronics, 51*(2), 110–122.
5. Abramowitz, M., & Stegun I. A., (Eds). (1994). In *Handbook of mathematical functions*. National Bureau of Standards, Applied Mathematics, Series 55.
6. Vukovic, V. (1986). *Surface waves in plasmas and solids*. World Scientific.
7. Shkaropsky, I. P., Johnston, T. W., & Bachynski, M. P. (1966). *The particle kinetics of plasmas*. Addison-Wesley Publishing Company.
8. Alexandrov, A. F., Bogdankevich, L. S., & Rukhadze, A. A. (1984). *Principles of plasma electrodynamics*. Springer.
9. Akhiezer, A. I., Akhiezer, I. A., Polovin, R. V., Sitenko, A. G., & Stepanov, K. N. (1975). *Plasma electrodynamics*. Pergamon Press.
10. Krall, N. A., & Trivelpiece, A. W. (1973). *Principles of plasma physics*. MacGraw-Hill.
11. Girka, O. I., Girka, I. O., & Girka, V. O., et al. (2008). Propagation of azimuthal waves in magneto-active waveguides filled with a current-carrying plasma. *Technical Physics, 53*(7), 905–912.
12. Girka, V. O., & Girka, I. O. (2002). Effect of toroidal magnetic field variations on the spectra of azimuthal surface waves in metal waveguides entirely filled with plasma. *Plasma Physics Reports, 28*(3), 190–195.
13. Lifshits, Ye. M., & Pitaevsky, L. P. (1981). *Course of theoretical physics. Physical kinetics. 10*. Pergamon Press.
14. Landau, L. D., & Lifshits, Y. M. (1960). *Course of theoretical physics. Electrodynamics of continuous media. 8*. Pergamon Press.
15. Girka, O. I., Girka, I. O., & Girka, V. O., et al. (2008). Coupled transverse modes of coaxial metal waveguides completely filled with magneto-active current-carrying plasma. *Plasma Physics Reports, 34*, 901–910.

Chapter 5
Surface Flute Waves in Inhomogeneous Magnetized Plasma Waveguides

This Chapter is devoted to the theory of surface waves propagating across the axis of symmetry in cylindrical metal waveguides with non-uniform magneto-active plasma filling. The presented theoretical results illustrate, how spatial non-uniformity of both the plasma particle density and the external static magnetic field can affect the dispersion properties of these waves. These problems are solved analytically as well as numerically. The analytical study of the obtained equations is carried out by the method of successive approximations; adequacy of this approach is proved here as well. Numerical evaluations of the possibilities to experimentally observe the phenomena, which accompany the propagation of these waves in non-uniform plasma filled waveguides, are carried out.

Experimental studies of non-uniform plasma waveguides are important for solving different problems of plasma physics. For instance, the plasma produced during a gas discharge affects the radiation efficiency of plasma antennas or an array of plasma antennas [1–3]. Inhomogeneity of a magneto-active plasma strongly affects the spatial distribution of the fields of axially non-symmetric waves, which are radiated by antennas with such plasma coating. Results of broad theoretical and experimental investigation into excitation of quasi-electrostatic and helicon waves by a ring antenna immersed into magneto-active plasma were presented in [4]. The radiated power was found there to resonantly depend on the plasma density profile. The HF surface wave (SW) radar was shown in [5] as a highly efficient tool, which can provide over-the-horizon surveillance of ships and aircrafts, which move close above sea level. From this point of view studying the features of surface flute waves (SFWs) in non-uniform waveguides seems to be important for the development of new types of powerful antennas.

SW properties are intensively studied for sustaining microwave gas discharges, which have broad practical applications for development of different plasma technologies [6–9]. The parameters of microwave gas discharges depend on different characteristics, such as: geometry of the discharge chamber, type of the operating gas and its pressure, magnitude of an external static magnetic field, type of the operating electromagnetic mode and so on. Utilization of an external static magnetic field

© The Author(s), under exclusive license to Springer Nature Switzerland AG 2022

I. Girka and M. Thumm, *Surface Flute Waves in Plasmas*, Springer Series on Atomic, Optical, and Plasma Physics 120, https://doi.org/10.1007/978-3-030-98210-2_5

and axially non-symmetrical SWs were found to allow one to increase the volume of the produced plasma and make the gas discharge longer. Plasma generators based on application of transverse waves excited by slot antennas [8, 9] seem to be efficient devices because the operating modes, which propagate along the azimuthal direction, do not lose their energy to the discharge chamber. Application of SWs has a lot of advantages as compared with bulk wave application [6]. This is an additional reason, that determines the importance of studying the properties of SWs, which propagate in plasmas with non-uniform particle density.

External static magnetic fields, which are usually applied for confining plasma structures in laboratory devices, are also non-uniform. This is connected with the methods utilized for production of such magnetic fields. To confine plasmas with a small volume one can apply either stationary solid ferromagnets or long solenoids. Only for small samples of semiconductor plasmas one can consider that the produced magnetic field is uniform. However, in the case of gaseous plasmas one has to apply a complicated chamber as trap for the produced plasma and discrete magnetic coils for generation of the confining magnetic field (e.g. toroidal tokamaks and stellarators, linear mirror magnetic traps, reversed field pinches, etc.). Thus, any large-scale plasma device has a non-uniform external static magnetic field. Even if this non-uniformity is small, it can affect the properties of the electromagnetic eigenwaves, whose frequency depends on the utilized magnetic field. That is why the problem of non-uniformity of an external static magnetic field is also under consideration in the present Chapter.

5.1 Surface Flute Waves in Waveguides Filled with Plasma Whose Particle Density is Non-uniform

The model of plasmas with uniform particle density describes only the case of solid-state plasmas in a sufficiently good manner [10–13]. Non-uniformity of the plasma particle density affects the frequencies of both bulk and surface waves. In the following, a three-component waveguide structure is considered, which consists of a cylindrical metal waveguide with inner radius R_2 which has a dielectric coating in the region $R_1 < r < R_2$ with dielectric constant ε_d and a concentric plasma cylinder with outer radius R_1 (see Fig. 2.1). The plasma particle density is supposed to be spatially non-uniform and the magnitude of its density in the central part is assumed to be sufficiently large, so that the inequality $\Omega_e^2(r \to 0) \gg \omega_e^2$ is valid, here Ω_e and ω_e are the electron plasma and electron cyclotron frequencies, respectively. An external static magnetic field is directed along the axial direction, $\vec{B}_0 \| \vec{z}$.

The dispersion properties of SFWs, whose fields depend on coordinates and time in the following form: $E, H \propto f(r)\exp[i(m\varphi - \omega t)]$, are studied. Then starting from solving the set of Maxwell equations, one can derive the following equation for the axial magnetic component of the surface flute X-modes (SFXMs) wave field in the region of the non-uniform plasma:

$$\frac{1}{r}\frac{d}{dr}\frac{r}{k_\perp^2}\frac{dH_z}{dr} - \left[1 + \frac{m^2}{k_\perp^2 r^2} - \frac{m}{r}\frac{d}{dr}\left(\frac{\mu}{k_\perp^2}\right)\right]H_z = 0, \tag{5.1}$$

here $\mu = \varepsilon_2/\varepsilon_1$, $k_\perp^2 = k^2(\mu^2 - 1)\varepsilon_1$, and $k = \omega/c$, where ε_j $(j = 1, 2)$ are the components of the dielectric permittivity tensor for magneto-active plasmas in hydrodynamic approximation [10, 14–16], and m is the azimuthal mode index. These notations are the same as in previous chapters.

The plasma particle density is assumed to be weakly non-uniform. Applying the method of etalon equations [17] for solving Eq. (5.1) in the plasma region, $r < R_1$, one can derive:

$$H_z = A(r) I_m(z); \tag{5.2}$$

where $A(r) = const \cdot \sqrt{z k_\perp^2/(r z')}$; $z' = k_\perp \sqrt{1 - (m/r) d(\mu/k_\perp^2)/dr}$, and $I_m(z)$ is a modified Bessel function [18]. The correctness of this solution depends on the satisfaction of the inequality:

$$\left|\frac{m}{r}\frac{d}{dr}\left(\frac{\mu}{k_\perp^2}\right)\right| \ll 1, \tag{5.3}$$

which is true in the case of weak non-uniformity of the plasma density.

The dispersion relation for SFXMs can be derived as result of the application of the boundary conditions indicated in Chap. 1 (see Sect. 1.4.4). In this case, the dispersion equation looks as follows [17]:

$$\frac{m}{R_1}\mu + \frac{1}{H_z}\frac{dH_z}{dR_1} = \frac{k_\perp^2}{k}\frac{J'_m(kR_1)N'_m(kR_2) - J'_m(kR_2)N'_m(kR_1)}{J'_m(kR_2)N_m(kR_1) - J_m(kR_1)N'_m(kR_2)}, \tag{5.4}$$

where $J_m(z)$ and $J'_m(z)$ are Bessel functions of the first kind and their derivatives over the argument, respectively; $N_m(z)$ and $N'_m(z)$ are Neumann functions and their derivatives, respectively [18]. The analysis of Eq. (5.4) allows one to determine the frequency ranges where SFXMs can propagate. There are two possible ranges, which can be referred here in the same manner as it is done in Sect. 1.4.1, namely the LF range (1.13) and the HF range (1.14), respectively. Applying asymptotes [18] of the Bessel functions in the limiting case $|m| \gg 1$, one can obtain analytical expressions for the SFXMs frequency in the HF and LF ranges, respectively:

$$\omega \approx \omega_1\left[1 + \frac{\delta^2}{2R_1^2}\left(\Delta\frac{R_1^2}{\delta^2}\frac{|\omega_e|}{\omega_1} + \frac{m\Omega_e^2}{\omega_1^2} - \frac{\Omega_e^2\xi}{\omega_1|\omega_e|}\right)^2\right], \tag{5.5}$$

$$\omega \approx \frac{|\omega_e|}{\sqrt{1+k_{ef}^{-2}}}\left[1-\Delta\frac{\Omega_e^2}{\omega_e^2}|m|\sqrt{1+k_{ef}^{-2}}-\frac{\xi R_1^2\omega_e^2}{m^3c^2(1+k_{ef}^{-2})^{3/2}}\right]^{-1}, \tag{5.6}$$

here $\omega_{1,2} = \mp 0.5|\omega_e| + \sqrt{\Omega_e^2 + 4^{-1}\omega_e^2}$ are the cut-off frequencies, ω_{LH} and ω_{UH} are the lower hybrid and upper hybrid frequencies, $\delta = c/\Omega_e$, $k_{ef} = m\delta R_1^{-1}$, $\xi = (R_1\delta^{-1})(d\delta/dR_1)$ is a parameter which describes the non-uniformity of the plasma density, and the parameter $\Delta = R_2R_1^{-1} - 1$ is the relative thickness of the dielectric layer, which separates the non-uniform plasma column from the metal wall of the waveguide.

Using the solutions (5.5) and (5.6) it is possible to obtain approximate expressions for the frequency ω of SFXMs in other limiting cases, namely: large depth of wave penetration into the plasma, $z(R_1) \ll 1$, and small penetration depth, $z(R_1) \gg |m|$. The results of a numerical analysis of dispersion relation (5.4) are presented in [17]. There it was found that the condition of weak non-uniformity of the plasma density slightly changes the SFXM frequency as compared with the case of uniform plasma. As far as these extraordinary polarized modes are non-reciprocal (propagate with different frequencies in mutually opposite directions, under the same conditions), increasing the radial plasma particle density gradient $d\Omega_e^2/dR_1$ affects the SFXMs frequency in the cases of positive and negative signs of the azimuthal mode index m in different way. For instance, the frequency of LF SFXMs with $m < 0$ decreases more slightly as compared with the case of positive azimuthal mode index $m > 0$.

It should be pointed out that the dispersion relation (5.4) once more describes also two other waveguide structures: first, a waveguide, which is partially filled with a cold non-magnetized plasma ($B_0 = 0$), and second, a waveguide which is entirely filled with a cold magneto-active plasma ($\Delta = 0$) . In the first case one can obtain the expression for the frequency of SFXMs within an accuracy up to the summands of the second order of smallness, $\Delta^2 \ll 1$:

$$\omega^2 \approx \frac{\Delta(1-\Delta/2)m^2\Omega_e^2}{\sqrt{m^2 + R_1^2/\delta^2 + \Delta(1+\Delta/2)R_1^2/\delta^2 - \xi}}. \tag{5.7}$$

In the second case, it is possible to derive the SFXM frequencies just from the expressions (5.5) and (5.6) putting there $\Delta = 0$. However, it should be underlined that if $\Delta = 0$ then SFXMs become unidirectional waves (i.e. they can propagate only along one direction). Thus, only SFXMs with $m > 0$ can propagate in the LF range, $\omega < \omega_1$, and only a negative sign of the azimuthal mode index $m < 0$ can be realized for SFXMs in the HF range, $\omega > \omega_{UH}$. Therefore, analyzing the expressions (5.6) and (5.7) one can make the conclusion: if the plasma particle density decreases in the direction towards the plasma boundary, the SFXM frequency increases compared with the case of uniform plasma particle density profile.

In the case of arbitrary type of the plasma density profile, there is a unique method to determine the SFXM spectrum—it is a numerical analysis. However, there is a special limiting case, when their property can be studied analytically. This is the case of SFXM propagation in a wide plasma cylinder in which the plasma radius is considerably larger than the SFXM penetration depth into the plasma. It can be shown that if nearby the plasma boundary its density is characterized by a linear profile then dispersion relation (5.4) can be solved analytically.

The further considerations in the present Section are restricted only to the case of LF SFXMs propagating in a metal waveguide filled with cold magneto-active plasma. It is assumed that the plasma, which is located in the region $r < R_1$, is characterized by the following particle density profile [19]:

$$n(r) = n(R_1) + (r - R_1)dn/dr\big|_{r=R_1}. \tag{5.8}$$

In the case of a large plasma cylinder ($R_1 \gg |m|\delta_1$, with $\delta_1 = c\Omega_e^{-1}(R_1)$) the inequality $|q(r)| \equiv \left|m^2 k_\perp^{-2} r^{-2} - (m/r)d\left(\mu k_\perp^{-2}\right)/dr\right| \ll 1$ is valid. Then, describing the SFXM field as a series: $H_z = H_z^{(0)} + H_z^{(1)} + \ldots$, where $\left|H_z^{(n+1)}\right| \ll \left|H_z^{(n)}\right|$, one can solve Eq. (5.1) with the aid of the method of successive approximations. The expressions for the zeroth and first order approximations are, respectively:

$$H_z^{(0)} = C_1 U'(\xi) + C_2 V'(\xi), \tag{5.9}$$

$$H_z^{(1)} = U'(\xi)\int_{\xi_1}^{\xi} V'(\xi) g d\xi - V'(\xi)\int_{\xi_2}^{\xi} U'(\xi) g d\xi. \tag{5.10}$$

Here is $g = -\frac{\delta_0^2}{r}\frac{dH_z^{(0)}}{dr} + q(r)H_z^{(0)} - \frac{k^2}{r}\frac{d}{dr}\left[\delta_0^4 r\left(1 + \frac{\omega_e^2}{\omega^2}\right)\frac{dH_z^{(0)}}{dr}\right]$, $\left|\frac{g}{H_z^{(0)}}\right| \ll 1$, $\delta_0 = c/\Omega_e(r < R_1)$, $\xi(r) = (\delta_1 k_0)^{-2} + k_0(R_1 - r)$, $k_0^3 = -d\delta_0^{-2}/dR_1$, $U'(\xi)$ and $V'(\xi)$ are derivatives of Airy functions over their arguments [18]. Applying the asymptotes of the Airy functions for large arguments and choosing the solution of Eq. (5.1) in the form of SWs, one finds that $C_1 = 0$, $\xi_1 = \xi(0)$ and $\xi_2 = \xi(R_1)$ if the plasma particle density increases along the direction from the plasma boundary towards the waveguide axis (this means that $k_0 > 0$). In the opposite case, if $k_0 < 0$, then $C_2 = 0$, $\xi_1 = \xi(R_1)$ and $\xi_2 = \xi(0)$. To find solutions of Eq. (5.1) in the form of the expressions (5.9) and (5.10) it is not sufficient to satisfy the inequalities $R_1 \gg |m|\delta$ and $\Omega_e^2 \gg \omega_e^2$. It is necessary also to provide a smooth variation of the plasma density, so that the following inequality should be realized:

$$|\delta_0 d\ln[n(r)]/dr|_{|r=R_1} \ll 1. \tag{5.11}$$

The condition (5.11) is satisfied if the inequality $k_0^3 \leq (R_1\delta_1^2)^{-1}$ is valid. In this case, the dispersion equation for SFXMs looks like in the previous case (see Eq. (5.4)).

Dispersion relation (5.4) can be solved approximately for the case of the plasma particle density profile (5.8) and narrow vacuum layer, $k(R_2 - R_1) \ll 1$. First, one has to substitute the solutions of the Maxwell equations in the forms (5.9) and (5.10) into Eq. (5.4) and apply the asymptotes of the Airy functions for large argument ($\xi \gg 1$) [18]. Then the eigenfrequency of LF SFXMs can be derived in the following form:

$$w \approx \frac{0.5m|\omega_e|}{R_1/\delta_2 + \Delta R_1^2/\delta_2^2} + \sqrt{\frac{0.25m^2\omega_e^2}{\left(R_1/\delta + \Delta R_1^2/\delta^2\right)^2} + \frac{\Delta m^2 \Omega_e^2(R_1)}{R_1/\delta + \Delta R_1^2/\delta^2}}, \tag{5.12}$$

where $\delta_2 = \delta_1\left(1 + k_0^3\delta_1^3/4\right)$.

Summarizing the results obtained in the present Section one can make the following conclusions. First, application of the approximation of the plasma particle density profile by a linear function allows one to solve the problem of the influence of plasma density non-uniformity on the LF SFXM frequency spectrum analytically. Second, the type of plasma density profile can be a nonlinear one in the plasma core, for example, it can change from a linear shape nearby the plasma interface into a parabolic form in the plasma core, but such a behavior does not affect the obtained solution. Third, if the skin-depth is $\delta = \delta_1$, formula (5.12) coincides with the corresponding expression, which was obtained for uniform plasmas in Sect. 2.1 (see Eq. (2.10)). Fourth, analyzing (5.12) one can see that if the plasma particle density decreases along the direction from the plasma boundary towards the plasma core, the effective penetration depth δ_2 of LF SFXMs into the plasma is smaller than in a uniform plasma. Fifth, in the case of the density profile shape indicated above, the frequency of LF SFXMs is lower than it was in the case of a uniform plasma as well.

One more example of non-uniform plasma particle density profile allows one to solve the dispersion relation of SFWs analytically. Studying the propagation of electromagnetic waves in thermonuclear fusion devices filled with gaseous plasma, the following form for the radial dependence of the plasma particle density nearby the boundary of the plasma is often applied:

$$n(r) = (r - R_1)dn/dR_1. \tag{5.13}$$

Here the gradient of the plasma density is assumed to have the following approximate value: $|dn/dr|_{|r=R_1} \sim n(0)/R_1$. The plasma particle density nearby the axis is sufficiently high, so that the inequality $|\varepsilon_1(0)| \gg 1$ is valid. Then, for the plasma core the following inequality $R_1 - r \gg k^2k_0^{-3}$ is satisfied. Therefore, it is possible to apply the solution of Eq. (5.1) in the form of Eqs. (5.9) and (5.10) using the following replacement $\xi \rightarrow k_0(R_1 - r)$, if the following inequalities: $k_0^2 \gg k^2$, $k_0R_1 \gg |m\omega_e|/\omega$, and $k_0R_1 \gg 1$ are satisfied. After that in the peripheral plasma region (nearby the plasma interface $0 \leq R_1 - r \ll k_0^{-1}$), the solution of Eq. (5.1) can be found with the aid of the method of a narrow layer [19]:

$$H_z = H_z(R_1) + H_z(R_1) \int\limits_{R_1}^{r} dr' k_\perp^2 \left[A_d + \int\limits_{R_1}^{r'} dr \frac{H_z(r)}{H_z(R_1)} (1 + q(r)) \right]. \tag{5.14}$$

The constant of integration A_d can be determined via application of the boundary conditions on the plasma-vacuum interface, which are considered in details in Sect. 1.4.1:

$$A_d = \frac{1}{k} \frac{N'_m(kR_1)J'_m(kR_2) - N'_m(kR_2)J'_m(kR_1)}{N'_m(kR_2)J_m(kR_1) - N_m(kR_1)J'_m(kR_2)}. \tag{5.15}$$

In the case of a narrow dielectric layer $R_2 - R_1 \ll R_1$ one can calculate an approximate value of the constant A_d using an expansion procedure over the small parameter $\Delta \ll 1$: $A_d \propto \Delta$.

The expressions for the SFXM fields obtained in the form of (5.9) and (5.10) in the radius region $R_1 - r \gg k^2 k_0^{-3}$ and those obtained in the form (5.14) in the region $k^2 k_0^{-3} \le R_1 - r \ll k_0^{-1}$ can be joined. This allows one to derive a dispersion equation, whose solution in the considered case (plasma density profile (5.13) is linear nearby the plasma boundary) has the following form:

$$\omega \approx \frac{\sqrt{9k_0^4 \Delta m^2 c^2 \left(\Delta R_1^2 + V_1 R_1 k_0^{-1}\right) + V_1^2 m^2 \omega_e^2}}{3k_0^2 \left(\Delta R_1^2 + V_1 R_1 k_0^{-1}\right)} + \frac{V_1 m |\omega_e|}{3k_0^2} \left(\Delta R_1^2 + \frac{V_1 R_1}{k_0}\right)^{-1}, \tag{5.16}$$

here $V_1 = -V(0)/V'(0) \approx 1.37$. From a comparison of the expressions (5.12) and (5.16) one can make the conclusion that in the case of a linear plasma particle density profile (5.13), the obtained expression for the SFXM frequency as function of the plasma parameters is similar to that one, which was obtained for a uniform plasma when doing the replacement of the SFXM penetration depth δ_1 by the parameter k_0^{-1} [20].

In this Section, the frequency ν of collisions between plasma particles has not been taken into the account so far, but under experimental conditions its value can be often not so small. That is why one should take into account the collision frequency ν in the expressions for the components of the plasma dielectric permittivity tensor, which leads to the appearance of SFXM damping. The damping rate of SFXMs caused by plasma particle collisions can be calculated from dispersion relation (5.4), which describes the case of a non-uniform plasma. Making the replacement $\omega \to \omega - i\gamma_c$, the following expression for the collisional damping rate can be obtained:

$$\gamma_c \approx \frac{\nu}{3} \left[1 + \frac{2m|\omega_e|}{3R_1 k_0 \omega}\right] \left[\frac{2m|\omega_e|}{3R_1 k_0 \omega} + \frac{2k_0 m^2 \Delta}{V_1 k^2 R_1}\right]^{-1}. \tag{5.17}$$

Analyzing expression (5.17) one can make the conclusion that the collisional damping rate of SFXMs propagating along a plasma-metal boundary is larger than that one calculated in the case of a plasma-dielectric boundary. It is possible to explain such behavior of the SFXMs by the circumstance that in the first case the SFXM energy propagates only in the plasma region, while in the second case the wave energy propagates as well in the region of the dielectric layer, which separates the plasma column from the waveguide metal wall.

There are some more reasons for SFXM damping in the case of their propagation in non-uniform plasmas. If the SFXM frequency exceeds the electron cyclotron frequency, the resonance $\mathrm{Re}(\varepsilon_1) = 0$ can be located nearby the plasma boundary. This leads to substantial increasing of the SFXM damping. In this case, the SFXM damping rate consists of two terms, namely the collisional and resonant damping rates $\gamma_c + \gamma_r$. To determine the resonant damping rate γ_r, whose appearance is connected with the presence of the resonance, one can take into the account the imaginary part of the component ε_1 of the dielectric permittivity tensor of the plasma filling in the parameter $k_\perp^2$, which enters into the first summand of Eq. (5.1). It will allow one to calculate the value of the resonant damping rate in the following form:

$$\gamma_r \approx \frac{\pi\omega\, m^2 k_0}{\varsigma\, k^2 R_1^2}\left(1 - \frac{m|\omega_e|}{3R_1\omega k_0}\right)\left[V_1 + \frac{\Delta k_0}{k^2 R_1}\left(k^2 R_1^2 + m^2 - \frac{2m^3|\omega_e|}{3R_1\omega k_0}\right)\right]^{-1}, \tag{5.18}$$

where $\varsigma = (d\varepsilon_1/dr)_{|r=R_1} > 0$. Inserting the expression for k_0, one can conclude that the value of the SFW resonant damping rate is proportional: $\gamma_r \propto (dr/d\varepsilon_1)^{2/3}$. In the case of SFWs propagating in a waveguide immersed into a weak external azimuthal magnetic field $\vec{B}_0 = B_{0\varphi}\vec{e}_\varphi$ (see Sect. 4.1) the resonant damping rate depends on the gradient of the plasma density in a stronger way, but there was as well a squared small parameter of that problem ($\beta = |\omega_e|/\omega$), so its value was proportional: $\gamma_r \propto \beta^2(dr/d\varepsilon_1)^{5/3}$ (see expression (4.8) for details).

Expression (5.18) is also valid for the cases of non-magnetized plasmas and SFXM propagation along the plasma-metal interface. Its analysis confirms that resonant damping of SFXMs becomes larger with moving the resonance from the plasma boundary towards its interior and with moving the plasma surface towards the metal wall of the waveguide. This corresponds to the results, which were obtained for other types of SWs [11, 21, 22].

Thus, if the plasma density is characterized by a linear profile nearby the plasma periphery the dispersion properties of SFXMs are similar to those determined for the case of a uniform plasma. The influence of the plasma non-uniformity can be taken into account by replacement of the SFXM penetration depth into the plasma by its effective value. The expressions for the SFXM damping rates correspond to the results obtained for the other SWs. Their values increase with increasing magnitude of the ratio $|m|/R_1$.

Possibilities to apply a numerical method to study the dispersion properties of SFWs using an analytical expression of the radial plasma density profile, which is

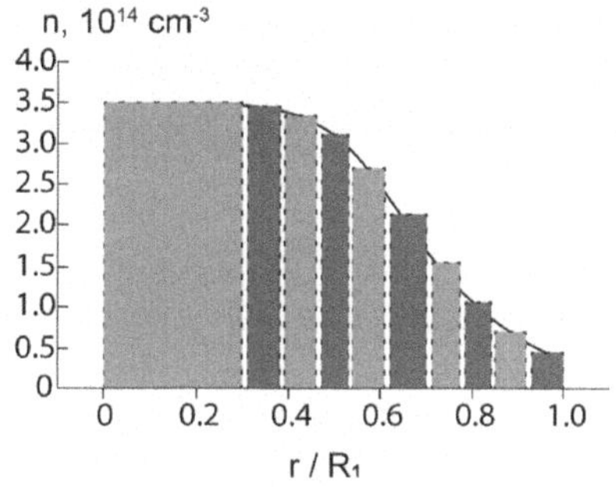

Fig. 5.1 Dependence of the plasma density on the normalized radius of the plasma column (the analytical dependence [23] is drawn by the *solid curve* and the modeled profile is represented by the *dashed lines*)

calculated in [23] for the conditions of a HELIAS reactor, can be illustrated. As far as the magnitude of the applied external static magnetic field in this case is substantially large (20–45 kG), SFXMs can propagate in the LF range ($\omega < |\omega_e|$), moreover their frequency is close to harmonics of the ion cyclotron frequency.

The plasma is considered as a composition of N uniform cylindrical plasma layers, which have different thickness and density values. The inner (first) plasma layer has the density n_1 and is located in the region $0 \le r \le r_1$. The second layer with density $n_2 < n_1$ is located in the region $r_1 < r < r_2$. The other layers with densities: $n_3, n_4, \ldots, n_N$ are located in the following regions, $r_2 < r < r_3$, $r_3 < r < r_4$, ..., $r_{N-1} < r < r_N$, respectively. Real (calculated accordingly to the paper [23]) and modelled plasma density profiles are shown in the Fig. 5.1.

In all the plasma layers, the magnetic component of the SFXMs is described by Eq. (5.1) where $\frac{\partial}{\partial r}\left(\frac{\mu}{k_\perp^2}\right) = 0$ due to uniformity of the plasma particle density within each layer.

Thus, the equation takes the form of a modified Bessel equation. Therefore, the H_z component of the SFXM wave field is described by a superposition of modified Bessel functions and MacDonald functions [18]. However, taking into account that the SFWs fields have finite magnitudes on the axis of the plasma column one can equate the MacDonald functions' multiplier to zero for the first plasma layer with density n_1. The expression for the tangential component of the SFXM electric wave field can be obtained from the set of Maxwell equations (see expression (1.10)).

After calculation of explicit expressions for H_z and E_φ it is easy to find the impedance $Z_{pl}^{(1)}(r_1) = E_\varphi(r_1)/H_z(r_1)$ of the first plasma layer on its interface $r = r_1$:

$$Z_{pl}^{(1)}(r_1) = \frac{iI_m'(k_{\perp 1} r_1)}{\psi_1 I_m(k_{\perp 1} r_1)} + \frac{im\mu_1}{k_{\perp 1} r_1 \psi_1^2}, \tag{5.19}$$

here the prime superscript means the derivative of the corresponding modified Bessel function with respect to its argument and the subscript 1 indicates that the corresponding parameter is calculated just for the first plasma layer, which means: $\mu_1 = \varepsilon_2(n_1)/\varepsilon_1(n_1)$, $\psi_1 = \psi(n_1) = \sqrt{(\mu_1^2 - 1)/\varepsilon_1(n_1)}$, and $k_{\perp 1} = k_\perp(n_1)$. As far as MacDonald and modified Bessel functions have finite values in all other plasma layers, both wave fields H_z and E_φ are expressed by both functions $I_n(z)$ and $K_n(z)$. Thus, the impedances for the rest of the plasma layers can be written in the similar manner:

$$Z_{pl}^{(j)}(r) = \frac{i\left[A_1^{(j)} I_m'\left(k_{\perp j} r\right) + A_2^{(j)} K_m'\left(k_{\perp j} r\right)\right]}{\psi_j\left[A_1^{(j)} I_m\left(k_{\perp j} r\right) + A_2^{(j)} K_m\left(k_{\perp j} r\right)\right]} + \frac{im\mu_j}{k_{\perp j} r \psi_j^2}, \tag{5.20}$$

here $j = 2, 3, \ldots, N$ is the number of the plasma layer.

As it has been mentioned above, the SFXM fields in the vacuum layer are described by Bessel functions of the first kind and Neumann functions. Therefore, in the vacuum region, one can apply Eq. (2.1) which was found in Sect. 2.1 for H_z^{vac}. Then E_φ^{vac} can be expressed in the following way: $E_\varphi^{vac} = dH_z^{vac}/d(ikr)$. There are two boundary conditions for the impedance $Z_v(r) = E_\varphi^{vac}(r)/H_z^{vac}(r)$ of the vacuum layer: first, it is equal to zero on the metal wall $Z_v(R_2) = 0$; second, there is continuity of the plasma impedance and vacuum impedance on the boundary between the plasma column and the vacuum layer $Z_{pl}^{(N)}(R_1) = Z_v(R_1)$. Therefore, one can obtain the following expression for $Z_v(R_1)$:

$$Z_v(R_1) = \frac{-i\left[J_m'(kR_2)N_m'(kR_1) - J_m'(kR_1)N_m'(kR_2)\right]}{J_m(kR_1)N_m'(kR_2) - J_m'(kR_2)N_m(kR_1)}. \tag{5.21}$$

It should be noted that the relative value of the vacuum layer thickness is small, so that the inequality $\Delta = \frac{R_2}{R_1} - 1 \ll 1$ is satisfied here.

After that, one can apply the boundary conditions for $Z_{pl}^{(j)}(r)$ on the interfaces between all plasma layers. This allows one to calculate the impedances for the neighboring plasma layers on the mutual boundaries. Doing this, one can find the following set of $N + 1$ equations for the coefficients B_j, which can be considered as equivalent of the SFXM dispersion equation for the case of a non-uniform plasma density profile [24]:

$$B_1\left\{\psi_1\psi_2 I_m'(r_1k_{\perp 2}) - I_m(r_1k_{\perp 2})\left[\frac{\psi_2^2 I_m'(r_1k_{\perp 1})}{I_m(r_1k_{\perp 1})} + \frac{m\left(\mu_1\psi_2^2 - \mu_2\psi_1^2\right)}{r_1k_{\perp 1}}\right]\right\}$$
$$= K_m(r_1k_{\perp 2})\left[\frac{\psi_2^2 I_m'(r_1k_{\perp 1})}{I_m(r_1k_{\perp 1})} + \frac{m\left(\mu_1\psi_2^2 - \mu_2\psi_1^2\right)}{r_1k_{\perp 1}}\right] - K_m'(r_1k_{\perp 2})\psi_1\psi_2,$$

$$B_N\left\{I_m'(k_{\perp N}R_1) + \psi_N I_m(k_{\perp N}R_1)\left[\frac{J_m'(kR_1)N_m'(kR_2) - J_m'(kR_2)N_m'(kR_1)}{J_m(kR_1)N_m'(kR_2) - J_m'(kR_2)N_m(kR_1)}\right.\right.$$
$$\left.\left. + \frac{m\mu_N}{k_{\perp N}R_1\psi_N}\right]\right\} = -K_m'(k_{\perp N}R_1) - \psi_N K_m(k_{\perp N}R_1)$$

$$\times\left[\frac{J_m^{'}(kR_1)N_m^{'}(kR_2)-J_m^{'}(kR_2)N_m^{'}(kR_1)}{J_m(kR_1)N_m^{'}(kR_2)-J_m^{'}(kR_2)N_m(kR_1)}+\frac{m\mu_N}{k_{\perp N}R_1\psi_N}\right],$$

$$\frac{B_{j-1}I_m^{'}(r_{j-1}k_{\perp j-1})+K_m^{'}(r_{j-1}k_{\perp j-1})}{B_{j-1}I_m(r_{j-1}k_{\perp j-1})+K_m(r_{j-1}k_{\perp j-1})}+\frac{m\left(\mu_{j-1}\psi_j^2-\mu_j\psi_{j-1}^2\right)}{\psi_j^2 r_1 k_{\perp j-1}}$$
$$=\frac{B_jI_m^{'}(r_{j-1}k_{\perp j})+K_m^{'}(r_{j-1}k_{\perp j})}{B_jI_m(r_{j-1}k_{\perp j})+K_m(r_{j-1}k_{\perp j})}\frac{\psi_{j-1}}{\psi_{j+1}}, \tag{5.22}$$

where the subscript j in the last equation of the set (5.22) changes in the limits from 2 to N. Thus, the last equation really represents $N-1$ equations obtained from equating the impedances of the neighboring plasma layers beginning from the impedance of the second layer.

The results of a numerical analysis of the set of Eq. (5.22) are plotted in Figs. 5.2, 5.3 and 5.4. The value of the SFXM frequency decreases with increasing plasma radius more strongly as compared with that one calculated for a uniform plasma (see Fig. 5.2). Increasing the vacuum layer thickness leads to decreasing SFXM frequencies, under the same other conditions. The results of this numerical analysis prove that their frequency decreases with increasing strength of the confining magnetic field (see Fig. 5.3).

The decrease of the SFXM frequency with increasing plasma column radius is found to happen faster in the case of a non-uniform plasma density profile than

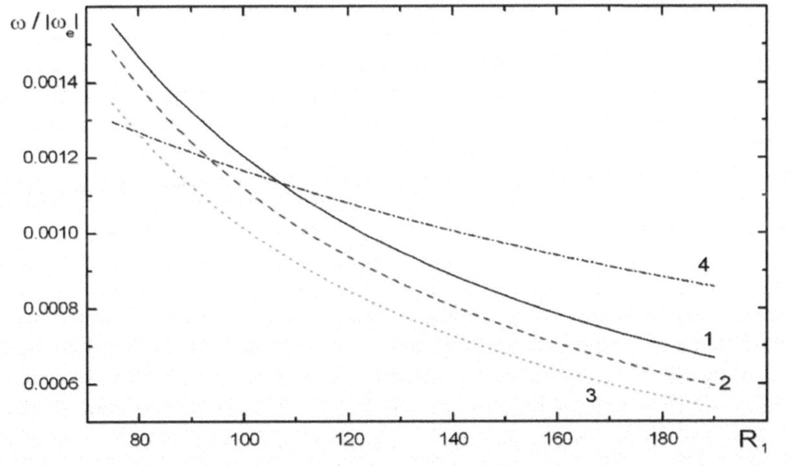

Fig. 5.2 Normalized frequency of the SFXMs versus plasma column radius (in cm) for $m = 3$, $B_0 = 30$ kG. The *numbers* 1, 2 and 3 relate to the cases $\Delta = 0.05$; 0.25 and 0.45, the *curve* marked by 4 relates to the case of a uniform plasma, $\Delta = 0.25$

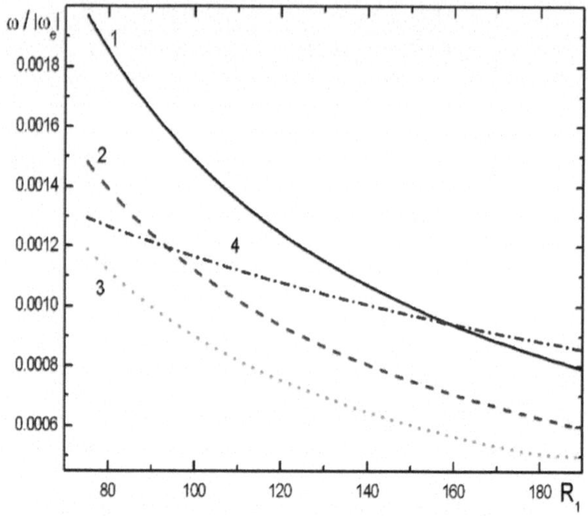

Fig. 5.3 Normalized frequency of the SFXMs versus plasma column radius (in cm) for $m = 3$, $\Delta = 0.25$. The *numbers* 1, 2 and 3 relate to the cases $B_0 = 30$; 40 and 50 kG, respectively. The curve marked by 4 relates to the case of a uniform plasma with $B_0 = 40$ kG

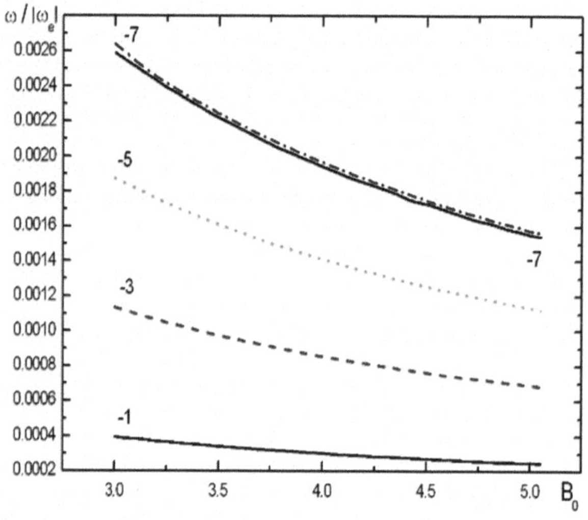

Fig. 5.4 Normalized frequency of the SFXMs versus confining magnetic field (in *Tesla*) for $R_1 = 130$ cm, $\Delta = 0.25$. The *numbers* indicate the values of the azimuthal mode index *m*. The *solid line* marked by -7 relates to the case of a uniform plasma

for a uniform plasma. Therefore, there is a crossing point of the dispersion curves related to the cases of uniform and non-uniform plasmas. The results of a numerical analysis of the SFXM frequency dependence on the azimuthal mode index m confirm the conclusion obtained analytically (see Fig. 5.4). It is confirmed numerically that there are two possible propagation directions of SFXMs with positive and negative values of the azimuthal mode index (see Chap. 2). It is found that $\omega(+m) < \omega(-m)$, under the same conditions, but the relative difference between these frequencies is small. For instance, for a uniform plasma, the order of magnitude of this relative difference is $\sim 10^{-2}$ and in the case of a non-uniform plasma, it is $\sim 10^{-1}$. Features of SFXMs with opposite values of the azimuthal mode indices are similar to each other. Therefore, the figures related to the case of $m < 0$ are not presented.

Analyzing the results shown in Fig. 5.4, one can see that non-uniformity of the plasma density leads to some increase of the SFXM frequency as compared with the case of uniform strongly magnetized plasma. Thus, numerical investigations confirm the conclusions of the analytical studies on the increase of the SFXM frequency with decreasing values of confining magnetic field, plasma radius and thickness of the vacuum layer.

Modelling of the radial non-uniformity of the plasma density by a set of radial layers applied here faces the following problem. Introducing additional interfaces between plasmas of different densities unavoidably results in arising additional roots in the dispersion relation (5.22). To distinguish which root has physical sense, the following trick is applied. One has to start from the uniform profile with the density n_1, and calculate the eigenfrequency. Then the density profile should be gradually modified to the desired shape with keeping the root, which is most close to that obtained in the previous step.

Such a numerical problem is well-known also in other fields of plasma electrodynamics. In particular, it is common to use discontinuous plasma–vacuum interfaces as a simple way of avoiding issues associated with the lower hybrid resonance in numerical calculations in the range of ion cyclotron resonance frequency (ICRF). Non-physical surface waves are indicated in [25] as a side effect of this common trick. These waves not only introduce non-physical roots of the dispersion relation but also prevent convergence. Introducing losses can certainly help in suppressing such surface-wave excitation. However, the effects of such interventions remain an open question.

5.2 Influence of Radial Non-uniformity of External Static Magnetic Field on Frequency Spectrum of Surface Flute Waves

Plasmas are confined in various magnetic traps with the help of an external static magnetic field $\vec{B}_0(r)$ with non-uniform radial profile. As a rule, this non-uniformity is relatively weak. That is why in the present Section just the case of a weak radial non-uniformity of the magnetic field and its influence on the SFXM frequency spectrum are considered. As far as the influence of the non-uniformity of the plasma density on SFXMs properties has been examined in the previous Section, here the plasma particle density radial profile is assumed to be uniform.

It is interesting that in this case the magnetic component of the extraordinary polarized SFXM field is also described by Eq. (5.1). That is why the latter can be solved using the method of etalon equations [17]. Applying this method one can find that in the plasma region, the axial magnetic wave field of SFXMs is described by the following expression:

$$H_z = A(r) I_m(q). \tag{5.23}$$

In (5.23), $A(r) = \text{Const}\sqrt{qk_\perp^2/(rq')}$, $q = \int_0^r k_\perp \sqrt{1-(m/r)d\left(\mu/k_\perp^2\right)/dr} dr$, and $q' = dq/dr$. This solution is valid if condition (5.3) is satisfied. Substituting the expression (5.23) into Eq. (5.1) one can calculate the correction ω_1 to the SFXM frequency, which is connected with radial non-uniformity of an external stationary magnetic field $\vec{B}_0$ in the limiting case of a wide plasma waveguide $R_1 \gg |m|\delta$ or a large azimuthal mode index $|m| \gg 1$:

$$\omega_1 \approx \frac{-m\xi\omega^2}{2\sqrt{m^2 + R_1^2/\delta}} \left[m\omega + 2\Delta \frac{m^2 \Omega_e^2}{|\omega_e(R_1)|} \right]^{-1}. \tag{5.24}$$

Here the characteristic parameter of the radial non-uniformity of the external magnetic field ξ is determined by the expression $\xi = r d \ln B_0/dr$, which is calculated at the radius $r = R_1$, $\delta = c/\Omega_e$. Under the considered conditions, the following inequality for the parameter of the magnetic field non-uniformity is valid $\xi \ll R_1^2\omega/\left|m\omega_e(R_1)\delta^2\right|$. Expression (5.24) can be applied as well for obtaining the frequency corrections ω_1 of these modes in other limiting cases, namely narrow plasma waveguide $k_\perp R_1 \ll 1$ or small azimuthal mode index $|m| \ll k_\perp R_1$.

If the thickness of the dielectric layer between plasma and wall is sufficiently small, the inequality: $\Delta < |\omega_e(R_1)|\, \omega/\left|m\Omega_e^2\right|$ is valid and the SFXMs become unidirectional waves (in this case the azimuthal mode index can be only positive: $m > 0$). In an external static magnetic field, whose strength increases along the radial direction to the plasma boundary, SFXM frequency is less than in the case of a uniform magnetic field. This conclusion agrees with the results of Ref. [17] concerning the SFXM frequency as function of the magnitude of the applied magnetic field B_0. If the thickness of the dielectric layer, which separates the confined plasma column from the metal wall of the waveguide, is so large that the following inequality $\Delta > |\omega_e(R_1)|\omega/\left|m\Omega_e^2\right|$ is valid, the influence of an external static magnetic field is more pronounced for SFXMs with negative sign of the azimuthal mode index $m < 0$ as compared with the case of positive sign. This can be explained as demonstration of the phenomenon of non-reciprocity for SFXMs (this means that in the case of propagation of these waves in opposite directions their frequencies are different, under the same conditions).

Summarizing the results presented above, one can make the following conclusions. First, radial non-uniformity of an external magnetic field affects much stronger those SFXMs, which propagate along the direction of electron gyration in the magnetic field B_0. Second, for fixed magnitude of the characteristic parameter $\xi = r d \ln B_0/dr$of the radial non-uniformity of the external magnetic field, the correction of the SFXM frequency decreases with increasing thickness of the dielectric layer, as well as with increasing absolute value of azimuthal mode index, plasma radius, plasma particle density in the waveguide and strength of the applied magnetic field $B_0(R_1)$.

5.3 Influence of Toroidal Magnetic Field Non-uniformity on Frequency Spectrum of Surface Flute Waves in Metal Waveguides Entirely Filled with Plasmas

In a gaseous plasma, which is confined in a toroidal experimental device, the influence of the toroidicity of the applied magnetic field on the transverse electromagnetic eigenwaves is revealed, first, via a spatial non-uniformity of the external static toroidal magnetic field; second, via a toroidal shift of the magnetic surfaces in these experimental devices; and third, via deviation of the shape of the poloidal cross-sections of the magnetic surfaces from circular cross-section. The influence of the second and third factors on the dispersion properties of the electromagnetic eigenwaves, which propagate across the external static uniform axial magnetic field along a small azimuthal angle, is particularly examined in the next chapter [26].

The present Section is devoted to the study of the influence of the first feature mentioned above on the properties of SFXMs, which propagate in a toroidal waveguide entirely filled with plasma. Doing that the method of successive approximations is applied. In the framework of this theory, the SFXM propagation in a uniform magnetic field is applied as zeroth approximation. As it is shown in the previous section, an external static magnetic field very strongly affects the SFXM dispersion properties in the case of the absence of a dielectric layer, which separates the plasma column from the metal wall of the utilized waveguide. That is why here just the case of a toroidal metal waveguide with circular cross-section and radius R_1, which is entirely filled with cold plasma is examined.

To solve the problem, one has to choose the suitable co-ordinate system. In the right-hand quasi-toroidal co-ordinate system (this means that the vector product of $\vec{e}_r$ with $\vec{e}_\vartheta$ is equal to $\vec{e}_\zeta$), the chamber is assumed to be symmetric along the circular axis, so that $\partial/\partial\zeta = 0$. In this case the poloidal angle ϑ is counted from the direction of the center of symmetry of the torus (see Fig. 5.5, where this co-ordinate system is also shown). It is assumed as well that the plasma filling is uniform and the applied external static toroidal magnetic field is described by the following expression:

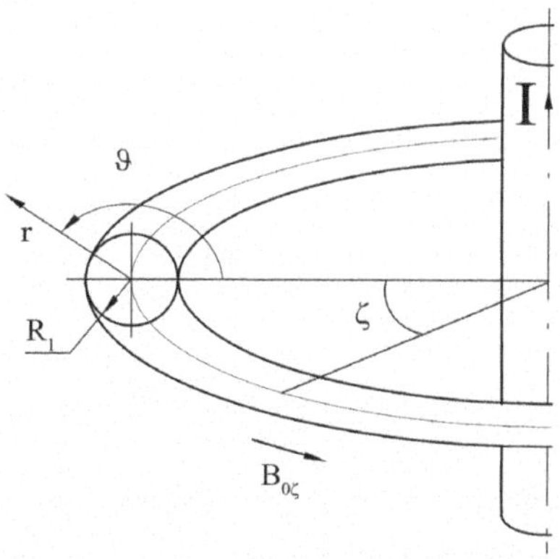

Fig. 5.5 Scheme of the waveguide structure consisting of a metal torus filled by magneto-active plasma. The toroidal (azimuthal) magnetic field is generated by the central internal, perpendicular current *I*

$$B_{0\zeta} = B_0/[1 - (r/R)\cos\vartheta], \tag{5.25}$$

here R is the major radius of the torus. Extraordinary polarized SFW eigenwaves with the components of the electromagnetic field E_r, E_ϑ, H_ζ can propagate in such toroidal waveguides. In electrodynamics of bounded plasma systems, propagation of SWs across an external stationary magnetic field oriented parallel to the plasma-metal interface is entitled wave propagation in Voigt geometry [12, 22].

In the framework of magnetic hydrodynamics, the vectors of electric induction and electric field are connected by the tensor ε_{ij} of the dielectric permittivity of the cold magneto-active plasma without taking into account collisions between plasma particles. In the case of small toroidicity (the parameter of toroidicity $\varepsilon_t = R_1/R \ll 1$) one can represent expressions for the components of the plasma permittivity tensor ε_{ij} as a series over the small parameter of toroidicity:

$$\varepsilon_{1,2} = \varepsilon_{1,2}^{(0)} + \varepsilon_{1,2}^{(1)}\cos\vartheta + \varepsilon_{1,2}^{(2)}. \tag{5.26}$$

The main summands in the series (5.26) do not depend on the applied co-ordinates, therefore, $\varepsilon_{1,2}^{(0)} = \varepsilon_{1,2}\left(B_{0\zeta} = B_0\right)$. The summands of the first order of smallness over the minor radius of the torus r are as follows:

$$\varepsilon_1^{(1)} = -\frac{2r}{R}\sum_\alpha \frac{\Omega_\alpha^2 \omega_\alpha^{(0)2}}{\left(\omega^2 - \omega_\alpha^{(0)2}\right)^2}, \quad \varepsilon_2^{(1)} = -\frac{r}{R}\sum_\alpha \frac{\omega_\alpha^{(0)}\Omega_\alpha^2\left(\omega^2 + \omega_\alpha^{(0)2}\right)}{\omega\left(\omega^2 - \omega_\alpha^{(0)2}\right)^2}, \tag{5.27}$$

here the sub-script α designates the type of plasma particle ($\alpha = i$ for ions, and $\alpha = e$ for electrons), Ω_α and ω_α are the plasma and cyclotron frequencies of the α-type particles, respectively. The second order corrections are quadratic with respect to the minor radius r of the torus. In the series (5.26) the summands of the second order of smallness, which are proportional to $\exp(2i\vartheta)$ are not taken into account because they make a contribution to the correction of the SFXM frequency whose order of smallness is higher than second order. In zeroth approximation, the cyclotron frequencies ω_α in the expressions (5.27) are determined by the strength of the toroidal magnetic field (5.25): $\omega_\alpha^{(0)} = \omega_\alpha(\varepsilon_t = 0)$.

The toroidal component of the SFXM magnetic wave field can be calculated from the following equation, which one can derive from the set of Maxwell equations:

$$\begin{aligned}\frac{1}{r}\frac{\partial}{\partial r}\left(\frac{r\varepsilon_1}{\varepsilon_\perp}\frac{\partial H_\zeta}{\partial r}\right) + \frac{i}{r}\frac{\partial}{\partial\vartheta}\left(\frac{\varepsilon_2}{\varepsilon_\perp}\right)\frac{\partial H_\zeta}{\partial r} + \frac{\omega^2}{c^2}H_\zeta \\ + \frac{1}{r^2}\frac{\partial}{\partial\vartheta}\left(\frac{\varepsilon_1}{\varepsilon_2}\frac{\partial H_\zeta}{\partial\vartheta}\right) - \frac{i}{r}\frac{\partial}{\partial r}\left(\frac{\varepsilon_2}{\varepsilon_\perp}\right)\frac{\partial H_\zeta}{\partial\vartheta} = 0,\end{aligned} \tag{5.28}$$

where $\varepsilon_\perp = \varepsilon_1^2 - \varepsilon_2^2$. In zeroth approximation, Eq. (5.28) describes the independent propagation of SFXMs with different values of the azimuthal mode index. However, in the case of toroidal non-uniformity of an external stationary magnetic field one has to take into consideration satellite harmonics of the SFXM field. Taking into account the symmetry of the problem (see Eqs. (5.26), (5.27)) the solution of Eq. (5.28) can be represented in the form of the following wave packet:

$$H_\zeta = \left[H_\zeta^{(0)}(r) + H_\zeta^{(2)}(r) + H_\zeta^{(+1)}(r)\, e^{i\vartheta} + H_\zeta^{(-1)}(r)\, e^{-i\vartheta}\right]\exp(im\vartheta - i\omega t). \tag{5.29}$$

The components of the SFXM electric wave fields can be written in a form which is similar to expression (5.29), where the next (with respect to the main harmonic $\propto \exp(im\vartheta)$) two satellite harmonics $\propto \exp[i(m \pm 1)\vartheta]$ are taken into account. The field correction $H_\zeta^{(2)}$, which is determined by the waveguide toroidicity, appears to be a summand of the second order smallness.

For further analysis, it is necessary to determine the relationship between the poloidal component of the electric wave field E_ϑ and its magnetic component H_ζ:

$$E_\vartheta^{(0)}(r) = \frac{c}{i\omega\varepsilon_\perp^{(0)}}\left[\varepsilon_1^{(0)}\frac{dH_\zeta^{(0)}}{dr} + \frac{m\varepsilon_2^{(0)}}{r}H_\zeta^{(0)}\right], \tag{5.30}$$

$$\begin{aligned} E_\vartheta^{(\pm 1)}(r) = \frac{c}{i\omega\varepsilon_\perp^{(0)}}\Bigg[&\varepsilon_1^{(0)}\frac{\partial H_\zeta^{(\pm 1)}}{\partial r} + \frac{m \pm 1}{r}\varepsilon_2^{(0)}H_\zeta^{(\pm 1)} + \frac{\varepsilon_\perp^{(0)}\varepsilon_1^{(1)} - \varepsilon_\perp^{(1)}\varepsilon_1^{(0)}}{2\varepsilon_\perp^{(0)}}\frac{\partial H_\zeta^{(0)}}{\partial r} \\ &+ \frac{m\left(\varepsilon_\perp^{(0)}\varepsilon_2^{(1)} - \varepsilon_\perp^{(1)}\varepsilon_2^{(0)}\right)}{2r\varepsilon_\perp^{(0)}}H_\zeta^{(0)}\Bigg], \end{aligned} \tag{5.31}$$

$$\begin{aligned} E_\vartheta^{(2)}(r) = \frac{1}{\varepsilon_\perp^{(0)}}\Bigg\{&-\frac{\varepsilon_\perp^{(2)}}{2}E_\vartheta^{(0)} - \frac{\varepsilon_\perp^{(1)}}{2}\left(E_\vartheta^{(+1)} + E_\vartheta^{(-1)}\right) + \frac{c\varepsilon_1^{(1)}}{2i\omega}\frac{\partial}{\partial r}\left(H_\zeta^{(+1)} + H_\zeta^{(-1)}\right) \\ &+ \frac{c\varepsilon_1^{(0)}}{i\omega}\frac{\partial H_\zeta^{(2)}}{\partial r} + \frac{c\varepsilon_1^{(2)}}{2i\omega}\frac{\partial H_\zeta^{(0)}}{\partial r} + \frac{cm\varepsilon_2^{(0)}}{i\omega r}H_\zeta^{(2)} + \frac{cm\varepsilon_2^{(2)}}{2i\omega r}H_\zeta^{(0)} \\ &+ \frac{cm\varepsilon_2^{(1)}}{2i\omega r}\left[(m+1)H_\zeta^{(+1)} + (m-1)H_\zeta^{(-1)}\right]\Bigg\}. \end{aligned} \tag{5.32}$$

Here

$$\begin{aligned} &\varepsilon_\perp^{(0)} = \varepsilon_1^{(0)2} - \varepsilon_2^{(0)2}, \quad \varepsilon_\perp^{(1)} = 2\varepsilon_1^{(0)}\varepsilon_1^{(1)} - 2\varepsilon_2^{(0)}\varepsilon_2^{(1)}, \\ &\varepsilon_\perp^{(2)} = \varepsilon_1^{(1)2} + 2\varepsilon_1^{(0)}\varepsilon_1^{(2)} - \varepsilon_2^{(1)2} - 2\varepsilon_2^{(0)}\varepsilon_2^{(2)}. \end{aligned} \tag{5.33}$$

To derive the dispersion relation for SFXMs propagating in a toroidal chamber, which is entirely filled with magneto-active plasma, one can apply the standard boundary conditions (see Sect. 1.4.1 for details).

In zeroth approximation one can employ the theory of SFXMs propagating along the azimuthal direction in a cylindrical metal waveguide with inner radius R_1, which is entirely filled with uniform plasma (see Sect. 2.2). The amplitudes of the satellite harmonics can be found as solutions of the considered problem in first approximation ($H_\zeta^{(\pm 1)}(r)$ are assumed to be parameters of the first order of smallness). They can be substituted into Eq. (5.28), taking into account the expressions (5.26). Then summands of the first order of smallness appear there with a factor $\propto \exp[i(m \pm 1)\vartheta]$. That is why accounting for summands of the first order of smallness does not change the amplitudes of the main harmonic and consequently the solution of the considered problem in first order approximation does not determine the SFXM frequency correction. These summands form non-uniform Bessel equations for the satellite harmonics $H_\zeta^{(\pm j)}(r)$ with the known right-hand parts. These equations have the following solutions, which satisfy the boundary condition that the wave fields should have a finite value on the waveguide axis:

$$H_\xi^{(\pm 1)}(r) = -I_{m\pm 1}(k_\perp r) \int_{R_1}^{r} K_{m\pm 1}(k_\perp r) \hat{L}_\mp^{(1)} I_m(k_\perp r) r dr + C_\pm I_{m\pm 1}(k_\perp r)$$
$$+ K_{m\pm 1}(k_\perp r) \int_0^r I_{m\pm 1}(k_\perp r) \hat{L}_\mp^{(1)} I_m(k_\perp r) r dr, \tag{5.34}$$

where the operator $\hat{L}_\pm^{(1)}$ has the following form:

$$\hat{L}_\pm^{(1)} = \left(2\varepsilon_1^{(0)}\varepsilon_\perp^{(0)}\right)^{-1} \left\{ \frac{\varepsilon_\perp^{(1)}}{r} \frac{\partial}{\partial r}\left(r\varepsilon_1^{(0)} \frac{\partial}{\partial r}\right) + \frac{\varepsilon_\perp^{(0)}}{r} \frac{\partial}{\partial r}\left(r\varepsilon_1^{(1)} \frac{\partial}{\partial r}\right) \right.$$
$$- \varepsilon_1^{(0)} \frac{\partial \varepsilon_\perp^{(1)}}{\partial r} \frac{\partial}{\partial r} - \frac{\varepsilon_\perp^{(0)}}{r} \frac{\partial \varepsilon_2^{(1)}}{\partial r} \hat{\Theta} + \frac{\varepsilon_2^{(0)}}{r} \frac{\partial \varepsilon_\perp^{(1)}}{\partial r} \hat{\Theta} - \frac{\varepsilon_\perp^{(1)}}{r^2} \varepsilon_1^{(0)} \hat{\Theta}^2$$
$$- \frac{\varepsilon_\perp^{(0)} \varepsilon_1^{(1)}}{r^2} \hat{\Theta}^2 + 2\frac{\omega^2}{c^2} \varepsilon_\perp^{(0)} \varepsilon_\perp^{(1)} \mp \frac{\varepsilon_\perp^{(0)} \varepsilon_\perp^{(1)}}{r^2} \hat{\Theta} \pm \frac{\varepsilon_1^{(0)} \varepsilon_\perp^{(1)}}{r^2} \hat{\Theta}$$
$$\left. \pm \frac{\varepsilon_2^{(1)} \varepsilon_\perp^{(0)}}{r} \frac{\partial}{\partial r} \mp \frac{\varepsilon_2^{(0)} \varepsilon_\perp^{(1)}}{r} \frac{\partial}{\partial r} \right\}, \quad \hat{\Theta} I_m = -m I_m. \tag{5.35}$$

The integration constants $C_\pm$ can be found from the boundary condition that the amplitudes of the first satellite harmonics of the poloidal electric wave field vanish at the metal surface of the waveguide, $E_\vartheta^{(\pm 1)}(R_1) = 0$:

$$C_{\pm} = -\left[\varepsilon_1^{(0)} k_\perp K'_{m\pm1} \int_0^{R_1} I_{m\pm1} \hat{L}_{\mp}^{(1)} I_m r dr + \frac{m \pm 1}{R_1} \varepsilon_2^{(0)} K_{m\pm1} \int_0^{R_1} I_{m\pm1} \hat{L}_{\mp}^{(1)} I_m r dr \right.$$
$$\left. + \frac{\varepsilon_\perp^{(0)} \varepsilon_1^{(1)} - \varepsilon_\perp^{(1)} \varepsilon_1^{(0)}}{2\varepsilon_\perp^{(0)}} k_\perp I'_m + \frac{m\left(\varepsilon_\perp^{(0)} \varepsilon_2^{(1)} - \varepsilon_\perp^{(1)} \varepsilon_2^{(0)}\right)}{2R_1 \varepsilon_\perp^{(0)}} I_m \right]$$
$$\times \left[\varepsilon_1^{(0)} k_\perp I'_{m\pm1} + \frac{m \pm 1}{R_1} \varepsilon_2^{(0)} I_{m\pm1}\right]^{-1} \Bigg|_{\substack{r = R_1 \\ \omega = \omega_m^{(0)}}} . \tag{5.36}$$

The correction of the SFXM frequency caused by the toroidicity of the applied static magnetic field can be calculated in second order approximation. The summands of second order smallness in Eq. (5.29) are proportional to one of the following phase multipliers: either $\exp(im\vartheta)$ or $\exp[i(m \pm 2)\vartheta]$. The summands, which have the factor $\exp[i(m \pm 2)\vartheta]$ arrange the equation for the amplitudes of the second satellite harmonics. Thus, these summands in series (5.29) can be neglected because their contributions to the SFXM frequency corrections have values of such order of smallness, which are higher than the second order. The summands of the second order of smallness, which are proportional to the factor $\exp(im\vartheta)$ lead to the non-uniform differential Bessel equation for the correction $H_\zeta^{(2)}(r)$ to the amplitude of the basic harmonic. A solution of this equation, which satisfies the boundary condition of finite magnitude of the wave field on the circular axis of the torus, can be found using the method of constant variation:

$$H_\xi^{(2)} = K_m(k_\perp r) \int_0^r I_m \left(\hat{L}_-^{(1)} H_\xi^{(-1)} + \hat{L}_+^{(1)} H_\xi^{(+1)} + \hat{L}^{(2)} H_\xi^{(0)}\right) r dr$$
$$- I_m(k_\perp r) \int_a^r K_m \left(\hat{L}_-^{(1)} H_\xi^{(-1)} + \hat{L}_+^{(1)} H_\xi^{(+1)} + \hat{L}^{(2)} H_\xi^{(0)}\right) r dr + C_2 I_m(k_\perp r). \tag{5.37}$$

The solution (5.37) is the sum of the particular solution of the inhomogeneous differential equation (the first two integral terms) and the general solution of the corresponding homogeneous differential equation (the third term $C_2 I_m(k_\perp r)$). In the general expression (5.29) for the axial component of the magnetic wave field, the term $C_2 I_m(k_\perp r)$ and the amplitude $H_\zeta^{(0)}(r)$ are similar terms. Therefore, the constant of integration C_2 plays the role of the correction to the normalized multiplier, which in zeroth order approximation was put to be unit. That is why the constant C_2 cannot be found from the above-mentioned boundary conditions, but with the aid of the following condition, which is similar to the condition of normalizing the wave function in quantum mechanics [27]:

$$\int_0^{R_1}\left[2H_\zeta^{(2)}I_m(k_\perp r)+\left(H_\zeta^{(+1)}\right)^2+\left(H_\zeta^{(-1)}\right)^2\right]rdr=0. \tag{5.38}$$

Equation (5.38) expresses the following physical condition for the value of the wave field energy: the value of the magnetic field energy calculated taking into account the summands of second order of smallness coincides with the value, which is calculated in zeroth order approximation. At the same time, it can be underlined that the type of the dispersion equation, which is derived by taking into account the summands of second order of smallness does not depend on the value of the integration constant C_2.

Application of the boundary condition on the metal wall of the considered toroidal waveguide for the basic harmonic of the poloidal component of these electric wave fields and taking into account the summands of second order of smallness lead to deriving the dispersion equation in the form $D^{(0)}+D^{(2)}=0$, where $D^{(0)}$ describes SFXMs in the case of uniform stationary magnetic field (see Sect. 2.2) and the second term is the correction $D^{(2)}\propto E_\vartheta^{(2)}(R_1)$, which is determined by the toroidicity of the applied magnetic field $\vec{B}_{0\zeta}$. That is why the correction to the eigenfrequency $\omega_m^{(0)}$ of these SFWs can be calculated by expanding $D^{(0)}$ in Taylor series. So, the value of this correction shows to be of the second order of smallness: $\omega_m^{(2)}\propto\varepsilon_t^2$. The condition of applicability of the applied method for determination of the dispersion properties of SFXMs, which propagate in toroidal metal waveguides, is given by the validity of the inequality $\left|\omega_m^{(2)}\right|\ll\omega_m^{(0)}$.

Therefore, the basic mathematical achievement of this result can be formulated in the following way: the problem of two-dimensional non-uniformity of an external stationary magnetic field over the waveguide poloidal cross-section has been reduced to a one-dimensional non-uniformity (this means that finally one has only a radial distribution of the considered external magnetic field). However, of course the obtained solutions have a complicated form.

In the limiting case of wide waveguides $R\gg R_1\gg|m|k_\perp^{-1}$, the frequency correction of SFXMs with an eigenfrequency given by Eq. (2.12) can be represented in the LF range by the following asymptotic expression [28]:

$$\omega_m^{(2)}=-0.25\varepsilon_t^2\omega_m^{(0)}. \tag{5.39}$$

In the HF range the eigenfrequency of SFXMs is given by Eq. (2.13) and their frequency correction can be written with the same accuracy in the following way:

$$\omega_m^{(2)}=\frac{\varepsilon_t^2}{4}\frac{c^2m^2}{\omega_m^{(0)2}R_1^2}\left(1+\frac{3|\omega_e|k_\perp R_1}{\omega_m^{(0)}|m|}\right)\omega_m^{(0)}. \tag{5.40}$$

For SFXMs which propagate in narrow $|m|\gg R_1k_\perp$ toroidal waveguides in the LF range one can derive the following asymptotic expression for the frequency correction:

$$\omega_m^{(2)} = \frac{2m^2 - m + 1}{4m^2(m+1)} \varepsilon_t^2 \omega_m^{(0)}. \tag{5.41}$$

In this limiting case, the frequency correction for HF SFXMs is approximately equal to:

$$\omega_m^{(2)} = \frac{2m^2 + m + 1}{2m^2} \varepsilon_t^2 \left(\frac{m\delta}{R_1}\right)^4 \frac{\Omega_e}{|\omega_e|} \omega_m^{(0)}. \tag{5.42}$$

It should be mentioned that the spatial distributions of the satellite harmonics in these two limiting cases are substantially different, for instance in the limiting case of a narrow waveguide: $H_\zeta^{(\pm 1)}(r) \propto r^{m\pm 1}$. However, their values on the metal waveguide wall are approximately equal to each other.

To carry out a comparative analysis of the obtained results one can apply asymptotical expressions for the correction $\omega_{el}^{(2)}$ to the eigenfrequency of SFXMs which is due to an ellipticity of the transverse cross-section of the metal waveguide, completely filled with plasma. For SFXMs, which propagate in wide waveguides in the LF range, the frequency correction $\omega_{el}^{(2)}$ is as follows:

$$\omega_{el}^{(2)} = -0.25\varepsilon_{el}^2 (4 - m^2 + 2R_1/\delta)\omega_m^{(0)}. \tag{5.43}$$

In this case the shape of the cross-section of the chamber is determined by the ellipticity parameter ε_{el} in the following way: $r(\vartheta) = R_1(1 + \varepsilon_{el}\cos(2\vartheta - \pi))$. In the limiting case of a narrow waveguide, the frequency correction is given by:

$$\omega_{el}^{(2)} = -\frac{\varepsilon_{el}^2}{2m} \frac{2 - 4m^2 - 4m^2}{2 - 4m^2} \frac{R_1^2}{\delta^2} \omega_m^{(0)}. \tag{5.44}$$

Thus, after comparing expressions (5.41) and (5.42) with expressions (5.43) and (5.44), respectively, one can make the following conclusions: first, the influence of a toroidal non-uniformity of the external magnetic field on the spectra of transversal SWs is stronger than the influence of an ellipticity of the waveguide cross-section in the case of low plasma density; second, comparison of the obtained results on the influence of the toroidal change of the magnetic surfaces is redundant, because in experimental practice this change is too small nearby the surface of the metal chamber, where the considered waves are localized [28].

Consequently, the toroidal non-uniformity of the external static magnetic field leads to the fact that SFXMs propagate in toroidal waveguides in the form of wave packets. The amplitudes A_{m+N} of the satellite harmonics in such packets are very small compared with the amplitude of the basic harmonic A_m so that $A_{m\pm N} \sim \varepsilon_t^N A_m$. The influence of the toroidal non-uniformity of the external stationary magnetic field on the amplitude of the basic harmonic shows up in the second order of smallness over the parameter of toroidicity ε_t, it can be stronger than the effect of non-circularity of the fusion chamber cross-section under the determined conditions. The presented results can be useful for practical application in the branch of plasma electronics, and also for interpretation of experimental results concerning phenomena in the peripheral plasma of toroidal fusion devices with magnetic confinement.

5.4 Propagation of Surface Flute Waves Around a Metal Ring in Non-uniform Toroidal Magnetic Fields

In the present Section, extraordinary polarized electromagnetic SFWs with the field components E_r, E_ϑ, H_ζ propagating across a static toroidal magnetic field around a metal ring is studied within the framework of the method of successive approximations. The scheme of the plasma-waveguide structure is similar to that shown in Fig. 5.5. The minor radius of the metal ring with ideal conductivity is assumed to be small compared with the major radius of the ring, so that the inequality $\varepsilon_t = R_1/R \ll 1$ is valid. The constant toroidal magnetic field $\vec{B}_0 = B_{0\zeta}\vec{e}_\zeta$ is created by a conductor with a DC current, which is located at the center-line of the ring, perpendicularly to the plane of symmetry of the ring, and the explicit expression for $B_{0\zeta}$ is given by formula (5.25). The plasma, which in the present Section surrounds this metal ring, is assumed to be cold and uniform [29]. All other assumptions, which have been done in Sect. 5.3, are valid in this Section as well.

The components of the dielectric permittivity tensor ε_{ik} can be written in the form of series (5.26) over the small parameter of toroidicity ε_t. In the present case, solution for the toroidal $H_\zeta(r)$ component of the magnetic wave field can be obtained from Eq. (5.28) in the form (5.29). The solution of Eq. (5.28) should satisfy the boundary conditions, which are particularly discussed in Sect. 1.4.1. In this case the amplitude $H_\zeta^{(0)}(r)$ of the basic harmonic of the considered SFWs in zeroth order approximation can be expressed in terms of MacDonald function [18] $K_m(k_\perp r)$.

The amplitudes $H_\zeta^{(\pm 1)}(r)$ of the satellite harmonics can be determined by a method, which is similar to the method applied in the previous section, because of the similarity of these problems. Thus, solving the present problem in the first order approximation does not change the amplitude of the main harmonic. Consequently, the toroidal variations of the external static magnetic field do not contribute to the SFW eigenfrequency in the first order approximation. These first-order terms are summands of the inhomogeneous Bessel equations for $H_\zeta^{(\pm 1)}(r)$ with known right-hand side. The noted equations have the following solutions (compare with (5.34)), which satisfy the boundary condition for the SFXM wave fields at infinite distances from the metal ring [29]:

$$H_\zeta^{(\pm 1)}(r) = K_{m\pm 1}(k_\perp r) \int_{R_1}^{r} r I_{m\pm 1}(k_\perp r) \hat{L}_\mp^{(1)} K_m(k_\perp r) dr$$
$$+ C_\pm K_{m\pm 1}(k_\perp r) - I_{m\pm 1}(k_\perp r) \int_{\infty}^{r} r K_{m\pm 1}(k_\perp r) \hat{L}_\mp^{(1)} K_m(k_\perp r) dr, \quad (5.45)$$

where the operator $\hat{L}_{\pm}^{(1)}$ was defined by Eq. (5.35). The integration constants $C_{\pm}$ can be found from the boundary condition that the amplitudes of the first satellite harmonics of the poloidal electric field vanish on the metal surface of the ring, $E_{\vartheta}^{(\pm 1)}(R_1) = 0$ (compare with Eq. (5.36)):

$$
\begin{aligned}
C_{\pm} = & \left[\frac{m+1}{R_1} \varepsilon_2^{(0)} K_{m\pm 1}(k_{\perp} R_1) - \varepsilon_1^{(0)} k_{\perp} K'_{m\pm 1}(k_{\perp} R_1) \right]^{-1} \\
& \left\{ \varepsilon_1^{(0)} k_{\perp} I'_{m\pm 1}(k_{\perp} R_1) \int_{\infty}^{R_1} r K_{m\pm 1}(k_{\perp} r) \hat{L}_{\mp}^{(1)} K_m(k_{\perp} r) dr \right. \\
& + \frac{m \pm 1}{R_1} \varepsilon_2^{(0)} I_{m\pm 1}(k_{\perp} R_1) \int_{\infty}^{R_1} r K_{m\pm 1}(k_{\perp} r) \hat{L}_{\mp}^{(1)} K_m(k_{\perp} r) dr \\
& \left. + k_{\perp} K'_m(k_{\perp} R_1) \frac{\varepsilon_{\perp}^{(0)} \varepsilon_1^{(1)} - \varepsilon_{\perp}^{(1)} \varepsilon_1^{(0)}}{2\varepsilon_{\perp}^{(0)}} - \frac{m\left(\varepsilon_{\perp}^{(0)} \varepsilon_2^{(1)} - \varepsilon_{\perp}^{(1)} \varepsilon_2^{(0)}\right)}{2R_1 \varepsilon_{\perp}^{(0)}} K_m(k_{\perp} R_1) \right\} \Bigg|_{\omega = \omega_m^{(0)}} .
\end{aligned} \tag{5.46}
$$

The corrections for the SFXM eigenfrequencies in this case can be found by the same method, as it was done in Sect. 5.3. The non-uniform differential Bessel equation for the second order correction $H_{\zeta}^{(2)}(r)$ to the amplitude of the basic harmonic of the SFXMs and its solution are similar to the corresponding equation and its solution, which have been obtained in the previous section. This is the reason why one can apply the method of constant variation once again for the present problem (compare with Eq. (5.37)):

$$
\begin{aligned}
H_{\zeta}^{(2)}(r) = & K_m(k_{\perp} r) \int_{R_1}^{r} r I_m(k_{\perp} r) \left(\hat{L}_{-}^{(1)} H_{\xi}^{(-1)} + \hat{L}_{+}^{(1)} H_{\xi}^{(+1)} + \hat{L}^{(2)} H_{\xi}^{(0)} \right) dr \\
& + C_2 K_m(k_{\perp} r) \\
& - I_m(k_{\perp} r) \int_{\infty}^{r} r K_m(k_{\perp} r) \left(\hat{L}_{-}^{(1)} H_{\xi}^{(-1)} + \hat{L}_{+}^{(1)} H_{\xi}^{(+1)} + \hat{L}^{(2)} H_{\xi}^{(0)} \right) dr.
\end{aligned} \tag{5.47}
$$

The boundary conditions mentioned above in this Section are insufficient to determine the integration constant C_2 in the solution (5.47). This constant can be found from a condition similar to (5.38) implying that the energy of the magnetic wave field calculated in the second order approximation should be equal to the energy calculated in the zeroth order approximation. This requirement is analogous to the normalization condition for the wave function in quantum mechanics [27].

In the limiting case of thick rings $R \gg R_1 \gg |m|/k_\perp$, the LF SFXM frequency correction $\omega^{(2)}$ is the same as expression (5.39) with the exception of the sign. In the HF range their frequency correction becomes negative, but it absolute value is the same as given in (5.40). In the waveguide structure of the present geometry (see Fig. 5.5), the studied SFXMs can exist in the HF range if the following inequality is valid for the considered plasma parameters $(\delta m/R_1)^2 < |\omega_e|\Omega_e^{-1}$.

For the eigenfrequencies of SFXMs, which propagate around a thin metal ring with $R_1 \ll |m|k_\perp^{-1} \ll R$ in the LF range, it is possible to derive the following asymptotic expression for the frequency correction:

$$\omega^{(2)} = -\varepsilon_t^2(2m^2 + m + 3)\omega_m^{(0)}/8m^2. \quad (5.48)$$

In this case, the SFXM frequency $\omega_m^{(0)}$ is determined by the formula (2.12). In the HF range, the value of the SFXM eigenfrequency in zeroth order approximation $\omega_m^{(0)}$ is determined by formula (2.13), and the frequency correction in this case is approximately equal to:

$$\omega^{(2)} = -\varepsilon_t^2(2m^2 - m + 2)(\delta/R_1)^4 m(m-1)\omega_m^{(0)}. \quad (5.49)$$

The radial dependences of the satellite harmonics for these SFWs in the limiting case of thin metal ring are substantially different from each other: $H_\zeta^{(\pm 1)}(r) \propto r^{-|m\pm 1|}$, but nevertheless these amplitudes appear to be of the same order of smallness on the surface of the metal ring because of the corresponding difference in the numerical factors $C_\pm$.

Consequently, the presence of a toroidal non-uniformity of the external magnetic field results in the fact that SFXMs propagate in the form of wave packets. The amplitudes of the satellite harmonics in such packets have small values compared with the amplitude of the main harmonic: $A_{m\pm N} \sim \varepsilon_t^N A_m$. The influence of the toroidal non-uniformity of the applied external stationary magnetic field $\vec{B}_0 = B_{0\zeta}\vec{e}_\zeta$ on the amplitude of the main harmonic appears to be of the second order of the small toroidicity parameter ε_t. The eigenfrequency corrections for these SFWs, which are determined by the toroidal non-uniformity of the $\vec{B}_0 = B_{0\zeta}\vec{e}_\zeta$ field, appear to have values of the second order of smallness. They are of the same order of smallness as it was found for the frequency corrections in the previous section, but they have the opposite sign compared with the results obtained there.

Concerning the practical application of the results presented in this Section, one can see that such type of a metal ring can be applied as an antenna, which is intended for excitation of transversal SWs, for example with the purpose to sustain a microwave gas discharge.

5.5 Surface Flute Waves in Weakly Rippled Magnetic Fields

An external stationary magnetic field, which confines the plasma in a laboratory device, is often rippled. For instance, magnetic mirror traps have a rippled magnetic field [30], the rippled confining magnetic field in tokomaks is connected with the discreteness of the toroidal magnetic field coils, and a rippled magnetic field was typical for such toroidal systems like the ELMO BUMPY TORUS [31, 32]. The so-called "mirror" non-uniformity is planned to be predominant in the confining magnetic field of the modular stellarator reactor HELIAS [33]. If in the expression of the confining magnetic field derived for HELIAS, which is given in [33], one replaces the flow co-ordinates by cylindrical co-ordinates then this expression will coincide with the expression of the rippled magnetic field applied in the present Section. In [34], the electromagnetic eigenwaves of the thermonuclear plasma of the modular stellarator HELIAS were investigated theoretically.

Thus, one can see that such types of problems were actively studied in the past and are under consideration at present time. It should be mentioned for instance, that the theory of propagation, conversion and absorption of magneto-hydrodynamic waves in a plasma column, which is located in a weakly rippled magnetic field has been developed in [35–39]. The waves in a rippled magnetic field propagate in the form of wave packets, which in addition to a basic harmonic contain an infinite number of satellite spatial harmonics.

Propagation of electromagnetic flute eigenwaves nearby the boundary of the rippled plasma column separated by a vacuum layer from circular cross-section metal chamber with infinite conductivity was considered in [40–42]. The external static magnetic field $\vec{B}_0 = B_{0z}\vec{e}_z + B_{0r}\vec{e}_r$ is assumed to be weakly rippled (see Fig. 5.6):

$$B_{0r} = B_{00}\varepsilon_m^{'} k_m^{-1} \sin(k_m z),\; B_{0z} = B_{00}[1 + \varepsilon_m(r)\cos(k_m z)], \tag{5.50}$$

where $\varepsilon_m^{'} \equiv d\varepsilon_m/dr$, $k_m = 2\pi/L$, and L is the ripple period. The characteristic parameter of the ripples ε_m is usually (for example, in all modern fusion reactors) small. For instance, nearby the boundary of the plasma confined in the tokamak ASDEX-Upgrade in Garching, Germany [39], it is about 0.05; in the HELIAS

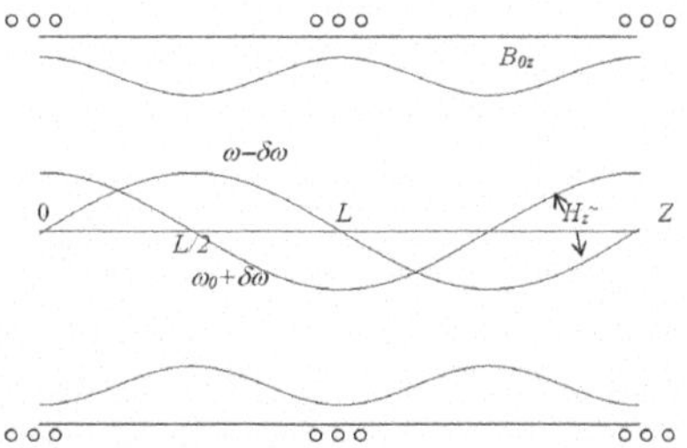

Fig. 5.6 Scheme of the absolute value of an external stationary magnetic field and standing eigenwaves versus the axial coordinate of a cylindrical metallic vessel in the case of a weakly rippled magnetic field

configuration [34] the "mirror" non-uniformity is planned to be small as well: $\varepsilon_m \sim 0.13$.

One can see that the expressions (5.50) do not automatically provide the implementation of the fundamental equation, namely, $div\ \vec{B}_0 = 0$. Therefore, one can use expressions (5.50) together with this fundamental equation and then obtain an equation like a modified Bessel equation, which determines the dependence ε_m on the radial co-ordinate. Its solution is proportional to the modified Bessel function [18] of zeroth order. Consequently, if the period of the ripples is larger compared to the minor radius of chamber of the fusion device then ε_m can be considered as a practically constant value. In the opposite case, if the ripple period is small, the ε_m value decreases approximately exponentially with increasing distance from the plasma interface going to the plasma center. Thus, the non-uniformity of the confining magnetic field is substantial only in a narrow layer within k_m^{-1} thickness nearby the metal wall of the fusion chamber.

As it is known from the basics of electromagnetic theory, the magnetic field fluxes are parallel to the vector of the confining magnetic field induction in every point of the co-ordinate space. The vector form of this condition allows one to derive the following equation for the flux lines in cylinder co-ordinates:

$$dr/B_{0r} = dz/B_{0z}. \tag{5.51}$$

Integrating Eq. (5.51) after inserting the explicit expressions (5.50) for the confining magnetic field components, one can obtain the equation for the magnetic surface in a more convenient way. Taking into account the properties of the ripple parameter $\varepsilon_m(r)$, which is described by modified Bessel functions, one can derive the expression for the radius of the magnetic surface:

$$r_0 \approx r + \frac{\cos(k_m z)}{r}\int_0^r r\varepsilon_m dr = r - \cos(k_m z)\left(1 - \frac{\varepsilon_m}{2}\cos(k_m z)\right)\varepsilon_m' k_m^{-2} + O(\varepsilon_m^3). \tag{5.52}$$

In this case, the equilibrium plasma density is a function of the magnetic surface and one can introduce the plasma density as function of the magnetic surface radius, $n(z, r) = n(r_0)$.

To simplify the consideration of the problem it is suitable to choose a convenient co-ordinate system. The system of co-ordinates $\vec{e}_1, \vec{e}_2, \vec{e}_3$ which is related to the flux lines of the magnetic field induction $\vec{B}_0$ is applied. Namely, the first vector is perpendicular to the magnetic surface in every point of the co-ordinate space $\vec{e}_1 = \nabla_{r_0}/\left|\nabla_{r_0}\right|$, the second vector coincides with the azimuthal basic vector in cylinder co-ordinates, and the third vector is parallel to the magnetic field flux lines $\vec{e}_3 = \vec{B}_0/\left|\vec{B}_0\right|$. In this system of co-ordinates, the components of the electric field induction $\vec{D}$ and the electric field strength $\vec{E}$ of the wave are coupled by the permittivity tensor of a magneto-active cold plasma without collisions in the simplest way:

$$\vec{D} = \varepsilon_1(E_1\vec{e}_1 + E_2\vec{e}_2) + \varepsilon_3 E_3\vec{e}_3 - i\varepsilon_2\vec{e}_3 \times \vec{E}. \quad (5.53)$$

Here the components of the cold magneto-active plasma permittivity tensor ε_j have a standard form (1.3) [10, 14–16]. The values of the cyclotron frequencies, which are present in the expressions for the components of the ε_j tensor, are determined by the complete magnetic field $B_0(r, z)$, and the plasma frequencies, which are applied there as well, can be calculated on the magnetic surface at the radial co-ordinate r_0.

Using the symmetry of the problem, the components of the plasma permittivity tensor can be written as series over the small parameter ε_m of the magnetic field ripples. Those of them, which describe the extraordinary flute modes, taking into account the summands of second order of smallness over the characteristic ripple parameter, are given by:

$$\varepsilon_{1,2} = \varepsilon_{1,2}^{(0)}(r) + \varepsilon_{1,2}^{(1)}\cos(k_m r) + \varepsilon_{1,2}^{(2)}(r) + O\left(\varepsilon_m^3\right), \quad (5.54)$$

where $\varepsilon_{1,2}^{(0)}(r)$ are the basic summands, and $\varepsilon_{1,2}^{(1)}\cos(k_m r)$ are corrections of the first order of smallness $\left|\varepsilon_{1,2}^{(1)}\right| \sim \left|\varepsilon_m\,\varepsilon_{1,2}^{(0)}\right|$. In second order approximation over the small parameter one can take into account only corrections to the basic summands $\left|\varepsilon_{1,2}^{(2)}\right| \sim \left|\varepsilon_m^2\,\varepsilon_{1,2}^{(0)}\right|$, because just they make contributions to the dispersion relation [40] of these SFXMs. As a result of calculating the first-order and second-order approximations of the components of plasma permittivity tensor ε_j in fluid approximation one can obtain:

$$\varepsilon_1^{(1)} = \frac{\partial \ln\left(\Omega_e^2\right)}{\partial r_0}\left[\varepsilon_1^{(0)}(r) - 1\right]\frac{\varepsilon_m'}{k_m^2} - 2\varepsilon_m\frac{\Omega_e^2(r)\omega_e^{(0)2}}{(\omega_e^{(0)2} - \omega^2)^2}, \quad (5.55)$$

$$\varepsilon_1^2(r) = -\frac{\varepsilon_m'\varepsilon_m}{k_m^2}\frac{\partial \ln\left(\Omega_i^2\right)}{\partial r_0}\left[\frac{\Omega_e^2(r)\omega_e^{(0)2}}{(\omega_e^{(0)2} - \omega^2)^2} + \frac{\Omega_e^2(r)}{4\left(\omega_e^{(0)2} - \omega^2\right)}\right]$$
$$+ \frac{\varepsilon_m^2}{2}\frac{\Omega_e^2(r)(3\omega_e^{(0)2} + \omega^2)}{(\omega_e^{(0)2} - \omega^2)^3} - \frac{\varepsilon_m'^2}{2k_m^2}\frac{\Omega_e^2(r)\omega_e^{(0)2}}{(\omega_e^{(0)2} - \omega^2)^2} + \frac{\varepsilon_m'^2}{4k_m^2}\frac{\partial^2\Omega_e^2}{\partial r_0^2}\frac{1}{\omega_e^{(0)2} - \omega^2}, \quad (5.56)$$

$$\varepsilon_2^{(1)}(r) = \frac{\varepsilon_m'}{k_m^2}\frac{\partial \ln \Omega_e^2}{\partial r_0}\varepsilon_2^{(0)} + \varepsilon_m\frac{\omega_e^{(0)}\Omega_e^2(r)(\omega^2 + \omega_e^{(0)2})}{\omega(\omega^2 - \omega_e^{(0)2})^2}, \quad (5.57)$$

$$\varepsilon_2^{(2)}(r) = \frac{\varepsilon_m'\varepsilon_m}{2k_m^2}\frac{\omega_e^{(0)}\Omega_e^2(r)(\omega^2 + \omega_e^{(0)2})}{\omega\left(\omega^2 - \omega_e^{(0)2}\right)}\frac{\partial \ln \Omega_e^2}{\partial r} + \frac{\varepsilon_2^{(0)}}{4\Omega_e^2(r)}\left[\frac{\partial^2\Omega_e^2}{\partial r^2}\frac{\varepsilon_m'^2}{k_m^4} - \frac{\partial\Omega_e^2}{\partial r}\frac{\varepsilon_m\varepsilon_m'}{k_m^2}\right]$$
$$+ \frac{\varepsilon_m'^2}{4k_m^2}\frac{\omega_e^{(0)}\Omega_e^2(r)(\omega^2 + \omega_e^{(0)2})}{\omega\left(\omega^2 - \omega_e^{(0)2}\right)^2} + \frac{\varepsilon_m^2}{2}\frac{\omega_e^{(0)3}\Omega_e^2(r)(3\omega^2 + \omega_e^{(0)2})}{\omega\left(\omega^2 - \omega_e^{(0)2}\right)^3}. \quad (5.58)$$

Applying the symmetry of the problem, one can investigate the solution of the Maxwell equations for the SFXM field as a wave packet, which contains the two nearest satellite harmonics proportional to $\exp[i(m\vartheta \pm k_m z)]$ in addition to the main harmonic proportional to $\exp(im\vartheta)$. In the vacuum layer, which separates the plasma column from the metal wall of the waveguide, the radial dependence of the amplitudes of the main harmonic (if the period of magnetic field ripples is sufficiently large, i.e. if the inequality $k_m c < \omega$ is valid) of the wave is expressed through the Bessel functions of the first kind and the Neumann functions [18]. It should be mentioned that this statement concerns the satellite harmonics as well.

As it is known from the basic theory of magneto-active plasmas (see e.g. [10, 14–16]), the properties of O-modes do not depend on the applied stationary magnetic field (from mathematical point of view it is explained by the absence of an external magnetic field in the expression for the component of the permittivity tensor ε_3), that is why one only considers the influence of ripples of the confining magnetic field on the properties of SFXMs in a plasma cylinder with a non-uniform radial profile of the plasma density. Therefore, in zeroth approximation, it is assumed that only the mth harmonic of the axial component of the SFW magnetic field $H_z^{(0)}(r)$ differs from zero, the other components of the magnetic field are equal to zero as well as the axial electric wave field: $E_z^{(0)}(r) = 0$, $H_\vartheta^{(0)}(r) = H_r^{(0)}(r) = 0$. The dispersion relation for the surface flute X-modes can be derived by application of the boundary conditions considered in details above in Sect. 1.4.1. In the present Section, the value of the eigenfrequency of the waves in zeroth order approximation is supposed to be known.

The inclusion of summands of the first order of smallness in the solution of the Maxwell equations in the plasma region does not change the amplitudes of the basic harmonics of the SFXM wave field components. However, it results in the appearance of the satellite harmonics, which are proportional to $\exp[i(m\vartheta \pm k_m z - \omega t)]$. Weak ripples of the external confining magnetic field are the reason for the coupling between the amplitudes $E_z^{(\pm)}$ of the satellite harmonics and the amplitudes $E_r^{(0)}$ and $H_z^{(0)}$ of the fundamental harmonics of the surface flute X-modes. The differential equation, which links these field components of SFXMs, has the form of a non-uniform Bessel equation. This is the reason why it is convenient to search for its solution by the method of constant variation. The appearance of the axial electric field of the SFWs, which is not equal to zero, is connected also with its weak coupling to the radial electric field because of application of the boundary condition for the tangential electric wave field on the metal wall of the torus chamber. The amplitudes of the satellite harmonics of the axial magnetic, and also the radial and poloidal electric fields of SFXMs are symmetrical: $E_r^{(+)} = E_r^{(-)}$, $E_\vartheta^{(+)} = E_\vartheta^{(-)}$, $H_z^{(+)} = H_z^{(-)}$ whereas the satellite harmonic amplitudes of the axial electric and the radial and poloidal magnetic fields are anti-symmetrical: $H_r^{(+)} = -H_r^{(-)}$, $H_\vartheta^{(+)} = -H_\vartheta^{(-)}$, $E_z^{(+)} = E_z^{(-)}$. There is no need to further refine the spatial distribution of the SFW axial electric field, as well as of the radial and poloidal magnetic wave fields, because this will merely lead to third- and higher-order corrections to the wave eigenfrequency.

By including the second-order terms, one can find the second-order corrections to the amplitudes of the fundamental harmonics of the axial magnetic field of the flute wave and of its radial and poloidal electric fields. The solution of the Maxwell equations for the axial component of the magnetic wave field can be expressed in the following form:

$$H_z^{\sim} = \exp[i(m\vartheta - \omega t)] \times \left[H_z^{\sim}(r) + H_z^{(+)}(r)\exp(ik_m z) + H_z^{(-)}(r)\exp(-ik_m z)\right], \tag{5.59}$$

where $H_z^{\sim}(r) = H_z^{(0)}(r) + H_z^{(2)}(r)$, $\left|H_z^{(\pm)}\right| \sim \left|\varepsilon_m H_z^{(0)}\right|$, $\left|H_z^{(2)}\right| \sim \left|\varepsilon_m^2 H_z^{(0)}\right|$. It should be underlined that in the derivation of formula (5.59), one can ignore the second-order terms which are proportional to $\exp(\pm 2ik_m z)$, because their contribution to the eigenfrequency is of a value, whose order is higher than ε_m^2.

The second order correction $H_z^{(2)}(r)$ to the amplitude of the main axial harmonic can be obtained by solving the Maxwell equations for the second order of smallness over the small ripple parameter ε_m:

$$H_z^{(2)} = \psi_0\left[A_2^{(2)} - \int\limits_{R_1}^{r} \frac{\tilde{\psi}_0 N_X^{(0)2}\hat{Q}H_z^{(0)}dr}{W\left(\psi_0, \tilde{\psi}_0\right)}\right] + \tilde{\psi}_0 \int\limits_{0}^{r} \frac{\psi_0 N_X^{(0)2}\hat{Q}H_z^{(0)}dr}{W\left(\psi_0, \tilde{\psi}_0\right)}. \tag{5.60}$$

Here the square of the transverse refractive index for SFXMs is $N_X^{(0)2} = \left(\mu^{(0)2} - 1\right)\varepsilon_1^{(0)}$, with $\mu^{(0)} = \varepsilon_2^{(0)}/\varepsilon_1^{(0)}$, and $W\left(\psi_0, \tilde{\psi}_0\right)$ is the Wronskian of the functions $\psi_0(r)$ and $\tilde{\psi}_0$. The radial profile of the magnetic wave field in zeroth order approximation, $H_z^{(0)}(r) = A^{(0)}\psi_0(r)$ (herein $A^{(0)}$ being a normalizing factor), is assumed to be known from the theory of SFXMs presented in Chap. 2. It satisfies the boundary condition, that the SFW magnetic field is of finite value inside the waveguide and, in particular, on the axis of plasma column $r = 0$. The function $\tilde{\psi}_0$ is a solution of the same equation as the $\psi_0(r)$ function, but it is linearly independent from $\psi_0(r)$ and has a singularity on the axis. The operator $\hat{Q}$ in solution (5.60) has the following form:

$$\hat{Q}H_z^{(0)} = \frac{N_\vartheta}{N_X^{(0)2}}\left(-q_1 + i\mu^{(0)}q_2\right) + \frac{c}{\omega r}\frac{d}{dr}\left[\frac{r}{N_X^{(0)2}}\left(\mu^{(0)}q_1 - iq_2\right)\right], \tag{5.61}$$

where $N_\vartheta = cm/(r\omega)$ is the poloidal refractive index,

$$q_1 = -\varepsilon_m N_\vartheta H_z^{(+)} + \varepsilon_m N_m H_\vartheta^{(+)} + \frac{c\varepsilon_m'}{\omega r k_m}\left[\frac{d}{dr}\left(rH_\vartheta^{(+)}\right) - imH_r^{(+)}\right] - \varepsilon_m \varepsilon_1^{(0)} E_r^{(+)}$$
$$- \varepsilon_1^{(1)} E_r^{(+)} - \varepsilon_m \varepsilon_1^{(1)} E_r^{(0)}/2 - \varepsilon_1^{(2)} E_r^{(0)} + i\varepsilon_m' \varepsilon_1^{(0)} E_z^{(+)}/k_m - i\varepsilon_2^{(1)} E_\vartheta^{(+)} - i\varepsilon_2^{(2)} E_\vartheta^{(0)}$$

$$- i\varepsilon_m \varepsilon_2^{(1)} E_\vartheta^{(0)} \Big/ 2 - i\varepsilon_m \varepsilon_2^{(0)} E_\vartheta^{(+)} - i(\varepsilon_m)^2 \varepsilon_2^{(0)} E_\vartheta^{(0)} \Big/ 4k_m^2, \tag{5.62}$$

$$q_2 = -\varepsilon_1^{(1)} E_\vartheta^{(+)} - \varepsilon_1^{(2)} E_\vartheta^{(0)} + i\varepsilon_2^{(0)} E_r^{(0)} \left[\frac{\varepsilon_m^2}{2} - \frac{(\varepsilon_m^{'})^2}{4k_m^2} \right]$$

$$+ i\varepsilon_2^{(1)} E_r^{(+)} + i\varepsilon_2^{(2)} E_r^{(0)} + \varepsilon_2^{(0)} \frac{\varepsilon_m^{'}}{k_m} E_z^{(+)}. \tag{5.63}$$

Note that the constant of integration $A_2^{(2)}$ in solution (5.60) is a small value of second order of smallness. If one inserts the correction given by Eq. (5.60) into expression (5.59) describing a wave packet, one shall see that the term $A_2^{(2)} \psi_0(r)$ is similar to $H_Z^{(0)}$. That is why the constant $A_2^{(2)}$ should not be determined from the boundary conditions but rather from a condition analogous to the normalization condition for the wave function in perturbation theory in quantum mechanics [27]. In this case, the mentioned condition implies the equality of the electromagnetic wave energy calculated in zeroth order approximation and in the second one. On the other hand, the form of second order dispersion relation and the correction to the SFXM eigenfrequency are independent on the constant $A_2^{(2)}$.

The inclusion of summands of the second order of smallness into the consideration enables one to determine the values of the corrections of the second order of smallness to the amplitude of the fundamental harmonic of the axial magnetic wave field and also the radial and poloidal electric fields of the SFWs. Thus, the spatial distribution of the SFXM fields can be determined by using the theory of successive approximations from the spatial distribution of the axial magnetic field of the SFWs, which is known in the zeroth approximation, within an accuracy up to summands of the second order of smallness. Applying all the above-mentioned boundary conditions one can derive the dispersion equation for the surface flute X-modes as it has been done in the previous sections.

It can be analytically studied only in some limiting cases, for example, in the case of a uniform dense plasma $\left(\Omega_e^2 \gg \omega_e^2\right)$ with a large radius ($R_1 \gg |m|\delta$), which is separated from the metal wall of the torus chamber by a narrow vacuum layer ($\Delta = R_2/R_1 - 1 \ll 1$). If this plasma filled waveguide is immersed into an external stationary magnetic field with a large ripple period ($k_m R_1 \ll 1$), the SFXM eigenfrequency can be determined in the following way: $\omega = \omega^{(0)} + \Delta\omega$, where $\omega^{(0)}$ is the SFXM frequency in zero order approximation calculated with an accuracy up to summands of the order of $\Delta^2 \ll 1$:

$$\omega^{(0)} \approx \frac{(1-\Delta)\delta^2 m|\omega_e| + \sqrt{(1-2\Delta)\delta^4 m^2 \omega_e^2 - \left(2R_1^2\Delta - R_1^2\Delta^2 + 2R_1\delta\right) m^2 c^2 \Delta(3\Delta - 2)}}{2R_1^2\Delta - R_1^2\Delta^2 + 2R_1\delta}, \tag{5.64}$$

and $\Delta\omega$ is a correction to this value, which is connected with the weak ripples of the external static magnetic field:

$$\frac{\Delta\omega}{\omega^{(0)}} \approx -\frac{(\varepsilon_m^{'})^2}{4k_m^4}\left[\frac{k^2\delta}{a}\left(1-\frac{m\delta|\omega_e|}{a\omega^{(0)}}\right)\left(\frac{m^2}{a^2k^2}-1\right)+\frac{\Delta}{\delta^2}-\frac{\Delta m^2\delta^{-2}}{a^2k^2}\right.$$
$$\left.-\frac{3m^2}{k^2a^4}+\frac{\delta+a}{a^2\delta}-\frac{m|\omega_e|}{a^2\omega^{(0)}}\right]\left[(1-\Delta)\frac{m\delta^2|\omega_e|}{a^2\omega^{(0)}}+(2-3\Delta)\frac{\Delta m^2}{k^2a^2}\right]^{-1}. \quad (5.65)$$

Summarizing the material discussed in the present Section one can emphasize that a similar mathematical approach has been applied for investigation of beam-plasma instabilities, which occur in waveguides with periodic dielectric inserts and/or in waveguides with corrugated walls [43, 44]. In those cases, the dispersion relations, which describe the excitation of such waveguides by charged particle beams, contain small correcting summands, which are proportional to the square of the ripple depth of the device. A similar correction, but proportional to the squared amplitude of an external static electric field that varies periodically along the axis of a magnetized waveguide, enters into the dispersion relation describing the emission from an electrostatic free-electron laser. The fact that the same mathematical methods for solving different problems (electrodynamics of dense electron beams and of plasmas in modular stellarators) yield dispersion relations with similar structures provides one more independent confirmation of the validity and correctness of the results obtained in the present Sections. Experimental study of the processes of mode conversion in a waveguide with corrugated walls proves that for practical aims it is sufficient to apply analytical calculations, which have been done within the framework of the theory of successive approximations [43].

5.6 Additional Electron Cyclotron Resonance Heating of Plasmas by Surface Flute Waves in Weakly Rippled External Static Magnetic Fields

In the previous Sect. 5.5, it is proved theoretically that surface flute waves (SFWs) can propagate in metal cylindrical waveguides, which are partially filled with magnetoactive plasma that is confined by a rippled external stationary magnetic field. The scheme of the plasma filled cylindrical waveguide immersed into a weakly rippled external stationary magnetic field is shown in Fig. 5.6. The analytical expression for this confining magnetic field $\vec{B}_0 = B_{0z}\vec{e}_z + B_{0r}\vec{e}_r$ is assumed to be a superposition of axial and radial components, whose forms are given by the expression (5.50). The axial distribution of the components B_{0r} and B_{0z} is plotted in Fig. 5.7. To search for a solution of the Maxwell equations the new convenient local system of co-ordinates related to the flux lines of the magnetic field induction $\vec{B}_0$ was applied in [40, 43,

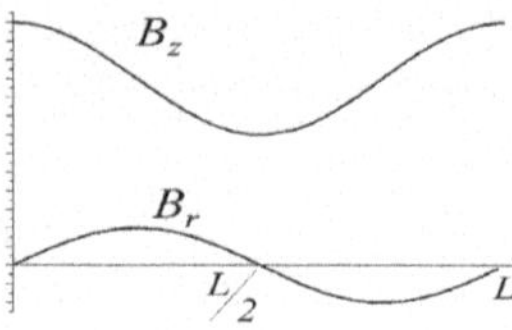

Fig. 5.7 Axial dependence of the components of the external stationary rippled magnetic field

44]. In this co-ordinate system the first basic vector is perpendicular to the magnetic surface, the second unit vector coincides with the azimuthal basic vector in cylinder co-ordinates, and the third basic vector is parallel to the magnetic field flux lines, $\vec{e}_3 = \vec{B}_0/\left|\vec{B}_0\right|$. As it was shown in the previous Sect. 5.5, the choice of this co-ordinate system allows one to express the components of the displacement vector $\vec{D}$ through the electric field $\vec{E}$ of the wave using the dielectric permittivity tensor of the magneto-active cold plasma without collisions in the simplest way (5.53). In this case the components of the dielectric permittivity tensor ε_j have the standard form (1.3) [10, 14, 15]. The values of the cyclotron and plasma frequencies are determined there by the complete magnetic field $B_0(r, z)$ and the value of the plasma particle density on the magnetic surface (see Eq. (5.52)), respectively.

The actuality of the topic, which is proposed for consideration here, is explained by the existence of essential losses of electromagnetic power in all types of magnetic fusion devices. These losses are considered as those connected with strong electromagnetic activity in the peripheral plasma. Numerous types of surface electromagnetic waves are recognized as the most possible cause of undesirable heating of the plasma periphery, strengthening of plasma-wall interaction and impurity generation. This has been clearly shown in experiments with ion cyclotron heating [45–47]. This proves the necessity of studying problems of surface waves for magnetically confined fusion plasmas and also for their technological applications. SFWs, which propagate perpendicular to a stationary axial magnetic field along the interface of a cylindrical plasma column, are proved to be eigenmodes of such type waveguides [28, 48] with non-uniform plasma filling (see Sect. 5.1). In the case of dense plasmas $\left(\Omega_e^2 > \omega_e^2\right)$, these modes propagate in particular in the LF range (1.13): $\omega_{LH} < \omega < \sqrt{\Omega_e^2 + \omega_e^2/4} - |\omega_e|/2$ (see Sect. 1.4.1). Here ω_{LH} is the lower hybrid frequency; ω_e and Ω_e are the electron cyclotron and electron plasma frequencies.

In the present Section, the possibility of applying SFW propagation in the case of a weakly rippled confining magnetic field for plasma heating is investigated. Doing that, both the geometry of the magnetic trap with the rippled confining magnetic field and the initial assumptions are the same as they were in the previous Sect. 5.5. The spatial distribution of the confining magnetic field (5.50) along the axial direction for this case is shown in Fig. 5.7.

It is assumed that the absorption of SFWs happens nearby the interface of the plasma column with average radius $r = R_1$ separated by a vacuum layer from the ideally conductive cylindrical metal chamber with radius $r = R_2$, whose value is a little bit larger than the plasma radius: $R_2 > R_1$. Here the equilibrium plasma density $n_e(r, z)$ is considered to be a function of only one variable r_0 that is the label of the magnetic flux surface: $n_e(r, z) = n_e(r_0)$. The variable r_0 is determined by the expression (5.52).

The Maxwell equations, which take into account the periodic non-uniformity of the stationary rippled magnetic field (5.50), are solved here by the method of successive approximation using the ripple parameter $\varepsilon_m \ll 1$ as the small parameter of the problem [40–42]. Because of the non-uniformity of the confining magnetic field one cannot separate the fields of SFWs into two independent subsets, which describe X- and O-polarized modes, respectively. Moreover, SFWs with different values of azimuthal mode index cannot propagate independently; they propagate in the form of wave packets, which are superpositions of the fundamental harmonic that is proportional to $\exp[i(m\varphi - \omega t)]$ (this is the same dependence as it was applied for description of the extraordinary flute mode, see Chap. 2) and the two nearest satellite harmonics, whose dependences on time and co-ordinates are as follows $\exp[i(m\varphi - \omega t \pm k_m z)]$. Here $k_m = 2\pi/L$, and L is the period of ripples (or in other words, L is the axial period of the stationary magnetic field inhomogeneity). As example of such presentation one can see the expressions (5.59) and (5.60) for the axial magnetic field of the SFW derived for the plasma region, which are obtained in the previous section.

It should be mentioned that the satellite harmonics are of the first order of smallness as compared with the fundamental harmonic: for example, satellites of the axial magnetic field $H_z^{(\pm)}(r) \propto \varepsilon_m H_z^{(0)}$. The second order correction of this axial magnetic wave field $H_z^{(2)}(r) \propto \varepsilon_m^2 H_z^{(0)}$ is of the second order of smallness over the small parameter of the problem $\varepsilon_m \ll 1$. In the studied case all six components of the SFW electromagnetic field are coupled with each other [40–42]. To find the spatial distribution of the fields one has to apply a complete set of corresponding boundary conditions for both fundamental harmonics and satellite harmonics. Adequate application of these conditions allows one to derive analytical expressions for the satellite harmonics of all components of the SFW electromagnetic field. In the previous Sect. 5.5, the dispersion properties of SXFMs are studied in the limiting case of a dense uniform cylindrical plasma with a relatively large radius. An evident expression for the SFXM frequency correction is found (see expression (5.65)) for this case. Unlike that, the case of cylindrical waveguides filled with non-uniform plasma is considered here.

In Ref. [28], the propagation of surface flute modes in plasmas with non-uniform density profile is studied. Since in a rippled stationary magnetic field, the fields of SFXMs have the form of an envelope, one can expect the possibility of resonant damping of either their fundamental harmonic or their satellite harmonics. If the frequency of SFXMs is higher than the electron cyclotron frequency, there is the possibility of electromagnetic power absorption in the peripheral plasma, where the following resonant condition is valid [19]:

$$\varepsilon_1(r_{res}) = 0. \tag{5.66}$$

In (5.66) ε_1 is the diagonal component of the plasma permittivity tensor and the radial co-ordinate r_{res} indicates the place in the plasma column, where the upper hybrid resonance is realized, $\omega = \omega_{UH}(r_{res}) = \sqrt{\omega_e^2 + \Omega_e^2(r_{res})}$.

The condition (5.66) can be realized for SFWs with angular frequencies $\omega > |\omega_e|$, in the case of high plasma density in the plasma core, $\Omega_e^2(r = 0) > \omega_e^2$. There is no upper hybrid resonance for the basic harmonic of the surface flute modes with a frequency $\omega < |\omega_e|$. Nevertheless both of their satellite harmonics experience the following local resonance:

$$\varepsilon_1(r_*) = N_m^2. \tag{5.67}$$

In (5.67) the physical sense of the parameter N_m is the axial refractive index, its value is $N_m = ck_m/\omega > 1$, and r_* is the radial co-ordinate determining the position of this local resonance. An illustration of the correlation between the radial plasma density profile and the corresponding variation of $\varepsilon_1(r)$ is plotted in Fig. 5.8.

Analyzing the expressions, which are derived in the previous section for the fields of the SFW satellite harmonics, one can see that nearby the radius r_*, that determines the place of the local resonance (5.67), a resonant growing up of these satellite harmonics happens. For example, the amplitudes of the satellite harmonics $H_\varphi^{(\pm)}(r)$ of the poloidal magnetic field component of these waves strongly grows up along the radial direction nearby the radius r_* of the local resonance (5.67). The same phenomenon is realized for the amplitudes of the satellite harmonics of the radial electric field component $E_r^{(\pm)}(r)$ of the SFWs:

$$H_\varphi^{(\pm)}(r \to r_*);\ E_r^{(\pm)}(r \to r_*) \propto [\varepsilon_1(r) - N_m^2]^{-1}. \tag{5.68}$$

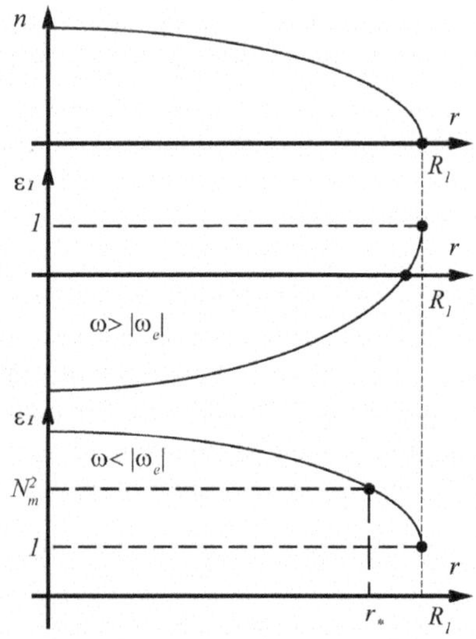

Fig. 5.8 Radial profile of the plasma density and corresponding variation of the first diagonal component of the plasma permittivity tensor along the plasma radius. The horizontal dashed lines show specific values $\varepsilon_1 = 1$, and $\varepsilon_1 = N_m^2$

The radial singularity of these amplitudes is stronger than that of the satellite harmonics of the radial and axial magnetic field components and of the poloidal electric field component of the SFWs field:

$$H_r^{(\pm)}(r \to r_*);\ H_z^{(\pm)}(r \to r_*);\ E_\varphi^{(\pm)}(r \to r_*) \propto \ln[\varepsilon_1(r) - N_m^2]. \tag{5.69}$$

Unlike satellite harmonics of these five components (see expressions (5.68) and (5.69)), the amplitudes of the satellite harmonics of the axial electric field component of the SFW have no singularity nearby the local resonance (5.67); their behaviour is continuous [41, 42].

It should be emphasized that in spite of the logarithmic singularity of the satellite harmonics of both the magnetic $H_z^{(\pm)}$ and electric $E_\varphi^{(\pm)}$ wave fields nearby the radial co-ordinate, the following combination of these satellites has a finite value:

$$\left[i\varepsilon_2^{(0)} E_\varphi^{(\pm)}(r) + N_\vartheta H_z^{(\pm)}(r)\right]_{|r=r_*} \ll \infty, \tag{5.70}$$

where $N_\vartheta = cm/(\omega r)$ is the poloidal refractive index of the SFW. The combination (5.70) changes slowly nearby the local resonance (5.67). This combination is a continuous function of the radial co-ordinate in the whole plasma column [41, 42]. This circumstance is very useful for both analytical and numerical investigation of the spatial distribution of the SFW fields nearby the resonance (5.67).

The solution of the Maxwell equations for the SFWs in the form of a wave packet, that contains the next satellite harmonics which are proportional to $\exp[i(m\varphi \pm jk_m z - \omega t)]$, with natural numbers $j = 2, 3, 4, \ldots$, shows that the amplitudes of these harmonics have small values with higher order of smallness than the amplitudes of the two nearest satellite harmonics. For example, out of the regions of local resonances, which are determined by the radial co-ordinates r_j, correspondingly, the amplitudes of the satellite harmonics of the axial magnetic field of the SFW have the following values: $H_z^{(\pm j)}(r) \propto (\varepsilon_m)^j$. These satellite harmonics have local resonances nearby the definite radii r_j, where the following resonant condition is valid:

$$\varepsilon_1(r_j) = (jN_m)^2, \tag{5.71}$$

with the natural numbers $j = 2, 3, 4\ldots.$

Nearby these radii r_j resonant growing up of the satellite harmonics $H_z^{(\pm j)}(r \to r_j)$ takes place and a small part of the electromagnetic power of the SFW, namely $\propto (\varepsilon_m)^{2j}$ is absorbed there as compared with the case of the local resonance realized for the nearest satellite harmonics. Because of that one can neglect the absorption of these satellite harmonics nearby the resonances (5.71) as compared with the absorption of the nearest satellite harmonics [41, 42].

Now one has to calculate the electromagnetic wave power P_{res} that can be absorbed within the local resonance (5.67) per unit length of the plasma cylinder.

Its physical sense is determined as the work that is done by the electric fields of the SFW satellite harmonics over the radial radio frequency (RF) currents $j_r^{(\pm)} \propto \exp[i(m\varphi \pm k_m z - \omega t)]$:

$$P_{res} = 2\pi r_* \cdot \mathrm{Re}\{ \int_{r_*-0}^{r_*+0} [j_r^{(+)}(r)]^* E_r^{(+)} dr\} = \frac{\pi\omega}{2} \left(\left| \frac{\partial \varepsilon_1^{(0)}}{\partial r} \right|^{-1} r \right)_{r=r_*} \left| A^{(+)} \right|^2, \quad (5.72)$$

where

$$A^{(+)} = \Big\{ 0.5 \Big[\varepsilon_m N_\vartheta H_z^{(0)} + \varepsilon_1^{(1)} E_r^{(0)} + \varepsilon_m \varepsilon_1^{(0)} E_r^{(0)} + i\varepsilon_2^{(1)} E_\varphi^{(0)} + i\varepsilon_m \varepsilon_2^{(0)} E_\varphi^{(0)} \Big] - \frac{ic^2}{\omega^2} k_m \frac{dE_z^{(+)}}{dr} + N_\vartheta H_z^{(+)} + i\varepsilon_2^{(0)} E_\varphi^{(+)} \Big\}_{r=r_*} \quad (5.73)$$

The expression (5.72) is derived under the assumption that the radial plasma density profile is linear near the resonant radius $r = r_*$, so that:

$$\varepsilon_1^{(0)}(r) = N_m^2 + \frac{d\varepsilon_1^{(0)}}{dr}_{|r_*} (r - r_*). \quad (5.74)$$

It should be underlined that the behaviour of the density profile is assumed to be linear only in the close vicinity of the resonant region where the electromagnetic wave is absorbed. Thus, in the region of the plasma column core, the radial profile of the plasma density can be of arbitrary form.

The possibilities of experimental observation of the considered phenomenon of plasma heating in a microwave plasma source can be estimated. This device can be presented as a cylindrical applicator, which is excited with the help of an RF antenna operating at the frequency $f = 2.386$ GHz. To avoid plasma heating due to the electron cyclotron resonance one has to utilize a confining steady magnetic field $B_0 = 1$ kG. In this case the difference between the electron cyclotron frequency $|\omega_e| = 1.76 \times 10^{10}$ s^{-1}and the operating frequency $\omega = 1.51 \times 10^{10}$ s^{-1} is sufficiently large, so that one can experimentally distinguish these different types of heating. In order to realize the inequality $\Omega_e^2 > \omega_e^2$, that is assumed to be valid for the considered SFWs, one has to provide a plasma with a core density of $n_e = 10^{12}$ cm^{-3}. This plasma density allows one to obtain the electron plasma frequency $\Omega_e = 5.61 \times 10^{10}$ s^{-1}, which is higher than the electron cyclotron frequency.

Because of the peculiarity of the utilized antenna (its azimuthal size is supposed to be $\Delta\varphi = \pi/2$) an electromagnetic wave with the azimuthal mode index $m = 2$ is mainly excited in this applicator. In the plasma cylinder with radius $R_1 = 3.1$ cm, which is separated from the metal cylindrical chamber by a vacuum layer with 1 cm thickness, the SFW with azimuthal mode index $m = 2$ propagates under these experimental conditions with the frequency $\omega = 1.5 \times 10^{10}$ s^{-1} [41, 42]. This circumstance allows one to identify the wave excited by the antenna as SFW.

The longitudinal size of the applicator $L_{res} = 50$ cm determines specific values of the axial wavenumbers of other waves (this means that they are not SFWs), which can be excited by the antenna: $k_z = \pi/L_{res} \approx 0.06$ cm^{-1}. For the electromagnetic waves with such wavenumbers (and as well for the waves with axial wavelength, that is larger than the vacuum wavelength $\lambda_{vac} = 12.56$ cm at $f = 2.386$ GHz), the condition of the local resonance $\varepsilon_1 = (ck_z/\omega)^2$, which is analogous to the condition (5.67), cannot be realized in the proposed waveguide.

In the following, the waveguide equipped with thin magnetic coils, which are located at the distance of 5 cm from each other, is considered. Then from the resonant condition (5.67) one can find that the power of the SFW satellite harmonics has to be absorbed nearby the radial co-ordinate, where the plasma density is $n_e = 1.4 \times 10^{11}$ cm^{-3}. For fixed magnitude of the confining magnetic field, a decrease of the period of the magnetic field ripple leads to a decreasing value of the radial co-ordinate, where the SFW power is absorbed. If under the same conditions the magnetic coils are modified in such a way that there is no ripple (for example, by changing the distance between the magnetic coils and/or the value of electric current in one of these coils), no resonant absorption is observed. Thus this action allows one to identify that the plasma under these experimental conditions is heated under the regime of additional electron cyclotron heating of the radially non-uniform plasma because of absorption of the satellite harmonics of the SFWs which propagate in a weakly rippled confining magnetic field.

Summarizing the presented results one can conclude that in contrast to the case of a homogeneous axial stationary magnetic field $\vec{B}_0$, due to the local resonance (5.67) additional absorption (5.72) of electromagnetic energy P_{res} can take place for SFWs with frequency values that are lower than the electron cyclotron frequency in a plasma with a radially inhomogeneous density profile. Analysis of the dependence of P_{res} proves that under the same other conditions the value of the absorbed power is larger for the case of low plasma density gradient at the resonance radius r_* and for larger amplitude of the satellites of the radial electric field $A^{(+)}$. For a fixed frequency of the RF generator and a magnetic field configuration such that $\omega < |\omega_e|$, one can obtain larger values of $\left|A^{(+)}\right|^2$ and therefore obtain enhanced magnitude of absorbed RF power in the following two cases.

First, for a given plasma density profile, the SFMs with poloidal mode index m and zero axial wavenumber $k_z = 0$ can be eigenmodes (this means that their damping rate is weak) for the considered plasma filled waveguide. In this case the amplitudes of the fundamental harmonics of the fields $H_z^{(0)}$, $E_r^{(0)}$, $E_\varphi^{(0)}$, and the amplitudes of satellite harmonics $H_z^{(+)}$, $E_r^{((+)}$, $E_\varphi^{(+)}$ become larger, leading to an enhanced magnitude of absorbed power at the radius r_*.

Second, the absorption of the satellite harmonics at the radius r_* can be enhanced also if for a specific plasma density profile the wave with poloidal mode index m and axial wavenumber $k_z = k_m$ (i.e. the mode $\propto \exp(im\varphi \pm ik_m z - i\omega t)$) is an eigenmode. In this case the last three terms $N_\varphi B_z^{(+)} + i\varepsilon_2^{(0)} E_\varphi^{(+)} - (icN_m/\omega)\, dE_z^{(+)}/dr$ in the expression for $A^{(+)}$ increase, which leads to enhanced absorbed power.

Surface waves with frequency $\omega < |\omega_e|$ and axial wavenumber are also resonantly absorbed nearby the radius r_*. However, the penetration depth of these modes decreases with increasing k_m value. Therefore, plasma heating caused by their absorption is less efficient as compared with absorption of SFWs, whose penetration depth into the plasma for the fundamental harmonic of the SFW is larger. It should be underlined also that a decrease of the scale of the axial magnetic field non-uniformity $L = 2\pi/k_m$ leads to a decreasing thickness of the layer in the plasma periphery where the inhomogeneity of $\vec{B}_0$ is essential. Because of that the interaction between the fundamental and satellite harmonics becomes weaker.

Since the dimensions of the plasma chamber and the strength of the axial stationary magnetic field in fusion devices are huge (so that the inequality $L|\omega_e| > 2\pi c$ is valid) the studied mechanism of plasma periphery heating due to SFWs does not pose hazard to fusion plasmas. However, it is known that the tokamak chamber has to be cleaned periodically employing RF antennas in order to remove impurities from the chamber wall. First wall conditioning is a necessary procedure for tokamak operation because it is needed to control the particle recycling, to reduce plasma impurities and to prevent tritium inventory in the plasma facing components of fusion reactors like ITER [49, 50]. This conditioning can be done without changes in hardware of ion and electron cyclotron resonance heating systems, under the influence of the residual magnetic field. These conditions could be suitable for SFW propagation.

The considered mechanism can be applied as well for technical processing plasmas. It should be mentioned that the properties of surface waves propagating under the regime of electron cyclotron resonance are actively utilized in the physics of gas discharges [6, 7]. The possibility to apply SFWs for sustaining gas discharges has been proved theoretically in [51].

5.7 Conclusions

The results of theoretical studies of the properties of SFXMs under the conditions of their propagation in plasma filled cylindrical non-uniform waveguides of different constructional peculiarities are presented here. Real experimental conditions, namely, spatial non-uniformity of the confined plasma density and of the applied external magnetic field are taken into account. Possible practical applications of these theoretical results are indicated here as well.

The performed investigations allow one to determine the frequencies of the electromagnetic flute eigenwaves and spatial distribution of their fields analytically, calculate expressions for their damping rates, which are connected both with collisions between the plasma particles and resonant conversion of these modes into bulk modes in the case of a dense plasma. It is proved that if the plasma density increases in radial direction, the LF SFXM frequency decreases as compared with the case of a uniform plasma density profile. It is shown that approximation of the periphery plasma density profile by a linear function (regardless of its form in the central region of the plasma column) allows one to find the LF SFXM frequency spectrum analytically. In the case of a linear type of the periphery plasma density profile the obtained expression for their frequency as function of the plasma parameters is similar to that one, which was obtained in the case of a uniform plasma, when doing a replacement of the SFXM penetration depth by a definite effective value.

It is found that radial non-uniformity of an external axial magnetic field has an influence on the frequency of these modes, which is similar (from mathematical point of view) to the case of the influence of a radial plasma density inhomogeneity. A small toroidicity of an external magnetic field (parameter $\varepsilon_t \ll 1$) results in the feature that SFWs propagate as wave packets. Corrections to the wave field of their main harmonic and to their eigenfrequency, which are determined by this non-uniformity of the utilized magnetic field, are proportional to ε_t^2.

Ripples of the confining magnetic field in modern thermonuclear reactors and in special devices for plasma electronics are characterized by a small parameter $\varepsilon_m \ll 1$. It is shown that such ripples can influence conversion and absorption of the flute eigenwaves. In this case SFWs propagate as well as wave packets, in which all six field components are present. Nevertheless, this problem can be solved by employing the method of successive approximations. It has been proved that this problem is analogous (from mathematical point of view) to the problems of beam instabilities in a plasma, which is confined in a waveguide with a periodic dielectric insert or with corrugated walls and even to the theory of free electron lasers. The dispersion relations obtained for the SFXMs in this case contain small summands, which are proportional to ε_m^2. They can be adequately solved by the methods of theory of successive approximations.

The possibility to utilize the propagation of SFWs in waveguides filled with non-uniform plasma immersed into the weakly rippled stationary magnetic field for plasma heating is investigated theoretically using the method of successive approximations. It is proved that resonant damping of the satellite harmonics of the SFW with $\omega > |\omega_e|$ leads to plasma heating. The efficiency of this process is estimated. If the condition of the local resonance $\varepsilon_1(r_*) = N_m^2$ is realized in the confined plasma, the absorption of SFWs leads to plasma heating. Here N_m is the axial refractive index, its value should be relatively large, $N_m = ck_m/\omega > 1$, ε_1 is the diagonal component of the plasma permittivity tensor, and r_* is the radial coordinate determining the position of this local resonance. The value of RF power absorbed nearby this resonance is found to be proportional to the square of the small ripple parameter.

It is proved that SFWs with frequencies $\omega < |\omega_e|$ are also resonantly absorbed nearby the radius r_*, but the plasma heating caused by their absorption is less efficient as compared with the case of SFWs with $\omega > |\omega_e|$, because the penetration depth of their fundamental harmonic is larger just in the present case. Decreasing of the axial scale of the magnetic field non-uniformity leads to a decrease of the thickness of the plasma layer where the inhomogeneity of $\vec{B}_0$ is essential and because of that to suppression of the studied phenomenon of plasma heating.

The most prospective field of this regime of plasma heating is to sustain gas discharges in technological plasma devices and even conditioning of the first wall in fusion plasma devices, because such procedure can be done without changes in the hardware of ion and electron cyclotron resonance heating systems, under the influence of the residual magnetic field produced by the superconductive coils of the fusion machine.

References

1. Anderson, T. (2011). *Plasma antennas*. Artech House.
2. Jenn, D. C. (2003). *Plasma antennas: Survey of techniques and the current state of the art*. San Diego, CA. SPAWAR PMW 189.
3. Kumar, V., Mishra, M., & Joshi, N. K. (2011). Study of a fluorescent tube as plasma antenna. *Progress in Electromagnetics Research Letters, 24*, 17–26.
4. Kudrin, A. V., Petrov, E Yu., Kyriacou, G. A., et al. (2005). Insulated cylindrical antenna in a cold magneto-plasma. *Progress in Electromagnetics Research, 53*, 135-166.
5. Wu, M., Wen, B. Y., & Zhou, H. (2009). Ionospheric clutter suppression in HF surface wave radar. *Journal of Electromagnetic Waves and Applications, 23*(10), 1265–1272.
6. Aliev, Yu. M., Schlüter, H., & Shivarova, A. (2000). *Guided-wave-produced plasmas*. Springer.
7. Margot, J., & Moisan, M. (1992). *Surface wave sustained plasmas in static magnetic fields for study of ECR discharge mechanisms—Microwave excited plasmas*. In: Moisan, M., & Pelletier, J. (Eds.). Amsterdam: Elsevier
8. Winter, R., Korzec, D., & Engemann, J. (1997). Large area plasma cleaning with 26" microwave slot antenna plasma source SLAN II. *Surface and Coating Technol., 93*(1), 134–141.
9. Korzec, D., Werner, F., Winter, R., et al. (1996). Scaling of microwave slot antenna (SLAN): A concept for efficient plasma generation. *Plasma Sources Science and Technology, 5*, 216–234.
10. Alexandrov, A. F., Bogdankevich, L. S., & Rukhadze, A. A. (1984). *Principles of plasma electrodynamics*. Springer.
11. Vukovic, V. (1986). *Surface waves in plasmas and solids*. World Scientific.
12. Beletski, N. N., Bulgakov, A. A., Khankina, S. I., & Yakovenko, V. M. (1984). *Plasma instabilities and non-linear phenomena in semi-conductors*. Kyiv: Naukova dumka. (in Russian).
13. Agranovich, V. M., & Mills, D. L. (1982). *Surface Polaritons*. North-Holland.
14. Akhiezer, A. I., Akhiezer, I. A., Polovin, R. V., Sitenko, A. G., & Stepanov, K. N. (1975). *Plasma electrodynamics*. Pergamon Press.
15. Krall, N. A., & Trivelpiece, A. W. (1973). *Principles of plasma physics*. MacGraw-Hill.
16. Brambilla, M. (1998). *Kinetic theory of plasma waves. Homogeneous plasma*. Clarendon Press.
17. Girka, V., & Girka, I. (1990). Influence of plasma inhomogeneity on the spectra of azimuthal surface waves. *Radiofisika, 33*(4), 516–517. (in Russian).
18. Abramowitz, M., Stegun I. A. (Eds.) (1964). *Handbook of mathematical functions*. New York: National Bureau of Standards, Applied Mathematics, Series 55.
19. Girka, V., & Girka, I. (1991). Azimuthal surface waves in a non-uniform plasma cylinder. *Radiophysics and Quantum Electronics, 34*(4), 324–328.
20. Girka, V., Girka, I., Kondratenko, A., et al. (1988). Azimuthal surface waves of magnetoactive plasma waveguides. *Journal of Communication Technology and Electronics, 33*(8), 37–41.
21. Kondratenko, A. M., & Kuklin, V. M. (1988). *Basis of plasma electronics*. Energoatomizdat. (in Russian).
22. Azarenkov, N. A., & Ostrikov, K. N. (1991). Surface magnetoplasma waves at the interface between a plasma-like medium and a metal in a Voigt geometry. *Physics Reports, 308*, 333–428.
23. Beidler, C. D., Grieger, G., Harmeyer, E., et al. (1995). *Helias reactor studies*. Preprint of Max-Planck-Institut für Plasmaphysik, 85748 Garching bei München, FRG, IPP 2/330.
24. Girka, V. (2006). Theory of transverse surface electro-magnetic waves propagating along a plasma–metal boundary with a finite radius of curvature in a magnetic field. *Plasma Physics Reports, 32*(5), 401–410.

25. Tierens, W., Colas, L., & EUROfusion MST1 Team. (2021). Slab-geometry surface waves on steep gradients and the origin of related numerical issues in a variety of ICRF codes. *Journal of Plasma Physics, 87*(4), 905870405.
26. Girka, V., Girka, I., & Pavlenko, I. (1997). Surface waves propagating in the direction transverse to the axis of magnetized plasma—filled waveguides with noncircular transverse cross sections. *Plasma Physics Reports, 23*(11), 959–963.
27. Landau, L. D., & Lifshits, E. M. (1977). *Course of theoretical physics. Quantum mechanics.* Vol. 3. Oxford: Pergamon Press.
28. Girka, V., Girka, I., Girka, A., et al. (2011). Theory of azimuthal surface waves propagating in non-uniform waveguides. *Journal of Plasma Physics, 77*(4), 493–519.
29. Girka, I. (2002). Azimuthal surface waves at the plasma-metal boundary in a non-uniform toroidal magnetic field. *Technical Physics, 47*(7), 845–850.
30. Ryutov, D. D. (1988). Open-ended traps. *Physics-Uspekhi, 31*, 300–327.
31. Baity, F. W., Berry, L. A., Bighel, L., et al. (1983). Plasma properties and ion heating in EBT-S and hot electron rings at TRW. In: *Proceedings of 9th Conference on Plasma Physics and Controlled Nuclear Fusion Research*, Baltimore, 2, 185.
32. Fujiwara, M., Kamimura, R., Hosokawa, M., et al. (1983). Experimental and numerical studies on plasma confinement in Nagoya Bumpy Torus (NBT). In: *Proceedings of 9th Conference on Plasma Physics and Controlled Nuclear Fusion Research*, Baltimore, 2, 197.
33. Beidler, C. D., Kolesnichenko, Y. I., Marchenko, V. S., et al. (2001). Stochastic diffusion of energetic ions in optimized stellarators. *Physics of Plasmas, 8*, 2731.
34. Kolesnichenko, Y. I., Lutsenko, V. V., Wobig, H., et al. (2001). Alfven continuum and high-frequency eigenmodes in optimized stellarators. *Physics of Plasmas, 8*, 491.
35. Girka, I., & Stepanov, K. (1992). MHD plasma oscillations in a bumpy magnetic field. *Ukrainian Physical Journal, 37*(1), 69–75. (in Russian).
36. Girka, I., Lapshin, V., & Stepanov, K. (1998). Splitting of the spectra of MHD plasma oscillations in a rippled magnetic field. *Plasma Physics Reports, 24*(11), 948–955.
37. Girka, I., Lapshin, V., & Stepanov, K. (1994). Plasma heating near satellite Alfven resonances in confinement systems with a ripple magnetic field. *Plasma Physics Reports, 20*(11), 916–922.
38. Beljaev, M., Girka, & I., Gritsyna, V. (2003). Effect of the periodic ripple in an axial confining magnetic field on the Alfven heating of a cylindrical plasma. *Plasma Physics Reports, 29*(5), 399–406.
39. Girka, I. O. (1996). *How ripples of the steady magnetic field in tokamaks influence propagation, conversion and absorption of Alfven and fast magneto-sonic waves.* Preprint of Max-Planck-Institut für Plasmaphysik, 85748 Garching bei München, FRG, IPP 4/274.
40. Girka, V., & Girka, I. (2006). Surface flute modes in a rippled magnetic field. *Plasma Physics Reports, 32*(9), 750–758.
41. Girka, V., & Girka, I. (2006). Additional ECR heating of a radially inhomogeneous plasma via the absorption of a satellite harmonics of the surface flute modes in a rippled magnetic field. *Plasma Physics Reports, 32*(12), 1047–1051.
42. Girka, V., & Girka, I. (2006). ECR plasma heating caused by absorption of surface flute modes in a bumpy magnetic field. In: *Proceedings of the 14th Joint Workshop on Electron Cyclotron Emission and Electron Cyclotron Resonance Heating. Santorini*, Greece, 9–12 May, 2006, pp.278–282.
43. Kuzelev, V. M., & Rukhadze, A. A. (1990). *Electrodynamics of dense electron beams in plasma.* Nauka. (in Russian).
44. Shestopalov, V. P. (1976). *Diffraction electronics.* Kharkiv: Vyscha Shkola. (in Russian).
45. Murphy, A.B. (1990). Waves in the edge plasma during ion cyclotron resonance heating. *Fusion Engineering and Design, 12*(1, 2), 79–92.
46. Noterdaeme, J.-M., et al. (1990). Experimental results on the edge effects during ICRF heating. *Fusion Engineering and Design, 12*(1, 2), 127–137.
47. Ballico, M.J., & Cross, R.C. (1990). Probe measurements of ICRF surface waves in the torus tokamak. *Fusion Engineering and Design, 12*(1, 2), 197–201.

48. Girka, V., Girka, I., & Thumm, M. (2014). *Surface flute waves: Theory and Applications.* Springer.
49. Jotaki, E., & Itoh, S. (1997). ECR-discharge cleaning for TRIAM-1M. *Fusion Engineering and Design, 36*, 447–450.
50. Ushigusa, K., Seki, M., Suganuma, K., et al. (1999). Electron cyclotron resonance discharge cleaning by using LHRF system on JT-60 U. *Fusion Engineering and Design, 45*, 137–144.
51. Girka, V., Girka, I., & Pavlenko, I. (2001). Electrodynamic model of the gas discharge sustained by azimuthal surface waves. *Contributions to Plasma Physics, 41*(4), 393–406.

Chapter 6
Surface Flute Waves in Waveguides with Non-circular Cross-Section

The most efficient way to excite waveguide structures (including those which are either corrugated or contain phase velocity slowing down elements like combs, disks, etc.) can be realized in the range of their eigenfrequencies [1–3]. This fact explains the interest to studying frequency spectra of multi-component plasmas filled waveguides supplied with different types of slow-wave structures. Multi-component waveguide structures are widely used [4] in modern devices of plasma electronics. The presence of different semiconductor and/or dielectric insertions in a metal waveguide increases the number of eigenmodes and power of electromagnetic radiation that can be obtained from generators and amplifiers built up on the basis of these type of waveguides. It also leads to the appearance of some specific features both in the frequency spectra of the waves propagating in such waveguides and in the spatial distribution of their fields compared with the case of application of uniform waveguides. In particular, different types of surface waves (SWs) can propagate there [5]. Properties of SWs, which propagate along a planar boundary between two different media, and properties of SWs propagating along a boundary, which is characterized by a finite value of curvature radius, differ substantially. For instance, in the case of the planar shape of the plasma boundary most of the SWs are slow waves, they are either electrostatic (i.e. their magnetic field is negligible small compared with their electric field) or at least can be considered in the electrostatic approach, see e.g. [6–9], unlike the case of SWs propagating along the plasma boundary with finite curvature. The frequency ranges of their existence are different for these two types of plasma interfaces.

The shape of the cross-section of the plasma-vacuum interface in fusion devices also distinguishes from a circular one [10, 11]. For example, the shape of the poloidal cross-section of the boundary of closed magnetic surfaces in the stellarator Wendelstein 7-X [12], which is being operated in Germany, changes along the axial co-ordinate. It periodically (there are five periods) has quasi-triangular, bean-like and intermediate shapes. Thus, when carrying out research into the propagation of axially non-symmetrical SWs in fusion devices one has to take into account the non-circularity of their chamber cross-section. The authors of the references [13–16] have

© The Author(s), under exclusive license to Springer Nature Switzerland AG 2022
I. Girka and M. Thumm, *Surface Flute Waves in Plasmas*, Springer Series on Atomic, Optical, and Plasma Physics 120, https://doi.org/10.1007/978-3-030-98210-2_6

made the conclusion that propagation of different types of SWs can be considered as the main reason of undesirable phenomena observed in fusion devices during various ion RF heating scenarios: heating of peripheral plasma regions, enhancement of interaction between plasma particles and the inner walls of the chamber, contamination of the plasma by impurity ions and consequent damage of the chamber. Thus, investigation of the properties of transverse SWs propagating in magneto-active plasma filled metal waveguides, which have non-circular cross-section, is important as well for analyzing the experimental data obtained in fusion devices.

Using an expansion into Fourier series one can simulate any real shape of the cross-section, which characterizes any experimental plasma device. If the coefficients in this series are small, one can apply the method of successive approximations for studying the problem of SW propagation in such non-circular cross-section waveguides. The theory of extraordinary polarized surface flute modes (SFXMs), which is presented in Chap. 2, can be applied as zeroth approximation while carrying out this investigation. Spatially periodical non-uniformity of the waveguides, which are under consideration in the present Chapter, is the reason that in this case, electromagnetic waves propagate in the form of wave packets. The existence of a small parameter of the problem allows one: first, to assume that the considered wave packets can be represented in the form of a superposition of the fundamental harmonic and the two nearest satellite harmonics; second, to formulate nonlinear boundary conditions for the considered SFXM fields in the form of Taylor series over this small parameter. This investigation allows one: to calculate the correction to the SFXM frequency, whose value in zeroth approximation is obtained in Chap. 2 (for the case of circular cross-section), and to analyze the spatial distribution of the SFXM fields in the case of longitudinally corrugated waveguides.

It is reasonable to start the consideration from the description of the general geometrical features of the waveguides, where the SFXM can propagate and from the formulation of boundary conditions, which are applied for solving the problem. In the general case, the inner radius R_2 of the waveguide metal chamber with arbitrary shape of its cross-section can be determined by the following series:

$$R_2 = b \cdot \left[1 + \sum_{n=1}^{\infty} h_n \sin(n\varphi - \varphi_{2n}) \right], \tag{6.1}$$

where b is the mean radius of the considered waveguide metal chamber and $h_n \ll 1$ is the relative depth of the longitudinal corrugation of the chamber wall. Expression (6.1) is an example of a Fourier series, which can describe arbitrary waveguide cross-section. Such a corrugated waveguide, generally speaking, can be filled with a uniform plasma column with outer radius R_1, whose shape is described by the following expression:

$$R_1 = a \cdot \left[1 + \sum_{n=1}^{\infty} \delta_n \cos(n\varphi - \varphi_{1n}) \right], \tag{6.2}$$

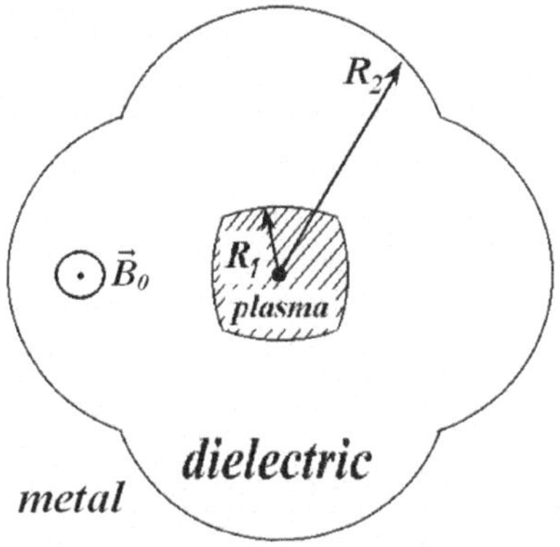

Fig. 6.1 Scheme of the magneto-active plasma filled waveguide, which has non-circular cross-sections of the chamber and plasma column

where a is the mean radius of the plasma column. The parameters $\delta_n \ll 1$ characterize the maximum relative deviation of the plasma column radius from its mean value a. The corrugation parameters in the expressions (6.1) and (6.2) satisfy the following inequality: $\delta_n,\ h_n < (b/a - 1) < 1$. The parameters $\varphi_{1n,2n}$ are the phases of the corrugation of the inner waveguide surface and the plasma interface, respectively.

In the present Chapter the plasma is assumed to be uniform, magneto-active and cold, and that it is separated by a thin $\left(\Delta = ba^{-1} - 1 \ll 1\right)$ dielectric layer with dielectric permittivity constant ε_d from the metal wall of the waveguide chamber. The dielectric permittivity of the plasma is described in fluid approximation [1, 17, 18]. An external static uniform magnetic field is oriented along the axis of the waveguide: $\vec{B}_0 || \vec{z}$. The general geometry of a waveguide chamber with corrugated shape of its metal wall and a plasma column, which also has non-circular cross-section, is shown in Fig. 6.1.

To derive the dispersion relation for SFXMs propagating in such a waveguide one can apply the following nonlinear (over both parameters h_n and δ_n, which determine the waveguide wall and plasma column interface corrugations) boundary conditions:

$$E_{r,\varphi}(r = 0) < \infty, \quad H_z(r = 0) < \infty, \tag{6.3}$$

(these conditions mean that the SFXM fields are of finite magnitudes in the volume occupied by the plasma column);

$$E_\tau(R_1 + 0) = E_\tau(R_1 - 0), \quad E_\tau(R_1) \propto R_1(\varphi)\, E_\varphi(R_1) + E_r(R_1)\, \frac{dR_1}{d\varphi},\ H_z(R_1 + 0) = H_z(R_1 - 0), \tag{6.4}$$

(these conditions mean continuity of the tangential electric and magnetic wave fields on the plasma-dielectric interface);

$$E_\tau(R_2) \propto R_2(\varphi)\, E_\varphi(R_2) + E_r(R_2)\, \frac{dR_2}{d\varphi} = 0, \tag{6.5}$$

(this condition means that the tangential electric wave field is zero on the metal wall).

For electromagnetic waves with zero axial wavenumber $\left(\frac{\partial}{\partial z} = ik_z = 0\right)$, the set of Maxwell equations splits into two subsets, which describe independent propagation of the X-mode with the field components E_r, E_φ, H_z and the O-mode with the other three field components. However, deviation of the cross-section of the plasma-dielectric interface from circular shape does not lead to coupling between X- and O-modes and surface flute O-modes cannot propagate in waveguides with a thin dielectric layer ($b - a \ll a;\ b - a \ll c/\omega$), as proved in Chap. 2. That is why the present investigation of propagation of eigenwaves in waveguides with non-circular cross-section is restricted only to the case of flute X-modes.

6.1 Surface Flute Waves in Corrugated Metal Waveguides Entirely Filled with Plasma

First, the case of SFXM propagation in longitudinally corrugated metal waveguides which are entirely filled with dense ($|\omega_e| \ll \Omega_e$) uniform cold magneto-active plasma is considered. Therefore, in this particular case, the shapes $R_1(\varphi)$ and $R_2(\varphi)$ in Fig. 6.1 are identical and there is no region of dielectric.

One can start this study from the limiting case, when the shape of the waveguide cross-section is described by (6.1), where only one arbitrary term $n = N$ (one type of corrugation) is taken into account from the complete summation over the index n. Then the obtained results can be generalized for the case of a waveguide with arbitrary cross-section. The components of the electric field of the considered SFXM can be expressed using the set of Maxwell equations through the axial component of the magnetic wave field in the following manner:

$$E_r = \frac{\omega}{ck_\perp^2}\left(\mu \frac{\partial H_z}{\partial r} - \frac{i}{r}\frac{\partial H_z}{\partial \varphi}\right), \quad E_\varphi = \frac{i\omega}{ck_\perp^2}\left(\frac{\partial H_z}{\partial r} - \frac{i\mu}{r}\frac{\partial H_z}{\partial \varphi}\right), \tag{6.6}$$

here $k_\perp^2 = k^2(\mu^2 - 1)\varepsilon_1$, and $\mu = \varepsilon_2/\varepsilon_1$, where ε_j are components of dielectric permittivity tensor of the cold magneto-active plasma [1, 17, 18]. The expressions (6.6) transform into Eq. (1.10) in a single-mode presentation.

In first order approximation with respect to h_N, the solution of the set of Maxwell equations for the axial component of the magnetic wave field has the following form:

$$\begin{aligned} H_z(r, \varphi, t) = \Big[& I_m(k_\perp r) + C_+^{(1)} I_{m+N}(k_\perp r) \exp(iN\varphi) \\ & + C_-^{(1)} I_{m-N}(k_\perp r) \exp(-iN\varphi) \Big] \exp(i(m\varphi - \omega t)), \end{aligned} \tag{6.7}$$

where $C_{\pm}^{(1)}$ are constants of integration. From the physical point of view, they are amplitudes of the SFXM satellite harmonics.

After using expression (6.7) for the calculation of E_r and E_{φ} in Eq. (6.6) these fields can be inserted in the boundary condition (6.5). This allows one to forecast that in linear approximation (with respect to h_N) the SFXM eigenfrequency does not change in comparison with the case of waveguides with circular cross-section. However, the fields of the SFXM even in linear approximation can be expressed in the form of wave packets, which contain the satellite harmonics. Their amplitudes $C_{\pm}^{(1)}$ have the following forms:

$$C_{\pm}^{(1)} = \frac{-h_N}{2D_{m\mp N}^{(0)}} I_m(k_{\perp}b)\left[k_{\perp}^2 b^2 + m(m \pm N)\left(1 - \mu^2\right)\right]_{|\omega^{(0)}}. \tag{6.8}$$

Thus, SFXMs propagate in the form of wave packets, which contain spatial harmonics with specified values of the azimuthal mode index $m = M \pm jN$, where j are natural numbers. These wave packets always have those satellites, which propagate in the direction, which is opposite to the direction of propagation of the fundamental harmonic. As one can see from formula (6.7) such a satellite harmonic appears already in the first approximation over h_N, if $|M| < N$. In the opposite case, if $|M| > N$, such a satellite harmonic appears in one of the higher order approximations. The part of the SFXM energy, which is transferred to these satellites, has a magnitude of the order h_N^{2k} (as compared with the main mode energy), where $k = 1$ for the case $|M|<N$ and $k = 2$ for $N < M < 2N$, and so on. Therefore, SFXM properties are different from those of SWs, which propagate in corrugated waveguides of planar geometry [7, 19].

In second order approximation with respect to the small parameter of the problem h_N, the boundary condition for the tangential component of the SFM electric wave field (6.5) allows one to derive the following equation:

$$\begin{gathered}(1 + h_N \sin N\varphi)E_{\varphi}(b) + h_N N \cos N\varphi \cdot E_r(b) + b\left(h_N \sin N\varphi + h_N^2 \sin^2 N\varphi\right)\frac{dE_{\varphi}}{db} \\ + h_N^2 b \sin N\varphi \cos N\varphi \cdot \frac{dE_r}{db} + \frac{b^2}{2} h_N^2 \sin^2 N\varphi \frac{d^2 E_{\varphi}}{db^2} = 0.\end{gathered} \tag{6.9}$$

Summands, which are proportional to $\exp(\pm 2Ni\varphi)$, can be neglected in Eq. (6.9) because of their smallness. Then one can derive a set of three equations for calculation of the factors, which are in front of the exponents for the different arguments. Its solution can be written in the following form:

$$D_0 + D_{\mathrm{mN}}^{(2)} = 0, \tag{6.10}$$

where $D_0(\omega_0) = 0$ is the dispersion relation of SFXMs propagating in a metal waveguide with circular cross-section of radius b, which is entirely filled with plasma (see Sect. 2.2), and $D_{mN^{(2)}}$ is a correction to D_0, which is connected with the spatial azimuthal non-uniformity of the waveguide interface. The superscript "(2)" means that it is proportional to the square of the small parameter of the problem. In the present case $D^{(2)}_{mn}$ has the following form:

$$\begin{aligned} D^{(2)}_{m\,n} &= 0.25h_n^2 I_m(k_\perp b)\big(m^2 + 3k_\perp^2 b^2 - \mu^2 m^2\big) \\ &+0.5h_n C_+^{(1)}\big[I'_{m+n}(k_\perp b)k_\perp b\,(1+\mu m) + I_{m+n}(k_\perp b)\big(k_\perp^2 b^2 + (m+n)(m+\mu)\big)\big] \\ &+0.5h_n C_-^{(1)}\big[I'_{m-n}(k_\perp b)k_\perp b\,(1+\mu m) + I_{m-n}(k_\perp b)\big(k_\perp^2 b^2 + (m-n)(m+\mu)\big)\big]_{|\omega=\omega_0} \end{aligned} \tag{6.11}$$

An analysis of the dispersion relation (6.10) for a wave packet, whose fundamental harmonic is characterized by the azimuthal mode index $m = M$, proves that in the limiting case of a narrow waveguide, $b \ll M\delta$, the SFXM frequency $\omega = \omega_0 + \Delta\omega_N$ becomes lower as compared with the case of circular cross-section, $\Delta\omega_N < 0$. Therefore, the spatial azimuthal non-uniformity of the surface of the waveguide wall studied here can be considered as slowing down structure (the frequency is lower). In the case of wide waveguides $b \gg M\delta$, the longitudinal corrugation of the plasma-metal interface decreases the value of the SFXM frequency only for SFXMs with small values of azimuthal mode index $M < N$. If the opposite inequality is valid, $M > N$, then in the mode index range $2M/\big(M^2 - N^2\big) < M\delta b^{-1} \ll 1$, deviation of the cross-section of the plasma-metal interface from a circular shape leads to an increase of the SFXM angular frequency, $\Delta\omega_N > 0$, in comparison with the case of waveguides with circular cross-section.

A numerical analysis proves that the SFXM frequency correction is with good accuracy proportional to the external magnetic field, $\Delta\omega_N \sim B_0$, and that the method of successive approximations cannot be applied in the limiting case of very wide waveguides $k_{ef} \to 0$, because of strong increasing of the $\Delta\omega_N$ value.

The investigation carried out above allows one to make the conclusion that the SFXM frequency correction caused by the deviation of the cross-section of the plasma-metal interface from a circular one does not depend in second order approximation with respect to the small parameter h_N on the presence of other small terms of the Fourier series in the right hand side of expression (6.1). Therefore, in the case of a waveguide with an arbitrary cross-section, the SFXM frequency correction caused by the deviation of the cross-section shape of the plasma-metal interface from a circular one can be found as the sum: $\Delta\omega = \sum_n \Delta\omega_n$, where the particular corrections $\Delta\omega_n$ can be independently calculated for different values of the index n (different types of corrugation). Every particular frequency correction can be calculated in the following way:

$$\Delta\omega_n = -D^{(2)}_{mn}\left(\frac{\partial D_0}{\partial\omega}\right)^{-1}_{|\omega=\omega_0}. \tag{6.12}$$

In limiting cases one can derive specific simple analytical expressions for $\Delta\omega_N$. In the case of wide waveguides, the correction of the frequency can be represented in the following form:

$$\frac{\Delta\omega_N}{\omega_0} \approx -\frac{h_N^2}{2}\left\{1 + \frac{2N|\omega_e|}{\omega_0}\right\}. \tag{6.13}$$

In the case of narrow waveguides one can obtain another expression for the correction of the frequency:

$$\frac{\Delta\omega_N}{\omega_0} \approx \frac{h_N^2}{2}\left\{1 - \frac{2N\omega_0}{|\omega_e|} + \frac{4N(\omega_0^2 - \omega_e^2)}{ka\Omega_e}\right\} < 0. \tag{6.14}$$

Analyzing the expressions (6.13) and (6.14) one can compare them with the case of waveguides with circular cross-sections, which is considered in Sect. 2.2. As one can see, the increase of the strength of an external magnetic field leads to a decrease of the SFXM frequency in comparison with the case considered in Sect. 2.2. This effect of an external magnetic field is more pronounced in the limiting case of narrow waveguides with non-circular cross-sections than in the case of wide waveguides.

Since SFXMs in longitudinally corrugated waveguides propagate in the form of wave packets, the spatial distribution of their fields is also different from the case of their propagation in waveguides with circular cross-section. Results of numerical analyses of the spatial distribution of the SFXM magnetic wave field in such type of waveguides are plotted in Fig. 6.2. There one can see the magnitude of the envelope of the SFXM magnetic wave field on the metal wall of the waveguide as function of the azimuthal angle for the following values of the waveguide parameters: $b = 2\delta$, $\Omega_e^2 = 50\omega_e^2$, $N = 4$, and $h_4 = 0.05$. The number of maxima of the SFXM magnetic wave field H_z located along the metal wall of the waveguide is equal to the azimuthal mode index. The numbers of the locations on the spatial distribution of H_z, where

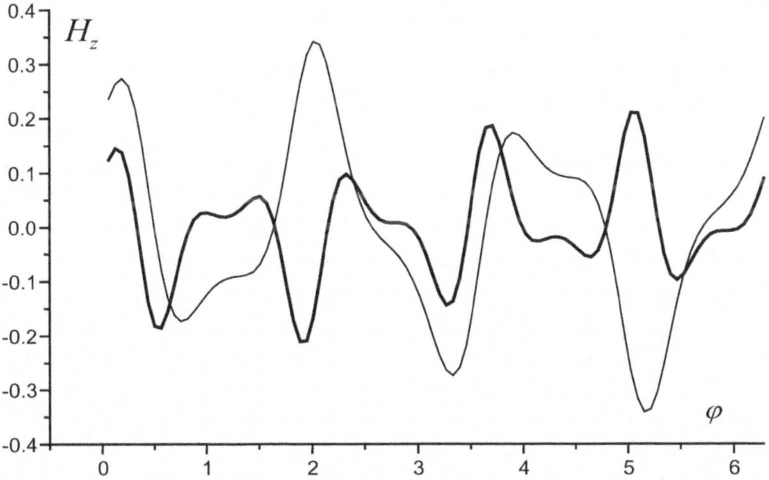

Fig. 6.2 Angular distributions of the axial magnetic field (in arb. units) of SFXMs on the metal wall of a corrugated waveguide for the cases $m = 5$ (*thick curve*) and $m = 3$ (*thin curve*)

one can see contributions of satellite harmonics, is equal to the corrugation parameter N, which determines the periodicity of the corrugation.

6.2 Surface Flute Waves in Square Metal Waveguides Entirely Filled with Plasma

From the practical point of view the results described above in this Chapter have one more important application. They can be used for studying SFXM propagation across an external axial magnetic field in a waveguide with square cross-section (square waveguide). It is interesting that in this case, application of the Cartesian system of co-ordinates seems to be more suitable: first, this is dictated by reasons of symmetry; second, it has allowed one already to solve similar problems in the cases of semi-bounded planar plasma layers (see e.g. [19]). Nevertheless, application of the Cartesian co-ordinates does not allow one to separate variables in the case of a square waveguide. Therefore, it is favorable to apply cylindrical co-ordinates.

The equation of the inner surface of a square waveguide with the side a in the cylindrical system of co-ordinates looks like:

$$R_2 = b(1 + h_4 \cos(4\varphi) + h_8 \cos(8\varphi) + h_{12} \cos(12\varphi) + ...), \tag{6.15}$$

where $b = (a/\pi)\ln\left(3 + 2\sqrt{2}\right)$, $h_4 = -0.1394$, $h_8 = 0.04395$, and $h_{12} = -0.02082$. In order to calculate the SFXM frequency correction according to expression (6.12) one has to neglect the terms of the third order of smallness. Then the applicability of expression (6.15) in the case of a square waveguide can be estimated by the value of the product $h_4^2 h_8 = 8{,}54 \times 10^{-4}$.

Now, the eigenfrequency of SFXMs in the square waveguide with $a = 1.8$ cm, which is filled by *n-Ge* semiconductor (density of conductivity electrons $n_e = 2.5 \times 10^{13}$ cm^{-3}, effective mass of electrons $m_e = 1.3 \times 10^{-27}$ g, dielectric permittivity constant $\varepsilon_d = 16$) is considered as an example [20]. It is immersed into a static magnetic field $B_0 = 100$ G. In physics of semiconductors such SWs are named as surface magneto-plasma polaritons, and such waveguide geometry is entitled as Voigt geometry [7, 8, 21]. The SFXMs with the azimuthal mode index $m = 1$ have the lowest eigenfrequency. Its value is given by: $\omega_0(m = 1) + \Delta\omega_4 + \Delta\omega_8 + \Delta\omega_{12} + ... \approx 2.23 \times 10^9 \text{s}^{-1}$. The penetration depth of the SFXM field into the semiconductor material *n-Ge* is $k_\perp^{-1} = 0.134$ cm, which is small compared with the transversal waveguide sizes. The life-time of conductivity electrons of $\tau = 10^{-5}$s in *n-Ge* is larger than the SFXM period, which justifies the application of the collisionless plasma model here.

The presented example proves that for practical needs to calculate the frequency of SFXMs, which propagate in a square waveguide, it is sufficient to take into account (in addition to the basic term $\omega_0(m)$) only the first order correction, because the contribution of the terms, which are neglected in this case, is about 4% only. However, such situation has no universal character. For example, this statement is not valid for waveguides, whose cross-sections have a rectangular shape with the sides b and d, respectively.

When the difference of these dimensions is relatively small, $|b - d| \ll b$, to estimate the value of terms, which are neglected in this case, one can calculate the asymptotic limit of expression (6.12) for large values of the summation parameter n. That means, if the inequality $n \gg k_{\perp}b \gg |m|$ is valid, one gets for low frequency SFXMs ($\omega < |\omega_e|$):

$$\Delta\omega_n = -0.5h_n^2 k_{\perp} b\, n\, \omega_0(m). \tag{6.16}$$

For high frequency SFXMs ($\omega_{UH} < \omega < \omega_2$, here $\omega_2 = |\omega_e|/2 + \sqrt{\Omega_e^2 + \omega_e^2/4}$ is the cut-off frequency and ω_{UH} is the upper hybrid resonance frequency), the frequency correction has another form:

$$\Delta\omega_n \approx -0.5h_n^2 k_{\perp} b\, n\, \omega_e^2/\omega_0(m). \tag{6.17}$$

An analysis of Eqs. (6.16) and (6.17) shows that the expression for $\Delta\omega_n$ in the form of series over h_n converges, if this small parameter decreases faster than n^{-1} with increasing parameter n.

Consequently, the SFXM frequency in a square waveguide with the side a is close to the SFXM frequency in a cylindrical waveguide will radius $R = (a/\pi)\ln\left(3 + 2\sqrt{2}\right)$. For determination of the SFXM frequency in zeroth approximation it is possible to apply the asymptotic formulas for SFXMs obtained in Sect. 2.2. Then its approximate value is: $\omega_0(m = 1) = 2.477 \times 10^9\ \mathrm{s}^{-1}$. This value differs from the value of the SFXM frequency calculated above by less than 2%. This is a good result, but the approximate value of the frequency correction calculated with expression (6.17) is $\Delta\omega_4 = -3.7 \cdot 10^7\ \mathrm{s}^{-1}$ which differs from the value obtained numerically from the SFXM dispersion relation by an order of magnitude. This noticeable difference is explained by the fact that in this case $k_{\perp}b = 7.5$, which means, that the inequality, which confirms the validity of the approximate expression (6.17), is only badly fulfilled [20].

Two differences of the properties of SFXMs, which propagate in a square waveguide from those SWs, which propagate in a planar plasma layer, should be pointed out. First, the SW frequency in a uniform planar plasma layer does not depend on the thickness of the planar waveguide (in hydrodynamic approach), but the SFXM frequency in a square waveguide explicitly depends on the waveguide sizes. This is explained by the fact that the frequency of SWs propagating in a plasma layer depends on the wave vector along the direction of their propagation. In the case of a plasma layer, which has an infinite size along the direction of SW propagation, this vector can be of arbitrary value. However, in the case of SFXMs propagating in a square waveguide the role of the wave vector is played by the ratio M/b for the fundamental harmonic. Second, SWs propagate along the opposite surfaces of the planar plasma layer independently from each other, whereas the SFXM phases on the opposite surfaces of a square waveguide exactly correlate (the difference of the phases for the fundamental harmonic is equal to $M\pi$). Thus, one can make the conclusion that due to the finite longitudinal sizes of the uniform planar plasma layer there exists a connection between SFXMs, which propagate along its opposite surfaces.

The spatial distributions of the magnetic and radial electric wave fields of the SFXM, whose fundamental harmonic is characterized by the azimuthal mode index $m = 1$, in a waveguide filled with $n - Ge$ are shown in Figs. 6.3 and 6.4. The graphs are calculated according to the formulas (6.6) and (6.7), in which small satellites $\propto \exp(i(m \pm 4)\varphi)$ and $\propto \exp(i(m \pm 8)\varphi)$ are taken into account. The magnitudes of the wave fields are given in arbitrary units. One can see that the SFXM magnetic wave field has maximal amplitude: this means that SFXMs are non-electrostatic, or in other words they are electromagnetic waves. Their field consists of a very large magnetic component and a very small electric component, which can be even neglected. The numerical analysis shows that the amplitude of the slow satellite $\propto \exp(i(M + N)\varphi)$ is larger than the amplitude of the fast satellite $\propto \exp(i(M - N)\varphi)$ if the inequality $M < N$ is valid, and vice versa, if the opposite inequality $M > N$ is fulfilled.

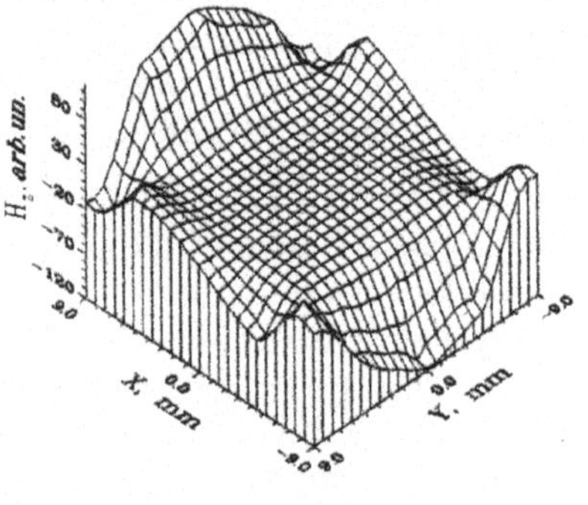

Fig. 6.3 Spatial distribution of axial magnetic wave field of the SFXM with azimuthal mode index $m = 1$ in the cross-section of a square waveguide, which is entirely filled with cold magneto-active plasma

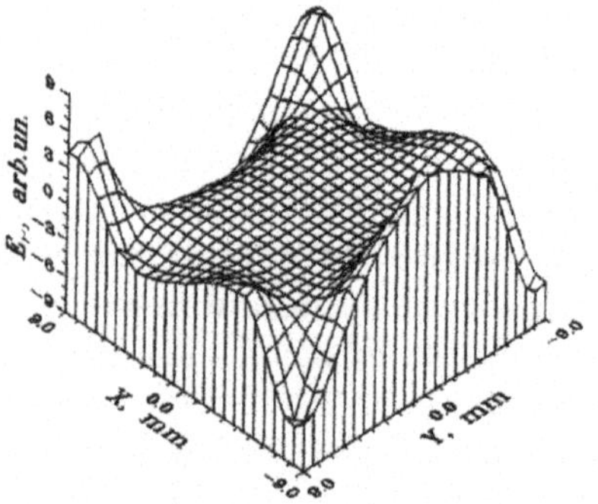

Fig. 6.4 Spatial distribution of radial electric wave field of the SFXM with azimuthal mode index $m = 1$ in the cross-section of a square waveguide, which is entirely filled with cold magneto-active plasma

The dependence of the SFXM frequency correction in a rectangular waveguide on the plasma density, constant magnetic field and linear dimensions of the waveguide is analogous to that for a square waveguide. To demonstrate the applicability of this method, LF SFXMs propagating in a waveguide with cross-section $d \times b = 1$ mm $\times$ 2 mm, which is filled with *n-Ge* semiconductor with the parameters mentioned above, are considered. It should be underlined that if one can take into account the first four terms on the right-hand side of Eq. (6.15), then it would be possible to describe rectangular waveguides, whose sides satisfy the equality $b = 2d$ with very high accuracy. For instance, the area of the waveguide cross-section in this approximation is only by 4% less than the area of the proper rectangle. In this case the frequency of LF SFXMs with the azimuthal mode index of the fundamental harmonic $m = 1$ can be calculated in the following way: $\omega \approx \omega_0(m = 1) + |\omega_e|(\Delta\omega_2 + \Delta\omega_4 + \Delta\omega_6 + \Delta\omega_8)$.. So, its value is $\omega \approx 1.78 \times 10^{10} - 1.95 \times 10^{10} \times \left(0.034 + 1.05 \times 10^{-4} + 2.842 \times 10^{-3} + 2.132 \times 10^{-3}\right)\,\mathrm{s}^{-1} = 1.707 \cdot 10^{10}\mathrm{s}^{-1}$. Therefore, the ratio of the frequency correction and the basic term is about 4.3%. One can see that the period of these LF SFXMs is smaller than the lifetime of conductivity electrons $\tau \approx 10^{-5}$ s in a typical *n*-semiconductor (this confirms the validity of the application of the collisionless plasma model for this problem).

The properties of SFXMs in such type of waveguides are similar to the properties of SFXMs in a waveguide with circular cross-section that has the radius $R \approx 0.0766$ cm. The penetration depth of the SFXM field into *n*-Ge is $k_\perp^{-1} \approx 0.142$ cm, which is larger than the radius *R*. The difference between the analytical value of the SFXM eigenfrequency (1.773×10^{10} s^{-1}) and the value calculated from the SFXM dispersion relation is less than 1%. In wide waveguides, where $k_{ef} < 0.3$, the eigenfrequency of LF SFXMs becomes small, and the absolute value of its correction increases. Therefore, in this limiting case the proposed method is not applicable.

The numerical analysis of the spatial distribution of SFXM fields shows that the amplitude of their magnetic component in a rectangular waveguide with the parameters considered above is one order of magnitude larger than the amplitudes of their electric fields. The amplitudes of their satellites $\propto \exp(-i\varphi)$ and $\propto \exp(-i3\varphi)$, which propagate in reverse direction as compared with the fundamental harmonic, relate to the amplitude of the fundamental harmonic as 4% and 0.1%, respectively. Propagation of the other group of satellites, which are proportional to $\propto \exp[i(m + 2j)\varphi]$, coincides with the propagation direction of the fundamental harmonic, but their phase velocity is less than the velocity of the fundamental harmonic. The amplitudes of these slow satellites are larger than the amplitudes of the satellites, which propagate in opposite direction. For example, the amplitudes of the satellites $\propto \exp(+i3\varphi)$ and $\propto \exp(+i5\varphi)$ relate to the amplitude of fundamental harmonic as 18% and 1.2%, respectively.

Finally, it should be noted that SFXMs can propagate as well nearby the surface of a rectangular metal bar, which is oriented parallel to a static external magnetic field and coated by a thick (as compared with the penetration depth of the SFXM field into this plasma) layer of *n*-semiconductor. Consequently, for this case one can also apply the described method and predict that the main results will be analogous to those presented here.

6.3 Surface Flute Waves in Metal Waveguides with Non-circular Cross-Section, Partially Filled with a Circular Cross-Section Plasma Column

In the present Section, a waveguide, whose radius is described by expression (6.1) is considered. It is filled with a cold uniform plasma column with radius $R_1 = a$; an external magnetic field is oriented along the $\vec{z}$ axis. Thus, to illustrate the geometry of the waveguide in this particular case one can also apply the scheme shown in Fig. 6.1, but the curve $R_1(\varphi)$, which describes the shape of the plasma column cross-section should be replaced by a circle. Solving the set of Maxwell equations for extraordinary polarized SFXMs in the plasma region, one can find expressions for their wave fields in the form of series (6.7). The components of their electric field are expressed by the formulas (6.6). In the dielectric region, $H_z(r,t)$ is determined by the superposition of Neumann functions and Bessel functions of the first kind, see expression (2.1) in Sect. 2.1. One should remind that in this region the components of the SFXM electric wave field are expressed through $H_z(r,t)$ by the formulas (6.6), where one can make the following replacements: $\mu \to 0$, $k_\perp^2 \to -\kappa_d^2$.

In zeroth approximation one can apply the theory presented in Sect. 2.1. [22]. Then the SFXM magnetic field can be written in the form of just one definite harmonic and the dispersion relation has the form of the SFXM dispersion relation (2.4) considered in Chap. 2. In the first order approximation one takes into account the terms $\propto h_n$ in the expressions for the SFXM wave fields. Application of the boundary conditions indicated above at the beginning of Sect. 6.1 allows to derive the dispersion relation in second order approximation. Its solution for the SFXM frequency can be written in the form of a sum as it was done in the previous section. Using the recurrent relations for cylindrical functions allows one to rewrite Bessel functions or their derivatives with the help of the Bessel functions of another order. Assuming the dielectric layer to be thin, it is possible to find [23] analytical expressions for $\Delta\omega_N$, which have a simple form in the limiting cases.

In a wide ($k_\perp a,\ \kappa_d a \gg |m|$) waveguide, the correction of the frequency is equal to:

$$\frac{\Delta\omega_N}{\omega_0} \approx -\frac{h_N^2}{16}\left\{1 - \frac{2k_\perp^2 a^2\left(k_\perp a + \varepsilon_2 m \varepsilon_1^{-1}\right)}{(k_\perp a + \varepsilon_2 m \varepsilon_1^{-1})^2 - \varepsilon_2^2 N^2 \varepsilon_1^{-2}}\right\}, \tag{6.18}$$

and in a narrow waveguide ($k_\perp a,\ \kappa_d a \ll 1$) it is:

$$\frac{\Delta\omega_N}{\omega_0} = \frac{h_N^2}{16}\left\{1 - \frac{2(\varepsilon_2 - \varepsilon_1)m^3}{\varepsilon_d\left(m^2 - N^2\right)}\right\}. \tag{6.19}$$

With decreasing radius of the metallic waveguide and increasing number N of its corrugations, the magnitude of the frequency correction increases, which leads to a decrease of the eigenfrequency. Numerical analyses show, that the correction increases with decreasing thickness of the dielectric layer Δ or with decreasing dielectric constant [23].

6.4 Low Frequency Surface Flute Waves in Metal Waveguides with Circular Cross-Section, Partially Filled with a Plasma Column of Non-circular Cross-Section

In the present Section, a cylindrical metal chamber with inner radius b and thin dielectric coating with dielectric permittivity ε_d is considered. An external static magnetic field is oriented along the z-axis: $\vec{B}_0 || \vec{z}$. The considered waveguide is uniform along its axis and is filled with a cold plasma column with sufficiently high density ($\Omega_e \gg |\omega_e|$). The shape of the plasma cross-section is assumed to be arbitrary (see expression (6.2)). Therefore, to illustrate the geometry of the plasma filled waveguide in this particular case one can apply Fig. 6.1 once again, but now the curve $R_2(\varphi)$, which describes the inner shape of the metal chamber, should be changed to be a circle.

In zeroth approximation, the plasma-dielectric interface is considered as circular and concentric with the metal chamber, so one can apply the theory of SFXMs presented in Chap. 2. However, periodic spatial non-uniformity of the plasma surface (6.2) leads to propagation of the SFXMs in the form of wave packets, which propagate not only along the azimuthal direction but along radial direction as well. Such a packet contains (in addition to the fundamental harmonic $\propto \exp(im\varphi)$) an infinite number of satellite harmonics $\propto \exp[i(m \pm jn)\varphi], j = 1, 2, 3\ldots$ According to the theory of electromagnetic waves propagating in waveguides with one-dimensional periodicity [24] the amplitudes of satellite harmonics are lower than the amplitude of the fundamental harmonic, their ratio is $\propto \delta_n^j$. To solve the problem of SFXM propagation in the considered waveguide in second order approximation over the small parameter δ_n, one can consider firstly a simpler definite shape of the plasma-dielectric interface, $R_1 = a[1 + \delta_N \cos(N\varphi + \varphi_N)]$, and then generalize the obtained results for the case of arbitrary shape of the plasma-dielectric interface (6.2). In this approximation it is sufficient to take into account only the fundamental harmonic and its two nearest satellites in the wave packet.

The radial dependence of the SFXM harmonics in the dielectric layer is expressed through Bessel functions of the first kind and Neumann functions. In the plasma region, they are determined by modified Bessel functions and their derivatives over the argument [25]. Within the accuracy of second order of the small parameter of the problem, the boundary conditions on the surface $r = R_1$ can be written in the following form:

$$\Big[E_\varphi(1-\delta_N^2N^2) - E_r\delta_N N\sin(N\varphi+\varphi_N) + a\delta_N\cos(N\varphi+\varphi_N)\tfrac{\partial E_\varphi}{\partial r} + \\ +\tfrac{a^2}{2}\delta_N^2\cos^2(N\varphi+\varphi_N)\tfrac{\partial^2 E_\varphi}{\partial r^2}\Big]_{|a} = 0, \tag{6.20}$$

$$\left[H_z + a\delta_N\cos(N\varphi+\varphi_N)\frac{\partial H_z}{\partial r} + \frac{a^2}{2}\delta_N^2\cos^2(N\varphi+\varphi_N)\frac{\partial^2 H_z}{\partial r^2}\right]_{|a} = 0. \tag{6.21}$$

The dispersion relation in zeroth approximation, $D_0(\omega_0) = 0$, can be derived by application of the boundary conditions on the circular plasma surface:

$$D_0 = I_m(k_\perp a)\beta_2 - \alpha_2(r=a)\beta_1, \tag{6.22}$$

where $\alpha_2(r) = J'_m(\kappa_d b)N_m(\kappa_d r) - J_m(\kappa_d r)N'_m(\kappa_d b)$; $\beta_1 = m\mu I_m(k_\perp a)/(ka\varepsilon_d) + I'_m(k_\perp a)\sqrt{\mu^2 - 1/\varepsilon_d}$; and $\beta_2 = -\alpha'_2(r=a)/\sqrt{\varepsilon_d}$.

Selecting only terms of the first order of smallness in the boundary conditions (6.20) and (6.21) allows to determine the amplitudes of SFXM satellite harmonics. An analysis proves that they have no symmetry: $H_z^{(+)} \neq \pm H_z^{(-)}$. Taking into consideration the terms $\propto \exp(im\varphi)$ in Eqs. (6.20) and (6.21), one can derive the SFXM dispersion relation within an accuracy of the second order smallness over δ_N: $D_0 + D_{\rm N}^{(2)} = 0$, where $D_{\rm N}^{(2)}$ is the second order correction. The explicit form of $D_{\rm N}^{(2)}$ is very cumbersome (see Ref. [26] for details), therefore, it is not shown here; instead of that the same equation, but for the case of HF flute X-modes is derived in Sect. 6.6.

The LF SFXM dispersion relation can be solved analytically only in some limiting cases. For example, in the case of wide plasma waveguides ($a/\delta \gg |m|, N$), their frequency is determined as the sum $\omega = \omega_0 + \Delta\omega_N$, where

$$\omega_0 \approx \frac{\sqrt{(1-2\Delta)\delta^4 m^2\omega_e^2 - (2a^2\Delta - a^2\Delta^2 + 2a\delta)m^2c^2\Delta(3\Delta-2)} + (1-\Delta)\delta^2 m|\omega_e|\sqrt{\varepsilon_d}}{(2a^2\Delta - a^2\Delta^2 + 2a\delta)\sqrt{\varepsilon_d}}, \tag{6.23}$$

$$\Delta\omega_N \approx -\delta_N^2\omega_0\frac{3m^2\kappa_d^2a^2 + 2\kappa_d^4a^4 - 3m^4 + m^2N^2}{4m^2(N^2 - 4m^2)\Delta^2}. \tag{6.24}$$

The expression (6.24) contains the resonant denominator $(N^2 - 4m^2)$, thus it cannot be applied in the case $N = 2|m|$. This is associated with the fact that the spectra of SFXMs propagating in waveguides with circular cross-section without external magnetic field are degenerate with respect to the sign of the azimuthal mode index. Analysis of this resonant case when $N = 2|m|$ requires additional studies; its results are presented in the next Sect. 6.6.

One can see from (6.24) that the frequency correction increases with decreasing dielectric layer thickness. However, this increase is limited by the inequality $\Delta > \delta_n$. Note that the SFXM frequency can be represented as a series over the small parameter δ_n, if δ_n decreases faster than $n^{-3/2}$ with increasing summation index n [26].

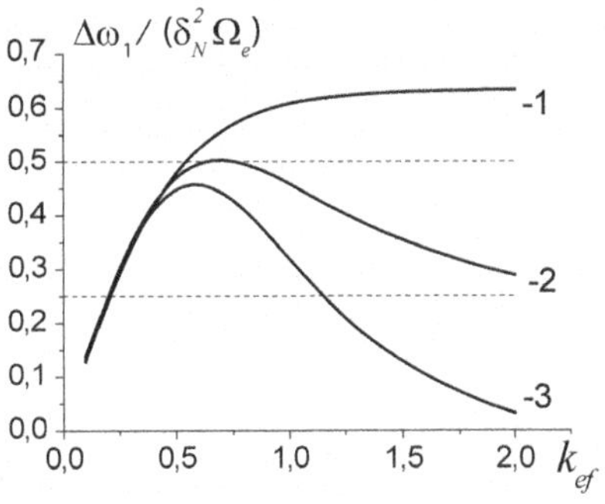

Fig. 6.5 Normalized corrections to the LF SFXM eigenfrequency in the non-resonant case for the azimuthal mode indices $m = -1, -2, -3$; $N = 1$ (case of eccentric plasma waveguide)

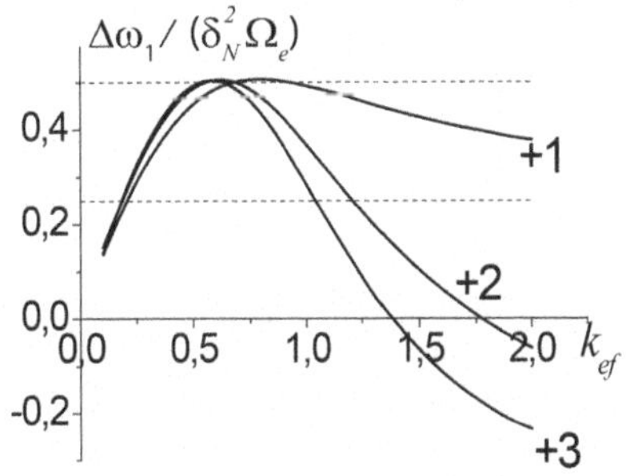

Fig. 6.6 Normalized corrections to the LF SFXM eigenfrequency in the non-resonant case for the azimuthal mode indices $m = 1, 2, 3$; $N = 1$ (case of eccentric plasma waveguide)

The results of numerical studies are shown in Figs. 6.5 and 6.6. They prove that for wide waveguides $\left(k_{ef} \ll 1\right)$ the dispersion curves $\Delta\omega_n\left(k_{ef}\right)$ weakly depend on m. Unlike in the case of narrow waveguides there exists a non-monotonous character of the curves. One can see on these plots, that there is no essential difference between the curves $\Delta\omega_1\left(k_{ef}\right)$ calculated for positive and negative values of the azimuthal mode index m in the region of small effective wavenumbers $k_{ef} \leq 0.5$. This numerical analysis confirms the conclusions of the analytical investigations that the value of $\Delta\omega_n$ increases with decreasing thickness and dielectric constant of the dielectric layer and with increasing B_0. The values of $\Delta\omega_n$ are different for SFXMs propagating in mutually opposite directions. It should be noted that in the considered case, the frequency of SFXMs with $m > 0$ is higher than the frequency of SFXMs with $m < 0$, the other parameters being the same [27]. This difference is small for wide waveguides, but it increases with increasing value of $|m|$ and for the case of waveguides with a narrow dielectric layer.

A study of the spectral composition of SFXM wave packets allows one to conclude the following. First, the amplitudes of fast (higher angular phase velocity) SFXM satellite harmonics are smaller than amplitudes of their slow harmonics (lower angular phase velocity). Second, the amplitudes of the satellite harmonics are larger in wide waveguides. Third, the magnitudes of satellite harmonics decrease with decreasing $|m|$. However, in the case of narrow waveguides for $N = 1$ (case of eccentric plasma waveguide) for SFXMs with $m = -1$, the amplitude of the standing wave satellite harmonic, which does not transfer its energy along the azimuthal angle, is larger than the amplitude of the slow satellite harmonic, which propagates along the azimuthal angle with the angular phase velocity $\omega/(m-N) = -\omega/2$, which is lower than the angular phase velocity of the fundamental harmonic [26]. In general, the effect of corrugation of the plasma column surface on the SFXMs is more pronounced in comparison with the effect of the metal wall corrugation on SFXMs propagating in such type of waveguides, which are partially filled with a plasma column, whose cross-section is of circular form.

6.5 Resonant Influence of Plasma-Dielectric Interface Shape on Surface Flute Waves in Magneto-active Waveguides

In the general case, which means that SFXMs propagate with azimuthal mode index m in cylindrical waveguide structures whose periodicity is characterized by the number of azimuthal periods $N \neq 2|m|$. The frequency corrections of SFXMs are shown in the previous section to be of second order of smallness, $\Delta\omega_n \propto \delta_n^2$. However, the results obtained there are invalid in the resonant case, $N = 2|m|$. Degeneration of the wave (or energetic) spectra happens in many branches of physics and its analysis has some specific features [28]. It was found in [29] that SFXMs propagating in isotropic waveguides with non-circular plasma cross-section exist as two standings waves, whose frequencies are close. That is why beating modes arise as the result of this standing wave interference and the correction to the eigenfrequency of SFXMs becomes larger than in the non-resonant case considered in the previous section.

Here the waveguide structure with the same geometry and in the same framework as in Sect. 6.4 is studied. But unlike the case $N \neq 2|m|$ considered in the previous section, now just the case of SFXM wave packets, whose fundamental harmonic satisfies the following resonant condition [30]: $|m| = N/2$, is studied. Then one can derive the secular dispersion relation $(D_0)^2 - \left(D_m^{(1)}\right)^2 = 0$, which is typical for the case of isotropic waveguides. Moreover, if the value of the external static magnetic field is sufficiently small one can apply the theory developed for isotropic plasma filled waveguides in [29]. This statement is valid if the following inequality is satisfied:

$$\frac{|\omega_e|}{\Omega_e} \ll \frac{\delta_{2|m|}}{2}\sqrt{\frac{|m|}{\Delta}}\left[1+\sqrt{\frac{\Omega_e a}{|m|c}}\varepsilon_d^{-1/4}\left(1+\frac{a\Delta}{c}\Omega_e\right)\right]. \tag{6.25}$$

If the inequality (6.25) is not valid, one can analyze the influence of the waveguide cross-section shape on the dispersion properties of SFXMs in the resonant case numerically. It was found that non-circularity of the plasma cross-section leads to a decrease of the SFXM frequency if $m < 0$ (except for the case of the wave with $m = -1$ propagating in a wide waveguide: $k_{ef} < 1$). The absolute value of $\Delta\omega_N$ is larger for the case of narrow waveguides than for wide waveguides (except for

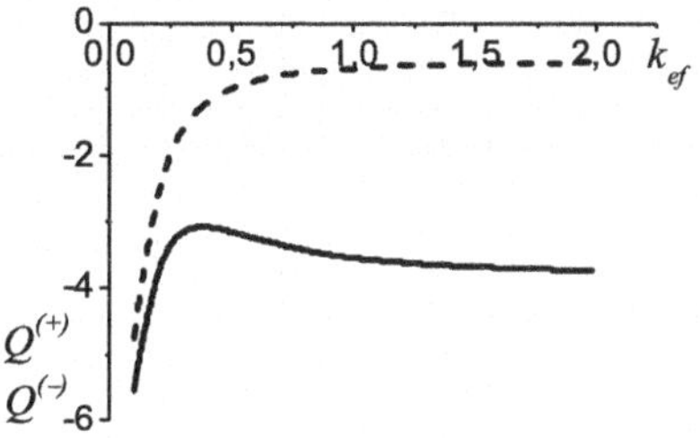

Fig. 6.7 The relative amplitudes of satellite harmonics versus k_{ef} in the resonant case, m = -1, $Z = 9$, $\varepsilon_d = 1$, $\Delta = 0.3$, and $N = 2$. The fast harmonic, $Q^{(+)}$, is depicted by the *solid curve*, the slow one, $Q^{(-)}$, is given by the *dashed curve*

the case of the mode index $m = -1$, when it is very small), its value increases with decreasing magnitude of B_0, thickness and permittivity of the dielectric layer. Changing the sign of the azimuthal mode index (from negative to positive) leads to changing sign of the frequency correction and also to some increase of its absolute value.

For the case $m = -1$ the results of numerical analyses of the SFXM spectral composition are plotted in Fig. 6.7. Curves calculated for the fast and slow satellites are marked by solid and dashed curves, respectively. Their analysis proves that the relative amplitudes of the slow and fast satellite harmonics sufficiently differ from each other unlike in the non-resonant case studied in the previous section. This can be explained by resonant increasing of the satellite harmonic, which propagates with the same angular phase velocity as the fundamental harmonic, but in the opposite direction (just this satellite harmonic is entitled here as resonant satellite harmonic). The magnitude of the relative amplitude of the resonant satellite harmonic essentially increases with increasing absolute value of the azimuthal mode index for the case $m < 0$. Decreasing the dielectric permittivity of the layer, which separates the plasma column from the waveguide metal wall, or a decrease of the strength of the external magnetic field lead to an increase of the absolute value of the relative amplitude of the resonant satellite harmonic.

Numerical analyses of the spectral composition of the SFXM wave packet, whose fundamental harmonic is characterized by $m < 0$, allow one to make the following conclusions on their peculiarities. First, the amplitudes of the fast satellites are smaller than the amplitudes of the slow satellites. Second, the amplitudes of satellite harmonics are larger in wide waveguides as compared with the case of narrow waveguides. Third, the amplitudes of satellite harmonics decrease with decreasing value of $|m|$. Fourth, in the case of positive sign of the azimuthal mode index m, the amplitude of the resonant satellite harmonic becomes firstly equal to zero and then becomes positive while the radius of the plasma column decreases until such waveguide becomes sufficiently small and can be already considered as a narrow one.

The possibility to control the SFXM frequency spectrum and spectral composition of their wave packet by choosing the shape of the plasma column cross-section is shown in [30] for the resonant case. In this case the LF SFXM frequency correction is $\Delta\omega_n \propto B_0^{-1}$ and its value becomes much larger, $\Delta\omega_n \propto \delta_n$, than in the non-resonant case. In general, the influence of the plasma column cross-section non-circularity is stronger than in the case when LF SFXMs propagate in waveguides with longitudinally corrugated metal wall (see previous Sect. 6.3). This is explained by the fact that the amplitude of SFXMs is maximal just on the plasma-dielectric interface and decreases if one goes away from the surface.

6.6 High Frequency Surface Flute Waves in Metal Waveguides with Circular Cross-Section, Partially Filled with a Plasma Column of Non-circular Cross-Section

In the present Section, the propagation of surface flute waves with extraordinary polarization in a perfectly conducting cylindrical metal chamber with radius b is studied. In the chamber, there is a column of a cold plasma separated from the chamber wall by a dielectric layer with the dielectric constant $\varepsilon_d \geq 1$. An external uniform static magnetic field is directed along the waveguide axis, $\vec{B}_0 || \vec{z}$. It is assumed as well that the SFXMs propagate strictly transversely to the magnetic field and the waveguide is uniform along the cylinder axis, $\partial/\partial z \equiv 0$. The cross-section of the plasma column is described by expression (6.2). To illustrate the geometry of the plasma filled waveguide, which is under consideration in the present Section one can apply Fig. 6.1 again, where the curve $R_2(\varphi)$, which describes the inner radius of the metal chamber should be replaced by a circle. The displacement vector and electric field vector are assumed to be related by the dielectric permittivity tensor of a weakly collisional cold magneto-active plasma (see, e.g. [1, 17, 18]). Unlike in the two previous sections, here the case of HF SFXMs in the frequency range (1.14), $\omega_{UH} < \omega < \omega_1$, is under consideration.

The dispersion properties of HF SFXMs are investigated here by the method of successive approximations. In zeroth approximation, the plasma-dielectric interface is circular and is concentric with the metal chamber. The properties of SFXMs in this approximation can be described by using the results obtained in Chap. 2. When the radius of curvature of the plasma-dielectric interface is constant in azimuthal direction, SFXMs with different azimuthal mode index propagate independently of each other. That is why, in zeroth approximation, one can assume that HF SFXMs propagate only with a definite azimuthal mode index m.

Because of the periodic spatial non-uniformity of the plasma-dielectric interface (see expression (6.2)), SFXMs propagate as wave packets. Each such wave packet contains the fundamental harmonic with a field varying as $\exp(im\varphi - i\omega t)$ and an infinite number of satellite harmonics with fields varying as $\exp[i(m + jN)\varphi - i\omega t]$ ($j = 1, 2, 3...$). From the theory of wave propagation in media whose properties vary periodically in one definite direction [24], the amplitudes of satellite harmonics are known to be small quantities (their values are of the order $(\delta_N)^j$) in comparison with the fundamental harmonic amplitude. With the objective to determine how the dispersion properties of HF SFXMs depend on the shape of the non-circular cross-section of the plasma-dielectric interface, one can consider further a wave packet with only two nearest satellite harmonics, which are proportional to $\exp[i(m \pm N)\varphi - i\omega t]$. The reason for this approximation is that the higher order satellite harmonics make only third and higher order contributions to the correction of the eigenfrequency of SFXMs. Using the expression (6.2) for the plasma-dielectric interface, one can see that, within second order terms, the boundary conditions (6.4) for the tangential components of the SFXM fields take the form of continuity conditions at the radius $r = a$, which are similar to expressions (6.20) and (6.21), respectively.

The spatial distributions of the azimuthal electric and axial magnetic wave fields of a SFXM in a dielectric layer ($R_1 < r < b$) are expressed in terms of Bessel functions of the first kind and Neumann functions [25]:

$$E_{\varphi}^{(V)}(r,\varphi) = \frac{-i}{\sqrt{\varepsilon_d}}[S^{(0)}P_m'(\kappa r) + \sum_{\pm} S^{(\pm)}P_{m\pm N}'(\kappa r)\exp(\pm iN\varphi)]\exp(im\varphi - i\omega t), \tag{6.26}$$

$$H_z^{(V)}(r,\varphi) = [S^{(0)}P_m(\kappa r) + \sum_{\pm} S^{(\pm)}P_{m\pm N}(\kappa r)\exp(\pm iN\varphi)]\exp(im\varphi - i\omega t), \tag{6.27}$$

where $P_m(\kappa r) = J_m(\kappa r)N_m'(\kappa b) - J_m'(\kappa b)N_m(\kappa r)$ with $\kappa^2 = \varepsilon_d k^2$. $S^{(0)}$ is a normalizing factor, which characterizes the amplitudes of the fundamental harmonics of the SFXM fields in the dielectric layer. The prime index denotes the derivative of the function over its argument, the symbol $\sum_{\pm}$ means summation over the upper and lower signs, the constants $S^{(\pm)}$ are determined by the boundary conditions (6.3)–(6.5), and the functions $P_m(\kappa r)$ and $P_{m\pm N}(\kappa r)$ are constructed in such a way that the azimuthal component of the electric wave field of the SFXMs satisfies the boundary condition (6.5).

The radial dependence of the SFXMs fields in the plasma region ($0 < r < R_1$) is expressed in terms of the modified Bessel functions and their derivatives [25]:

$$\begin{aligned} E_{\varphi}^{(p)}(r,\varphi) = \tfrac{ik}{k_{\perp}^2}\exp(im\varphi - i\omega t)\big\{A^{(0)}\big[\tfrac{m\mu}{r}I_m(k_{\perp}r) + k_{\perp}I_m'(k_{\perp}r)\big] \\ + \sum_{\pm} A^{(\pm)}\Big[\tfrac{(m\pm N)\mu}{r}I_{m\pm N}(k_{\perp}r) + k_{\perp}I_{m\pm N}'(k_{\perp}r)\Big]\exp(\pm iN\varphi)\Big\}, \end{aligned} \tag{6.28}$$

$$\begin{aligned} H_z^{(p)}(r,\varphi) = \exp(im\varphi - i\omega t) \\ \Big[A^{(0)}I_m(k_{\perp}r) + \sum_{\pm} A^{(\pm)}I_{m\pm N}(k_{\perp}r)\exp(\pm iN\varphi)\Big], \end{aligned} \tag{6.29}$$

where $A^{(0)}$ is a normalizing factor, which characterizes the amplitude of the fundamental harmonics of the SFXM fields in the plasma column. The constants $A^{(\pm)}$, which characterize the amplitudes of the satellite harmonics of these fields in the plasma, can be determined from the boundary conditions given above. The expressions (6.28) and (6.29) for the field components $E_\varphi^{(p)}(r,\varphi)$ and $H_z^{(p)}(r,\varphi)$ satisfy the boundary condition (6.3).

In first order approximation, using the mentioned boundary conditions, one can derive relations between the amplitudes of the satellite harmonics of the HF SFXM fields and the amplitudes of their fundamental harmonics for the dielectric and plasma regions, respectively. They are given by the following expressions:

$$A^{(\pm)} = p_1^{(\pm)} A^{(0)} + q_1^{(\pm)} S^{(0)}, \quad S^{(\pm)} = p_2^{(\pm)} A^{(0)} + q_2^{(\pm)} S^{(0)}, \tag{6.30}$$

$$p_1^{(\pm)} = \left(\alpha_2^{(\pm)}\beta_3^{(\pm)} - \alpha_3\beta_2^{(\pm)}\right)/D^{(\pm)}, \quad p_2^{(\pm)} = \left(\alpha_3\beta_1^{(\pm)} - \alpha_1^{(\pm)}\beta_3^{(\pm)}\right)/D^{(\pm)}, \tag{6.31}$$

$$q_1^{(\pm)} = \left(\alpha_2^{(\pm)}\beta_4^{(\pm)} - \alpha_4\beta_2^{(\pm)}\right)/D^{(\pm)}, \quad q_2^{(\pm)} = \left(\alpha_4\beta_1^{(\pm)} - \alpha_1^{(\pm)}\beta_4^{(\pm)}\right)/D^{(\pm)}, \tag{6.32}$$

$$D^{(\pm)} = \alpha_1^{(\pm)}\beta_2^{(\pm)} - \alpha_2^{(\pm)}\beta_1^{(\pm)}, \tag{6.33}$$

$$\alpha_1^{(\pm)} = I_{m\pm N}(k_\perp a), \quad \alpha_2^{(\pm)} = -P_{m\pm N}(\kappa a), \tag{6.34}$$

$$\alpha_3 = 0.5\delta_N k_\perp a I_m'(k_\perp a), \quad \alpha_4 = -0.5\delta_N \kappa a P_m'(\kappa a), \tag{6.35}$$

$$\beta_1^{(\pm)} = \frac{k}{k_\perp^2}\left[\frac{(m\pm N)}{a}\mu I_{m\pm N}(k_\perp a) + k_\perp I_{m\pm N}'(k_\perp a)\right], \quad \beta_2^{(\pm)} = \frac{P_{m\pm N}'(\kappa a)}{\sqrt{\varepsilon_d}}, \tag{6.36}$$

$$\beta_3^{(\pm)} = -\frac{\delta_N k}{2k_\perp}\left[\frac{(m\mu \mp mN - k_\perp^2 a^2 - m^2) I_m(k_\perp a)}{k_\perp a} + I_m'(k_\perp a)(1 - m\mu \mp \mu N)\right], \tag{6.37}$$

$$\beta_4^{(\pm)} = 0.5\delta_N k a^{-1}\kappa^{-2}\left[(m^2 \pm mN - a^2\kappa^2)P_m(\kappa a) + \kappa a P_m'(\kappa a)\right]. \tag{6.38}$$

An analysis of the expressions for the SFXM fields shows that the amplitudes of their satellite harmonics do not possess any symmetry: $H_z^{(+)} \neq \pm H_z^{(-)}$.

The dispersion relation for HF SFXMs propagating along the longitudinally corrugated plasma interface has the same form as it was obtained in [26] for LF SFXMs (see Sect. 6.3), $D_0 + D_N^{(2)} = 0$, where $D_0(\omega_0) = 0$ is the dispersion relation of SFXMs in zeroth approximation (see Sect. 2.1). The expression for $D_N^{(2)}$ (super-script "(2)" means that it is a second order correction over the small parameter of the problem) reads as follows:

$$D_N^{(2)} = \alpha_1^{(0)}\beta_2^{(2)} - \alpha_2^{(2)}\beta_1^{(0)} + \alpha_1^{(2)}\beta_2^{(0)} - \alpha_2^{(0)}\beta_1^{(2)}, \tag{6.39}$$

where

$$\alpha_1^{(2)} = 0.5\delta_N a\{k_\perp\Big[p_1^{(+)}I'_{m+N}(k_\perp a) + p_1^{(-)}I'_{m-N}(k_\perp a)\Big]$$
$$-\kappa\Big[p_2^{(+)}P'_{m+N}(\kappa a) + p_2^{(-)}P'_{m-N}(\kappa a)\Big]\} + 0.25\delta_N^2 k_\perp^2 a^2 I_m''(k_\perp a);$$
$$\alpha_2^{(2)} = 0.5\delta_N k_\perp a\Big[q_1^{(+)}I'_{m+N}(k_\perp a) + q_1^{(-)}I'_{m-N}(k_\perp a)\Big]$$
$$-0.5\delta_N \kappa a\Big[q_2^{(+)}P'_{m+N}(\kappa a) + q_2^{(-)}P'_{m-N}(\kappa a)\Big] - 0.25\delta_N^2\kappa^2 a^2 P_m''(\kappa a);$$
$$\beta_1^{(2)} = 0.5\delta_N N\Big[p_1^{(-)}\beta_5^{(-)} - p_1^{(+)}\beta_5^{(+)}\Big] + 0.5a\delta_N\Big[p_1^{(-)}\beta_6^{(-)} + p_1^{(+)}\beta_6^{(+)}\Big]$$
$$+0.25\delta_N^2 a^2\beta_7 + 0.5\delta_N Nk\kappa^{-2}a\Big[(m-N)p_2^{(-)}P_{m-N}(\kappa a) - (m+N)p_2^{(+)}P_{m+N}(\kappa a)\Big]$$
$$+0.5ak\delta_N\Big[p_2^{(-)}P''_{m-N}(\kappa a) + p_2^{(+)}P''_{m-N}(\kappa a)\Big];$$
$$\beta_2^{(2)} = 0.5\delta_N\Big[N\Big(q_1^{(-)}\beta_5^{(-)} - q_1^{(+)}\beta_5^{(+)}\Big) + a\Big(q_1^{(-)}\beta_6^{(-)} + q_1^{(+)}\beta_6^{(+)}\Big)\Big]$$
$$+\delta_N^2 a^2\kappa^2 P_m'''(\kappa a)/\big(4\sqrt{\varepsilon_d}\big)$$
$$+\delta_N Nk\Big[(m-N)q_2^{(-)}P_{m-N}(\kappa a) - (m+N)q_2^{(+)}P_{m+N}(\kappa a)\Big]/\big(2a\kappa^2\big)$$
$$+0.5ak\delta_N\Big[q_2^{(-)}P''_{m-N}(\kappa a) + q_2^{(+)}P''_{m-N}(\kappa a)\Big];$$
$$\beta_5^{(\pm)} = \frac{k}{k_\perp^2}\Big[\frac{(m\pm N)}{a}I_{m\pm N}(k_\perp a) + \mu k_\perp I'_{m\pm N}(k_\perp a)\Big];$$
$$\beta_6^{(\pm)} = \frac{k}{k_\perp^2}\Big[\frac{(m\pm N)\mu}{-a^2}I_{m\pm N}(k_\perp a) + \frac{(m\pm N)\mu}{a}k_\perp I'_{m\pm N}(k_\perp a) + k_\perp^2 I''_{m\pm N}(k_\perp a)\Big];$$
$$\beta_7 = \frac{k}{k_\perp^2}\Big[\frac{2m\mu}{a^3}I_m(k_\perp a) - \frac{2m\mu}{a^2}k_\perp I_m'(k_\perp a) + \frac{m\mu}{a}k_\perp^2 I_m''(k_\perp a) + k_\perp^3 I_m'''(k_\perp a)\Big].$$

The solution of the dispersion relation for HF SFXMs can be represented in the form $\omega = \omega_0 + \Delta\omega_N$, where ω_0 is the eigenfrequency of the HF SFXMs (see Sect. 2.1). The frequency correction $\Delta\omega_N$ can be calculated for arbitrary number N (type of corrugation) according to formula (6.12) as it was done in Sect. 6.1.

With an accuracy of up to the second order of smallness over the small parameter δ_n, the different small terms $\delta_n\cos(n\varphi + \varphi_{1n})$ in expression (6.2) for the plasma column radius make their contribution independently to the correction of the HF SFXM frequency. That is why the dependence of the eigenfrequency of these waves on the deviation of the cross-section of the plasma-dielectric interface from circular shape can be described by representing the complete correction $\Delta\omega$ to the eigenfrequency as the sum of particular corrections $\Delta\omega_n$, which can be calculated according to formula (6.12).

Application of a thin dielectric layer, which separates the plasma column from the metal chamber wall, causes a shift of the dispersion curve $\omega = \omega\big(k_{ef}\big)$ of HF SFXMs towards larger values of the effective wavenumber $k_{ef} = |m|\delta/a$ within the considered frequency range, unlike in the case considered in Sect. 6.1:

$$\omega_0 \approx \sqrt{\Omega_e^2\Big(1 + k_{ef}^2\Big) + \omega_e^2} - |m|\omega_e\Delta\big(\varepsilon_d - k_{ef}^2\big). \tag{6.40}$$

Nearby the lower limit of the HF range, the SFXM penetration depth into the plasma becomes small, so that the inequality $k_\perp a \gg 1$ becomes valid. This means that in this case $\varepsilon_1 \to 0$. As far as a dispersion curve of the HF SFXMs asymptotically approaches to the lower range limit for smaller values of effective wavenumber $\left(k_{ef} \ll 1\right)$, the derivative, which is located in the denominator of the expression for the SFXM frequency correction $\Delta\omega_n$, goes to infinity, or in other words $(\partial D_0/\partial\omega)^{-1}_{|\omega=\omega_0} \to 0$. Therefore, in this case, one can expect a vanishing effect of the deviation of the cross-section of the plasma-dielectric interface from circular shape on the dispersion properties of HF SFXMs. It should be mentioned, that in this band of k_{ef} one does not observe efficient excitation of HF SFXMs by gyrating electron beams [31]. This property of HF SFXMs makes them distinctive from LF SFXMs, whose frequency is almost proportional to k_{ef} in wide waveguides.

Unlike that, nearby the upper HF range limit (or, in other words, nearby the right limit of the k_{ef} band, where HF SFXMs can propagate), SFXMs well penetrate into the plasma, $k_\perp a \ll 1$, which means validity of the relation $\varepsilon_1 \approx \varepsilon_2$. Nearby the upper HF range limit, the frequency correction of SFXMs, caused by the deviation of the cross-section of the plasma-dielectric interface from circular shape, strongly increases, $\Delta\omega_N \propto k_\perp^{-2}$. Thus, the usability condition, $\Delta\omega_N \ll \omega_0$, of the method of successive approximations applied here is violated.

The results of numerical investigations of the waveguide cross-section non-circularity on dispersion properties and spatial distribution of HF SFXMs are shown in Figs. 6.8, 6.9, 6.10, 6.11, 6.12 and 6.13. The figures are given in terms of dimensionless parameters, the correction to the frequency of HF SFXMs, $\Delta\omega_N/\left(\Omega_e\delta_N^2\right)$, normalized to the electron plasma frequency and the squared small parameter of the problem versus the effective wavenumber $k_{ef} = |m|\delta/a$. The electron plasma frequency can conveniently be chosen as a normalizing divisor for the following two reasons. First, the plasma particle density and, accordingly, the plasma frequency are constant in the model under consideration. Second, in the plots, one can immediately see how the correction to the eigenfrequency of SFXMs scales with such a characteristic parameter like the electron plasma frequency. The choice of the skin depth δ (penetration depth of electromagnetic fields into the plasma) as a normalizing factor appears to be convenient when using the fluid approximation. The numbers in these figures denote the azimuthal mode indices m.

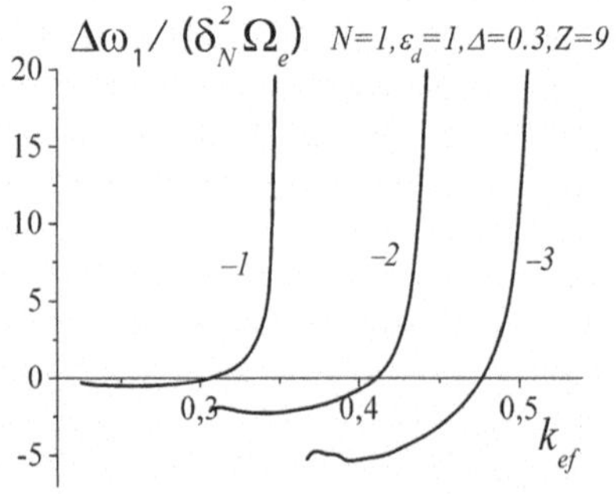

Fig. 6.8 Normalized HF SFXMs frequency correction versus k_{ef}; $N = 1$; $Z = 9$; $\varepsilon_d = 1$; $\Delta = 0.3$; $m = -1, -2, -3$

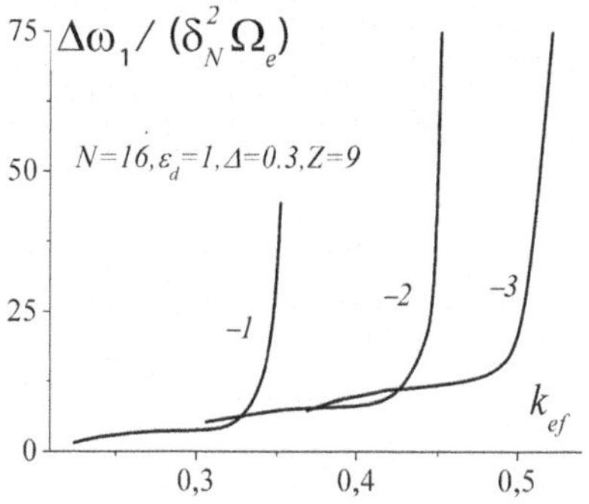

Fig. 6.9 The same as in Fig. 6.8, but for $N = 16$

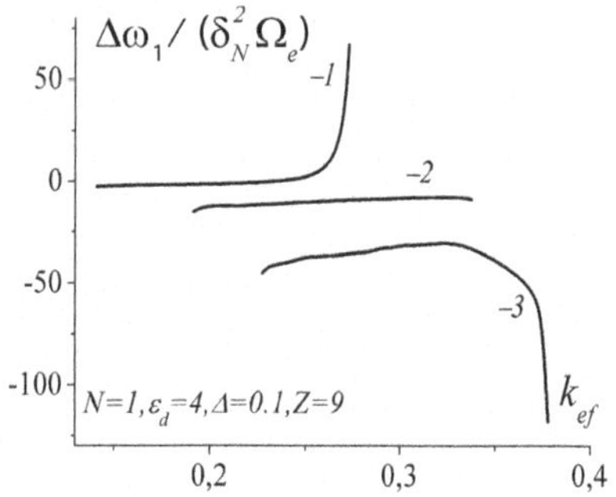

Fig. 6.10 The same as in Fig. 6.8, but for $\varepsilon_d = 4$; and $\Delta - 0.1$

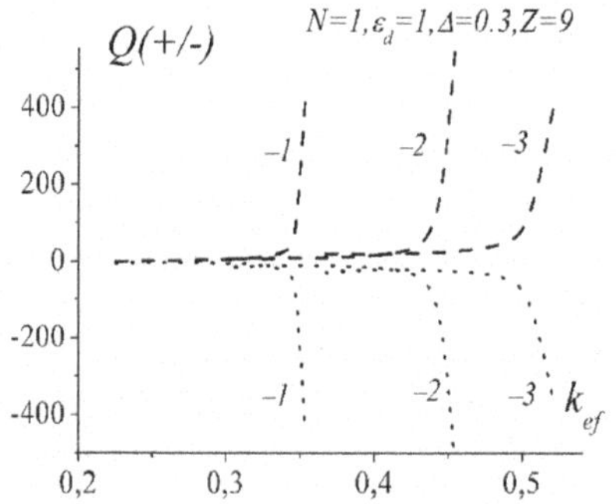

Fig. 6.11 Relative amplitudes of satellite harmonics versus the effective wavenumber k_{ef}. The fast harmonics, $Q^{(+)}$, are depicted by the *dotted curves*, the slow ones, $Q^{(-)}$, are given by the *dashed curves*. The plasma waveguide parameters are the same as in Fig. 6.8. The *numbers* correspond to the azimuthal mode index m

The case under consideration should be compared with the case of LF SFXMs. Note, that the deviation of the cross-section of the plasma-dielectric interface from circular shape causes a decrease of the eigenfrequency of LF SFXMs with negative mode index (except for the wave with $m = -1$, if they propagate in wide waveguides $(k_{ef} \ll 1)$). Thus, a cylindrical metal waveguide with non-circular plasma-dielectric

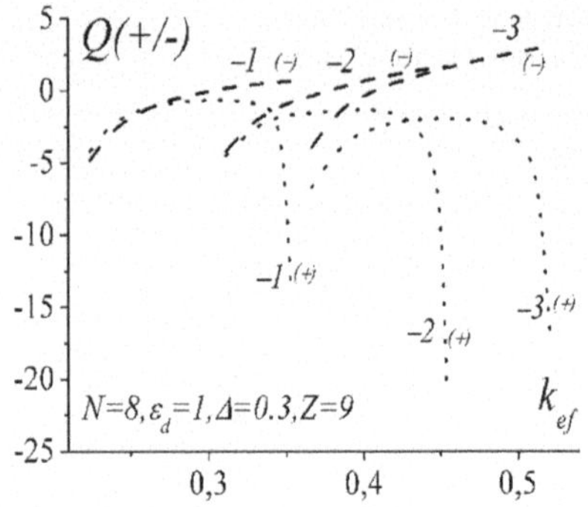

Fig. 6.12 The same as in the Fig. 6.11, but for $N = 8$. The fast harmonics, $Q^{(+)}$, are depicted by the *dotted curves*, the slow ones, $Q^{(-)}$, are given by the *dashed curves*. The *numbers* correspond to the azimuthal mode index m

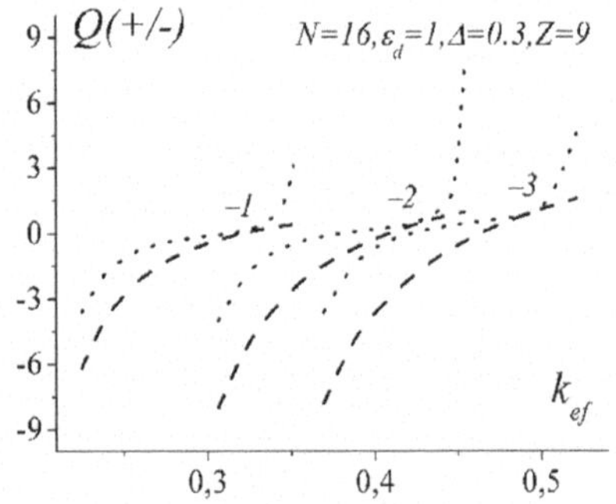

Fig. 6.13 The same as in the Fig. 6.11, but for $N = 16$. The fast harmonics, $Q^{(+)}$, are depicted by the *solid curves*, the slow ones, $Q^{(-)}$, are given by the *dashed curve*. The *numbers* correspond to the azimuthal mode index m

interface is a slowing-down structure for such waves. On the other hand, the frequency correction for HF SFXMs caused by the deviation of the cross-section of the plasma-dielectric interface from circular shape, in its turn, can be both negative and positive depending on the values of the waveguide parameters.

Figure 6.8 illustrates how the frequency correction $\Delta\omega_N/\left(\Omega_e\delta_N^2\right)$ depends on k_{ef} for $N = 1 \leq |m|$, $\varepsilon_d = 1$, $\Delta = 0.3$, $Z = \Omega_e/|\omega_e| = 9$. Analyzing the curves plotted in this figure one can see that the minimum of the frequency correction $\Delta\omega_1$ strongly depends on the azimuthal mode index m: $\Delta\omega_1 \propto |m|^{\beta}$, where $\beta \approx 2.102$. The dependence of $\Delta\omega_N/\left(\Omega_e\delta_N^2\right)$ on k_{ef} is shown for the case $N = 16 > |m|$ in Fig. 6.9. Increase of N leads to reduction of the wavebands of k_{ef}, where the frequency corrections $\Delta\omega_N$ are negative. For $N = 8$, the frequency correction for SFXMs with $m = -1$ is positive for the whole range of k_{ef} under the consideration here, whereas the curves $\Delta\omega_N\left(k_{ef}\right)$ for the SFXMs with $m = -2, -3$ still contain the bands of k_{ef}, where the corrections are negative. For $N = 12$ the frequency corrections are positive for all three values of the azimuthal mode index. Increasing N is not followed by any essential increase in the absolute value of $\Delta\omega_N$. An analysis of the expression for $\Delta\omega_N\left(k_{ef}\right)$ shows that it converges if the small parameters δ_n decrease with the index of summation n faster than $n^{-1.5}$. This requirement is not too restrictive for the study because, when, e.g., the plasma-dielectric interface has a square cross-section, the small parameters of the corrugation δ_n decrease inversely to the square of the summation index n:

$$\delta_n = 4\sqrt{2}(-1)^{\frac{n}{4}} / \left[n^2 \ln\left(3 + 2\sqrt{2}\right) \right]. \tag{6.41}$$

Increasing B_0 leads to an essential decrease of the correction $\Delta\omega_l$. It should be underlined that the dispersion properties of SFXMs, which propagate in magneto-active waveguides, differ strongly from the properties of SFXMs propagating in non-magnetized waveguides. For instance, if an external magnetic field is applied, SFXMs can propagate along the cold plasma-metal interface [27]. This is impossible in the case of $B_0 = 0$. It should be also pointed out that the spectra of SFXMs are degenerate with respect to the sign of the azimuthal mode index in circular plasma waveguides without constant magnetic field: $\omega^{(0)}(+|m|) = \omega^{(0)}(-|m|)$.

However, in a waveguide filled with cold plasma, whose interface has a non-circular cross-section, SFXMs with $n = 2|m|$ are generally non-degenerate [29]. One can remind for comparison, that even in a weak $\left(\Omega_e^2 \gg \varepsilon_0\omega_e^2\right)$ static magnetic field, the dispersion properties of LF SFXMs with these azimuthal mode index ($|m| = n/2$) are more sensitive to the plasma shape than those of LF SFXMs with other azimuthal mode indices ($|m| \neq n/2$), for the same other conditions [30]. The reason is, that for LF SFXMs with azimuthal mode index $|m| = n/2$, the denominators $D^{(\pm)}$ in expressions (6.31) and (6.32) for $p_1^{(\pm)}$, $p_2^{(\pm)}$, $q_1^{(\pm)}$, and $q_2^{(\pm)}$ are very small. If $m > 0$, the denominator $D^{(+)}$ is not small, but the denominator $D^{(-)}$ is:

$$D^{(-)} = -P_m(\kappa a)\frac{2m\mu k}{k_\perp^2 a} I_m(k_\perp a) \ll D^{(+)}. \tag{6.42}$$

Vice versa, if $m < 0$ the denominator $D^{(+)}$ is small (in this case it is expressed by the Eq. (6.33)), but $D^{(-)}$ has no small value. It should be reminded that in the HF range, SFXMs can propagate in waveguides with narrow dielectric layer only with negative azimuthal mode indices. That is why the deviation of the cross-section of the plasma-dielectric interface from circular shape does not cause any resonant influence on the dispersion properties of HF SFXMs. Moreover, if there is no external magnetic field, the HF range does not exist at all.

The influence of the dielectric layer thickness parameter Δ on the behavior of the curves representing the dependence of the frequency correction on k_{ef} appears to be not essential, unlike in the case of LF SFXMs [26], for which $\Delta\omega_1 \propto \Delta^{-2}$ with high accuracy. It also should be mentioned, that an infinite increase of frequency correction $\Delta\omega_1$ value starts from smaller values of the parameter k_{ef}, if Δ decreases.

Figure 6.10 illustrates the influence of the dielectric constant of the dielectric layer on the normalized frequency correction $\Delta\omega_1/\left(\Omega_e\delta_1^2\right)$. Here the following parameters of the waveguide were chosen: $N = 1$, $\Delta = 0.1$, $Z = 9$, dielectric permittivity $\varepsilon_d = 4$. An increase of the ε_d value not only strongly effects the absolute value of $\Delta\omega_1$, but also changes its sign, which determines the sign of the infinity of the frequency corrections nearby the right band limit of k_{ef}, where HF SFXMs with given azimuthal mode index exist. For instance, for HF SFXMs with azimuthal mode index $m = -3$, the normalized frequency correction is approximately equal to $\Delta\omega_1/\left(\Omega_e\delta_1^2\right) \approx -200$. To satisfy the usability condition of the method of successive approximations, $\Delta\omega_1/\omega_0 \ll 1$, the small parameter of the problem should be sufficiently small, namely, the following rigid condition should be valid: $h_l \ll 0.07$.

Figure 6.11 shows how the relative amplitudes $Q^{(\pm)}$ of the satellite harmonics, varying as $\exp[i(m \pm N)\varphi - i\omega t]$, depend on the effective wavenumber,

$$Q^{(\pm)} = A^{(\pm)} I_{m\pm N}(k_\perp a)/\left[h_1 A^{(0)} I_m(k_\perp a)\right]. \tag{6.43}$$

The calculations are carried out for the same plasma waveguide parameters as in Fig. 6.7. For the wave packet of SFXMs with azimuthal mode index $m = -3$ the fundamental harmonic and its two satellite harmonics propagate in the same direction. The angular phase velocities of these satellite harmonics $\omega/(m \pm N)$ are equal to $(-\omega_0/2)$ and $(-\omega_0/4)$, respectively; i.e., one of the satellite harmonics, varying as $\propto \exp[i(m+N)\varphi - i\omega t]$, propagates faster than the fundamental harmonic, whose angular phase velocity is equal to $-\omega_0/3$, and the other satellite harmonic, varying as $\propto \exp[i(m-N)\varphi - i\omega t]$, propagates slower than the main one. For the wave packet of SFXMs with the fundamental harmonic $m = -1$, one of the satellites, varying as $\propto \exp[i(m+N)\varphi - i\omega t]$, is a standing wave in azimuthal direction. The other, varying as $\propto \exp[i(m-N)\varphi - i\omega t]$, has the angular phase velocity $-\omega_0/2$; i.e., it propagates slower than the fundamental harmonic, which has the angular phase velocity $-\omega_0$.

Unlike the LF SFXMs, whose amplitudes of the fast satellite harmonics are lower than the amplitudes of the slow satellite harmonics [26], the amplitudes of HF SFXM satellite harmonics are approximately symmetric: $Q^{(+)} \approx Q^{(-)}$ in the case of waveguides with $N = 1$. This symmetry does not change with variation of the parameters Δ, Z and ε_d. For wide waveguides (if $k_{ef} \to 0$) the amplitudes of satellite harmonics of HF SFXMs turn to zero which confirms the usability condition of the method of successive approximations. However, approaching the right limit of the k_{ef} range, where HF SFXMs with a given value of m can exist, the amplitudes of HF SFXMs satellite harmonics become infinitely large and the usability condition of the method of successive approximations, $Q^{(\pm)} \sim h_N \ll 1$, is violated. In the case of $N = 1$, the absolute values of $Q^{(\pm)}$ increase with increasing Z and ε_d, but weakly depend on the thickness Δ of the dielectric layer. The plots of the dependencies of SFXM satellite harmonics amplitudes calculated for the values of the waveguide parameters, which correspond to Figs. 6.9 and 6.10, are not presented since they do not differ qualitatively from those presented in Fig. 6.11.

For satellite harmonics in the case $N \gg |m|$, the character of the dependence of $Q^{(\pm)}$ on the effective wavenumber differs from that in the case of $N = 1 \leq |m|$. Plots of the dependence of $Q^{(\pm)}$ on k_{ef} calculated for the same values of the waveguides parameters as done in Fig. 6.8, but for the satellite harmonics with $N = 8$ and16 are presented in Figs. 6.12 and 6.13, respectively. The curves calculated for the amplitudes $Q^{(-)}$ of slow satellite harmonics, which propagate in the same direction as the fundamental harmonic, are shown by dashed curves. The curves which illustrate the dependencies of the satellite harmonics amplitudes $Q^{(+)}$, which propagate in the opposite direction as compared to the fundamental harmonics, are shown by dotted curves. Analyzing these dependences, the asymmetry of the dependencies of $Q^{(\pm)}$ should be noted, first of all. The amplitudes $Q^{(-)}$ are smaller than the amplitudes $Q^{(+)}$. In the case $N = 8$, $Q^{(-)}$ reaches its maximum absolute value 4.82 (see Fig. 6.12) already nearby the left limit of the k_{ef} range, where SFXM modes with $m = -1$ can exist. Increasing number N leads to widening the k_{ef} ranges, where the derivative $\partial Q^{(+)}/\partial k_{ef}$ is positive. In the case $N = 16$ (see Fig. 6.13), the amplitudes $Q^{(+)}$ of all first three surface flute modes turn to positive infinity when k_{ef} approaches to its right limit.

The results obtained above can be employed for estimating how the dispersion properties of SFXMs depend on a shift of the axes of plasma column and metal waveguide (that means $N = 1$). For example, for a gaseous plasma with a density of $n = 10^{11}$ cm^{-3}, the electron plasma frequency is equal to 1.8×10^{10} s^{-1} and the skin depth is about 1.7 cm. The relationship $\Omega_e = 9|\omega_e|$ between electron plasma and electron cyclotron frequency can be achieved with a moderate external magnetic field, $B_0 \approx 0.161$ kG. For the SFXM with azimuthal mode index $m = -1$, which propagates in a plasma cylinder of radius $a \approx 8.5$ cm, the effective wavenumber is equal to $k_{ef} = 0.2$. The condition $\Delta = 0.1$ in this case implies that the thickness of the vacuum layer is about $b-a \approx 0.85$ cm. This is less than 9% of the vacuum wavelength, $\lambda \approx 0.52c/\omega$, so the usability condition for the method of successive approximations is satisfied. In this case, the SFXM propagates at the frequency 2.9 GHz, which is 1.02 times higher than the electron plasma frequency. If the axes of the plasma column and the metal waveguide are shifted by 1 mm, the small parameter of the problem is approximately equal to $\delta_l \approx 0.012$. In this case, the correction to the SFXM frequency is as small as 662 kHz, which is much less than the eigenfrequency, as it is required by the method of successive approximations. In such a waveguide, the amplitude of the satellite harmonic $\propto \exp[-i2\varphi]$ is lower than 2.4% of the amplitude of the fundamental harmonic. The amplitude of the standing harmonic in the azimuthal direction (for which $m + N = 0$) is somewhat larger: it is equal to 2.9% of the fundamental harmonic amplitude.

6.7 Conclusions

The present Chapter is devoted to the theory of SFXMs propagating in magnetoactive plasma filled waveguides with non-circular cross-sections. Periodic change of the curvature of the plasma column interface with a dielectric layer and/or a metal chamber along the azimuthal angle is shown to cause the propagation of electromagnetic surface waves in the form of wave packets. As a rule, the SFXM eigenfrequency correction, which is caused by the deviation of the waveguide cross-section from circular shape, is found to be a small value of second order over the small parameter of the problem. In the special (resonant) cases associated with degeneration of the frequency spectrum, the value of the frequency correction for the LF SFXMs can be linearly proportional to the small parameter of the problem.

Periodic azimuthal non-uniformity of the plasma waveguide is shown to result in splitting the spectra of SFXM eigenwaves, if the angular period of the non-uniformity of the dielectric layer thickness is equal to half the angular period of the wave. The value of this splitting is found to be of the first order of smallness. The correction to the SFXM eigenfrequency, caused by the non-uniformity of the dielectric layer thickness, is more sensitive to a corrugation of the plasma-dielectric layer interface (or its deviation from circular shape) than to a corrugation of the metal chamber. Thus, designing a waveguide with a special shape of the cross-section can be used for control of the eigenfrequency and spectral contents of the electromagnetic eigenwaves propagating there.

The results reported here can be essentially interesting both for studying wave processes in fusion devices with magnetic confinement and for plasma electronics. For example, corrugated waveguide lines are utilized as interaction circuits of free electron masers [32]. One can strongly enhance the pulse power (about 20 times) of HF electromagnetic radiation [33] by sectioned waveguide-compressors with screw corrugated surface. This is associated with the well-known advantages of plasma waveguides [2, 4, 34] in comparison with vacuum waveguides, namely: enhanced value of limiting current, optimized conditions of beam-wave interaction, possibility to enhance the power of the generated signal and to tune smoothly its frequency. The applicability of the method of successive approximations has been proved not only analytically in theoretical papers referenced here, but there is also an experimental confirmation [35] of its sufficient accuracy and usability for practical calculations.

The frequency correction of HF SFXMs propagating in waveguides, which are partially filled with plasma, is shown to increase with decreasing external magnetic field and increasing dielectric constant of the dielectric layer. Varying the thickness of the dielectric layer (under the initial assumption that it is thin) weakly affects the frequency correction. The sign and absolute value of $\Delta\omega_N$ are very sensitive to the type of the corrugation of the plasma-dielectric interface (shape parameter N).

Examination of the spectral content of HF SFXMs propagating along the surface of a plasma column with non-circular cross-section makes it possible to reveal the following properties of the wave packet. In the case $N = 1$, the amplitudes of the satellite harmonics of HF SFXMs are almost anti-symmetric, $Q^{(-)} \approx -Q^{(+)}$. In the case $N \gg |m|$, the amplitudes $Q^{(-)}$ are small compared with those of $Q^{(+)}$.

References

1. Alexandrov, A. F., Bogdankevich, L. S., & Rukhadze, A. A. (1984). *Principles of plasma electrodynamics*. Springer.
2. Kuzelev, V. M., & Rukhadze, A. A. (1990). *Electrodynamics of dense electron beams in plasma*. Nauka. (in Russian).
3. Miller, R. (1982). *Introduction to the physics of strong-current beams of charged particles*. Plenum Press.

4. Kondratenko, A. M., & Kuklin, V. M. (1988). *Basis of plasma electronics*. Energoatomizdat. (in Russian).
5. Landau, L. D., & Lifshits, Ye.M. (1960). *Course of theoretical physics. Electrodynamics of continuous media* (Vol. 8). Oxford: Pergamon Press.
6. Gradov, O. M., & Stenflo, L. (1983). Linear theory of a cold bounded plasma. *Physics Reports (Review Section Physics Letter), 94*, 111–137.
7. Azarenkov, N. A., & Ostrikov, K. N. (1991). Surface magnetoplasma waves at the interface between a plasma-like medium and a metal in a Voigt geometry. *Physics Reports, 308*, 333–428.
8. Beletski, N. N., Bulgakov, A. A., Khankina, S. I., & Yakovenko, V. M. (1984). *Plasma instabilities and non-linear phenomena in semi-conductors*. Naukova dumka. (in Russian).
9. Gradov, O. M., & Stenflo, L. (2001). Theory of non-linear plasma surface waves. *Journal of Plasma Physics, 65*, 73–77.
10. Wesson, J. (2004). *Tokamaks* (3rd ed.). Clarendon Press.
11. Volkov, E. D., Suprunenko, V. A., & Shyshkin, O. O. (1983). *Stellarators*. Naukova dumka. (in Russian).
12. Beidler, C. D., Grieger, G., Harmeyer, E., et al. (1995). *Helias reactor studies*. Preprint of Max-Planck-Institut für Plasmaphysik, 85748 Garching bei München, FRG, IPP 2/330.
13. Ballico, R. C., & Cross, M. J. (1990). Probe measurements of ICRF Alfven surface waves in the TORTUS tokamak. *Fusion Engineering and Design, 12*(1–2), 197–201.
14. Murphy, A. B. (1990). Waves in the edge plasma during ion cyclotron resonance heating. *Fusion Engineering and Design, 12*(1–2), 79–92.
15. Porkolab, M. (1990). Parametric instabilities in the tokamak edge plasma in the ion cyclotron heating regimes. *Fusion Engineering and Design, 12*(1–2), 93–103.
16. Noterdaeme, J. -M., ICRH-team, ASDEX-team, & NI-team. (1990). Experimental results on edge effects during ICRF heating of ASDEX plasmas. *Fusion Engineering and Design, 12*(1–2), 127–201
17. Akhiezer, A. I., Akhiezer, I. A., Polovin, R. V., Sitenko, A. G., & Stepanov, K. N. (1975). *Plasma electrodynamics*. Pergamon Press.
18. Krall, N. A., & Trivelpiece, A. W. (1973). *Principles of plasma physics*. MacGraw-Hill.
19. Azarenkov, N. A., & Kondratenko, A. N. (1989). Surface waves in gyrotropic plasma layers bounding a metal. *Soviet Journal of Communications Technology and Electronics, 34*(15), 101–103.
20. Girka, I., & Zolotukhin, O. (1994). Transversal surface magneto-plasma waves in metal waveguide of rectangular cross-section filled with n-semiconductor. *Soviet Journal of Communications Technology and Electronics, 39*(1–2), 1961–1968.
21. Agranovich, V. M., & Mills, D. L. (1982). *Surface polaritons*. North-Holland.
22. Girka, V., Girka, I., Girka, A., et al. (2011). Theory of azimuthal surface waves propagating in non-uniform waveguides. *Journal of Plasma Physics, 77*(4), 493–519.
23. Girka, V., Girka, I., & Pavlenko, I. (1997). Surface waves propagating in the direction transverse to the axis of magnetized plasma-filled waveguides with noncircular transverse cross sections. *Plasma Physics Reports, 23*(11), 959–963.
24. Karpov, S. Y., & Stolyarov, S. N. (1993). Propagation and transformation of electromagnetic waves in one-dimensional periodic structures. *Physics-Uspekhi, 36*(1), 1–22.
25. Abramowitz, M., & Stegun, I. A. (Eds.). (1964). *Handbook of mathematical functions*. National Bureau of Standards, Applied Mathematics, Series 55.
26. Girka, O. I., Girka, V. O., Girka, I. O., et al. (2007). Effect of the shape of the cross-section of a plasma-dielectric interface on the dispersion properties of azimuthal surface modes. *Plasma Physics Reports, 33*(2), 91–101.
27. Girka, V., Girka, I., Pavlenko, I., et al. (2012). Coupled azimuthal modes propagating in current-carrying plasma waveguides. *Journal of Plasma Physics, 78*(2), 105–123.
28. Landau, L. D., & Lifshits, E. M. (1977). Course of theoretical physics. In *Quantum mechanics* (Vol. 3). Pergamon Press.
29. Girka, I. O. (1997). Multi-peak structure of the spectra of azimuthal surface waves in isotropic plasma waveguides with noncircular transverse cross-sections. *Plasma Physics Reports, 23*(3), 246–251.

30. Girka, O., Girka, V., Girka, I., et al. (2007). Resonant effect of the noncircular shape of the plasma surface on the dispersion properties of extraordinary azimuthal surface modes in magneto-active waveguides. *Plasma Physics Reports, 33*(7), 543–552.
31. Girka, V., Girka, I., & Pavlenko, I. (2011). Excitation of azimuthal surface modes by relativistic flows of electrons in the high-frequency range. *Plasma Physics Reports, 37*(5), 447–454.
32. Ginzburg, N. S., Zaslavskii, V. Y., Peskov, N. Y., et al. (2006). Theory of a planar free-electron maser with transverse electromagnetic flux circulation in a 2D Bragg mirror. *Technical Physics, 51*(12), 1618–1623.
33. Mishakin, S. V., & Samsonov, S. V. (2009). Optimization of frequency-modulated pulse compression in a sectioned waveguide with a helically corrugated surface. *Technical Physics, 79*(11), 1655–1662.
34. Onishchenko, I. N., Sidorenko, D. Y., & Sotnikov, G. V. (1995). Electron beam interaction with a plasma-filled rippled coaxial waveguide. *Plasma Physics Reports, 21*(8), 670–676.
35. Aksenova, L. A., Keier, A. P., Nemirovskaya, L. L., et al. (1988). Experimental analysis of mode conversion in waveguides with corrugated walls. *Radiophysics and Quantum Electronics, 31*(10), 873–878.

Chapter 7
Excitation of Surface Flute Waves in Electron Cyclotron Frequency Range by Relativistic Electron Beam Gyrating Along Large Larmor Orbits

Interaction of a charged particle beam with plasma eigenoscillations is widely used in plasma electronics for generation and amplificaion of electromagnetic waves [1–6]. Doing that, great attention is paid to studying processes of eigenmode excitation in plasma filled waveguides with the aid of charged particle beams, which gyrate in an external magnetic field [7–19]. Devices of plasma electronics have a lot of advantages compared with vacuum devices, namely, a wider spectrum of eigenfrequencies, the possibility to control the frequency spectrum in a wide frequency range, the ability to generate and/or amplify waves in super-high frequency ranges and the possibility to enhance electromagnetic wave power generated by these devices due to the ability to increase the electric current, which flows through the plasmas.

An important issue of the general problem of beam excitation of plasma filled waveguides is the investigation of the interaction between surface waves with charged particle flows, because this process has a lot of peculiarities compared with the case of bulk (volume) wave excitation [1, 20, 21]. Taking into account that finite plasmas are characterized by a relatively large value of dissipation as compared with the case of infinite (large scaled) plasmas, there exists an additional interesting issue, which is connected with the experimental ability to realize a regime of dissipative instabilities [1, 22, 23]. Such a regime is possible in the case, when the effective particle collisional frequency becomes larger than the growth rate of beam instabilities.

The present Chapter is devoted to results of theoretical investigations on excitation of surface flute X-polarized modes (SFXMs) by a charged particle beam in a cylindrical metal waveguide supplied with a concentric plasma filling. The peculiarity of this problem is that the SFXM amplitude decreases exponentially from the plasma interface toward the waveguide axis, however a vacuum layer, which separates the plasma column from the metal waveguide wall, is naturally transparent for these waves. Therefore, the presence of some slot in the metal waveguide wall leads to leakages of the SFXMs power into the outer space that is typical for bulk waves. The dependence of the studied electromagnetic waves on time t and azimuthal angle φ is assumed in the present Chapter as follows: $E,\ H \propto \exp\left(im\varphi - i\omega t\right)$, while the space is assumed to be uniform along the $\vec{z}$ axis. The only exception is Sect. 7.7.

© The Author(s), under exclusive license to Springer Nature Switzerland AG 2022

I. Girka and M. Thumm, *Surface Flute Waves in Plasmas*, Springer Series on Atomic, Optical, and Plasma Physics 120, https://doi.org/10.1007/978-3-030-98210-2_7

where the initial stage of interaction between a gyrating relativistic electron beam and long-wavelength electromagnetic surface waves, $E,\ H \propto \exp(ik_z z + im\varphi - i\omega t)$, is studied. These considerations are restricted to the case of SFXMs propagating in the range of the electron cyclotron frequency. Thus, it is proposed to utilize an electron beam for their excitation. Excitation of SFXWs in the ion cyclotron frequency range by light ion beams gyrating around a plasma column in strongly magnetized plasma waveguides is studied later in Sect. 10.4.

Electronic devices based on a rotating beam have some significant advantages as compared to devices based on longitudinal (linear) ones. For example, the instability growth rates and the efficiency of the energy exchange in free electron lasers based on annular electron beams are $\gamma^{2/3}$ times larger (γ is the relativistic Lorentz factor) in comparison to those applying longitudinal beams [12]. Moreover, in linear theory, the device length limits the efficiency of generators based on longitudinal beams. In devices based on rotating beams, the particles travel along large Larmor orbits in the gap between the chamber wall and the plasma column. These particles transform their energy into electromagnetic waves until they (particles) reach the plasma surface due to deceleration. In this case, these particles are able to propagate along a path that is much longer than the size of devices based on longitudinal beams. Thus, application of such charged particle beams allows one to develop electronic devices, which would be more compact along the axial direction.

At present time, the achievement of very high power levels in gyrotrons based on traditional cylindrical resonators seems to be practically impossible from a technical point of view because of significant heating of the resonator walls, mode competition, influence of space charge of the applied beam and some other reasons. Combination of these influences restricts the output power of modern gyrotrons to a level of a little more than 1 MW. One way to overcome these difficulties is the application of coaxial resonators in order to develop new types of gyrotrons [24, 25]. Introduction of an inner conductor allows one to essentially decrease the influence of the beam eigenfields (space-charge voltage depression) [26], the mode competition (due to modification of the eigenfrequency spectrum) [27] and other influences, which also restrict the power of conventional resonator gyrotron. Currently, the record value of the output power of such coaxial-cavity gyrotrons is about 2.2 MW. This power level was obtained in the regime of short pulses (1 ms) at an operating frequency of 170 GHz with an output efficiency of ~30% without applying a depressed collector for energy recovery [28].

The present Chapter consists of eight sections, the last one summarizes the conclusions. Basics of the linear theory of SFXMs excitation by an electron beam gyrating around the plasma column in cylindrical metal waveguides are presented in Sect. 7.1. (The nonlinear stage of SFXM-beam interaction is investigated in Chap. 9). Both cases of resonant beam-plasma and beam-dissipative instabilities are considered. The next section is devoted to investigation of the resonant beam-plasma instability of higher radial modes of SFXMs. Two mechanisms of resonance overlapping in the excitation of SFXMs by rotating relativistic electron beams in the electron cyclotron frequency range are demonstrated in Sect. 7.3. Specific features of beam excitation of SFXMs in toroidal waveguides are studied in Sect. 7.4. The influence of a non-circularity of the plasma column cross-section on the interaction between the SFXM and the gyrating electron beam is investigated in Sect. 7.5. Transfer of a beam-plasma instability into a beam-dissipative one is well-known since long time. The question how to distinguish whether one or the other instability takes place is discussed in Sect. 7.6. The influence of small axial wavenumber on the surface wave beam excitation is studied in Sect. 7.7.

7.1 Linear Theory of Surface Flute Wave Excitation by Electron Beams Gyrating Around a Plasma Column

The geometry of the waveguide proposed for the present consideration is as follows. The cylindrical metal waveguide has the inner radius R_2, the outer radius of the plasma column, which fills the waveguide, is R_1 (see Fig. 7.1). Therefore, the relative thickness of the vacuum layer, which separates the plasma column from the metal wall, is $\Delta = R_2/R_1 - 1$. An external static magnetic field is oriented along the plasma axis, $\vec{B}_0 \| \vec{z}$. In solid state physics, the case when the external stationary magnetic field is parallel to the plasma interface and perpendicular to the direction of wave propagation is called Voigt configuration [29]. Therefore, SFXWs are eigenmodes of the Voigt configuration not in Cartesian but in cylindrical coordinates. The magnitude of $\vec{B}_0$ is assumed to be relatively small, so that the electron plasma frequency is larger than the electron cyclotron frequency: $\Omega_e^2 \gg \omega_e^2$. The electron beam is propagating in the region $R_1 < r < R_2$, its particle density is assumed to be sufficiently small in order to satisfy the condition that the influence of the electron beam on the properties of the plasma column can be neglected.

To develop a linear theory of SFXM excitation due to their interaction with the electron beam, the model of a flow of oscillators is applied here. It is assumed that all beam electrons move with the same value of transverse momentum $p_{\perp 0}$ and with zero axial momentum, $p_z = 0$. The beam-plasma system is compensated over both the

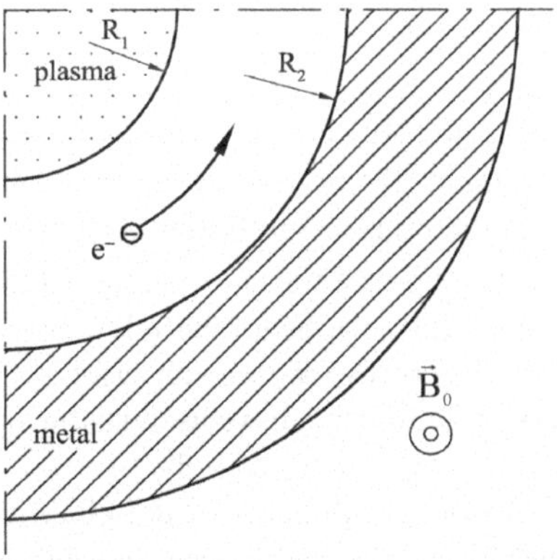

Fig. 7.1 Schematic of the considered electron beam-plasma waveguide structure immersed into a static axial magnetic field

electric current and charge. Such an electron beam is described by the equilibrium distribution function [21]:

$$f_0 = n_b \delta(p_\perp - p_{\perp 0})\delta(p_z)/(2\pi p_{\perp 0}), \tag{7.1}$$

where n_b is the beam particle density and the transverse momentum of the beam electrons is $p_{\perp 0} = m_e \upsilon_{\perp 0}\gamma$, with $\gamma = \sqrt{1 + p_{\perp 0}^2 m_e^{-2} c^{-2}}$ as relativistic Lorentz factor. It should be mentioned that just this form of the γ-factor is convenient for the numerical analysis of the studied surface flute extraordinary mode (SFXM) instability, because there is no calculation of a difference. Thus one can avoid the difficulty connected with "numerically dividing by zero". The presentation of the distribution function f_α of the beam particles in the form of the sum consisting of the equilibrium part f_0 described by expression (7.1) and the first correction to the distribution function δf_α allows one to derive the following general linearized kinetic Vlasov-Boltzmann equation:

$$-i\omega_1 \, \delta \, f_\alpha + ik_\perp \upsilon_\perp \cos\varphi \, \delta \, f_\alpha + e_\alpha \frac{\vec{E}}{m_\alpha} \frac{\partial f_0}{\partial \vec{\upsilon}} - \frac{\omega_\alpha}{\gamma} \frac{\partial \delta \, f_\alpha}{\partial \varphi} = 0, \tag{7.2}$$

where the integral of collisions is approximated by the effective collisional frequency ν and $\omega_1 = \omega + i\nu$. The value of ν is assumed to be small so that the inequality $\nu \ll \omega$ is satisfied. The subscript α relates to the type of beam particles (ions, electrons, etc.). Their cyclotron frequency ω_α is calculated using their "rest mass" m_α. In the case of relativistic particle beams one has to take into account the corresponding γ—factor. The angle φ is the azimuthal angle in the momentum phase space. The solution of the kinetic Eq. (7.2) is as follows:

$$f_1 = \exp(ia_1 \sin\varphi) \frac{e_b \gamma}{\omega_\alpha} \sum_{s=-\infty}^{+\infty} \left\{ E_z J_s(a_1) \frac{\partial f_0}{\partial p_z} \right.$$
$$\left. + \frac{\partial f_0}{\partial p_\perp} \left[\frac{s E_x}{a_1} J_s(a_1) + i E_y J_s'(a_1) \right] \right\} \frac{i \exp(-is\varphi)}{s - b_1}, \tag{7.3}$$

where $a_1 = k_\perp \upsilon_\perp \gamma / \omega_\alpha$, $b_1 = \gamma \omega_1 / \omega_\alpha$, and $J_s(a_1)$ are the Bessel functions of the first kind [30]. The superscript "prime" means the derivative of this function over its argument. Integration of the corresponding term of (7.3) over $d\vec{p} = p_\perp dp_\perp d\varphi dp_z$ with the relative multiplier $e_\alpha \upsilon_j$ (here υ_j is j th component of the velocity) allows one to calculate the components of the electric conductivity tensor $\sigma_{jn}^{(b)}$ for the space occupied by the beam particles. In the conductivity tensor the subscript n coincides with the corresponding subscript of the perturbed electric field, so the corresponding term mentioned above is just the multiplier before the related component of the electric wave field, namely E_n.

Those components of the electric conductivity tensor, which are applied in the further consideration, look as follows:

$$\sigma_{11}^{(b)} = -i\frac{\Omega_b^2}{4\pi\gamma}\sum_{s=-\infty}^{+\infty} s^2\left[\frac{\left(J_s^2(m)\right)'}{(s-q)k_\perp v_{\perp 0}} + \frac{\omega J_s^2(m)}{(s-q)^2 c^2 k_\perp^2}\right]; \tag{7.4}$$

$$\sigma_{12}^{(b)} = -\sigma_{21}^{(b)}$$

$$= \frac{\Omega_b^2}{4\pi\omega_b}\sum_{s=-\infty}^{+\infty} s\left[\frac{\left(J_s(m)J_s'(m)\right)'}{s-q} + \frac{J_s(m)J_s'(m)}{(s-q)m} + \frac{J_s(m)J_s'(m)\omega v_{\perp 0}}{(s-q)^2 c^2 k_\perp}\right]; \tag{7.5}$$

$$\sigma_{22}^{(b)} = -i\frac{\Omega_b^2}{4\pi\omega_b}\sum_{s=-\infty}^{+\infty}\left[\frac{2\left(J_s'(m)\right)^2}{s-q} + \frac{2mJ_s'(m)J_s''(m)}{s-q} - \frac{\left(J_s'(m)\right)^2 v_{\perp 0}^2 q}{(s-q)^2 c^2}\right], \tag{7.6}$$

where the ion terms are omitted taking into account that in the present problem just an electron beam is considered. Therefore, $\Omega_b^2 = 4\pi e^2 n_b m_e^{-1}$, $q = \omega_1\gamma/|\omega_e|$, and the other notations are the same as in (7.3).

The argument of the cylindrical functions in Eqs. (7.4)–(7.6) for the components of the dielectric permittivity tensor is given in [21] as $\tau = \frac{k_{perp}v_\perp}{|\omega_e|}\gamma$. Since the electrons of the flow do not move along the cylinder axis, $v = v_\perp$. SFXMs propagate along the azimuthal angle only, therefore, $k_{perp} = k_\varphi$. The Larmor radius of the electrons is given by: $\rho_e = \frac{v}{|\omega_e|\gamma}$, therefore $\tau = k_\varphi\rho_e$. The wave vector k_φ is defined by the condition of periodicity along the azimuthal angle: the mode with the azimuthal mode index m should consist just of m full perturbations along the azimuth for any radial coordinate. Therefore, $2\pi r = m\lambda = 2\pi/k_\varphi$. From this relation, k_{perp} for the dielectric permittivity tensor in [21] becomes immediately as $k_{perp} = m/r$. Finally, since an electron beam which rotates just along the Larmor radius is considered, $r = \rho_e$, and thus $\tau = m$.

The electrodynamic properties of the space region, which is occupied by the electron beam, are described with the dielectric permittivity tensor $\varepsilon_{ik}^{(b)}$ [21]. It should be mentioned that the expressions presented in the textbook [21] have some misprints, that is why one should apply the expressions for $\sigma_{ik}^{(b)}$ given above in order to derive the components of $\varepsilon_{ik}^{(b)}$, which are applied here for solving the problem of SFXM excitation. The relation between those tensors is as follows: $\varepsilon_{ik}^{(b)} = \delta_{ik} + 4\pi i\sigma_{ik}^{(b)}/\omega$, where δ_{ik} is the piecewise Kronecker delta-function.

Searching for solutions of the SFXM fields from the set of Maxwell equations in the region occupied by the beam particles, while the mentioned components of the tensor $\varepsilon_{ik}^{(b)}$ are applied, one can derive a Bessel equation for the axial magnetic component of the wave field. Thus the components of the SFXM field in the space region occupied by the electron beam are described by a superposition of the Bessel function $J_m(x)$ of the first kind and the Neumann function $N_m(x)$ [30]. The magnetic component of these modes propagating with arbitrary value of the azimuthal mode index m is given by:

$$H_z^{(b)} = C_1 J_m(\zeta) + C_2 N_m(\zeta), \tag{7.7}$$

where $\zeta = kr\sqrt{\psi_b}$, $\psi_b = \varepsilon_{22}^{(b)} + \left(\varepsilon_{12}^{(b)}\right)^2 \left(\varepsilon_{11}^{(b)}\right)^{-1}$, and $C_{1,2}$ are integration constants which can be found from the corresponding boundary conditions. In the plasma region, the SFXM field is described by a superposition of the modified Bessel function $I_n(x)$ and the MacDonald function $K_n(x)$ [30] as it was shown in the Sect. 1.4.1. So without detailed discussion, the expression for the SFXM magnetic wave field can be written as:

$$H_z^{(p)} = C_3 I_m(\xi) + C_4 K_m(\xi), \tag{7.8}$$

here $\xi = kr\sqrt{\psi}$, $\psi = \varepsilon_1(\mu^2 - 1) = \varepsilon_2^2/\varepsilon_1 - \varepsilon_1$, where ε_{ik} is the dielectric permittivity tensor of a cold magneto-active plasma [1, 21, 31], and $C_{3,4}$ are integration constants that can be found from the corresponding boundary conditions.

To obtain the dispersion relation which describes the SFXM excitation in the studied waveguide, that is filled with cold magneto-active plasma, one can apply the following boundary conditions.

- The electromagnetic field of the SFXM should have a finite magnitude on the axis of the plasma column:

$$E(r = 0);\ H(r = 0) \ll \infty. \tag{7.9}$$

- The tangential component of the SFXM electric wave field is equal to zero on metal surfaces:

$$E_\varphi(r = R_2) = 0. \tag{7.10}$$

- The tangential components of the electric and magnetic wave fields of the SFXM are continuous at the plasma-vacuum interface:

$$\begin{aligned} E_\varphi(r = R_1 + 0) &= E_\varphi(r = R_1 - 0), \\ H_z(r = R_1 + 0) &= H_z(r = R_1 - 0). \end{aligned} \tag{7.11}$$

In the limiting case of a uniform plasma particle density profile this dispersion relation has the following form [31]:

$$D_0 + \frac{J_n'(\zeta_1) - \Phi N_m'(\zeta_1)}{\sqrt{\psi_b}[J_n(\zeta_1) - \Phi N_m(\zeta_1)]} - \frac{im\varepsilon_{12}^{(b)}}{\varepsilon_{11}^{(b)}\zeta_1\sqrt{\psi_b}} = 0, \tag{7.12}$$

with

$$D_0 = \frac{I_m'(k_\perp R_1)k}{k_\perp I_m(k_\perp R_1)} + \frac{\mu m k}{R_1 k_\perp^2}, \tag{7.13}$$

where $D_0(m, R_1, \omega) = 0$ is the dispersion relation of SFXMs propagating in the magneto-active waveguide entirely filled by plasma, which is considered in Sect. 2.2, and

$$\Phi = \left[\frac{im\varepsilon_{12}^{(b)}}{\varepsilon_{11}^{(b)}\zeta_2} J_n(\zeta_2) - J_n'(\zeta_2)\right] \times \left[\frac{im\varepsilon_{12}^{(b)}}{\varepsilon_{11}^{(b)}\zeta_2} N_m(\zeta_2) - N_m'(\zeta_2)\right]^{-1}, \; \zeta_1 = \zeta(R_1), \; \zeta_2 = \zeta(R_2).$$

The dispersion relation (7.12) has to be analyzed for the LF range (1.13) which is lower than the cut-off frequency for the bulk waves, $\omega < \sqrt{\Omega_e^2 + \omega_e^2/4} - |\omega_e|/2$, if the following resonance condition is valid:

$$\omega = \omega_X + \delta\omega = N|\omega_e|\gamma^{-1} + \delta\omega, \tag{7.14}$$

where N is a natural number, ω_X is the eigenfrequency of the SFXM propagating in the considered waveguide (without any beam), $n_b = 0$, $|\delta\omega| \ll \omega_X$, and $\delta\omega$ is a correction for the eigenfrequency that is determined by the presence of the electron beam gyrating in the region $R_1 < r < R_2$.

In the case of resonant beam-plasma instability of the SFXM (which means that dispersion relation (7.12) is satisfied by substituting (7.14) therein), the wave growth rate is equal to [31]:

$$\mathrm{Im}(\delta\omega) = \frac{\sqrt{3}}{2}\left(\frac{\alpha\Delta\Omega_b^2 N^2 J_N^2(m)\upsilon_{\perp 0}^2}{\gamma\, c^2}\right)^{\frac{1}{3}}\left(\frac{\partial D^{(0)}}{\partial\omega}\right)^{\frac{-1}{3}}. \tag{7.15}$$

Hereinafter in this Chapter, the parameter $\alpha = n_b/n_e$ determines the relative particle density of the utilized electron beam. Remind, that the beam particle density is assumed to be much smaller than the plasma density, so that $\alpha \ll 1$, and that the influence of the beam on the properties of the plasma column is neglected.

The possibility of LF SFXW excitation by an annular electron beam was demonstrated for the first time numerically in [31]. Therein, numerical analysis was carried out for the following values of plasma and waveguide parameters: $Z = \Omega_e/|\omega_e| = 10$, $\Delta = R_2/R_1 - 1 = 0.1$, and $\alpha = 0.01$. It was shown, that the values of the growth rates have well expressed maxima at the effective wave numbers $k_{ef} = |m|\delta/R_1 \in 0.3 \div 0.6$ for the azimuthal mode index range $m \in 1 \div 6$,

As it is well-known [3–5, 32, 33], excitation of electromagnetic waves takes place in the most efficient way just nearby eigenfrequencies of plasma waveguides. Analysis of the dispersion relation (7.12) allows one to see, that the influence of a low-density electron beam ($\alpha \ll 1$) can be sufficient in the case of appearance of resonant denominators in the expressions for $\varepsilon_{ik}^{(b)}$. Just nearby the intersection points of the SFXM dispersion curves (constructed for the case $\alpha = 0$) with the curves

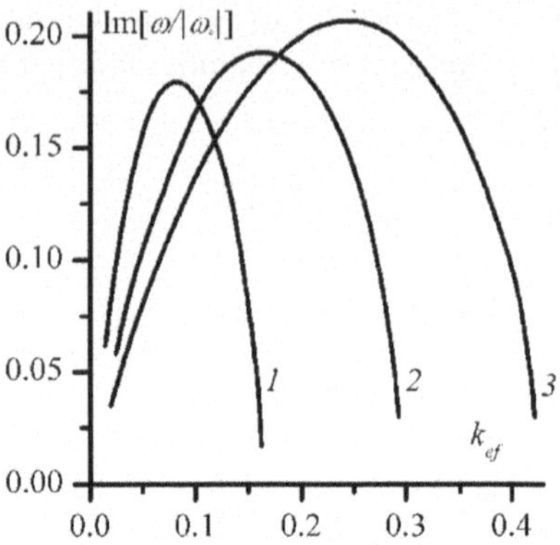

Fig. 7.2 Normalized growth rates of SFXWs with azimuthal mode indices $m = 1{,}2{,}3$ versus effective wavenumber k_{ef} for $Z = 10$, $\Delta = 0.2$, and $\alpha = 0.01$

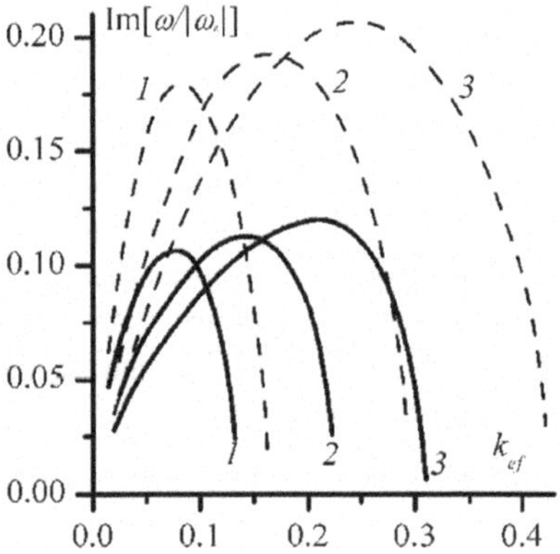

Fig. 7.3 Normalized growth rates of SFXWs with azimuthal mode indices $m = 1{,}2{,}3$ versus effective wavenumber k_{ef} for low particle beam density ($\alpha = 0.00125$, *solid curves*) and $Z = 10$, $\Delta = 0.2$. For comparison, the curves of Fig. 7.2 are also displayed here (*dashed curves*)

$q = s$, along which the components of the tensor $\varepsilon_{ik}^{(b)}$ have resonant denominators, excitation of these waves is possible.

Results of a detailed numerical analysis [34] of the dependence of the LF SFXW growth rates on the effective wavenumber k_{ef} are plotted in Figs. 7.2, 7.3, 7.4 and 7.5. The numbers in these figures designate the azimuthal mode indices m. For each value of the effective wavenumber (for each value of plasma column radius R_1), it is assumed that the relativistic electron beam rotates along Larmor orbits precisely in the middle part of the vacuum gap with the radius $r = R_1(1 + 0.5\Delta)$, which causes the dependence $\gamma = \gamma(k_{ef})$. This numerical analysis confirms that the most efficient excitation of SFXWs occurs just near the values of k_{ef}, for which the condition (7.14) is satisfied.

Results of studying the beam excitation of SFXWs with mode indices $m = 1, 2$, and 3 are shown in Fig. 7.2 for the beam-plasma system with the following parameters: $\alpha = 0.01$, $Z = 10$, and $\Delta = 0.2$. The curves shown in the Fig. 7.2 are also displayed in the Figs. 7.3, 7.4 and 7.5 as dashed curves for comparison, since these figures are calculated with changing one of the parameters: α, Z, or Δ. The maximum growth rate in Fig. 7.2 increases with increasing azimuthal mode index m, namely, the magnitude

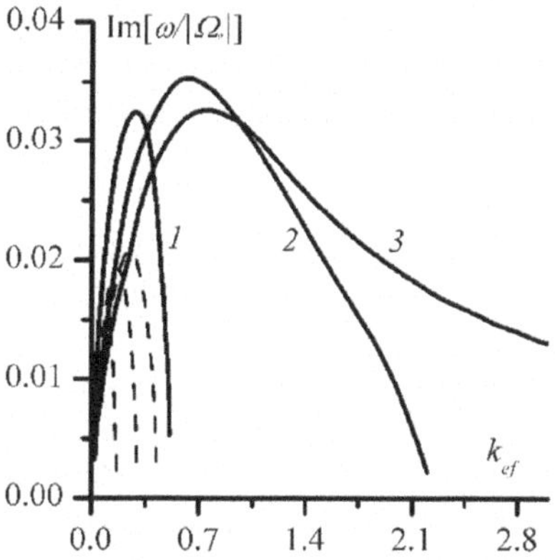

Fig. 7.4 Normalized growth rates of SFXWs with azimuthal mode indices $m = 1, 2, 3$ versus effective wavenumber k_{ef} for two values of constant external magnetic field and $\Delta = 0.2$, $\alpha = 0.01$. The *solid curves* correspond to stronger magnetic field ($Z = 3$), the *dashed curves* to weaker magnetic field ($Z = 10$)

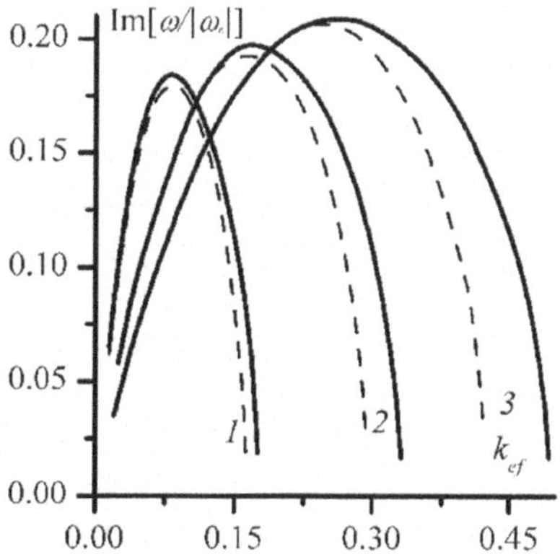

Fig. 7.5 Normalized growth rates of SFXWs with azimuthal mode indices $m = 1,2,3$ versus effective wavenumber k_{ef} for two values of width of the vacuum layer and $Z = 10$, $\alpha = 0.01$. The *dashed curves* correspond to a wider layer ($\Delta = 0.2$), the *solid curves* to a narrower layer ($\Delta = 0.1$)

of $\mathrm{Im}(\omega/|\omega_e|)$ increases from 0.18 for the SFXM with $m = 1$ up to 0.207 for the SFXM with $m = 3$. One can also conclude from Fig. 7.2, that the range of k_{ef}, where SFXM excitation is efficient, is wider for larger m.

The electron beam velocity that is required for LF SFXM excitation can be estimated as follows. For example, under the conditions of Fig. 7.2 the SFXM with azimuthal mode index $m = 1$ is excited in the most efficient way at $k_{ef} \approx 0.08$ which means that the beam velocity should be $\upsilon \approx 0.809\,c$ and the relativistic factor $\gamma \approx 1.7$ which is not too large (acceleration voltage is ≈360 kV).

As it follows from formula (7.15), the maximum value of the SFXM growth rate under the condition of beam instability is proportional to the cubic root of the beam particle density which is typical for beam instabilities [4, 5]. The results of studying the dependence of SFXM growth rates under the condition of beam instability with azimuthal mode indices $m = 1$, 2, and 3 on the effective wavenumber k_{ef} are presented in Fig. 7.3 for a beam density that is eight times smaller ($\alpha = 0.00125$) than in the case shown in Fig. 7.2. The other parameters of the beam-plasma system are the same as for Fig. 7.2. According to the theoretical predictions, a decrease of the beam particle density by a factor of 8 should be accompanied by the corresponding twice decrease of the growth rate. The numerical analysis shows satisfactory agreement of calculations and theoretical predictions. In the case of higher beam density, $\alpha = 0.01$, the maximum value of $\mathrm{Im}(\omega/|\omega_e|) = 0.18$ for the SFXM with azimuthal mode index $m = 1$, while at lower density of $\alpha = 0.00125$, it is 1.68 times smaller, namely, $\mathrm{Im}(\omega/|\omega_e|) = 0.107$.

Three phenomena more, associated with the influence of the beam particle density on SFXW excitation, should be pointed out as well. First, a decrease of the beam particle density leads to a decreasing real part of the correction to the SFXM frequency. Therefore, the frequency becomes closer to the value calculated for the case of the absence of the beam. Second, a decrease of the beam density causes narrowing of the k_{ef} range, where efficient excitation of SFXMs takes place. Third, decrease of the beam particle density is accompanied by transfer of the wave excitation from the regime of resonance overlapping to the regime of separate resonances. This phenomenon is analyzed in details in Sect. 7.3.

The magnitude of the stationary external magnetic field has appreciable influence on SFXM propagation [35]. This influence is demonstrated in Fig. 7.4. In this case normalization of the SFXM growth rates to $|\omega_e|$ that was done in the previous figures becomes inconvenient. That is why here the growth rates are normalized to the electron plasma (Langmuir) frequency Ω_e. Therefore, the physical observables shown along both axes of this figure are normalized to Ω_e. The solid curves correspond to the case of stronger magnetic field, $Z = 3$, and the dashed curves to the case of a weaker magnetic field, $Z = 10$. One can see, that decreasing magnetic field causes decreasing growth rates of the SFXM beam-plasma instability and significant narrowing of the k_{ef} range, where efficient beam-wave interaction is possible. In the case of stronger magnetic field, $Z = 3$, excitation of the SFXM with $m = 1$ happens due to interaction of the wave with the first cyclotron harmonic of the beam. Detuning of the SFXM frequency from the resonant value $|\omega_e|/\gamma$ at increasing k_{ef} leads to a corresponding decrease in growth rate. In the HF range (1.14) of SFXM propagation, the influence of the value of an external magnetic field on SFXM excitation is much stronger (see the Sect. 10.1): decreasing the magnetic field by 25% leads to a decrease by almost a factor 6 of the maximum value of the growth rate (for fixed value of the azimuthal mode index) [36].

The dependence of LF SFXM growth rates on the thickness of the vacuum gap, where the electron beam gyrates, is shown in Fig. 7.5. The solid curves were calculated for the gap width $\Delta = 0.1$ corresponding to a narrower gap, and the dashed curves to the wider gap with $\Delta = 0.2$. Decreasing width of the vacuum layer causes decreasing SFXM eigenfrequency as well as insignificant enhancement of the maximum growth rate and the range of k_{ef} where SFXMs efficiently interact with the electron beam.

Thus, the influence of the width of the vacuum layer on LF SFXM excitation essentially differs from its influence on HF SFXM excitation [36] (see the Sect. 10.1 below for details). Namely, in the case of HF SFXMs, a decrease of the width of the vacuum layer leads to an insignificant decrease of the maximum value of the growth rate, but to a marked expansion of the k_{ef} range, where efficient SFXM excitation is possible. This is explained by the fact that in the case of HF SFXMs, the narrower the vacuum layer is, the more often the dispersion curve crosses the resonant curves $\omega = N|\omega_e|/\gamma$ which is favorable for meeting the resonance conditions (7.14).

In the following, the possibility of experimental observation of the phenomenon studied here is estimated. As one can see, for example, from Fig. 7.2, the LF SFXM with azimuthal mode index $m = 1$ is efficiently excited at $k_{ef} \approx 0.09$. Thus for a laboratory device with plasma column dadius R_1 ~10 cm, the Langmuir frequency should be equal to $\Omega_e \approx 3.4 \times 10^{10}$ s^{-1}, which corresponds to the plasma density $n_e \approx 3.5 \times 10^{11}$ cm^{-3}. For the ratio $Z = 10$ (see the plot in Fig. 7.2), $|\omega_e| \approx 3.4 \times 10^9$ s^{-1}, that corresponds to an induction of the magnetic field $B_0 \approx 191$ G. In this case SFXMs propagate with the frequency $\omega \approx 2.5 \times 10^9$ s^{-1}. The beam instability is characterized by Im(ω) $\approx 6.1 \times 10^8$ s^{-1} caused by an electron beam with $n_b = 0.01$ $n_e \approx 3.5 \times 10^9$ cm^{-3} particle density. Using this data, it is possible to estimate the optimal current density of the electron beam to be $j = en_b\upsilon \approx 10.4$ A/cm^2.

In the present Section, the possibility to realize the condition of LF SFXM excitation by annular relativistic electron beams is studied in detail. Results of the numerical analysis are in satisfactory agreement with the analytical predictions. It is shown, that the magnitude of the growth rate Im(ω) of LF SFXM resonant beam-plasma instability increases with increasing external static magnetic field, azimuthal mode index, and electron beam particle density.

Excitation of electromagnetic waves by annular electron beams is characterised by the following feature. The beam rotates in the external magnetic field in the layer between the plasma column and the metallic wall. Therefore the beam velocity is determined by the induction of the magnetic field and the radii of the plasma column and the metal chamber $|\omega_e|R_1/\gamma < \upsilon < |\omega_e|\, R_2/\gamma$. Some conclusions are associated with this. First, the closer the beam radius is to the radius of the metal chamber, the more kinetic energy can be transferred to the electromagnetic wave before the beam reaches the plasma surface during deceleration. On the other hand, the closer the electron beam is to the metal chamber, the weaker is the electrical wave field, which interacts with the beam, and the lower is the efficiency of beam-wave interaction. Second, the role of the wave vector of SFXMs is played by the ratio of azimuthal mode index to the radius of the plasma column, $k_\varphi = m/R_1$. Thus, the condition (7.14) of resonant beam-plasma excitations, $\omega = |\omega_e|N/\gamma$, depends on the azimuthal mode index.

The character of dependence of the efficiency of LF SFXM excitation by electron beams on the radius R_1 of the plasma column is closely connected with the dispersion properties of these waves. In wide waveguides, $k_{ef} \to 0$, the SFXM frequency approximately depends inversely on the radius R_1, $\omega \propto k_{ef}$. However, in narrow waveguides, at $k_{ef} >> 1$, the frequency of LF SFXMs weakly depends on the radius R_1 [35]. Therefore, for excitation of LF SFXMs in the general case, a plasma column with not too small radius is preferable. However, the range of k_{ef}, for which there is efficient SFXM excitation for a given azimuthal mode index m in the resonator, can be prolonged in the direction of larger k_{ef}. For this purpose one has to choose the parameters of the resonator so that the SFXM frequency is close to the resonant value $\omega = |\omega_e|N/\gamma$.

The wider the vacuum layer between the plasma column and the metal wall of the waveguide is (larger Δ), the higher is the obtainable efficiency. Increasing Δ causes also an insignificant decrease in the growth rate of SFXW excitation unlike that observed in the HF range [36].

The value of an external static magnetic field is characterized here by the parameter $Z = \Omega_e/|\omega_e|$. The magnitude of the external axial magnetic field influences the excitation of LF SFXMs weaker as compared with the case of HF SFXMs [36]. Nevertheless, increasing B_0 causes a noticeable increase of the growth rate of LF SFXM beam excitation and a widening the k_{ef} ranges where excitation of these waves takes place.

Growth rates of LF SFXMs are larger as compared with those of HF SFXMs. This is explained not only by the better possibility of realizing the resonant beam-wave condition (7.14): for HF SFXMs at the best the resonance for the fifth ($N = 5$) cyclotron harmonic can be realized, while $\mathrm{Im}(\omega) \propto [J_N(|m|)]^{2/3}$. Besides, for LF SFXMs the increase in beam particle density causes the transition from the regime of beam instability due to resonant interaction of the wave with a separate cyclotron harmonic to the regime of overlapping resonances.

While the plasma system can be described by the model of an infinite medium, the frequency of collisions between plasma particles can be neglected (as a rule) compared with the frequency of the investigated electromagnetic waves. However, in the case of bounded plasmas the effective plasma particle collisional frequency ν can be relatively large compared with the growth rate of the resonant beam instability of the considered eigenwaves. Application of an effective collisional frequency allows one to model all real dissipative processes which can happen in the considered waveguide.

The damping rate of SFXMs is determined by the effective collisional frequency and directly proportional to its value ν [1, 37], see as well Chap. 2. Under the condition of validity of the inequality $\nu > \mathrm{Im}(\omega)$, the interaction between SFXM and the gyrating electron beam can lead to the development of a dissipative instability. The relative value of its growth rate is lower than the growth rate of the resonant beam instability, but for the cases of solid-state plasmas (e.g. semiconductor plasmas) and/or meta-plasmas (one applies this term for the case of special artificial environments) its value can be large enough.

In the following, the possibility to realize a regime of dissipative instability of the SFXM interacting with a gyrating electron beam is considered [38]. Geometry and design of the waveguide is the same as in the case of the resonant beam instability considered above. The basic equations of the SFXM fields and the related boundary conditions in the case of dissipative instability are the same as they are in the case studied in the first part of this Sect. 7.1 (see Eqs. 7.9–7.11). The dispersion relation that describes the initial stage of the interaction between SFXM and electron beam, which gyrates in the vacuum layer between plasma column and waveguide wall, $R_1 < r < R_2$, has a form that is similar to relation (7.12). There is only one distinguishing feature. Namely, one has to take into account the presence of the effective collisional frequency ν in the expressions (3.25) for the components of the dielectric permittivity tensor ε_{ij} for the plasma column. The dispersion relation is studied under the assumption of the validity of resonance condition (7.14). The growth rate of the dissipative instability of the SFXMs is as follows [38]:

$$\mathrm{Im}(\omega) \approx \sqrt{\frac{c\alpha\Omega_e^3}{2\gamma^2\nu|\omega_e|R_1}} N^{3/2} J_N(m). \tag{7.16}$$

Comparing the growth rates obtained for the cases of dissipative (7.16) and resonant beam (7.15) instabilities allows one to make the conclusion that the dissipative instability is characterized by a smaller value of the growth rate as compared with the resonant beam instability. This feature is mainly due to the increase of the degree to which the Bessel function of the first kind is rising up and the appearance of a small multiplier $\sqrt{\alpha}$.

Because of the smallness of the growth rate of the dissipative instability of SFXMs as compared with the growth rate of the resonant beam instability one can conclude that there is only one way for increasing the efficiency of the dissipative instability, namely, one can increase the plasma density. Therefore, utilization of a semiconductor-plasma as a plasma filling of the waveguide seems to be more prospective compared with the case of a gaseous plasma for applying the dissipative instability for some technological applications.

In the following, the case of semiconductor-plasma application is considered. Then the role of the electron flow, which gyrates around the plasma column, would be played by conduction band electrons of another semiconductor, which fills the space $R_1 < r < R_2$. The concentration of conduction band electrons in this outer semiconductor layer n_{e2} should be sufficiently low in order to satisfy the inequality: $\Omega_{e1}^2 \gg \varepsilon_{01,02}\omega_e^2 \gg \Omega_{e2}^2$. Here $\Omega_{e1,2}$ are the plasma (Langmuir) frequencies of the first (inner) and second (outer) semiconductors, respectively, and $\varepsilon_{01,02}$ are the dielectric constants of the lattice of the first and the second semiconductors. In this case conduction band electrons move with the velocity $\upsilon_0 \equiv \upsilon_d = cE_0/B_0$, where E_0 is an external radial electric field, which should be utilized for realization of the azimuthal motion of the conduction band electrons in the outer, second semiconductor. This electric field can be produced by application of an electric voltage between the metal wall of the waveguide and additional axial electrodes, which can be located around the circumference of the interface between the two semiconductors. As positive features of the proposed arrangement one should mention that there is no necessity to control a vacuum state in the utilized waveguide, to elaborate an electron beam injector (gun) and to install a collector for the electrons after passing the interaction region, etc.

Taking into account that the density of conduction band electrons in semiconductors is larger than in gaseous plasmas, solid state devices of HF electronics [32] have some evident advantages compared with vacuum electronic [4, 33] and plasma electronic [1–3] devices. The main advantage is the possibility to generate electromagnetic radiation in a relatively wide range of frequencies, namely from the electron cyclotron frequency up to the Langmuir frequency.

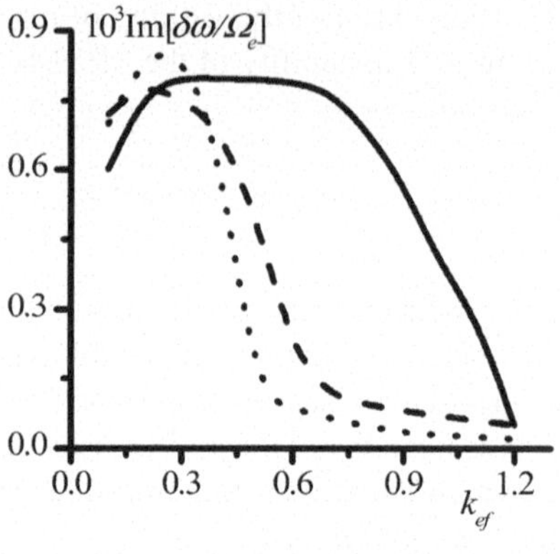

Fig. 7.6 Dependence of the normalized SFXW growth rates in the case of dissipative instability on the effective wavenumber for the azimuthal mode index $m = 2$, *Solid curve*: $\Delta = 0.1$, $B_0 = 100$ G; *dashed curve*: $\Delta = 0.1$, $B_0 = 200$ G; *dotted curve*: $\Delta = 0.2$, $B_0 = 100$ G

In Fig. 7.6 the solid curve shows the dependence of the growth rate $\mathrm{Im}(\delta\omega)$ on the effective wavenumber $k_{ef} = mc/(R_1\Omega_{e1})$ for the SFXM with azimuthal mode index $m = 2$ under the regime of dissipative instability. Doing that it is assumed that the inner semiconductor is *n-Ge* (density of conduction band electrons $n_1 = 2.5\times10^{13}\mathrm{cm}^{-3}$, dielectric constant of the lattice $\varepsilon_{01} = 16$, longitudinal effective mass of the conduction band electrons $M_{1\|} = 1.3m_e$, effective transverse electron mass $M_{1\perp} = 0.082m_e$, where m_e is the electron rest mass). The outer semiconductor is assumed to be similar to *n-Si* ($n_2 = 10^{10}\mathrm{cm}^{-3}$, ε_{02}, $M_{2\|} = 0.97m_e$, $M_{2\perp} = 0.19m_e$) [39]. The thickness of the outer semiconductor is assumed to be relatively small (parameter$\Delta = 0.1$) and the value of the applied magnetic field induction vector $B_0 = 100$ G can be obtained without any special effort (thus the relation between the frequencies $|\omega_{e1}| = 0.1\,\Omega_{e1}$ is a typical one for these conditions). The life-time of the quasi-particle "conduction band electrons" in chemically pure semiconductor *n-Ge* under the regime of electron–phonon interactions is large enough $\left(\tau_1 = 10^{-5}\mathrm{s}\right)$ [40]. That is why there is no necessity to search for a very pure semiconductor or to realize the special condition of low temperature, because one can obtain a collisional frequency ν, whose value is sufficiently lower than the SFW eigenfrequency in this case. Therefore, let it have approximately the value: $\nu \approx 1.74\times10^9\ s^{-1}$, this is four orders of magnitude larger than τ_1^{-1}. Such choice of the ν value is possible because it can be realized due to the following dependence of the effective collisional frequency on temperature in the range of ordinary room temperature: $\nu \propto T^{1.6}$ [40].

The solid curve in Fig. 7.6 shows the dependence of $\mathrm{Im}(\delta\omega)$ on k_{ef}. The dashed curve gives $\mathrm{Im}(\delta\omega)$ versus k_{ef} for the same values of the waveguide parameters as taken in the previous case, excluding the magnitude of the utilized magnetic field. It is chosen as $B_0 = 200$ G, so the value of the electron cyclotron frequency is twice compared with the previous case: $|\omega_{e2}| = 0.2\,\Omega_{e2}$. The dotted curve shows the dependence of $\mathrm{Im}(\delta\omega)$ on k_{ef} for the same values of the waveguide parameters as used for the case presented by the solid curve, excluding the relative thickness of the outer semiconductor sheeth, which here is chosen as $\Delta = 0.2$.

All these plots have a similar bell-type form where the maximum of the SFXM growth rate is for all these cases clearly located in the range $0.2 < k_{ef}0.7$.

To analyze the influence of the waveguide parameters on the value of the SFXM growth rate under the condition of dissipative instability one can conclude that an increasing external stationary magnetic field leads to a shift of the $\mathrm{Im}(\delta\omega)$ maximum to the region of smaller k_{ef} and makes the range of large values of the growth rate of the dissipative instability narrower. The similar influence on the form of the dependence $\mathrm{Im}(\delta\omega) = f(k_{ef})$ is obtained by increasing the thickness of the outer semiconductor layer with low density of the conduction band electrons. The maximum value of $\mathrm{Im}(\delta\omega)$ in the case of the dissipative instability for a fixed value of the azimuthal mode index m and the effective wavenumber k_{ef} weakly depends [38] on the thickness of the outer semiconductor sheeth and on the external magnetic field; it is approximately equal to $8 \times 10^{-3}\, \Omega_{e1}$ and $5 \times 10^{-3}\, \Omega_{e1}$ for the SFWs with $m = 2$ and $m = 3$, respectively. Increasing azimuthal mode index leads to decreasing range of effective wavenumber k_{ef}, where the value of the growth rate $\mathrm{Im}(\delta\omega)$ has its maximum. A comparison of this result with the case of the maximum of the resonant beam instability shows that in the present case the maximum of $\mathrm{Im}(\delta\omega)$ is shifted to the range of slightly smaller magnitudes of the effective wavenumbers k_{ef} [34, 38].

7.2 Excitation of Higher Radial Modes of Surface Flute Waves

In the previous section, the excitation of the zeroth radial mode of LF SFXMs was studied only. Explanation of the term "the zeroth radial mode" is as follows. For the zeroth radial mode of SFXMs, the axial magnetic wave field B_z has only one node within the full range of the radial co-ordinate: $f(r) = 0$ for $r = 0$. (Here $f(r)$ decribes the radial dependence of the RF axial magnetic wave field B_z within the complete dependence of B_z on the spatial co-ordinates and time t, $B_z(\vec{r}, t) = f(r)\exp(im\varphi - i\omega t)$.) The lth SFXM radial mode has l nodes within the range $0 < r < R_2$ (here $r = R_2$ is the inner radius of the cylindrical metal chamber) in addition to that for $r = 0$. The radial distribution of B_z with the azimuthal mode index $m = 2$ is plotted in Fig. 7.7 for the zeroth (solid curve), first (dashed curve) and second (dotted curve) radial modes. Increase of the radial mode index is accompanied by a successive increase of the

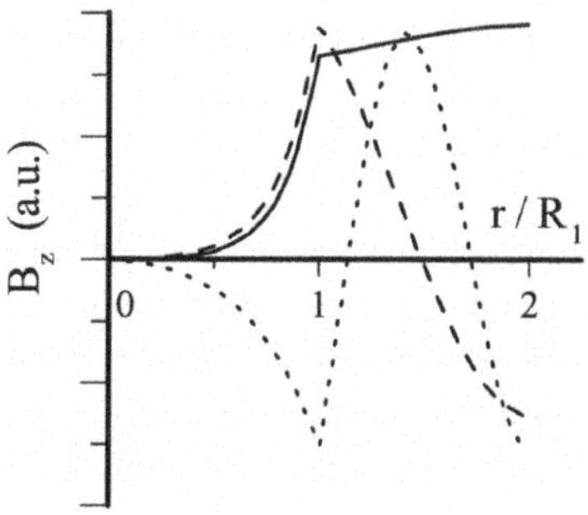

Fig. 7.7 Radial distribution of the SFXM axial magnetic field B_z in arbitrary units for $Z \equiv \Omega_e/\omega_e l$=5; $R_2/R_1 = 2.0$; $m = 2$; $l = 0$ (*solid curve*), $l = 1$ (*dashed curve*), $l = 2$ (*dotted curve*); effective wavenumber $k_{ef} \equiv mc/(\Omega_e R_1) = 0.3$

eigenfrequency; in these calculations the frequency increases from $\omega = 0.979|\omega_e|$ for the zeroth radial mode to $\omega = 2.319|\omega_e|$ for the first radial mode and further to $\omega = 4.157|\omega_e|$ for the second radial mode.

The reasons why the excitation of the zeroth radial SFXM mode only has been studied previously are as follows. First, the basic one, in this case the plasma parameters (as plasma column radius, strength of stationary magnetic field, and plasma particle density) cause the strongest impact on the SFXM dispersion properties. Second, the electrodynamic properties of the space where the electrons rotate, $R_1 < r < R_2$, can be described by the permittivity tensor $\varepsilon_{ik}{}^{(b)}$ [21, 34] whose weak dependence on the radial co-ordinate can be neglected. But in the case of propagation of the first and higher radial modes, the role of the plasma filling decreases, and just the parameters of the dielectric layer (its width and the value of its dielectric constant) mostly affect the SFXM dispersion properties. (In the studied case, the dielectric layer, $R_1 < r < R_2$, is replaced by the vacuum gap where the beam of electrons gyrates). Also in this case the velocity of the beam electrons cannot be assumed to be constant anymore.

The higher order radial modes of electromagnetic waves are characterized by their higher eigenfrequencies and shorter vacuum wavelengths than those in the case of the zeroth radial modes. This feature promises the possibility to expand the applicability of SFXM beam excitation to the case of higher frequencies by utilizing the first or even higher radial SFXM modes.

In the present Section, beam excitation of LF SFXMs in a cylindrical waveguide partially filled by plasmas is considered (see Fig. 7.1). The schematic of the problem is the same as in the previous Sect. 7.1. The electrodynamic properties of the uniform plasma column with radius R_1 are described in terms of the dielectric permittivity tensor ε_{ij} of a cold collisionless plasma [41]. The second order differential equation of Bessel type for the SFXM axial magnetic field B_z within the plasma column can be derived from the Maxwell equations. Its solution is proportional to the modified Bessel function $I_m(k_\perp r)$ (see Eq. (9.6.10) on p. 375 in [30]). The second solution of the Bessel equation, that is the MacDonald function $K_m(k_\perp r)$ (see Eq. (9.6.11) on p. 375 in [30]), does not contribute to the expression for B_z since it is known to turn to infinity at the axis, $lim_{r\to 0} K_m(k_\perp r) = \infty$, and the boundary conditions demand the wave fields to be finite inside the waveguide.

The wave penetration depth of the wave into the plasma $k_\perp{}^{-1}$ is determined as follows, $k_\perp^2 = k^2\varepsilon_1\left(\mu^2 - 1\right)$, where $\mu = \varepsilon_2/\varepsilon_1$, $k = \omega/c$, and $\varepsilon_{1,2}$ are the components of the dielectric permittivity tensor [21, 41]. The condition of positive $k_\perp{}^2$ gives the expressions for the boundaries of the frequency range (1.13) where LF SFXMs propagate.

The electron beam particles rotate in the stationary axial magnetic field $\vec{B}_0 || \vec{z}$ along large Larmor orbits around the plasma column within the gap between the column and the metal chamber, $R_1 < r < R_2$. These particles are assumed to have only an azimuthal component of the momentum, $p_{r,z} = 0, p_\varphi = m_e \upsilon \gamma$, where the relativistic factor γ is a function of the radial co-ordinate since electrons rotate with different velocities υ along the Larmor orbits with different radii.

The electrodynamic properties of this space can be presented in terms of the dielectric permittivity tensor ${\varepsilon_{ij}}^{(b)}$ [21, 34]. The following second order differential equation for the SFXM axial magnetic wave field B_z within the gap between the column and the metal chamber can be derived from the Maxwell equations:

$$\frac{1}{r}\frac{d}{dr}\left(\frac{r}{k_b^2}\frac{dB_z}{dr}\right)+\left(1-\frac{m^2}{k_b^2 r^2}+\frac{m}{r}\frac{d}{dr}\left(\frac{\mu}{k_b^2}\right)\right)B_z=0. \tag{7.17}$$

In (7.17), $k_b^2 = k^2\left(\varepsilon_{22}^{(b)} + \varepsilon_{12}^{(b)2}/\varepsilon_{11}^{(b)}\right)$,

$$\varepsilon_{11}^{(b)} = 1 + \frac{\Omega_b^2}{\omega\gamma}\sum_{s=-\infty}^{+\infty} s^2\left[\frac{\gamma\left(J_s^2(m)\right)'}{(s-q)m|\omega_e|} + \frac{kJ_s^2(m)\,r^2}{(s-q)^2 cm^2}\right]; \tag{7.18}$$

$$\varepsilon_{12}^{(b)} = \frac{i\Omega_b^2}{\omega|\omega_e|}\sum_{s=-\infty}^{+\infty} s\left[\frac{\left(J_s(m)J_s'(m)\right)'}{s-q} + \frac{J_s(m)J_s'(m)}{(s-q)m} + \frac{J_s(m)J_s'(m)\,kvr}{(s-q)^2 cm}\right] = -\varepsilon_{21}^{(b)}; \tag{7.19}$$

$$\varepsilon_{22}^{(b)} = 1 + \frac{\Omega_b^2}{\omega|\omega_e|}\sum_{s=-\infty}^{+\infty}\left[\frac{2\left(J_s'(m)\right)^2}{s-q} + \frac{\left(\boldsymbol{J_s'(m)}\right)^2\boldsymbol{v^2y}}{\boldsymbol{(s-q)^2c^2}} + \frac{2mJ_s'(m)J_s''(m)}{s-q}\right]. \tag{7.20}$$

In (7.18)–(7.20), $\Omega_b^2 = 4\pi e^2 n_b m_e^{-1}$ is the plasma frequency of the electron beam squared, $q = \omega\gamma/|\omega_e|$, and $J_s(x)$ is the Bessel function of the first kind (see Eq. (9.1.10) on p. 360 in [30]). A prime denotes the derivative in the respect of the argument.

Hereinafter one assumes the approximation of weak particle density of the electrons in the beam, $\alpha = n_b/n_e << 1$. The expressions (7.18)–(7.20) for the components of permittivity tensor correspond to the expressions (7.4)–(7.6) for the components of the conductivity tensor presented in the Sect. 7.1.

Since the dielectric permittivity tensor ${\varepsilon_{ik}}^{(b)}$ weakly depends on the radial coordinate one can get the asymptotic solution of Eq. (7.17) in the following form applying the method of standard equation [42]:

$$B_z = \sqrt{\frac{c\zeta}{\omega\, r}}\left[C_1 J_m(\zeta) + C_2 N_m(\zeta)\right]. \tag{7.21}$$

In (7.21), $\zeta(r) = \zeta_1 + \int_{R_1}^{r} k_b dr$, $\zeta_1 = \zeta(R_1) = kR_1$ and $C_{1,2}$ are the constants of integration. In the present Section, the considerations are not restricted to assuming the SFXM radial phase change, $\int_{R_1}^{R_2} k_b dr$, within the layer where the electrons gyrate, to have a small value, which is an attribute of the zeroth radial mode. For the higher radial modes studied here, on the contrary, the SFXM radial phase change is large, $\int_{R_1}^{R_2} k_b dr \approx \pi l$.

Application of the boundary conditions (7.9)–(7.11) makes it possible to derive the dispersion relation (7.12) which describes the initial stage of the electron beam excitation of LF SFXMs. Using the theory of residues one can calculate ζ_2:

$$\zeta_2 = kb + \pi i \sum res(k_b) = kb + \pi i \sqrt{\frac{\alpha|\omega_e|}{\omega}\frac{Z}{m}} \sum_N \left(N^{1.5} J_N(m)\right). \quad (7.22)$$

Since the value $q = \omega\gamma/|\omega_e|$ is a function of the radial coordinate the function k_b can have few resonances along the radial coordinate. Each of these resonances is caused by a resonance denominator in some terms of the sums of the components (7.18)–(7.20). That term in the component $\varepsilon_{22}{}^{(b)}$ (7.20) of the electron beam dielectric permittivity tensor is marked in bold which makes the main contribution to the residuals in (7.22). The number of terms in the sum (7.22) is equal to the number of resonances of the function k_b along the radial coordinate and those terms are taken into account in the sum (7.22) for which the resonance condition (7.14) happens somewhere along the radial coordinate.

Figure 7.8 shows the dependence of the radial refractive index $k_b(r)/k$ versus the radial coordinate for the case in which three resonances (7.14) (for $N = 4$, 5, and 6) exist within the interval $R_1 < r < R_2$. These resonances are situated at the radii $r_4 = 1.051R_1$, $r_5 = 1.429R_1$, and $r_6 = 1.785R_1$. These coordinates are marked in Fig. 7.8 with vertical dotted lines. The following beam-plasma parameters are chosen for the calculations: $m = 2, Z = 5.0, k_{ef} = 0.3, \alpha = 0.01$, and $R_2/R_1 = 2$. In this case, the first radial mode of the SFXM propagates with the frequency $\omega = (2.32 + 0.097\mathrm{i})|\omega_e|$. The dependence of the real part of the radial refraction index, $\mathrm{Re}(k_b/k)$, is shown by the dashed curve and that of the imaginary part, $\mathrm{Im}(k_b/k)$, by the solid curve. The higher the number N of the resonance is, the less is the contribution of the resonance to the SFXM excitation. For the resonance $N = 4$, the imaginary part of the radial refractive index $k_b(r)/k$ deviates from zero with a maximum $\mathrm{Im}[k_b/k] = 0.025$ at $r/R_1 = 1.022$ and a minimum $\mathrm{Im}[k_b/k] = -0.0072$ at $r/R_1 = 1.1$. For the next resonance $N = 5$, the extrema of $\mathrm{Im}[k_b/k]$ are much less pronounced: the maximum is $\mathrm{Im}[k_b/k] = 0.022$ at $r/R_1 = 1.4$, and the minimum $\mathrm{Im}[k_b/k] = 0.00019$ at $r/R_1 = 1.48$. After the r/R_1 ratio exceeds 1.5, $\mathrm{Im}[k_b/k]$ reaches to the local maximum, $\mathrm{Im}[k_b/k] = 0.00057$ ($N = 6$), at $r/R_1 = 1.734$, and then decreases monotonously to $\mathrm{Im}[k_b/k] = 0.00043$ at $r/R_1 = 2.0$.

From the mathematical point of view, the expression (7.22) is the novel result of this Section—just this expression distinguishes the present work from the previous studies [31, 34, 43, 44] on SFXM beam excitation in cylindrical waveguides partially filled by magneto-active plasmas with narrow vacuum gap where the electron beam rotates.

Figures 7.9, 7.10, 7.11, 7.12 and 7.13 present some results of numerical studies on the dispersion relation (7.12) with account for the newly obtained value (7.22) of the argument ζ_2 of the Bessel functions. The effective wave vector $k_{ef} = m\delta/R_1$ is used there as abscissa. For SFXMs the ratio m/R_1 plays the role of the azimuthal wave vector. It is normalized by the skin depth $\delta = c/\Omega_e$. Figures 7.9, 7.10, 7.12,

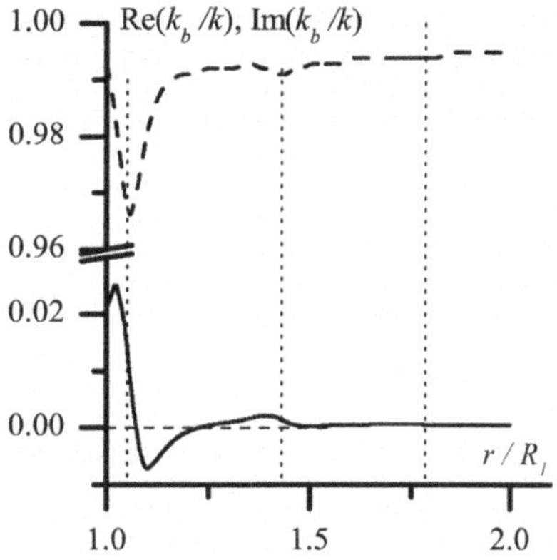

Fig. 7.8 Radial dependence of the SFXM radial refractive index k_b/k. $m = 2$, $Z = 5.0$, $k_{ef} = 0.3$, $\alpha = 0.01$, $R_2/R_1 = 2$, $l = 1$, $\omega = (2.32 + 0.097i)|\omega_e|$. Real part: *dashed curve*; imaginary part: *solid curve*; resonances for $N = 4$, 5 and 6: vertical *dotted lines*

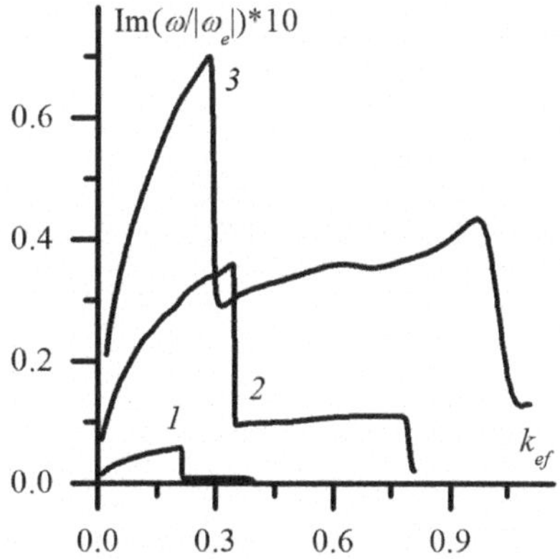

Fig. 7.9 Dependence of the normalized SFXM growth rate on effective wavenumber for $Z = 5$; $\alpha = 0.00125$; $R_2/R_1 = 2$; $m = 1, 2, 3$; $l = 1$. The *numbers* indicate the azimuthal mode indices *m*

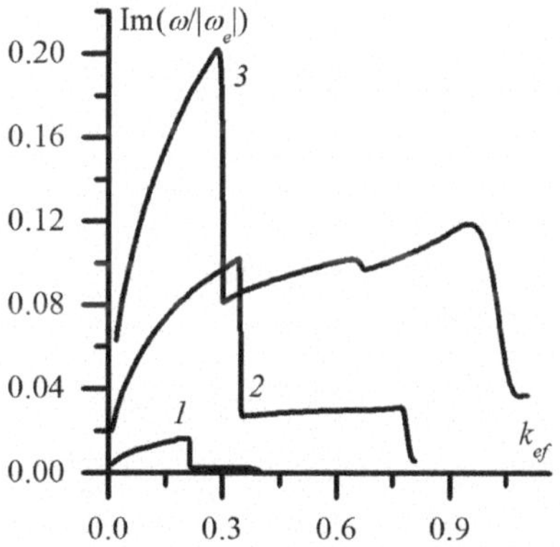

Fig. 7.10 Dependence of the normalized SFXM growth rate on effective wavenumber (to be compared with Fig. 7.9 to outline the effect of electron beam particle density) for $Z = 5$, $\alpha = 0.01$, $R_2/R_1 = 2$, $m = 1, 2, 3$; $l = 1$. The *numbers* indicate the azimuthal mode indices *m*

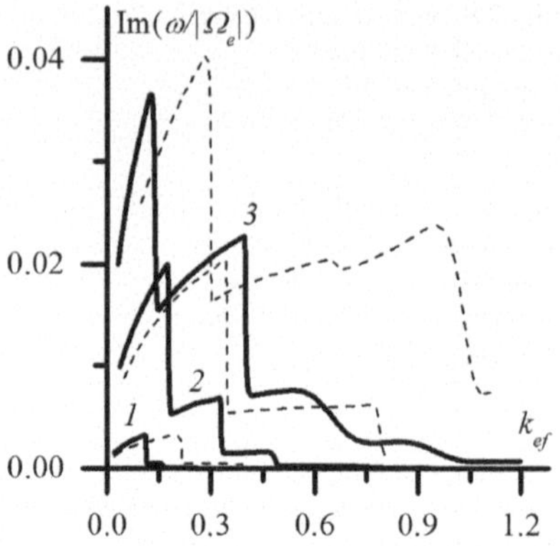

Fig. 7.11 Dependence of the normalized SFXM growth rate on effective wavenumber for $Z = 8$ (*solid curves*), $Z = 5$ (*dashed curves*), $\alpha = 0.01$, $R_2/R_1 = 2$, $m = 1, 2, 3$; $l = 1$. The *numbers* indicate the azimuthal mode indices m

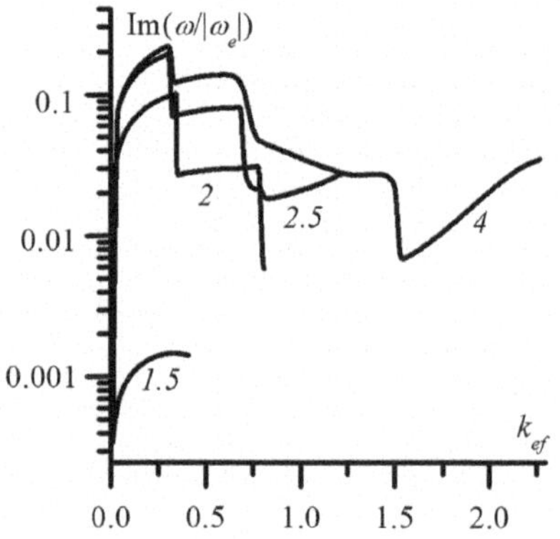

Fig. 7.12 Dependence of the normalized SFXM growth rate on effective wavenumber for $Z = 5$; $\alpha = 0.01$; $R_2/R_1 = 1.5, 2, 2.5, 4$; $m = 2$; $l = 1$. The *numbers* indicate the ratio of radii

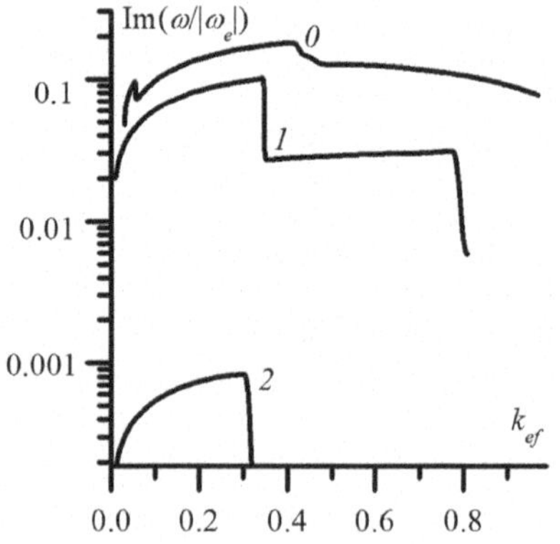

Fig. 7.13 Dependence of the normalized SFXM growth rate on effective wavenumber for $Z = 5$, $\alpha = 0.01$, $R_2/R_1 = 2$, $l = 0, 1, 2$, $m = 2$. The *numbers* indicate the radial mode indices l

and 7.13 demonstrate the dependence of the growth rate $\mathrm{Im}(\omega/|\omega_e|)$ on k_{ef}. When studying the effect of the stationary magnetic field B_0 on LF SFXM excitation one applies $\mathrm{Im}(\omega/|\Omega_e|)$ as ordinate in Fig. 7.11. In this case, the scales of both axes do not depend on B_0 and are inversely proportional to the same function, the plasma particle density, $\propto n_e^{-1/2}$. To control which radial mode is excited we calculate the radial phase change $\zeta_2 - \zeta_1$ within the layer $R_1 < r < R_2$ where the electrons rotate to get each point of the curves.

For the calculations, two values of relative electron beam particle density, $\alpha = n_b/n_e$, are chosen. In Fig. 7.9, the electron beam particle density is smaller, $\alpha = 0.00125$. In Figs. 7.10, 7.11 and 7.12, it is larger, $\alpha = 0.01$. The numbers in Figs. 7.9, 7.10 and 7.11 indicate the azimuthal mode indices m. The stationary magnetic field B_0 is characterized in this calculations by the dimensionless parameter $Z = \Omega_e/|\omega_e|$. It is chosen as $Z = 5$ for Figs. 7.9, 7.10, 7.12, and 7.13. However, in Fig. 7.11, the effect of the stationary magnetic field B_0 on LF SFXM excitation is shown, and growth rates for the cases $Z = 5$ (Fig. 7.10) and $Z = 8$ (Fig. 7.11) are compared. The dependence of the LF SFXM growth rate on k_{ef} for the case $R_2/R_1 = 2$ is studied numerically in Figs. 7.9, 7.10, 7.11 and 7.13. The influence of the width of the vacuum gap (of the parameter R_2/R_1) where the electrons rotate on LF SFXM excitation is demonstrated in Fig. 7.12.

In Fig. 7.9, the dependence of the normalized growth rate $\mathrm{Im}(\omega/|\omega_e|)$ of the first radial mode, $l = 1$, versus k_{ef} is shown for the following beam-plasma parameters: $Z = 5$, $\alpha = 0.00125$, $R_2/R_1 = 2$, $m = 1, 2, 3$. Increasing the azimuthal mode index m results in an increase of the eigenfrequency $\mathrm{Re}(\omega/|\omega_e|)$ and an expansion of the range of k_{ef} where higher radial modes of LF SFXM with this wavenumber can propagate. The right limits of the range of k_{ef} are as follows: $k_{ef} \approx 0.404$ for the SFXM with $m = 1$, $k_{ef} \approx 0.81$ for the SFXM with $m = 2$, and $k_{ef} \approx 1.11$ for the SFXM with $m = 3$. At these limits the eigenfrequency $\mathrm{Re}(\omega)$ reaches the right limit $\omega = \omega_1 - |\omega_e|$ of the LF range (1.13). In the following, the behavior of the curve in Fig. 7.9 which corresponds to the SFXM with $m = 3$ is explained as an example. For $0 < k_{ef} < 0.29$, the growth rate increases monotonously up to the value $\mathrm{Im}(\omega/|\omega_e|) \approx 0.071$. Within this range, the electrons which rotate in the layer $R_1 < r < R_2$ are four times in resonance (7.14) with the wave for $N = 4, 5, 6$, and 7. Further increase of k_{ef} (decrease of the plasma column radius R_1 and subsequent decreasing layer $(R_2 - R_1)$) excludes the resonance $N = 4$, and the SFXM beam excitation is caused by the resonances $N = 5, 6$, and 7 only for the range $0.29 < k_{ef} < 0.65$. Exclusion of the lower resonance $N = 4$ decreases the growth rate almost twice—to the value $\mathrm{Im}(\omega/|\omega_e|) \approx 0.036$ for $k_{ef} = 0.295$. Within the range $0.65 < k_{ef} < 1.0$ the resonances $N = 5$ and 6 contribute to SFXM excitation. Note that exclusion of the higher resonance $N = 7$ at $k_{ef} \approx 0.65$ causes an observable but not essential change of the growth rate. The latter decreases from $\mathrm{Im}(\omega/|\omega_e|) \approx 0.036$ at $k_{ef} = 0.65$ to $\mathrm{Im}(\omega/|\omega_e|) \approx 0.035$ at $k_{ef} = 0.7$. Within the range $1.0 < k_{ef} < 1.11$, the only observed resonance is $N = 6$. At $k_{ef} = 1.11$ the eigenfrequency reaches the upper limit of the LF range (1.13) and the calculations are stopped. The same resonances cause beam-wave interaction for the LF SFXM with $m = 2$. The resonance $N = 7$ does not contribute to the excitation of the LF SFXM with $m = 1$.

At a first glance it seems to be very difficult to use the electromagnetic wave interaction with higher cyclotron harmonics $N > 1$ for RF generation. Indeed, harmonic operation of gyrotrons with traditional annular small-orbit electron beam configuration is inevitably accompanied by significant mode competition from fundamental frequency modes. As a result, harmonic operation at or even beyond the third harmonic, $N \geq 3$, are rarely reported for small-orbit gyrotrons [45]. However, a team from the University of Strathclyde, UK, reported in [46] on designing and modeling of a ~390 GHz harmonic gyrotron operating at the seventh harmonic of the electron cyclotron frequency, $N = 7$. Moreover, a 0.42-THz eighth-harmonic large-orbit gyrotron was designed and theoretically studied in [47]. Experimental results on high-harmonic gyrotrons employing large-orbit, axis-encircling electron beams have been achieved with frequencies up to 1 THz using TE_{mn}-resonator modes where the azimuthal mode index m is equal to the harmonic number N ($m = N$) for $N \leq 5$ [45]. Interaction of an annular electron beam with surface waves and higher order modes in oversized G-band slow-wave structures were studied in [48]. Growth rates of the surface waves were shown there to be larger than those for the higher-order modes.

Figure 7.10 provides the possibility to study the effect of electron beam particle density on LF SFXM beam excitation. There, the shapes of the curves are very similar to those in Fig. 7.9. They have the same ranges of k_{ef}. Classical theory [49] of beam-wave interaction predicts the following dependence of the maximum growth rate on the beam particle density: $\mathrm{Im}(\omega) \propto n_b{}^{1/3}$. However, the beam particle density enters to the dispersion relation (see Eq. 7.22) as $n_b{}^{1/2}$. Numerical analysis of the dispersion relation confirms that an increase of α by a factor of eight results in an increase of $\mathrm{Im}(\omega)$ by $8^{\times}$ with $x = 0.488$, 0.501, and 0.512 for LF SFXMs with $m = 1$, 2, and 3, respectively. The difference between these results and those in [49] is explained as follows. In [49], all the beam particles are assumed to move with the same velocity and that they are in resonance with the wave simultaneously; whereas in the present study, only the electrons rotating along definite orbits, for which the condition (7.14) is valid, contribute to wave excitation.

It is well-known [31, 34, 43, 44] that the stationary magnetic field B_0 has the most significant influence on the SFXM beam excitation. The effect of B_0 on the dependence of the growth rate $\mathrm{Im}(\omega/\Omega_e)$ versus k_{ef} is demonstrated in Fig. 7.11. The dependences for the case $Z = 8$, which corresponds to a weaker B_0, are given by the solid curves. For comparison, those for the case $Z = 5$ are shown by the dashed curves. Decreasing stationary magnetic field results in narrowing the range of k_{ef} where higher radial modes of LF SFXMs propagate, moving the position of the growth rate maximum to smaller values of k_{ef}, and decreasing the value of the growth rate maximum. For instance, for the SFXM with $m = 3$, the growth rate maximum $\mathrm{Im}(\omega/\Omega_e) \approx 0.041$, which corresponds to $k_{ef} \approx 0.3$ in the case of $Z = 5$, decreases to $\mathrm{Im}(\omega/\Omega_e) \approx 0.038$ which is observed at $k_{ef} \approx 0.135$ in the case of $Z = 8$.

The influence of the width of the vacuum layer (where the electrons rotate) on LF SFXM beam excitation is demonstrated in Fig. 7.12 where the numbers nearby the curves indicate the dimensionless parameter R_2/R_1. The ordinate in Fig. 7.12 is given in logarithmic scale to make visible the curve which corresponds to the case $R_2/R_1 = 1.5$. From general considerations it is clear that the larger space occupied by the electron beam is the reason of the higher efficiency of wave excitation (larger

Table 7.1 Comparison of the parameters for LF SFXM beam excitation by relativistic electron beams for the conditions of Fig. 7.12

R_2/R_1	k_{ef0}	$\mathrm{Im}(\omega/|\omega_e|)$	$\mathrm{Re}(\omega/|\omega_e|)$	N
1.5	0.325	0.00147	3.928	7, 8
2.0	0.346	0.102	2.616	4–6
2.5	0.31	0.196	1.829	3–6
4.0	0.31	0.224	1.224	2–6

growth rate). The numerical analysis confirms that an increase of R_2/R_1 results in increasing growth rate but this increase is not linear. An increase of R_2/R_1 from $R_2/R_1 = 1.5$ to $R_2/R_1 = 2$ results in a remarkable increase of the maximum growth rate from $\mathrm{Im}(\omega/|\omega_e|) \approx 0.00147$ at $k_{ef} \approx 0.325$ to $\mathrm{Im}(\omega/|\omega_e|) \approx 0.102$ at $k_{ef} = 0.346$. This increase of the ratio R_2/R_1 is accompanied by a decrease of the LF SFXM eigenfrequency from $\mathrm{Re}(\omega/|\omega_e|) \approx 3.928$ to $\mathrm{Re}(\omega/|\omega_e|) \approx 2.616$. Further increase of R_2/R_1 from $R_2/R_1 = 2.5$ to $R_2/R_1 = 4$ results in a negligible increase of the growth rate maximum from $\mathrm{Im}(\omega/|\omega_e|) \approx 0.196$ at $k_{ef} \approx 0.31$ to $\mathrm{Im}(\omega/|\omega_e|) \approx 0.224$ at $k_{ef} = 0.31$. In this case the increasing parameter R_2/R_1 causes a reduction of the LF SFXM eigenfrequency from $\mathrm{Re}(\omega/|\omega_e|) \approx 1.829$ to $\mathrm{Re}(\omega/|\omega_e|) \approx 1.224$. To get shorter wavelengths of the excited waves is considered as the main objective of this research. An approximate formula for estimation of the wavelength is given by: $\lambda \approx 2(R_2 - R_1)/l$. Therefore, when designing a generator for electromagnetic radiation based on LF SFXMs one has to make a compromise between higher growth rate and shorter wavelength. Table 7.1 summarizes the main quantities which characterize the conditions of reaching the maximum LF SFXM growth rate: ratio R_2/R_1, value k_{ef0} of the effective wavenumber for which the maximum LF SFXM growth rate is observed, normalized maximum growth rate $\mathrm{Im}(\omega/|\omega_e|)$, corresponding eigenfrequency $\mathrm{Re}(\omega/|\omega_e|)$, and the harmonic numbers N of the resonances which contribute to the excitation.

The crucial point of the present investigations is that just the excitation of higher radial modes of LF SFXMs leads to reach shorter wavelengths in the same beam-plasma system. Figure 7.13 compares the results of numerical analysis of the dispersion relation for the second azimuthal mode index, $m = 2$, and the first three radial modes: $l = 0$, 1, and 2. In Fig. 7.13 the corresponding radial mode index is given nearby the curves. Logarithmic scale is chosen for the ordinate axis to make visible the curve which corresponds to the case $l = 2$. This figure also gives generator designers the chance to make a choice between larger growth rate and shorter wavelength. The second radial mode does not look as a good candidate for LF SFXM beam excitation: its maximum growth rate is $\mathrm{Im}(\omega/|\omega_e|) \approx 8.4 \times 10^{-4}$ while the eigenfrequency is $\mathrm{Re}(\omega/|\omega_e|) \approx 4.267$. The first radial mode looks much better when searching for optimal conditions of LF SFXM beam excitation. Its maximum growth rate $\mathrm{Im}(\omega/|\omega_e|) \approx 0.102$ is of the same order as for the zeroth radial mode, $\mathrm{Im}(\omega/|\omega_e|) \approx 0.183$. In addition the eigenfrequency of the first radial mode $\mathrm{Re}(\omega/|\omega_e|) \approx 2.616$ is sufficiently higher than that of the zeroth radial mode, $\mathrm{Re}(\omega/|\omega_e|) \approx 1.393$. The

Table 7.2 Comparison of the parameters for LF SFXM beam excitation by relativistic electron beams for the conditions of Fig. 7.13

l	k_{ef0}	$\mathrm{Im}(\omega/\|\omega_e\|)$	$\mathrm{Re}(\omega/\|\omega_e\|)$	N
0	0.42	0.183	1.393	2
1	0.346	0.102	2.616	4–6
2	0.31	8.4×10^{-4}	4.267	7–11

main quantities which make it possible to compare the conditions for reaching the maximum growth rates of the first three radial modes of LF SFXMs are summarized in Table 7.2. These quantities are as follows: radial mode index l, effective wavenumber k_{ef0} for the maximum LF SFXM growth rate, values of the maximal normalized growth rate $\mathrm{Im}(\omega/|\omega_e|)$, corresponding eigenfrequency $\mathrm{Re}(\omega/|\omega_e|)$, and the harmonic numbers N of the resonances which contribute to the excitation.

The numerical analysis should be completed by estimation of the parameters of the beam-plasma system in which higher radial modes of LF SFXMs can be efficiently excited by relativistic electron beams. Those parameters are used which relate to the maximum of the growth rate calculated in Fig. 7.10 for the SFXM with $m = 3$: $Z = 5$, $\alpha = 0.01$, $R_2/R_1 = 2$, $k_{ef} = 0.3$, $\mathrm{Re}(\omega/|\omega_e|) = 1.788$, and $\mathrm{Im}(\omega/|\omega_e|) = 0.201$. This growth rate is caused by the four resonances (7.14): $N = 4$, 5, 6, and 7. In this case, one can be sure that just the first radial mode is under study, $l = 1$, since the phase propagation is about 3.537. To demonstrate whether the electron beams under consideration are relativistic or not one can calculate the relativistic factor: it is in the range $\gamma = 2.236 \div 4.123$, which refers to the velocities $\upsilon \approx 0.894 \div 0.97c$ and acceleration voltages $0.63 \div 1.60$ MV.

In the present study, the question about the production of the electron beam was not discussed up to now. The rotating beams used in the large-orbit experiments carried out at the University of Maryland [50] were generated by passing annular, linearly streaming beams through non-adiabatic magnetic transitions (magnetic cusps, kickers). This technique had proven to be an efficient way to generate thin high-current beams with large perpendicular energies. Patent [51] proposes to produce an electron beam with radius 6.4 cm which is of the order of the typical radius of a plasma column in a University laboratory. In the following, the plasma column of radius $R_1 = 5$ cm is considered. Then the metal waveguide should have the radius $R_2 = 10$ cm. The indicated value of $k_{ef} \equiv mc/(R_1 \Omega_e) = 0.3$ dictates the electron plasma frequency $\Omega_e \approx 6 \times 10^{10}\ \mathrm{s}^{-1}$ which corresponds to the plasma particle density $n_e \approx 1.13 \times 10^{12}\ \mathrm{cm}^{-3}$ which is clearly achievable in a University laboratory. Then the electron beam particle density should be $n_b \approx 1.13 \times 10^{10} \mathrm{cm}^{-3}$. Since $Z = 5$ is applied, the electron cyclotron frequency is $|\omega_e| \approx 1.2 \times 10^{10}\ \mathrm{s}^{-1}$, which corresponds to the stationary magnetic field $B_0 \approx 682$ G, and an electromagnetic wave in the GHz range ($\approx$3.4 GHz) with the angular frequency $\mathrm{Re}(\omega) = 2.15 \times 10^{10}\ \mathrm{s}^{-1}$ is excited with the growth rate $\mathrm{Im}(\omega) = 2.47 \times 10^{9}\ \mathrm{s}^{-1}$.

In the present Section, the prediction that application of the first radial mode of LF SFXMs can provide higher frequency of the generated electromagnetic waves while keeping the same order of the growth rate as for the zeroth radial mode is confirmed [52]. Application of higher modes than the first radial mode does not seem to be reasonable. Transition from $l = 1$ to $l = 2$ only allows to increase the wave frequency by a factor which is close to one. However, to get this moderate result one has to accept that the growth rate decreases by two orders of magnitude.

Comprehensive numerical analysis shows that higher growth rates are reachable for denser electron beams (higher beam current), stronger stationary external magnetic field B_0, and wider vacuum layer where the electrons rotate [52].

The increase of the growth rate with increasing electron beam particle density is clear from a general point of view. The dependence of the peak value of the LF SFXM growth rate versus the electron beam particle density in the case of a wide vacuum layer, $\mathrm{Im}(\omega) \propto {n_b}^{1/2}$, differs from that of a narrow layer, $\mathrm{Im}(\omega) \propto {n_b}^{1/3}$, considered in Sect. 7.1. This controversy is explained by the difference in the initial assumptions of the models. In Sect. 7.1, the vacuum layer, where the electrons rotate, is assumed to be sufficiently narrow to provide simultaneous reaching the resonance (7.14) for all the electrons of the beam. In the present Section, a vacuum layer which is much wider than the radial range where the resonant electrons rotate is considered.

The increase of the LF SFXM growth rate with the stationary external magnetic field can be explained as follows. The stronger B_0 is, the lower the parameter Z and the harmonic number N of the resonance (7.14) are. The number N enters into the dispersion relation (7.12) via the combination $N^{1.5} J_N(m)$ (see 7.22). This combination is known [30] to decrease rapidly in the limit $N >> m$.

The increase of the LF SFXM growth rate with the width of the vacuum layer where the electrons rotate is not linear. Increase of the layer from $R_2/R_1 = 1.5$ to $R_2/R_1 = 2$ causes the increase of the peak growth rate by two orders of magnitude but only the decrease of the wave frequency just by the factor of 1.5. The further increase of the layer width from $R_2/R_1 = 2$ to $R_2/R_1 = 4$ results in the reduction of the wave frequency by a factor of two, accompanied by only an almost doubled growth rate.

7.3 Two Mechanisms of Resonance Overlapping in Excitation of Surface Flute Waves by Rotating Relativistic Electron Beams

In the 1970s ion cyclotron resonance plasma heating (ICRH) in thermonuclear fusion devices with radiofrequency (RF) antennas was considered very difficult because of the problem of antenna matching to the plasma with continuously changing density and shape. This problem was successfully solved due to taking into account the finite quality factor Q of electromagnetic wave resonances (non-zero damping rate of eigenmodes) in toroidal fusion devices (see e.g. Chaps. 6–7 in [53]). A comprehensive contribution to understanding the physics of overlapping RF resonances in large fusion devices was provided by Longinov and Stepanov [54]. Following experiments confirmed the realization of the regime of overlapping resonances at ICRH frequencies in large fusion machines with magnetically confined plasmas. The present Section deals with resonance overlapping in the electron cyclotron frequency range.

If a charge and current compensated particle beam, which causes electromagnetic wave excitation, is spatially separated from the plasma, the beam can be described as a flow of oscillators [21], which generally oscillate at an infinite number of cyclotron harmonics. Wave excitation can be caused by beam interaction with either a single harmonic or a number of harmonics, e.g. the wave-particle resonance (7.14) causes SFXM excitation by a gyrating electron beam.

Usually, the dependence of the wave growth rate on a physical parameter X is similar to a "bell-shaped" resonance curve. The "bell" curve reaches its maximum for a value X_N, which is close to that for which the condition (7.14) is satisfied exactly. The half-width δX_N of the "bell" is proportional to the wave growth rate, $\mathrm{Im}(\omega)$. Within the studied range of X values, the resonance (7.14) can be fulfilled for different values of N. If the distance between two resonance maxima $|X_{N+1} - X_N|$, is larger than the half-widths, $\delta X_N + \delta X_{N+1} < |X_{N+1} - X_N|$, then the resonances are well separated in the plot of the growth rate versus X. In this case one calls the "bell" curves as isolated resonances. If the "bell" curves are wide and close to each other, so that $\delta X_N + \delta X_{N+1} > |X_{N+1} - X_N|$, one resonance curve continuously develops into the other one without an obvious boundary between them and the two "bell" curves look like one wide "bell". This case is called resonance overlapping [55].

Beam matching with a plasma-filled waveguide for the purposes of SFXM excitation is better in the case of resonance overlapping. Such a regime was studied e.g. in [31, 34] where the model of a mono-energetic electron beam rotating along large Larmor orbits around the plasma column was studied. The permittivity tensor of the space occupied by such a beam is given in the previous section (see 7.18–7.20). To achieve the regime of resonance overlapping one has to provide a rather high growth rate of the waves, which in turn needs application of a rather high beam particle density.

The geometry of the beam-plasma structure studied here is the same as described in Sects. 7.1 and 7.2, and is shown in Fig. 7.1. Beam excitation of SFXMs is studied in Sect. 7.1 under the assumption of a narrow gap, $R_1 < r < R_2$ (see Fig. 7.1), which separates the plasma column from the metallic waveguide wall,

$$R_2 - R_1 \ll R_1. \tag{7.23}$$

Electrons injected into the gap rotate along their Larmor orbits with a specific radius, lose their energy due to interaction with an electromagnetic wave or because of synchrotron radiation and pass to orbits with smaller radii. This causes almost a linear dependence of the electron velocity and a weak dependence of the dielectric permittivity tensor on the radial coordinate. The assumption (7.23) allows the radial dependence of the permittivity tensor to be neglected and thus simplifies the problem. In this case, all the electrons of the flow are assumed to rotate with the same velocity.

For a given wave frequency ω, the condition (7.14) determines the position $r = r_N$ of the electrons which are in resonance with the wave. The beam electrons which mainly contribute to wave excitation rotate in an electron beam layer $r_N - \Delta r < r < r_N + \Delta r$, where Δr is the half-width of the layer. This can be derived from condition (7.14) by variation of eigenfrequency ω and radial coordinate r. The dependence of condition (7.14) on the radial coordinate is given by the relativistic factor, $\gamma = \gamma\ (r)$. Thereafter, the differential $d\omega$ of the frequency is replaced by the growth rate Im(ω) and the differential dr of the radial coordinate is replaced by the half-width Δr of the layer:

$$\Delta r \sim \frac{\mathrm{Im}(\omega)\gamma^2(R_1)c^2}{\omega R_1 \omega_e^2}. \tag{7.24}$$

If the growth rate Im(ω) is not very small, which can be achieved e.g. by appropriate choice of the beam particle density n_b, the width $2\Delta r$ of the electron beam layer is wider than the width $(R_2 - R_1)$ of the gap where the electron flow rotates,

$$2\Delta r \ \geq \ (R_2 - R_1). \tag{7.25}$$

In this case all the electrons of the beam contribute to wave excitation so that the applicability of the limit (7.23) (narrow gap) considered in the Sect. 7.1 is justified. The combined contribution of all the electrons of the flow results in an enhanced dependence of the wave growth rate Im(ω) on the beam particle density n_b: Im(ω) $\propto {n_b}^{1/3}$ [49].

To access higher frequencies (shorter wavelengths), which is of interest for modern plasma electronics, it is possible to take advantage of excitation of higher order radial modes [52]. SFXM higher radial modes propagate in waveguides with a wide gap, $(R_2 - R_1) \gtrsim R_1$ [56]. In this case, the assumption of uniform permittivity tensor components is not valid anymore to describe electrodynamic properties of the space where the electron beam gyrates. However, asymptotic methods [42] of solving the differential equations can be successfully utilized. In this case, two or even more resonant electron orbits $r = r_N$ can exist within the gap and all these resonances contribute simultaneously to SFXM excitation. This mechanism of achieving resonance overlapping is natural for the cylindrical geometry and distinguishes the physics of electromagnetic wave excitation in cylindrical geometry from that in Cartesian geometry.

A large width of the gap $(R_2 - R_1)$ can cause the violation of condition (7.25). Consequently, the beam electrons rotate with different velocities on different orbits and they cannot contribute to the particular resonances (7.14) all together. Only a small part of electrons which rotate within the narrow layers $r_N - \Delta r < r < r_N + \Delta r$ makes a contribution. Accordingly, the growth rate appears to be lower in this case than in the previous one when all electrons make their contribution: Im(ω) $\propto {n_b}^{1/2}$.

The objective of the present Section is to demonstrate two mechanisms of achieving resonance overlapping: first, by increasing the electron beam particle density n_b (thus by increasing the growth rate Im(ω)) in the case of a narrow gap and second, by increasing the gap width ($R_2 - R_1$) to get several resonance layers inside the gap.

The electrodynamic properties of the space $0 < r < R_1$ filled by the plasma (see Fig. 7.1) are described by the dielectric permittivity tensor (1.2) of a cold magneto-active plasma [1, 21, 41]. The metal wall is assumed to have infinite conductivity, which allows applying non-dissipative boundary conditions.

The electrons gyrate in the magnetic field $\vec{B}_0$ in the gap, $R_1 < r < R_2$, along large concentric axisymmetric Larmor orbits. The electrons are assumed to have no axial momentum, $p_z = 0$, but they have the same angular velocity, $\vec{\upsilon} = \upsilon_\perp \vec{e}_\varphi$, where the azimuthal component is a function of the radial coordinate, $\upsilon_\perp = r|\omega_e|\gamma(r)$. The electron beam is assumed to be compensated with respect to electric charge and current. The electrodynamic properties of the space $R_1 < r < R_2$ are described by the permittivity tensor $\varepsilon_{ik}{}^{(b)}$ [21]. The components of this tensor being used in the present Section are given by (7.18)–(7.20).

In the case of a longitudinal electron beam, $\vec{\upsilon} = \upsilon_{||}\vec{e}_z$, the beam interaction with plasmas due to the Cherenkov resonance is known to dominate over the interaction via the cyclotron resonance [21]. If the electrons have also a transversal component of the velocity, $\vec{\upsilon} = \upsilon_\perp \vec{e}_\varphi + \upsilon_{||}\vec{e}_z$, the interaction due to the cyclotron resonance becomes as strong as the Cherenkov one [21]. In the case of the rotating electron beam, $\vec{\upsilon} = \upsilon_\perp \vec{e}_\varphi$, considered here, only cyclotron interaction (7.14) is possible being represented by the second order poles in (7.18)–(7.20). These poles indicate that the most efficient beam-wave interaction takes place for the frequencies satisfying the resonance condition (7.14) [21].

In the present Section, beam excitation of LF SFMs with extraordinary polarization is considered. Their field components E_r, E_φ, and B_z depend on time t and coordinates as follows: $B_z(\vec{r}) = B_z(r)\exp(im\varphi - i\omega t)$. The radial behavior of the SFXM axial magnetic wave field amplitude $B_z(r)$ is governed by the Bessel equation, which is of the second kind if the wave frequency belongs to one of the two frequency ranges: (1.13) and (1.14) [35, 57]. The LF range (1.13) includes the electron cyclotron frequency and is considered here. The HF range (1.14), which is out of the scope of this Chapter, is above the upper-hybrid frequency.

Within the gap, $R_1 < r < R_2$, the electromagnetic field is of bulk nature. If the gap is narrow, $(R_2 - R_1) << R_1$, one can neglect the weak radial dependence of the components (7.18)–(7.20) of the permittivity tensor $\varepsilon_{ik}{}^{(b)}$ and assume the SFXM radial wave vector k_b to be a constant, where $k_b^2 = k^2\left(\varepsilon_{22}^{(b)} + \varepsilon_{12}^{(b)2}/\varepsilon_{11}^{(b)}\right)$. Therefore, if the electron beam is sufficiently intense to provide the condition (7.25), one can assume that all beam electrons contribute to SFXM excitation through the resonance (7.14). In this case the electrons can be assumed to rotate along the large orbits with the same radius which is located just in the middle of the gap, $r = (R_1 + R_2)/2$. Within the gap the radial dependence of the SFXM axial magnetic wave field is governed by the Bessel equation [31, 34]. Its solution can be written as a linear combination of Bessel functions of the first, $J_m(k_b r)$, and second, $N_m(k_b r)$, kind.

If the gap is not narrow, $(R_2 - R_1) \gtrsim R_1$, and the electron beam particle density is sufficiently small, such that $\Delta r << (R_2 - R_1)$, the solution of the Maxwell equations for the amplitude of the SFXM axial magnetic wave field can be found by asymptotic methods [42]. Since the dielectric permittivity tensor $\varepsilon_{ik}{}^{(b)}$ weakly depends on the radial coordinate one can get the asymptotic solution for the SFXM axial magnetic wave field in the following form applying the method of standard equation [42]:

$$B_z = \sqrt{\frac{c\zeta}{\omega\, r}}[C_1 J_m(\zeta) + C_2 N_m(\zeta)]. \tag{7.26}$$

In (7.26), $\zeta(r) = \zeta_1 \int_{R_1}^{r} k_b dr$, $\zeta_1 = \zeta(R_1) = kR_1$ and $C_{1,2}$ are the constants of integration.

Application of linear non-dissipative boundary conditions allows to derive the dispersion relation in the form (7.12). In the case of narrow gap, $(R_2 - R_1) << R_1, \Delta r$, one has in the dispersion relation:

$$\zeta_1 = k_b R_1,\ \zeta_2 = k_b R_2. \tag{7.27}$$

If the gap is not narrow, Δr, $R_1 << (R_2 - R_1)$, then $\zeta_1 = \zeta(R_1) = kR_1$, and

$$\zeta_2 = kR_2 + \pi i \sum res(k_b) = kR_2 + \pi i \sqrt{\frac{\alpha|\omega_e|}{\omega}\frac{Z}{m}} \sum_S \left(S^{1.5} J_S(m)\right). \tag{7.28}$$

The second term in (7.28) is calculated according to the theory of residues. The term $\frac{\left(J_s'(m)\right)^2 \upsilon_{\perp}^2 y}{(s-y)^2 c^2}$ in (7.20) makes the main contribution to $\pi\ i \sum res(k_b)$ providing a pole of the first order. Just this addendum $\pi\ i \sum res(k_b)$ causes the electromagnetic wave excitation if the gap is not narrow.

In the following, the influence of the beam particle density on the transition from the regime of resonance overlapping to that of isolated resonances in the case of a narrow gap, $(R_2 - R_1) << R_1$, Δr, is studied.

In the first step, the difference between the neighboring eigenfrequencies for which the resonant condition (7.14) is valid for SFXMs with azimuthal mode index $m = 2$ is calculated. The dependence of the SFXM eigenfrequency (dashed curve) versus the effective azimuthal wavenumber $k_{ef} = mc/(\Omega_e R_1)$ is plotted in Fig. 7.14. Just these values are chosen along the abscissa axis, since it is a common practice to give a dispersion relation as a function of the eigenfrequency versus wave vector. In the case of SFXMs, the characteristic azimuthal wave vector is m/R_1, and k_{ef} is m/R_1 normalized to the skin depth $\delta = c/\Omega_e$. The ordinate axis is chosen as eigenfrequency normalized to the electron cyclotron frequency: $\omega_0/|\omega_e|$. The following parameters of the plasma waveguide are chosen: $Z \equiv \Omega_e/|\omega_e| = 8$, $(R_2 - R_1)/R_1 = 0.1$, $\alpha = n_b/n_e = 0$, where n_e is the particle density of the plasma column. These parameters correspond to modest parameters of a University laboratory device.

The most efficient excitation of SFXMs is expected under the resonance condition (7.14). For this case the curves $\omega/|\omega_e| = N/\gamma$ which describe the behavior of the right-hand side of condition (7.14) are shown in Fig. 7.14. The solid line presents the

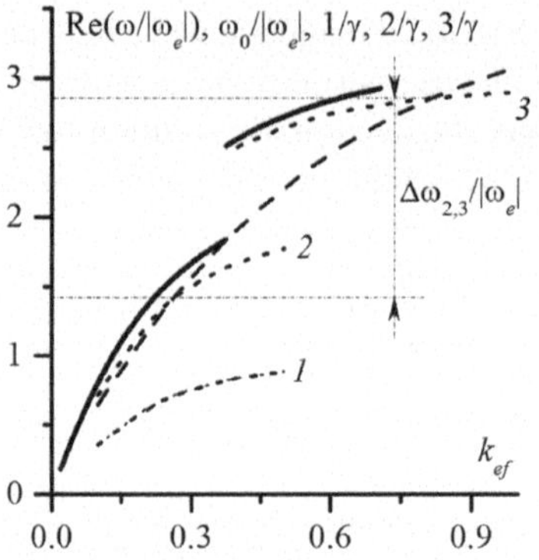

Fig. 7.14 Normalized SFXM eigenfrequency (*dashed curve*) versus effective wavenumber k_{ef} for the case $m = 2$, $Z \equiv \Omega_e/|\omega_e| = 8$ and $(R_2 - R_1)/R_1 = 0.1$ and resonant lines N/γ for $N = 1, 2$, and 3 (*dotted curves,* marked with these *numbers*). The *solid curve* presents the real part of the SFXM frequency in presence of an electron beam with the particle density $n_b = 0.01\ n_e$

real part of the SFXM frequency in presence of an electron beam with the particle density $n_b = 0.01\ n_e$. The dotted curves depict the dependence of N/γ on k_{ef} for $N = 1, 2, 3$. The intersections of these curves with the SFXM dispersion curve (dashed curve) identify the k_{ef} values, which are most suitable for an efficient SFXM excitation. The lowest curve $1/\gamma$ does not cross the SFXM dispersion curve. That is why there is no related curve in Fig. 7.15 as a result of the resonance (7.14) with $N = 1$. The lines $2/\gamma$ and $3/\gamma$ cross the dispersion line for $k_{ef} = 0.263$ and $k_{ef} = 0.812$. The eigenfrequency difference $\Delta\omega_{2,3}/|\omega_e|$ for these values of k_{ef} is equal to $\Delta\omega_{2,3}/|\omega_e| = \omega_0(0.812)/|\omega_e| - \omega_0(0.263)/|\omega_e| = 1.439$. The phenomenon of resonance overlapping is well known in physics of wave-particle interaction [53, 54] to exist if the growth rate is of the order or larger than the frequency difference between two resonances (see e.g. Chaps. 5 and 10 in [55]).

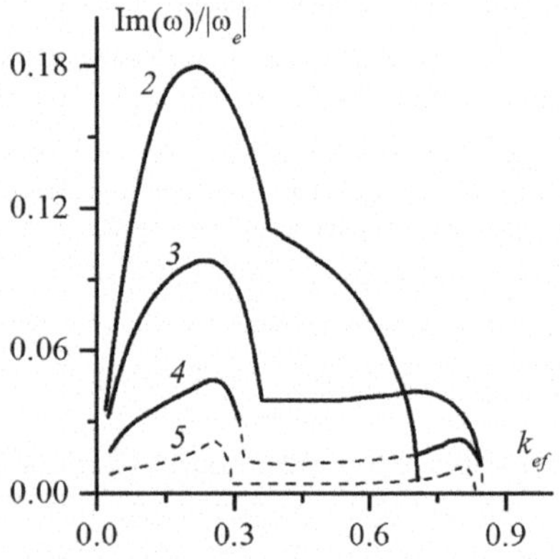

Fig. 7.15 Normalized SFXM growth rate versus effective wavenumber for $m = 2$, $Z \equiv \Omega_e/|\omega_e| = 8$, $(R_2 - R_1)/R_1 = 0.1$, $\alpha = 10^{-2}$ (2), $\alpha = 10^{-3}$ (3), $\alpha = 10^{-4}$ (4), and $\alpha = 10^{-5}$ (5). The *numbers* at the curves indicate $(-\lg(\alpha))$.The *dashed sections* of the curves 4 and 5 are calculated beyond condition (7.25) and can be considered as upper limit of the growth rate only

Now the SFXM growth rate is calculated in order to be compared with the already determined eigenfrequency difference $\Delta\omega_{2,3}/|\omega_e|$. The curve 2 in Fig. 7.15 demonstrates the resonance overlapping for the beam particle density $\alpha = 10^{-2}$. The numbers in Fig. 7.15 correspond to different beam particle densities, they indicate ($-\lg(\alpha)$). The curve 3 corresponds to a moderate value of the electron beam particle density, $\alpha = 10^{-3}$ and shows the transition from the regime of overlapping resonances to that of isolated resonances, however, it still can be considered as the case of resonance overlapping. Curve 4 ($\alpha = 10^{-4}$) and especially curve 5 ($\alpha = 10^{-5}$) clearly demonstrate just isolated resonances.

The dashed sections of the curves 4 and 5 indicate those values of the growth rate for which condition (7.25) is violated. In this case, those values can be considered as upper limit of the growth rate only.

Thus, Fig. 7.15 confirms that resonance overlapping takes place for sufficiently high beam particle densities, which correspond to higher magnitudes of the SFXM growth rate when $\mathrm{Im}(\omega)$ is comparable to the eigenfrequency difference $\Delta\omega_{2,3}/|\omega_e|$. Consequently, decrease of the beam particle density results in decreasing SFXM growth rate and transition from the regime of resonance overlapping to that of isolated resonances. The transition becomes pronounced when $\mathrm{Im}(\omega)$ becomes much lower than $\Delta\omega_{2,3}/|\omega_e|$.

Curve 2 in Fig. 7.15 is composed of two sections which correspond to resonances (7.14) with different harmonic numbers N. It demonstrates the case of resonance overlapping. To show this more clearly, the solid curve in Fig. 7.16 combines the solutions of the dispersion relation (7.12) under the condition (7.27). The dashed curve is obtained from the same dispersion relation when the sums in (7.18)–(7.20) are replaced by only one term ($N = 2$). The dotted curve is calculated in the similar way with taking into account only the term defined by $N = 3$. The dashed and dotted curves intersect at $k_{ef} \approx 0.36$ and the solid curve follows this correspondingly: it has a kink for $k_{ef} \approx 0.38$. The small difference between the two abscises ($k_{ef} \approx 0.36$ and $k_{ef} \approx 0.38$) is explained by the fact that the solid curve was calculated with account for more terms in addition to the resonant ones: $N = 2$ (for the dashed curve) and $N = 3$ (for the dotted curve). For $k_{ef} < 0.38$, the SFXM growth rate

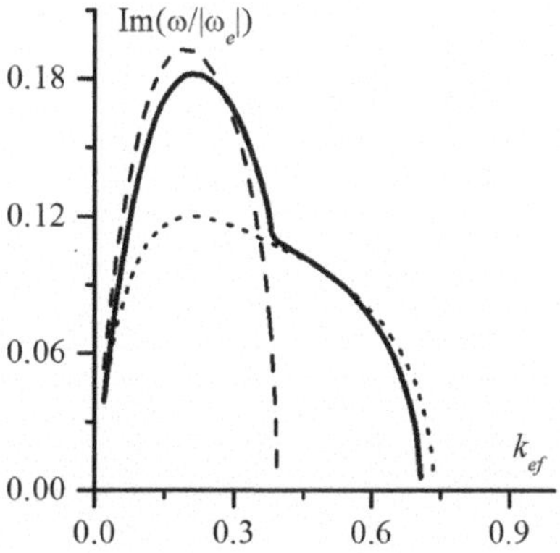

Fig. 7.16 Normalized SFXM growth rate versus effective wavenumber for $m = 2$, $Z \equiv \Omega_e/|\omega_e| = 8$, $(R_2 - R_1)/R_1 = 0.1$, and $\alpha = 0.01$ (*solid curve*). The model curves (*dashed and dotted curves*) are calculated for two partial cases described in the text

is caused almost completely by the SFXM interaction with the second cyclotron harmonic of the beam. Similarly, for $k_{ef} > 0.38$, the SFXM excitation is provided almost completely by the wave interaction with the third cyclotron harmonic of the beam. The following note can be considered as obvious, but it is of importance for comparing the two mechanisms of resonance overlapping under discussion in this Section. Curve 2 is calculated for the following range of effective wavenumber: $0.02 < k_{ef} < 0.74$. The dispersion relation (7.12) has two solutions for the azimuthal mode index $m = 2$ in this range: one corresponds to the SFXM interaction with the second cyclotron harmonic, $N = 2$, and the other to the third harmonic, $N = 3$ (see also solid lines in Fig. 7.14). The result of joint interaction of the SFXM with two harmonics is not the sum of two growth rates, one has to select the highest growth rate and draw it into the figure. In linear approximation with respect to the wave amplitude, both electromagnetic waves (which correspond to the two solutions of the dispersion relation) propagate independently. The exponential growth of the wave amplitude with higher growth rate is much more pronounced and just this determines the temporal evolution of the wave. Note, that a nonlinear analysis [58] did not disclaim this result for SFXMs.

The mode competition shown in Figs. 7.15 and 7.16 is explained above in terms of the Fourier space variables: wave vector-frequency. The similar explanation can be done in terms of the real space variables (see Fig. 7.17). The dotted curves in Fig. 7.17 corresponds to the solutions of Eq. (7.14) in respect of the radial position of the beam electrons for $N = 2$ (lower curve) and $N = 3$ (upper curve). The horizontal dash-dotted lines in Fig. 7.17 show the position of the plasma-vacuum (lower line) and vacuum-metal (upper line) boundaries. The dashed curves mark the position of the layer boundaries of the beam electrons which cause interaction of SFXMs with the third cyclotron harmonic. The solid curves outline the layer boundaries of the beam electrons which cause interaction of SFXMs with the second cyclotron harmonic. The vertical dash-dotted lines correspond to the values of the effective wavenumber for which mode competition is discussed below.

For the present plasma waveguide parameters and $k_{ef} = 0.02$, the electrons which are in resonance (7.14) with $N = 2$ (referred below as belonging to the second

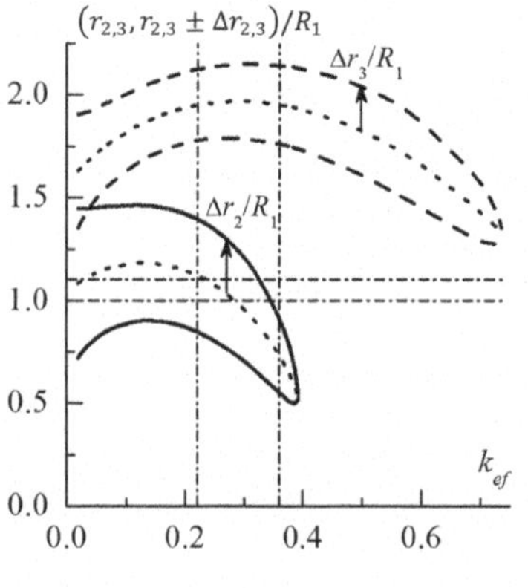

Fig. 7.17 Resonant orbits of beam electrons versus effective wavenumber

electron group) should rotate within the layer $r_2-\Delta r < r < r_2 + \Delta r$, where $r_2 \approx 1.085R_1$ and $\Delta r \approx 0.363R_1$ (as it follows from Eqs. 7.14 and 7.24). That is why the SFXM excitation is caused by its interaction just with the second cyclotron harmonic $N = 2$. For the same parameters the electrons in the resonance with $N = 3$ (referred below as belonging to the third electron group) should rotate within the radius layer $r_3-\Delta r < r < r_3 + \Delta r$ with $r_3 \approx 1.631R_1$ and $\Delta r \approx 0.274R_1$. But this group of electrons does not exist and does not contribute to the SFXM excitation.

When the effective wavenumber is equal to $k_{ef} \approx 0.22$ (here the growth rate in Fig. 7.16 has the maximum) the electrons of the second group should rotate on orbits with average radius $r_2 \approx 1.122R_1$ within a layer with half-width $\Delta r \approx 0.275R_1$. This means that just this group of electrons fills the gap $R_1 < r < R_2$ and determines the value of the SFXM growth rate. Meanwhile the third group of electrons should rotate with average radius $r_3 \approx 1.95R_1$ within a layer with half-width $\Delta r \approx 0.12R_1$ and thus cannot contribute to SFXM excitation.

For plasma waveguides with $k_{ef} \approx 0.36$ the growth rates caused by two different groups of electrons are equal. The second group should rotate with $r_2 \approx 0.734R_1$ and $\Delta r \approx 0.181R_1$ which means that the impact of the second group on SFXM excitation is much less than previously. The characteristics of the orbits of the third group of electrons are in this case $r_3 \approx 1.95R_1$ and $\Delta r \approx 0.19R_1$ which means that the third group also cannot dominate so far. The impacts of the two groups are comparable, and the modes compete.

If the effective wavenumber is larger than $k_{ef} \approx 0.36$, the half-width of the second group of electrons vanishes, and the growth rate of SFXM excitation is determined by the third group of electrons only.

In the following, the possibility to reach the regime of resonance overlapping via the other mechanism is investigated, which is increasing the width of the gap. When an SFXM is excited due to interaction with two or more cyclotron harmonics, the resonance (7.14) happens inside the gap for two or more N values. Figure 7.18 shows the dependences of the SFXM growth rate versus effective wavenumber for different relative widths of the gap: $(R_2 - R_1)/R_1 = 0.1$ (given by the dotted curve), $(R_2 - R_1)/R_1 = 0.4$ (given by the solid curve) and $(R_2 - R_1)/R_1 = 1.5$ (given by

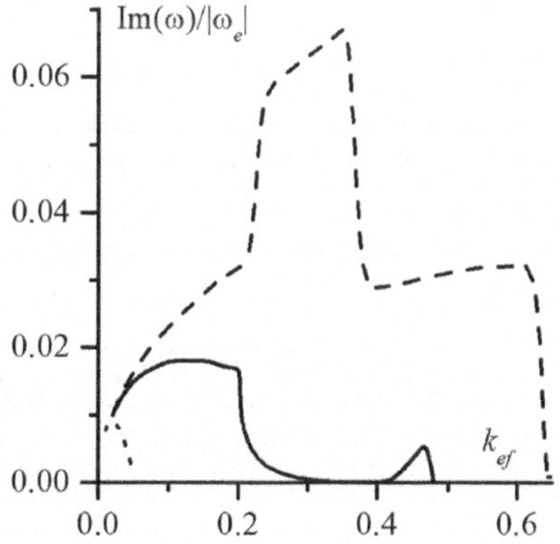

Fig. 7.18 Normalized SFXM growth rate versus effective wavenumber for $m = 2$, $Z \equiv \Omega_e/|\omega_e| = 8$, and $\alpha = 0.0001$: $(R_2 - R_1)/R_1 = 0.1$ (*dotted curve*), $(R_2 - R_1)/R_1 = 0.4$ (*solid curve*), $(R_2 - R_1)/R_1 = 1.5$ (*dashed curve*)

the dashed curve). These dependences are obtained from numerical solution of the dispersion relation (7.12) using Eq. (7.28). The following parameters of the beam-plasma system are chosen: $m = 2$, $Z \equiv \Omega_e/|\omega_e| = 8$, and $\alpha = 0.0001$.

One can see that widening of the layer from $(R_2 - R_1)/R_1 = 0.1$ to $(R_2 - R_1)/R_1 = 0.4$ causes widening of the well-separated resonances, which means expansion of the range of k_{ef} for which efficient SFXM excitation takes place. Further widening of the gap from $(R_2 - R_1)/R_1 = 0.4$ to $(R_2 - R_1)/R_1 = 1.5$ results in achieving such parameters of the system for which two orbits of resonant electrons (with $N = 2, 3$) coexist inside the gap, $R_1 < r < R_2$.

The dashed curve in Fig. 7.18 demonstrates the resonance overlapping in more detail. For small values of the effective wavenumber, $k_{ef} < 0.22$, excitation of the SFXM is caused by its interaction with the beam through the resonance (7.14) with $N = 2$ only. Increasing the k_{ef} value from $k_{ef} = 0.02$ to $k_{ef} = 0.22$ looks like a relay race. In this race, the resonance (7.14) with $N = 2$ is handed as a relay baton from electrons which rotate along the large Larmor radius $r_2 = 1.717R_1$ (where the half-width of the layer, $\Delta r = 0.109R_1$, is much smaller than the gap width $(R_2 - R_1) = 1.5\ R_1$) to those rotating at the radius $r_2 = 1.508R_1$ (where the half-width of the layer is $\Delta r = 0.06R_1$).

When k_{ef} exceeds the value 0.22 there are the conditions for one more resonance (7.14), that is the resonance with $N = 3$. Electrons which rotate just nearby the metal wall start to form the resonance layer with the half-width $\Delta r = 0.041R_1$.

For the range $0.22 < k_{ef} < 0.36$, two resonances contribute together to the SFXM excitation which is resonance overlapping. Applying the analogy with a relay race, one can consider the increase of the k_{ef} value from $k_{ef} = 0.23$ to $k_{ef} = 0.36$ as a relay race, in which two sport teams take part. One team is the layer of electrons which form the resonance (7.14) with $N = 2$. This team hands its relay baton from electrons which rotate along the large Larmor radius $r_2 = 1.508R_1$ to those rotating just nearby the plasma column (half-width of the layer is $\Delta r = 0.078R_1$). The other team is formed by the electrons which contribute to the resonance (7.14) with $N = 3$. They hand their relay baton from the metal wall to those electrons which rotate with the radius $r_3 = 2.248R_1$; the layer becomes wider reaching the value of $\Delta r = 0.051R_1$.

In the present Section, only the initial stage of SFXM excitation by a rotating electron beam is studied. In the framework of the linear (in respect of the wave amplitude) approach, one can expect that, as the beam electrons in a resonant layer lose their energy to the fields, their orbit radii are reduced and the beam electrons may find themselves in a different resonant layer. However, the time period which the electrons need to pass from one resonant layer to another seems to be much larger than that of the transition to the stage of nonlinear beam-wave interaction.

For $k_{ef} > 0.36$, the radius of the electrons which rotate along the large Larmor orbit corresponding to resonance condition (7.14) with $N = 2$ becomes less than the plasma column radius R_1, and this resonance stops to contribute to SFXM excitation. Then, only the resonance with $N = 3$ remains and isolated resonances are observed instead of resonance overlapping.

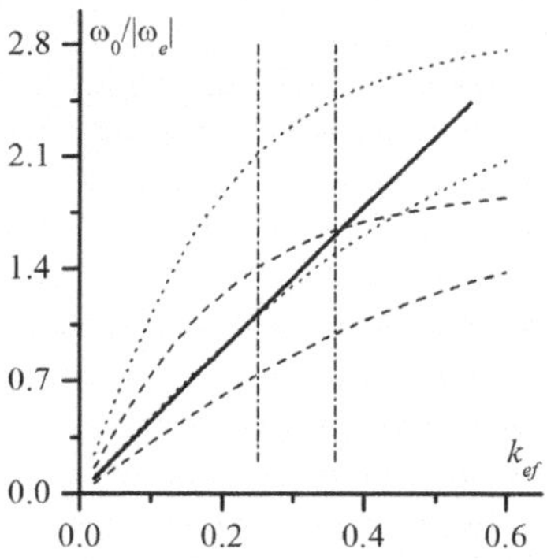

Fig. 7.19 Normalized SFXM eigenfrequency (*solid line*) versus effective wavenumber for $m = 2$, $Z \equiv \Omega_e/|\omega_e| = 8$, $\alpha = 0.0$, and $(R_2 - R_1)/R_1 = 1.5$. The other curves and the vertical lines are explained in the text

To the right of the range, $0.23 < k_{ef} < 0.36$, where the resonance overlapping is observed, one more isolated resonance, caused by SFXM interaction with the fourth cyclotron harmonic, $N = 4$, takes place for $0.67 < k_{ef} < 0.9$. This resonance is less pronounced: its growth rate is about $\mathrm{Im}(\omega) \lesssim 0.013|\omega_e|$.

This explanation is confirmed by Fig. 7.19, which shows the dependence of the SFXM eigenfrequency (solid curve) on the effective wavenumber. The vertical dash-dotted lines are $k_{ef} = 0.25$ and $k_{ef} = 0.36$. The dashed curves correspond to the conditions $2/\gamma$ (R_1) and $2/\gamma$ (R_2). The dotted curves depict the boundaries of the same range but for $N = 3$. For small values of the effective wavenumber, $k_{ef} < 0.22$, the eigenfrequency belongs to the range $2/\gamma\ (R_1) < \omega/|\omega_e| < 2/\gamma\ (R_2)$ only. Here the resonance (7.14) with $N = 2$ contributes to SFXM excitation. One can see that for the range $0.25 < k_{ef} < 0.36$, the eigenfrequency belongs to both ranges: $2/\gamma\ (R_1) < \omega/|\omega_e| < 2/\gamma\ (R_2)$ and $3/\gamma\ (R_1) < \omega/|\omega_e| < 3/\gamma\ (R_2)$. Consequently, both resonances ($N = 2$ and $N = 3$) contribute together to SFXM excitation, and the resonances in Fig. 7.18 overlap.

In the case of a wide gap, $(R_2 - R_1) >> R_1$, in the region Δr, where two or even more resonant electron layers exist, one does not deal with any mode competition. Electrons of all the resonant layers (7.14) which rotate inside the gap $(R_2 - R_1)$ contribute to the excitation of the same mode. The corresponding solid line in Fig. 7.19 is the only one, in the contrast to Fig. 7.14, where two solid lines are present.

In the present Section, two physics mechanisms are investigated to reach the conditions of resonance overlapping in plasma electronics devices based on electron beam excitation of surface type electromagnetic waves of extraordinary polarization propagating across an external static magnetic field in cylindrical metallic waveguides. The regime of resonance overlapping is of interest for practical applications since it provides better matching of the beam to the plasma waveguide during the beam excitation of electromagnetic waves.

One of the mechanisms is also known for Cartesian geometry. It consists of increasing the beam particle density in the narrow gap between plasma column and waveguide wall where the electrons move such that the growth rate becomes of the order or greater than the distance between the nearest resonance frequencies.

In this case one assumes that all the beam electrons contribute to wave excitation, $\Delta r > R_2 - R_1$. The gap which separates the plasma column from the metal wall is so narrow that there is no place for two or more resonant electron layers with different harmonic number N. Therefore, the modes with different N have to compete.

The other mechanism is natural for cylindrical beam-plasma structures only. It consists in widening the gap where the electron beam of low particle density rotates such that more than one resonant electron orbit appears to belong to the gap. In this case the width Δr of the layer where the beam electrons are in resonance with the wave is assumed to be narrower than that of the gap $(R_2 - R_1)$ between the plasma column and metal wall, $\Delta r << R_2 - R_1$. This mechanism is not related with mode competition.

For both these mechanisms, just increasing the number of electrons which take part in wave excitation is the physics reason of the resonance overlapping. However, the first one looks like more preferable due to its higher efficiency: in this case all the electrons of the beam take part in the excitation, and the SFXM growth rate $\mathrm{Im}(\omega)$ increases with the beam particle density n_b as $\mathrm{Im}(\omega) \propto n_b^{1/3}$. The second mechanism runs such that a small part of electrons of the beam transfers its energy to the SFXM. These are only those whose orbit radii are close to the resonant values determined by the condition (7.14). In this case the SFXM growth rate is naturally lower, $\mathrm{Im}(\omega) \propto n_b^{1/2}$. Nevertheless, the second mechanism was demonstrated for the first time in [59] to take place. It is of interest for the general theory of beam-plasma interaction in cylindrical geometry and can be applied in plasma electronics devices. This mechanism can also be observed in Cartesian geometry (in plane plasma waveguides) when a velocity shear is allowed in a direction normal to the interface.

7.4 Excitation of Surface Flute Waves by Electron Beams in Toroidal Waveguides

In the previous Sects. 7.1–7.3 results of theoretical studies on excitation of surface flute X-modes (SFXMs) by a rotating electron beam in a plasma, which is immersed into a uniform stationary axial magnetic field, are presented. Of course, in general, an external static magnetic field is spatially non-uniform. The influence of magnetic field non-uniformity on the dispersion properties of SFXMs is investigated in Chap. 5. A magnetic field non-uniformity influences the features of charged particle beams propagating in a waveguide with such a magnetic field and on its interaction with eigenmodes of the waveguide as well. Thus, the main goal of this Section is to prove theoretically that the beam-plasma interaction in a toroidally non-uniform static magnetic field is not weaker than in the case of a uniform magnetic field (see Sect. 7.1).

The effect of an azimuthally non-uniform external magnetic field on wave excitation was experimentally investigated in a combined axial and wiggler magnetic field [60]. Toroidal traps as tokamaks and stellarators are widely used to conduct controlled thermonuclear fusion plasma experiments. That is why the toroidal non-uniformity of the external static magnetic field is chosen in the present Section to study its effect on SFXM excitation.

Propagation of SFXMs across the circular axis of a toroidal metal waveguide with magnetized plasma filling was studied in Sect. 5.3 using the theory of successive approximations [61]. The spatial distribution of the wave fields was obtained there analytically. When the external static magnetic field has a toroidal non-uniformity, then the SFXMs propagate in the waveguide as a wave packet. If the toroidicity is weak, $\varepsilon_t = R_1/R \ll 1$ (here R_1 is the small plasma radius and R is the major radius of the torus), the amplitudes of the satellite harmonics of the wave packet are much smaller than the amplitude of the fundamental harmonic. An influence of the toroidal non-uniformity of the external magnetic field on the amplitude of the fundamental harmonic is found to be of the second order of smallness with respect to the toroidicity parameter ε_t. The correction to the SFXM eigenfrequency, which belongs to the electron cyclotron frequency range, caused by the toroidal non-uniformity of the external magnetic field is calculated in Sect. 5.3. It is proved that its value is of the second order of smallness. The present Section shows the results of studying the effect of the toroidal non-uniformity of the external static magnetic field on the initial stage of the resonance beam-plasma instability when the excited SFXMs propagate inside a toroidal metal waveguide partially filled by a cold magneto-active plasma.

A toroidal metal chamber with torus radius R and circular cross-section with inner radius R_2 is considered. The waveguide is filled with a circular plasma with radius R_1 (see Fig. 7.20). The waveguide is assumed to be symmetric along its circular axis, so that $\partial/\partial\zeta = 0$, in the right-handed quasi-toroidal coordinate system. The poloidal angle ϑ is counted from the direction to the symmetry center of the torus. The analytical expression for the external static toroidal confining magnetic field is given by:

$$B_{0\zeta} = B_0/[1 - (r/R)\cos\vartheta], \tag{7.29}$$

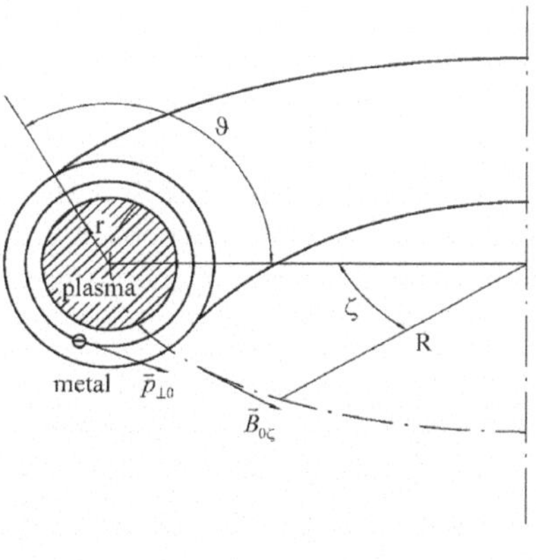

Fig. 7.20 Geometry of toroidal metal chamber with torus radius R and circular cross-section with inner radius R_2. The waveguide is filled with a circular plasma with radius R_1

where r is radial coordinate. Here surface waves with extraordinary polarization (with the electromagnetic field components E_r, E_ϑ, H_ζ) are under consideration, because ordinary surface waves (with the electromagnetic field components E_ζ, H_ϑ, H_r) do not propagate in waveguides with a narrow gap between the plasma column and the metallic chamber wall [62, 63].

In the present case, the electromagnetic properties of the plasma are described by two components (1.3) of the dielectric permittivity tensor obtained in the fluid model approximation, namely: $\varepsilon_{11} = \varepsilon_1$ and $\varepsilon_{12} = i\varepsilon_2$ [1, 21, 41]. The ordinary form of these expressions is inconvenient for the following study because one should apply therein the expression for the complete magnetic field. In the case of weak toroidicity ($\varepsilon_t = R_1/R \ll 1$) studied here, the form of the components ε_1 and ε_2 can be simplified by expanding them into a series over the small toroidicity parameter ε_t (see Eq. 5.26):

$$\varepsilon_{1,2} = \varepsilon_{1,2}^{(0)} + \varepsilon_{1,2}^{(1)} \cos\vartheta + \varepsilon_{1,2}^{(2)}. \tag{7.30}$$

The plasma particle density is assumed to be uniform in the present Section, that is why the main contributions to (7.30) do not depend on the coordinates, $\varepsilon_{1,2}^{(0)} = \varepsilon_{1,2}|_{B_{0\zeta}=B_0}$. Therefore, they can be written explicitly:

$$\varepsilon_1^{(0)} = \varepsilon_0 - \frac{\omega_{pe}^2}{\omega^2 - \omega_{ce}^{(0)2}}, \quad \varepsilon_2^{(0)} = \frac{\omega_{ce}^{(0)}\omega_{pe}^2}{\omega\left(\omega^2 - \omega_{ce}^{(0)2}\right)}. \tag{7.31}$$

The first order terms in (7.30) linearly depend on the small radius r of the torus (see Eq. (7.5.27), whereas the second order terms are quadratic with respect to r:

$$\varepsilon_1^{(2)} = \frac{r^2}{2R^2}\frac{\omega_{pe}^2\omega_{ce}^{(0)2}\left(\omega_{ce}^{(0)2} + 3\omega^2\right)}{\left(\omega_{ce}^{(0)2} - \omega^2\right)^3}, \quad \varepsilon_2^{(2)} = \frac{r^2}{2R^2}\frac{\omega_{ce}^{(0)}\omega_{pe}^2\omega\left(\omega^2 + 3\omega_{ce}^{(0)2}\right)}{\left(\omega_{ce}^{(0)2} - \omega^2\right)^3}. \tag{7.32}$$

The expansion (7.30) does not contain second order terms, which are proportional to $\exp(2i\vartheta)$ because they cause a correction to the SFXM eigenfrequency, which is at least by an order of magnitude smaller than the second one. The electron cyclotron frequency in the expressions (7.31) and (7.32) is defined by the magnetic field value B_0 neglecting its non-uniformity (calculated in zeroth approximation).

The components of the SFXM electric wave field are calculated with the Maxwell equations from the magnetic wave field component [61], which corresponds to a solution of the inhomogeneous Bessel equation [30]. The components of the SFXM wave field have to satisfy the following boundary conditions.

1. The wave fields have to be of a finite value inside the waveguide.
2. The tangential (poloidal in this consideration of toroidal waveguides) component of the electric wave field should be equal to zero on the inner surface of the metal chamber ($E_\vartheta(r = R_2) = 0$).

3. The tangential components of the electric and magnetic wave fields should be continuous on the plasma-vacuum interface including the satellite harmonics.

The SFXMs with different azimuthal mode indices propagate independently in the zeroth approximation over the small parameter of toroidicity ε_t. This allows to set in zeroth order approximation all the amplitudes of SFXM harmonics to zero except the amplitude of the definite harmonic with poloidal mode index m. The amplitude $H_\xi^{(0)}(r)$ of the main harmonic of the axial magnetic wave field in zeroth order approximation is proportional to the modified Bessel function $I_m(\xi)$ [30], where $\xi = rk_\perp$. The SFXM wave field penetration depth into the plasma $1/k_\perp$ is defined from the following relation,

$$k_\perp^2 = -(\omega/c)^2 \varepsilon_\perp^{(0)}/\varepsilon_1^{(0)} > 0, \tag{7.33}$$

where $\varepsilon_\perp^{(0)} = (\varepsilon_2^{(0)})^2 - (\varepsilon_1^{(0)})^2$. The inequality (7.33) defines the frequency ranges where the SFXM waves can propagate. In the present Section, the consideration is restricted to the low-frequency (LF) range (1.13): $\omega_{LH} < \omega < \omega_1 - |\omega_{ce}|$, where $\omega_1 = 0.5|\omega_{ce}| + \sqrt{\omega_{pe}^2 + \omega_{ce}^2/4}$ is the cut-off frequency, and ω_{LH} is the lower hybrid frequency.

Due to the periodic spatial non-uniformity of the external magnetic field (see expression (7.30)) the SFXMs propagate as wave packets. Each of the packets contains the fundamental harmonic (the wave field of this main harmonic is proportional to $exp(im\vartheta - i\omega t)$) and an infinite number of satellite harmonics (the fields of the satellite harmonics change proportionally to $exp[i(m \pm j)\vartheta - i\omega t]$, where $j = 1, 2, 3...$). It is known from the theory of wave propagation in media with periodic changes of the properties along the direction of propagation [64], that the amplitudes of the satellite harmonics are smaller (proportional to $(\varepsilon_t)^j$) as compared with the amplitude of the fundamental harmonic. The wave packet with the two nearest satellite harmonics (their wave fields change proportionally to $\exp[i(m \pm 1)\vartheta - i\omega t]$) is considered in Sect. 5.3 to study the effect of the toroidal magnetic field non-uniformity on the SFXM dispersion properties [61]. The consideration of the higher order satellite harmonics causes corrections to the SFXM eigenfrequency, which are of the third and higher orders of smallness.

Taking into consideration the first order terms (5.27) with the respect to ε_t in the expressions (7.30) for the components of the dielectric permittivity tensor does not change the amplitude of the main harmonic. However, these terms are the reason for the appearance of the small satellite harmonics of the field which are proportional to $\exp[i(m \pm 1)\vartheta - i\omega t]$. Second order considerations with the respect to ε_t show that the amplitude of the fundamental harmonic is corrected by the second order term, which together with the amplitudes of the first satellite harmonics gives the second order contribution to the dispersion relation $D^{(2)}(\omega, m...) \propto \varepsilon_t^2$. As a result, instead of the zeroth-order dispersion relation $D^{(0)}(\omega, m...) = 0$, a second-order dispersion relation describes the present problem: $D^{(0)}(\omega, m...) + D^{(2)}(\omega, m...) = 0$. The explicit expression for $D^{(2)}(\omega, m...)$ is too bulky and therefore it is not presented here, but it is described in detail in Sect. 5.3 and in [61].

The rotating electron beam is assumed to be injected in the vacuum gap in the radii range $R_1 < r < R_2$ between the plasma column and the waveguide metal wall. The beam is described as a set of oscillators with the same transverse momentum $p_{\perp 0}$ and zero axial momentum, $p_\zeta = 0$. This model is applied in the Sects. 7.1 and 7.2. The electrodynamic properties of the waveguide part with the electron beam are described by the dielectric permittivity tensor $\varepsilon_{ik}^{(b)}$ (see expressions (7.18)–(7.20) in Sect. 7.2).

Solving the Maxwell equations in the space occupied by the electron beam provides the expressions for the SFXM field as a linear combination of Bessel functions of the first kind $J_m(\zeta)$ and Bessel functions of the second kind $N_m(\zeta)$ [30], and their derivatives with respect to their argument $\zeta = kr\sqrt{\psi_b}$, where $\psi_b = \varepsilon_{22}^{(b)} + \left(\varepsilon_{12}^{(b)}\right)^2 \left(\varepsilon_{11}^{(b)}\right)^{-1}$.

The application of the boundary conditions discussed above allows one to derive the following dispersion relation, which is the main object of our study in this Section:

$$\frac{I_m'(k_\perp R_1)k}{k_\perp I_m(k_\perp R_1)} + \frac{\mu m k}{k_\perp^2 R_1} + D^{(2)}(\omega, m) = \frac{im\varepsilon_{12}^{(b)}}{\varepsilon_{11}^{(b)}\zeta_1\sqrt{\psi_b}} - \frac{J_m'(\zeta_1) - \Phi N_m'(\zeta_1)}{\sqrt{\psi_b}[J_m(\zeta_1) - \Phi N_m(\zeta_1)]}, \tag{7.34}$$

where

$$\Phi = \left[\frac{im\varepsilon_{12}^{(b)}}{\varepsilon_{11}^{(b)}\zeta_2} J_m(\zeta_2) - J_m'(\zeta_2)\right]\left[\frac{im\varepsilon_{12}^{(b)}}{\varepsilon_{11}^{(b)}\zeta_2} N_m(\zeta_2) - N_m'(\zeta_2)\right]^{-1},$$

with $\zeta_1 = \zeta(R_1)$, $\zeta_2 = \zeta(R_2)$. Equation (7.34) is analyzed here, assuming that the LF SFXM frequency is close to the Nth electron cyclotron harmonic, so that the following resonance condition is satisfied:

$$\omega = \omega_0 + \Delta\omega_t = N|\omega_e|\gamma^{-1} + \Delta\omega_t, \tag{7.35}$$

where N is a natural number, ω_0 is the eigenfrequency of the studied SFXMs calculated for the case when the electron beam is absent and when the external magnetic field is uniform, and $\Delta\omega_t$ is the correction to the wave frequency caused by the wave-electron beam interaction in the non-uniform magnetic field.

The results of numerical analyses of Eq. (7.34), which describes the excitation of LF SFXMs by the electron beam in the case of toroidal non-uniformity of the static external magnetic field, are plotted in Figs. 7.21, 7.22, 7.23, 7.24, 7.25 and 7.26. Each of the three pairs of Figs. 7.21, 7.22, 7.23 and 7.24, as well as Figs. 7.25 and 7.26 is computed for the same parameters of the electron beam-plasma system. The ratio of the beam particle density to the plasma density is the same for all these calculations, $\alpha = n_b/n_p = 10^{-3}$. It is proved numerically that an increase of the beam particle density leads to the transfer from the excitation regime due to the resonant interaction of the wave with separated cyclotron harmonics to the regime of

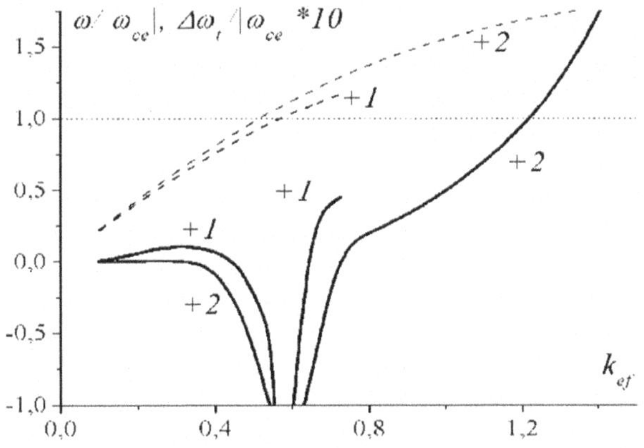

Fig. 7.21 Normalized SFXM frequency in zeroth approximation (*dashed curves*) and its correction caused by the toroidicity of the stationary external magnetic field (*solid curves*) versus the effective wavenumber for different azimuthal mode indices *m*, the *numbers* correspond to the values of *m*. The parameters of the beam-plasma system are $Z = 3$, $\Delta = 0.3$, $\varepsilon_t = 0.1$

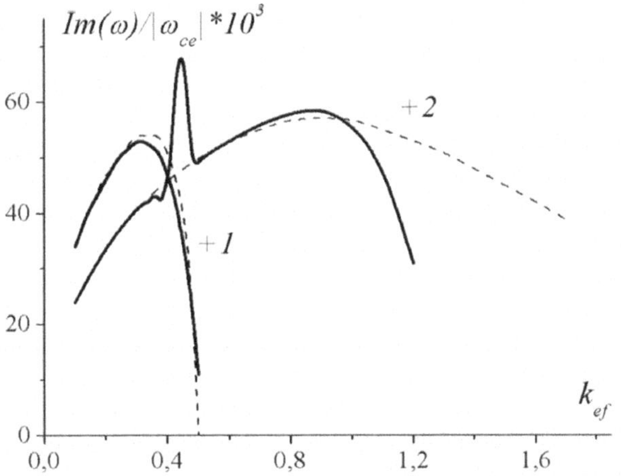

Fig. 7.22 Normalized growth rate of the SFXM resonance beam instability versus the effective wavenumber for different values of the azimuthal mode index *m*. The parameters of the beam-plasma system are the same as in Fig. 7.21. Zeroth approximation (*dashed curves*)

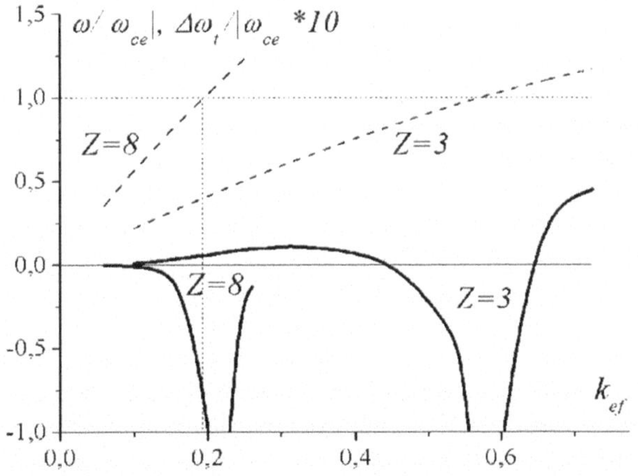

Fig. 7.23 Normalized SFXM frequency in zeroth approximation (*dashed curves*) and its correction caused by the toroidicity of the external static magnetic field (*solid curves*) versus the effective wavenumber for different values of the external magnetic field; $m = +1$, $\Delta = 0.3$, $\varepsilon_t = 0.1$

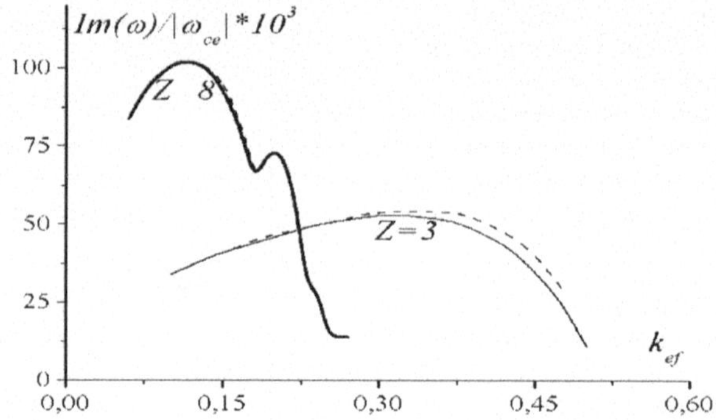

Fig. 7.24 Normalized growth rate of the SFXM resonance beam instability versus k_{ef} for different values of the external magnetic field ($Z = 8{,}3$). The parameters of the beam-plasma system are the same as in Fig. 7.23. Zeroth approximation (*dashed curves*)

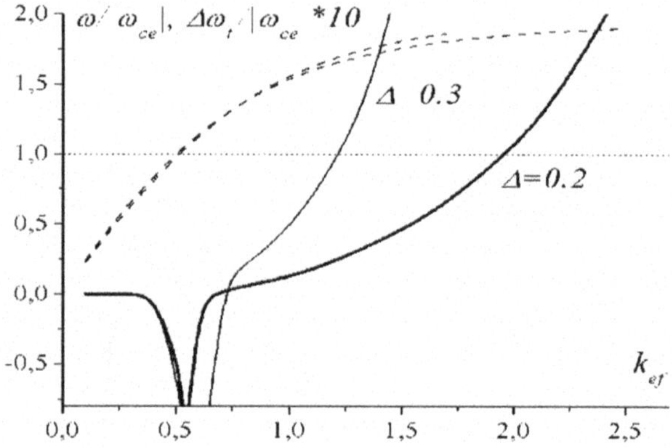

Fig. 7.25 Normalized SFXM frequency (*dashed curves*) and it correction caused by the toroidicity of the static external magnetic field (*solid curves*) versus k_{ef} for different widths Δ of the vacuum gap; $m = +2$, $Z = 3$. In the case of $\Delta = 0.3$, the parameter of toroidicity is $\varepsilon_t = 0.1$, in the case of $\Delta = 0.2$, $= 0.05$

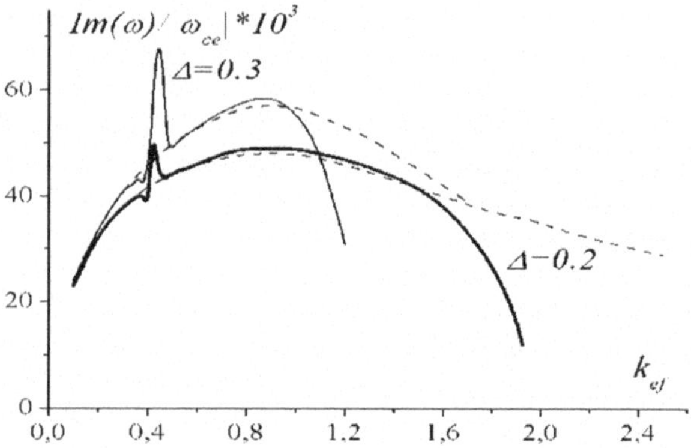

Fig. 7.26 Normalized growth rate of the SFXM resonance beam instability versus k_{ef} for different widths Δ of the vacuum gap. The parameters of the beam-plasma system are the same as in Fig. 7.25. Zeroth approximation (*dashed curves*)

resonance overlapping. The curves depicting the dependence of LF SFXM growth rates $\mathrm{Im}(\omega/|\omega_e|)$ on the effective wavenumber $k_{ef} = |m|c/(\Omega_e R_1)$ in Fig. 7.27 illustrate just this peculiarity of the resonant beam instability of the studied waves.

The curves presented in Fig. 7.27 are calculated for $\Delta = R_2/R_1 - 1 = 0.2$ and $Z = \Omega_e/|\omega_e| = 10$. The numbers indicate the value of the azimuthal mode index

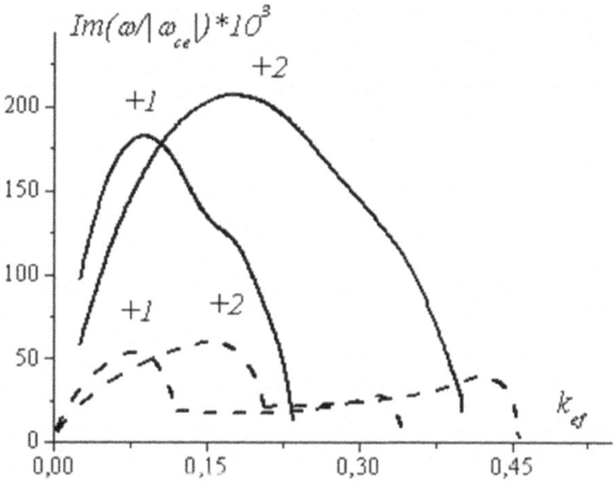

Fig. 7.27 Normalized growth rate of the SFXM resonance beam instability versus k_{ef} for different electron beam particle densities for $Z = 10$, $\Delta = 0.2$. *Solid curves* correspond to a dense beam, $\alpha = 10^{-2}$, *dashed curves* to a low-density beam, $\alpha = 10^{-4}$

m, the solid and dashed curves relate to the cases of a higher electron beam density, $\alpha = 10^{-2}$, and a lower beam density, $\alpha = 10^{-4}$, respectively. Maximum values of $\mathrm{Im}(\omega/|\omega_e|)$ are obtained in the case of higher beam density. For the SFXM with azimuthal mode index $m = +1$ this maximum is $\mathrm{Im}(\omega/|\omega_e|) = 0.184$ for $k_{ef} = 0.075$. The maximum growth rate of the $m = +2$ SFXM is even larger: $\mathrm{Im}(\omega/|\omega_e|) = 0.207$. Doing that, resonant condition (7.35) is satisfied for the first beam cyclotron harmonic $N = 1$ if $m = +1$ and for the second harmonic $N = 2$ if the azimuthal mode index is $m = +2$. Analyzing the solid curves in Fig. 7.27 one can see that there are two points of inflection (at k_{ef} equal to 0.13 and 0.28 for the first and second azimuthal mode indices, respectively). They indicate the transfer from interaction between the SFXM with definite m value and beam cyclotron harmonic N to the interaction between this eigenwave and the beam cyclotron harmonic with the next number $N + 1$. In the present case, it is the transfer from $N = 1$ to $N = 2$ for the LF SFXM with $m = +1$, and from $N = 2$ to $N = 3$ for the LF SFXM with $m = +2$. Thus, these solid lines show the "single bell"-shape $\mathrm{Im}(\omega/|\omega_e|)$ curves, which describe the regime of resonance overlapping.

However, if the electron beam particle density is decreased, the $\mathrm{Im}(\omega/|\omega_e|)$ curves have "double bell" shapes. In the present study, transfer from the case $n_b/n_p = 10^{-2}$ to the case of lower beam density $n_b/n_p = 10^{-4}$ (see dashed curves in Fig. 7.27) is accompanied not only by a decrease of the maximum $\mathrm{Im}(\omega/|\omega_e|)$ value, for instance from 0.187 to 0.054 for the LF SFXM with $m = +1$, but also by separation of the maxima of $\mathrm{Im}(\omega/|\omega_e|)$ related to the interaction with different numbers N of the beam cyclotron harmonics in the space of effective wavenumbers k_{ef}. The maximum values of $\mathrm{Im}(\omega/|\omega_e|)$ are realized for k_{ef} which are different by a factor of four. Namely, the most efficient interaction with the first cyclotron harmonic takes place for $k_{ef} \approx 0.08$ and with the second cyclotron harmonic for $k_{ef} \approx 0.32$. Thus, this regime of the LF SFXM instability can be called as regime of interaction between this wave and different beam cyclotron harmonics.

The solid lines in Figs. 7.21, 7.23 and 7.25 show the dependence of the correction to the LF SFXMs frequency caused by the space non-uniformity of the external stationary magnetic field, $\Delta\omega_t = -D^{(2)}\left(\partial D^{(0)}/\partial\omega\right)^{-1}_{|\omega=\omega_0}$, as functions of the k_{ef}. The correction $\Delta\omega_t$ is normalized there to the absolute value of the electron cyclotron frequency $|\omega_e|$. The dependence of the normalized SFXM frequency $\omega/|\omega_e|$ is shown there by the dashed lines for comparison. To make the comparison convenient for readers in the same graph, the value of the correction was multiplied by a factor of 10.

The solid lines in Figs. 7.22, 7.24 and 7.26 show the dependence of the growth rate of the SFXM resonant beam instability $\mathrm{Im}(\omega/|\omega_e|)$ (normalized to the absolute value of the electron cyclotron frequency) on the k_{ef} value, taking into consideration the toroidal non-uniformity of the external magnetic field. The dependence of the growth rate in the case of a uniform magnetic field is shown there by the dashed lines for comparison.

The numbers +1, +2 in Figs. 7.21 and 7.22 indicate the values of the azimuthal mode index m. The curves for the SFXMs with $m = +1$ have been plotted in Fig. 7.21 for the smaller range of k_{ef} which corresponds to the range where the growth rate dependence is studied. The curves of the correction caused by the toroidicity of the static external magnetic field diverge near k_{ef1}, which corresponds to $\omega \approx |\omega_e|$. This divergence comes from the ratio $\left(\varepsilon_1^{(2)}/\varepsilon_1^{(0)}\right)\Big|_{\omega\to|\omega_e|} \propto (\omega - |\omega_e|)^{-2}$. The "gap" in the dependence of $\Delta\omega_t(k_{ef})$ is wider for larger values of $|m|$. This discontinuity in the graphs of the correction to the wave frequency is accompanied by a feature (but not a discontinuity) in the graphs of the instability growth rate near the same singular values of k_{ef1}.The physical reason for the frequency spectrum discontinuity in the vicinity of the electron cyclotron frequency is the application of the plasma fluid model, which is not adequate in this frequency range [21].

The toroidicity has little effect on the maximal value of the growth rate, as can be seen from Fig. 7.22, especially for small values of k_{ef}. For the SFXMs with $m = +1$, the maximal value of the growth rate $\mathrm{Im}(\omega/|\omega_e|) = 0.054$ in the case of uniform magnetic field is reached when $k_{ef} = 0.35$, and exceeds the maximal value of the growth rate in the case of toroidal non-uniformity of the magnetic field by 3.7% only. For the SFXMs with $m = +2$, the maximal value of the growth rate $\mathrm{Im}(\omega/|\omega_e|) = 0.057$ in the case of a uniform magnetic field is reached when $k_{ef} = 0.9$ and, oppositely, the maximal value of the growth rate in the case of toroidal non-uniformity of the magnetic field is reduced by 3.4%. In addition to that it is found that the toroidal non-uniformity has an essential effect on the width of the k_{ef} range where SFXM excitation by electron beams is efficient. This range broadens in the direction of larger values of k_{ef}, or in other words to smaller values of the plasma column radius and lower plasma densities [43].

The effects of the magnetic field toroidicity on SFXM excitation are compared for different values of the external magnetic field in Figs. 7.23 and 7.24. A decrease of the stationary external magnetic field $B_{0\zeta}$ leads to increasing SFXM eigenfrequency (see Fig. 7.23), to higher growth rates (see Fig. 7.24), to a shift of the frequency correction to the LF SFXM eigenfrequency caused by the space non-uniformity of $B_{0\zeta}$ into the region of smaller values of k_{ef}, and to a shift of the "bell" of the dependence of $\mathrm{Im}(\omega/|\omega_e|)$ on k_{ef} into the region of lower values of k_{ef} as well.

In Figs. 7.21, 7.23 and 7.25, one can see a singularity in the dependence of the correction $\Delta\omega_t$ on k_{ef}. It happens when the wave frequency is close to $|\omega_e|$. Correlations between the positions of the points of intersection between the curves $\omega/|\omega_e|$ and the asymptotic lines $\omega = |\omega_e|$ and the "gaps" in the curves can be easily seen in these figures. The correction to the LF SFXM frequency caused by the toroidicity of the $B_{0\varsigma}$ is calculated here by employing the theory of successive approximations. Therefore, in the second approximation the correction is $\Delta\omega_t = -D^{(2)}\left(\partial D^{(0)}/\partial\omega\right)^{-1}_{|\omega=\omega_0} \propto (\omega - |\omega_e|)^{-2}$, and one should keep in mind that the validity of this approximation requires the satisfaction of the inequality $\omega >> |\Delta\omega_t|$. Thus, this approximation shows only a tendency of increasing $|\Delta\omega_t|$ for $\omega \to |\omega_e|$. To obtain a more correct expression for $|\Delta\omega_t|$ one can calculate the next higher order approximations. To show the tendency of $\Delta\omega_t$ and to indicate the position of such type resonance in the space of wave numbers it is sufficient to apply the plasma fluid model. However, one should always be aware of the fact that in the vicinity of the ion and electron cyclotron resonances just the kinetic model is correct rather than the fluid model. However, from the mathematical point of view the kinetic model is more difficult than the fluid model.

The difference in the dependences of the growth rates on k_{ef} in the cases of uniform and non-uniform external magnetic field becomes smaller when the parameter $Z = \Omega_e/|\omega_e|$ increases (see Fig. 7.24). It correlates with the result presented in Fig. 7.22 since this difference becomes negligible just for the range of small values of k_{ef}. Unlike that, near the right limit of the k_{ef} range where SFXM excitation is efficient, this difference becomes larger and is accompanied by broadening of this range of k_{ef} [43].

The influence of the toroidicity of the external static magnetic field on the value of the SFXM growth rates is mostly pronounced if their eigenfrequencies are close to the electron cyclotron resonance frequency, because in the second approximation $\Delta\omega_t \propto (\omega - |\omega_e|)^{-2}$. That is why one can observe in Figs. 7.22 and 7.26 a sharp increase of the growth rate of the SFXM with $m = +2$ nearby $k_{ef} = 0.5$ in the case $Z = 3$. In Fig. 7.24, where the results related to the case of $m = +1$ are plotted, one can see that since the curve $\omega/|\omega_e|$ crosses the asymptotical line of the electron cyclotron resonance for $Z = 3$ at $k_{ef} = 0.6$, where the resonance condition (7.35) is not satisfied, the related growth rate curve has no peak. However, if $Z = 8$ (lower magnetic field), the crossing point mentioned above is located nearby the effective wavenumber $k_{ef} = 0.2$ (see Fig. 7.23). Then the curve $\mathrm{Im}(\omega/|\omega_e|)$ has an additional peak just near this value of the effective wavenumber (see Fig. 7.24).

The effect of the toroidicity on the excitation of SFXMs is also studied for different widths of the gap between the plasma column and the metal wall. These results are presented in Figs. 7.25 and 7.26. Since the gap width weakly affects the SFXM frequency for small values of the parameter Z, the singularity in the dependence of the frequency correction $\Delta\omega_t$ caused by the toroidicity is almost at the same value of the effective wavenumber: $k_{ef} = 0.55 \div 0.6$ (see Fig. 7.25). In the case of a large gap, the related "hole" in the curve becomes wider. On the left side from the "hole", the dependencies are similar, but on the right side the correction to the frequency is smaller than in the case of the wider gap. This decrease of the correction is only partly due to the difference in Δ, but also due to the difference in ε_t ($\varepsilon_t = 0.1$ for the graph in the case $\Delta = R_2/R_1 - 1 = 0.3$ and $\varepsilon_t = 0.05$ for the case $\Delta = 0.2$). The important issue is that the strong inequality $\varepsilon_t \ll \Delta$ should be satisfied when solving the problem. The dependencies of the SFXM growth rate $\mathrm{Im}(\omega/|\omega_e|)$ on the effective wavenumber for different gap widths are plotted in Fig. 7.26. The differences in the graphs for the cases of uniform and non-uniform magnetic fields are qualitatively the same for both gap widths. But increasing gap width leads to increasing $\mathrm{Im}(\omega/|\omega_e|)$ [43].

The inequality $\varepsilon_t \ll 1$ should be satisfied in the present consideration to utilize the expansion (7.30) and dispersion equation (7.34). But the toroidal non-uniformity of the external magnetic field has an effect not only on the electrodynamic plasma properties, but also on the dynamics of the electron beam motion. The beam electrons rotating along Larmor radii in the toroidal non-uniform external magnetic field move with the drift velocity

$$\vec{\upsilon}_D = \frac{\upsilon_\perp \rho_{L\perp}}{2B_0^2}\left[\vec{B}_0, \nabla B_0\right], \tag{7.36}$$

where $\rho_{L\perp} = cp_\perp/(eB_0)$ is the Larmor radius. Due to the drift motion, the electron beam hits the metal wall or the plasma surface after the time $\tau = (R_2 - R_1)/\upsilon_D$. The excitation of SFXMs is only possible when the time τ is much larger than the SFXM temporal period $2\pi/\omega$. This request leads to the stricter inequality $\varepsilon_t \ll \Delta$, which is satisfied in all the calculations in this Section.

Summarizing the obtained results one can see that the toroidal non-uniformity of $\vec{B}_0$ affects most essentially the dispersion properties of the LF SFXMs when their eigenfrequency is close to the absolute value of the electron cyclotron frequency. This spatial non-uniformity of $\vec{B}_0$ weakly affects the maximal (for the complete range of the effective wavenumbers where the instability can be realized) value of the SFXM beam instability growth rate. However, in the studied case, the efficiency of power transfer from the electron beam into the excited waves becomes restricted due to the electron drift in the non-uniform magnetic field [43].

The range of k_{ef}, where the instability can be realized, becomes larger by shifting the right limit of the range towards the side of larger values of k_{ef} as compared with the case of spatially uniform confining magnetic field. This shift is larger for larger azimuthal mode indices m. The effect of the $\vec{B}_0$ toroidicity on the SFXM growth rates is weaker for a plasma filling with larger plasma particle density [43].

The toroidal non-uniformity of $\vec{B}_0$ leads to a drift of the electron beam. That is why these electrons disappear on the metal wall or the plasma surface faster. But if the toroidicity is weak so that $R_1/R \ll (R_2 - R_1)/R_1 < 1$, the beam has enough time during the drift to excite the LF SFXMs. Since the correction to the SFXM eigenfrequency caused by the toroidal non-uniformity of $\vec{B}_0$ is quadratic with respect to the small parameter of the toroidicity, $\Delta\omega_t \propto \varepsilon_t^2$, the effect of the toroidicity on the SFXM excitation is larger for larger gaps between the plasma column and the metallic toroidal chamber wall.

7.5 Excitation of Surface Flute Waves in Cylindrical Waveguides with Non-circular Cross-Section of Plasma Filling

Nowadays, the theory of beam-plasma instabilities is still incomplete since the interaction of charged particle beams with the eigenwaves of different waveguides depends essentially on a large number of factors including the dispersion properties of the waves, the wave polarization, the spatial distribution of the wave field, and the geometry and design features of the waveguides. Therefore, studying the interaction of an annular beam of the charged particles with the eigenwaves of plasma-filled metallic waveguides is an actual problem in plasma physics [65, 66].

A periodic spatial inhomogeneity of the medium along the direction of electromagnetic wave propagation can enhance the efficiency of plasma electronic devices. The results of the theoretical study of LF SFXW propagation in magnetized waveguides with non-circular cross-section, partially filled with plasmas [67], are presented in Sect. 6.3. The analysis is carried out with the assumptions that the external static magnetic field is directed along the waveguide axis and that there is a narrow dielectric gap between the plasma column and the metallic wall of the cylindrical waveguide. It is shown that in this case the frequency spectrum and the spectrum composition of the wave packet are predetermined by the shape of the cross-section of the plasma column. The correction $\Delta\omega_N$ to the eigenfrequency caused by the cross-section non-circularity is shown to increase with decreasing width of the dielectric layer and magnitude of its dielectric constant. An increase of the stationary magnetic field leads only to an insignificant increase of $\Delta\omega_N$. A difference in the corrections $\Delta\omega_N$ for waves propagating in opposite directions has also been found.

In addition, the effect of the shape of the plasma cross-section on the LF SFXW dispersion properties [68] has been investigated in Sect. 6.5 for the case when the angular period of the wave perturbation is exactly twice larger than the inhomogeneity period of the plasma-dielectric interface. In this case, the correction $\Delta\omega_N$ to the frequency is much larger than in the general case [67]. It is shown that changing the sign of the azimuthal mode index m of the SFXW main harmonic (from negative to positive) leads to a change of the sign of the correction and some increase of the absolute value of $\Delta\omega_N$.

The propagation of electromagnetic waves with extraordinary polarization (the non-zero components are E_r, E_φ, B_z) inside a cylindrical chamber of circular cross-section with inner radius R_2 (Fig. 7.28) is considered here. The chamber is assumed to be made of a perfect conductor (metal). The external homogeneous stationary magnetic field is directed along the waveguide axis, $\vec{B}_0 || \vec{z}$. The waveguide is assumed to be uniform along the cylinder axis, $\partial/\partial z \equiv 0$. These extraordinary modes

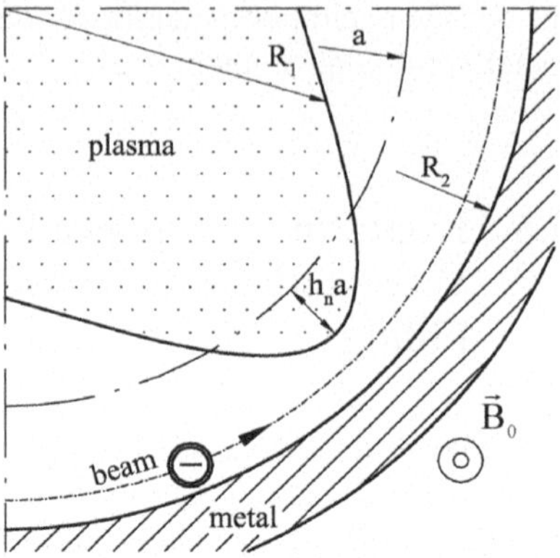

Fig. 7.28 Schematic of the electron beam-plasma structure with a non-circular cross-section of plasma filling in a magnetized circular waveguide

(X-modes) propagate strictly perpendicular to the magnetic field, for example, $H_z(r,\varphi) = H_z(r)\exp(im\varphi - i\omega t)$.

The plasma column with radius $R_1(\varphi)$ is placed inside the metallic chamber. The plasma particle density is assumed to be homogeneous to study just the effect of the cross-section shape of the plasma-vacuum interface on the beam-plasma interaction. The radial coordinate of the plasma-vacuum interface $R_1(\varphi)$ depends on the azimuthal angle according to formula (6.2).

The individual terms in the expression (6.2) for $R_1(\varphi)$ are proportional to $\cos(n\varphi + \varphi_{n1})$, that is why they contribute independently to the SFXW dispersion properties in the second order approximation with respect to the small parameters δ_N (the higher order terms are neglected). That is why the following analysis is carried out only for one term (one specific type of plasma deformation) in (6.2):

$$R_1(\varphi) = a \cdot [1 + \delta_N \cos(N\varphi + \varphi_N)]. \tag{7.37}$$

This allows excluding the effects of other terms. However, one should mention that there are no fundamental problems to consider any finite number of terms from (6.2) in the studied dispersion equation. Moreover, some values of the parameter N in (7.37) are of special physical interest: the term with $N = 1$ describes a violation of the coaxiality between the plasma column and the metal chamber, the term with $N = 2$ takes into account an ellipticity of the plasma column cross-section, and a cross-section triangularity is taken into account by the term with $N = 3$, etc.

It is assumed that the displacement vector and the electric field are related by the permittivity tensor (1.2) of the weakly collisional cold magnetized plasma (see, e. g., [21, 41]). The two components of the permittivity tensor which are used in the following considerations are ε_1 and ε_2. Their explicit form (1.3) is presented in Sect. 1.4.1.

The components of the SFXW electric field can be expressed in terms of the magnetic field component H_z in the following way:

$$E_r = \frac{k}{k_\perp^2}\left(\mu\frac{\partial H_z}{\partial r} - \frac{i}{r}\frac{\partial H_z}{\partial \varphi}\right), \quad E_\varphi = \frac{k}{k_\perp^2}\left(i\frac{\partial H_z}{\partial r} + \frac{\mu}{r}\frac{\partial H_z}{\partial \varphi}\right). \tag{7.38}$$

It should be noted that the derivation of these expressions does not coincide with (1.10) given in Sect. 1.4.1. The presentation (7.38) is based on the substitution $(\partial/\partial t) \to -i\omega$, but does not apply the replacement $(\partial/\partial\varphi) \to im$ because one cannot solve this problem in the framework of a single-mode approach. In the Eqs. (7.38), $k = \omega/c$ is the free-space wavenumber, the value $k_\perp^{-1}$ determines the penetration depth of the SFXW electromagnetic field into the plasma, $k_\perp^2 = k^2\varepsilon_1\left(\mu^2 - 1\right)$, and $\mu = \varepsilon_2/\varepsilon_1$. The considered waves can propagate just in the frequency ranges (1.13) and (1.14) where the $k_\perp^2$ value is positive.

The plasma particle density is assumed to be sufficiently high to ensure that the inequality $\Omega_e^2 > \varepsilon_r\omega_e^2$ is valid. This is surely realized in semiconductor plasmas but this condition can be also fulfilled in laboratory gas plasmas when the external static magnetic field is sufficiently weak. As mentioned in Sect. 1.4.1, in this case $k_\perp^2 > 0$ is valid in the LF range (1.13) and HF range (1.14). In the present Section the SFXWs propagating in the low frequency range (1.13), $\omega_{LH} < \omega < |\omega_e|$ and $|\omega_e| < \omega < \omega_1 = -|\omega_e|/2 + \sqrt{\Omega_e^2 + \omega_e^2/4}$, are under investigation.

It is assumed that the annular electron beam is injected into the vacuum layer $R_1 < r < R_2$ between the plasma column and the metal wall of the waveguide. The model of a gyrating electron beam is described in detail in Sect. 7.1. The electrodynamic properties of the waveguide inner part occupied by the beam are described by the permittivity tensor $\varepsilon_{ik}^{(b)}$ [21, 34]. The three components of the tensor $\varepsilon_{ik}^{(b)}$ which are used in the calculations below are expressed by (7.18)–(7.20) in Sect. 7.1.

Solving the set of Maxwell equations in the region occupied by the electron beam with the mentioned components of the permittivity tensor $\varepsilon_{ik}^{(b)}$ leads to the expressions for the SFXW fields as linear combinations of Bessel functions of the first kind $J_m(\zeta)$, Bessel functions of the second kind $N_m(\zeta)$ and their derivatives with respect to the argument $\zeta = kr\sqrt{\psi_b}$, where $\psi_b = \varepsilon_{22}^{(b)} + \left(\varepsilon_{12}^{(b)}\right)^2\left(\varepsilon_{11}^{(b)}\right)^{-1}$.

The dispersion properties of the considered hybrid waveguide structure can be studied using the method of successive approximations. In zeroth approximation, the plasma-vacuum interface is assumed to be circular and coaxial with the metallic chamber. The SFXW properties in this approximation can be described using the results presented in [63]. SFXWs with different azimuthal mode index m propagate independently when the curvature radius of the plasma-vacuum interface does not depend on the azimuthal angle. That is why one can assume that in zeroth approximation with respect to the small parameter δ_N the electromagnetic waves propagate with a single azimuthal mode index m. As a result of the periodic spatial inhomogeneity of the plasma-vacuum interface (see expression 7.37) SFXWs propagate as wave packets. Each such packet consists of the fundamental harmonic with a field which is proportional to $\exp(im\varphi - i\omega\, t)$ and of an infinite number of satellite harmonics with fields which are proportional to $\exp[i(m \pm jN)\varphi - i\omega t]$ ($j = 1, 2, 3\ldots$). It is well-known from the theory of wave propagation in media with periodical properties along a definite direction [64] that the amplitudes of the satellite harmonics have small values (of the order $(\delta_N)^j$) as compared with the amplitude of the fundamental harmonic. A wave packet consisting of the fundamental harmonic and the two nearest satellite ($j = 1$) harmonics with fields which are proportional to $\exp[i(m \pm N)\varphi - i\omega t]$ has been considered in [67, 68] to study the dependence of the SFXW dispersion properties on the shape of the non-circular cross-section of the plasma-vacuum interface. Such approach is based on the fact that taking into consideration the higher satellite harmonics gives a contribution to the correction of the SFXW eigenfrequency which is in third and higher orders of the small value δ_N.

Taking into account only those terms which are of the first order in δ_N does not change the amplitude of the fundamental harmonic but leads to the appearance of non-zero small satellite harmonics of the wave field which vary proportionally to $\exp[i(m \pm N)\varphi - i\omega t]$. In the framework of the second order approximation with respect to the small parameter δ_N, the amplitude of the fundamental harmonic obtains a correction of second order in δ_N which together with the amplitude of the first satellite harmonic gives a correction of the second order in δ_N to the dispersion equation, $D^{(2)}(\omega,m,N,\ldots) \propto \delta_N{}^2$. In such a way the dispersion relation of the zeroth order approximation, $D^{(0)}(\omega,m,\ldots) = 0$, transforms to the dispersion relation of the second order approximation: $D^{(0)}(\omega,m,\ldots) + D^{(2)}(\omega,m,N,\ldots) = 0$. The expression for the term $D^{(2)}(\omega,m,N,\ldots)$ is too cumbersome to be presented here, one can find it in details in Sect. 6.6 (see Eq. 6.39) [67].

The following boundary conditions for the wave fields should be used to solve the problem of LF SFXW propagation in the described waveguide.

- The tangential component of the SFXW electric field is equal to zero at the metal surface:

$$E_\varphi(r = R_2) = 0. \tag{7.39}$$

- The tangential components of the electric and magnetic fields of the SFXWs are continuous at the non-circular plasma-vacuum interface:

$$E_\tau(r = R_1 + 0) = E_\tau(r = R_1 - 0), \quad H_z(r = R_1 + 0) = H_z(r = R_1 - 0). \tag{7.40}$$

- The SFXW fields have finite values everywhere inside the waveguide, particularly on the waveguide axis:

$$H_z(r = 0) < \infty. \tag{7.41}$$

Applying these boundary conditions allows to obtain the dispersion relation that is under consideration here:

$$\begin{aligned} &\frac{I'_m(k_\perp a)k}{k_\perp I_m(k_\perp a)} + \frac{\mu m k}{k_\perp^2 a} + D^{(2)}(\omega, m, N, \ldots) \\ &= \frac{im\varepsilon_{12}^{(b)}}{\varepsilon_{11}^{(b)}\zeta_1\sqrt{\psi_b}} - \frac{J'_m(\zeta_1) - \Phi N'_m(\zeta_1)}{\sqrt{\psi_b}[J_m(\zeta_1) - \Phi N_m(\zeta_1)]}, \end{aligned} \tag{7.42}$$

where $\Phi = \left[\frac{im\varepsilon_{12}^{(b)}}{\varepsilon_{11}^{(b)}\zeta_2} J_m(\zeta_2) - J_m'(\zeta_2)\right]\left[\frac{im\varepsilon_{12}^{(b)}}{\varepsilon_{11}^{(b)}\zeta_2} N_m(\zeta_2) - N_m'(\zeta_2)\right]^{-1}$, with $\zeta_1 = \zeta(a)$, $\zeta_2 = \zeta(R_2)$ [69]. Equation (7.42) describes the excitation of LF SFXWs by an electron beam in the case of non-circular cross-section of the plasma column. It is analyzed under the resonance beam condition (7.14).

The results of the numerical analysis of Eq. (7.42) and some explanation of the results are plotted in Figs. 7.29, 7.30, 7.31, 7.32 and 7.33 where the dependences of the growth rates of the resonance beam instability, normalized by the absolute value of the electron cyclotron frequency, on the effective wavenumber $k_{ef} = |m|c/(\Omega_e a)$ for different plasma waveguide parameters are presented. Since Fig. 7.31 is devoted to studying the effect of external static axial magnetic field on the excitation of LF SFXMs, the growth rate is normalized by the electron plasma (Langmuir) frequency and shown along ordinate.

The dependence of the growth rate on k_{ef} for the LF SFXW with the azimuthal mode index $m = 2$ is shown in Fig. 7.29. The following parameters of the plasma-beam system are used for the calculations: $Z \equiv \Omega_e/|\omega_e| = 8$, $R_2 - a = 0.1a$, $\delta_N = 0.1$, $\alpha = 10^{-3}$. The numbers near the solid curves indicate the value of the

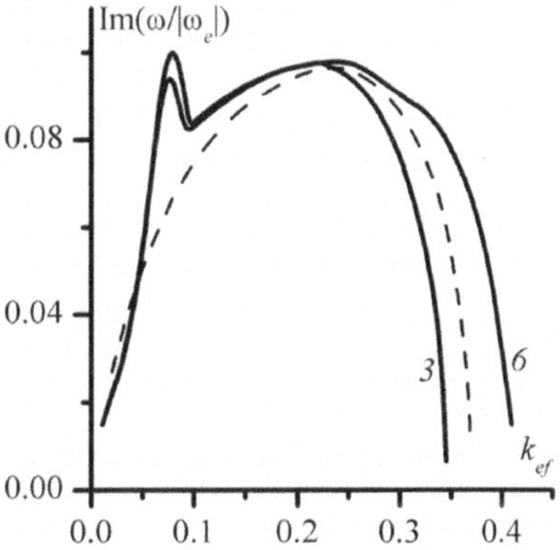

Fig. 7.29 Normalized beam-plasma instability growth rate of the LF SFXW with $m = 2$ versus effective wavenumber for $Z = 8$, $R_2 - a = 0.1a$, $\alpha = 10^{-3}$, $\delta_N = 0.05$, $N = 3$ and $N = 6$ (*solid curves*). The *dashed curve* is for circular plasma cross-section

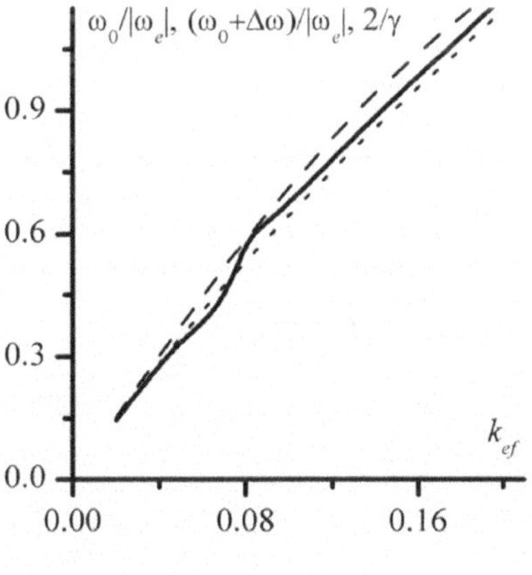

Fig. 7.30 Normalized eigenfrequency of LF SFXM versus effective wavenumber for the conditions of Fig. 7.29. The dependence $\omega_0/|\omega_e|$ is given by the *dotted curve*; $(\omega_0+\Delta\omega_N)/|\omega_e|$ by the *solid curve*; and the resonance curve $2/\gamma$ is presented by the *dashed curve*

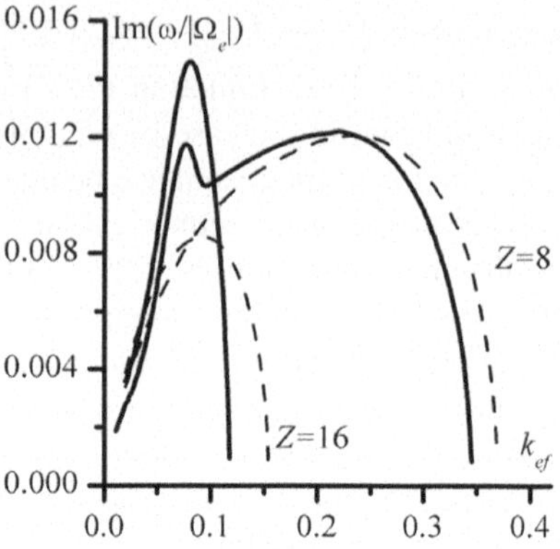

Fig. 7.31 The same as in Fig. 7.29, however for $N = 3$, $Z = 8$ (*dotted curves*) and $Z = 16$ (*solid curves*). The *dashed curves* are for circular plasma cross-section

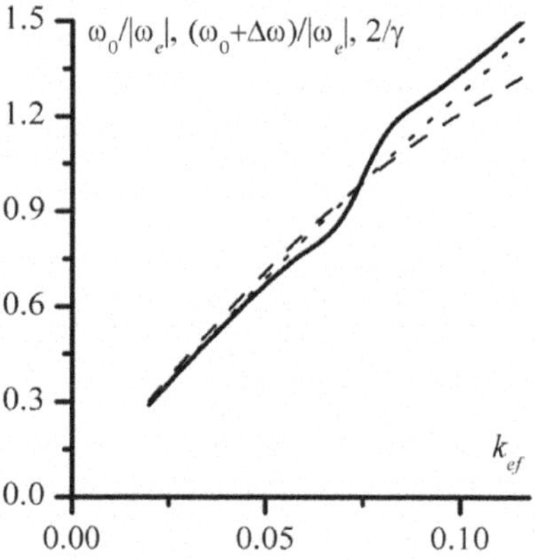

Fig. 7.32 Normalized eigenfrequency of LF SFXM versus effective wavenumber as in Fig. 7.30, however for the conditions of Fig. 7.31

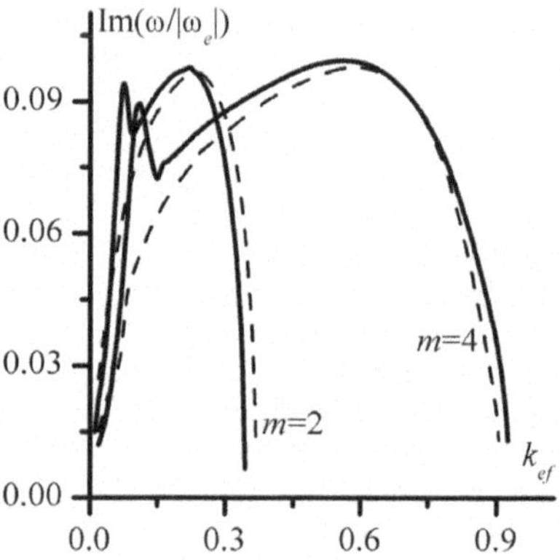

Fig. 7.33 The same as in Fig. 7.29, however for $m = 2$ and $m = 4$. The *dashed curves* are for circular plasma cross-section

corrugation parameter $N = 3,6$, which defines the number of the angular periods of the plasma-vacuum interface. The dashed curves in Figs. 7.29, 7.31 and 7.33 show the dependence of the LF SFXW growth rate in the case of a circular cross-section of the plasma-vacuum interface, $\delta_N = 0$. The plasma column cross-section non-circularity causes a change in the range of k_{ef} for which beam excitation is essential. An additional local maximum appears at $k_{ef} \approx 0.08$ due to the non-circularity. Its appearance is explained by the behavior of the curves in Fig. 7.30. The dotted curve presents the dependence of $(\omega_0/|\omega_e|)$ versus k_{ef}. The dashed curve corresponds to the dependence of the resonant curve (7.14), $\omega_0/|\omega_e| = 2/\gamma$. The account for small non-circularity with $N = 3$ causes weak change in the dependence of $((\omega_0 + \Delta\omega_N)/|\omega_e|)$ presented by the solid curve in the vicinity of $k_{ef} \approx 0.08$. However, this small change results in approaching of the curve to the resonant curve $\omega_0/|\omega_e| = 2/\gamma$. Just this small change in the dispersion properties of LF SFXMs provides significant change in the growth rates. Note that application of the non-circularity with $\delta_N = 0.1$ satisfies the conditions of applicability of the method of successive approximations. It increases the LF SFXM eigenfrequency by 11% only.

It was shown in [35] that the strongest effect on the dispersion properties of the LF SFXW is caused by an external static axial magnetic field. The solid curves in Fig. 7.31 show the dependence of the growth rate of the LF SFXW with $m = 2$ for the case of weaker external static magnetic field, $Z = 16$, in comparison with the data presented in Fig. 7.29, $N = 3$. In both cases, the non-circularity makes the range of k_{ef} for which the SFXM excitation is efficient narrower. Therefore, the larger magnitudes of k_{ef} are excluded from this range.

Figure 7.31 clearly demonstrates that for a definite range of k_{ef}, SFXM growth rate can be larger in the presence of non-circularity than in a plasma with circular cross-section. Explanation for this phenomenon can be found in considering the LF SFXM dispersion properties. The influence of the non-circularity on the LF SFXM dispersion properties is plotted in Fig. 7.32. The dotted, dashed and solid curves denote the same as in Fig. 7.30, but for $Z = 16$. The larger magnitude of the growth rate in the case of $\delta_N \neq 0$ correlates with larger angle between the tangents to the solid and dashed curves for $k_{ef} \approx 0.08$ as compared to the angle between the tangents to the dotted and dashed curves.

The growth rates of LF SFXMs with azimuthal mode index $m = 2$ and $m = 4$ are compared in Fig. 7.33, $N = 3$. The non-circularity in the case of $m = 4$ also leads to a change in the range of k_{ef} for which excitation of LF SFXMs is efficient as it is for $m = 2$ in Fig. 7.29. The “bell”, which presents the dependence of the LF SFXM with $m = 4$, is wider than that for $m = 2$. The corresponding position of the local maximum for $m = 4$ ($k_{ef} \approx 0.11$) is located to the right of that for $m = 2$ ($k_{ef} \approx 0.08$).

The initial stage of the resonance beam instability of LF SFXWs in a circular metallic waveguide with the non-circular plasma-vacuum interface is studied in the present Section. The non-circular shape of the plasma column cross-section does not essentially affect the LF SFXW excitation by the annular electron beam [69]. This is explained by the fact that the effective wavenumber range (the range of the beam-plasma system parameters) wherein the excitation of LF SFXWs is efficient [34] differs from the range where the non-circularity of the plasma-vacuum interface essentially influences the dispersion properties of these waves [67, 68]. This fact can be useful for technological purposes: small non-circularities of gas discharge tubes do not decrease the efficiency of LF SFXW excitation by the electron beam. This is correct even in the case when the angular period of the wave perturbations is just twice larger than the period of the inhomogeneity of the plasma-dielectric interface [68] and the non-circularity of the plasma column cross-section most essentially affects the LF SFXW dispersion properties. For some cases, the non-circularity of the plasma column cross-section even increases the growth rates of the resonance beam instability of LF SFXWs [69].

Beam excitation of LF SFXWs in corrugated metal chambers (waveguides) is not studied here since it is well-known that the effect of the corrugation of the plasma-vacuum interface on the wave dispersion properties is much larger than the effect of a non-circularity of the metal chamber. This feature of the LF SFXW dispersion properties is related to the fact that the LF SFXW power is concentrated just near the plasma-vacuum interface but not near the metal wall.

7.6 Transition Between Beam-Plasma and Beam-Dissipative Instability Regimes in Interaction of Relativistic Large Larmor Orbit Electron Beams and Surface Flute X-Modes

Excitation of electromagnetic waves in plasmas caused by wave interaction with electron beams is a matter of interest for a long time, both in the mode of beam-plasma instability [49] and beam-dissipative instability [70]. Nevertheless, numerous applications of these physical mechanisms initiate their further active investigation [71–74].

Surface flute waves (SFWs) are intensively studied in recent years [57] due to their possible application for usage in high-power plasma electronics devices [57], quickly on/off-switchable plasma-antenna systems [75–79], plasma nano-science using metal-powder plasmas [80] and plasma production for various technological applications in materials processing and plasma chemistry [81]. The introduction of plasmas in high-power (up to GW level) vacuum microwave devices often allows one to enhance output power and efficiency due to higher injected beam currents and improved control of wave-beam interaction. These effects have been observed in plasma-filled relativistic large-orbit gyrotrons and in backward-wave oscillators, where the plasma-filled smooth circular metallic waveguide interaction circuit allows electron-beam excitation of slow-waves.

SFWs are electromagnetic eigenmodes of cylindrical plasma-filled waveguides and propagate along the plasma column interface across an axial external static magnetic field. SFWs are known to propagate within two frequency ranges if the plasma is sufficiently dense, so that $\Omega_e^2 >> \varepsilon_r \omega_e^2$, where Ω_e is the electron plasma (Langmuir) frequency, ω_e is the electron cyclotron frequency, and ε_r is the relative dielectric constant of a crystal or the crystal lattice of a semiconductor, or of a metamaterial ($\varepsilon_r > 1$). For gaseous plasmas one has $\varepsilon_r = 1$. These two ranges are the electron cyclotron frequency range (1.13, LF range) and the frequency range above the upper hybrid frequency (1.14, HF range). The first one is under study in this Section.

Surface flute modes of extraordinary polarization (SFXMs with non-zero components E_r, E_φ, B_z) efficiently interact with electron beams rotating along the Larmor orbits in the gap which separates the plasma column from the loss-free metal chamber wall [31, 34]. The initial stage of the beam-plasma instability of SFXMs caused by their interaction with the electron beam is studied well in the previous sections with taking into account specific features of the beam-plasma system, like deviation of the shape of the plasma column cross-section from circular one [69], weak toroidal inhomogeneity of the external magnetic field [43], etc. Advantages of plasma electronics devices based on surface type waves and rotating electron beams as compared to those based on volumetric waves and longitudinal beams are discussed in [34].

In the studies of SFXM beam excitation in the previous Sects. 7.1–7.5, dissipation was neglected. Exclusively, excitation of SFWs in a cylindrical semiconductor plasma column in the presence of drift motion of the electrons flow was investigated in Sect. 7.1 [38]. In the present Section, the influence of dissipation on SFXM beam excitation is studied in detail and the transition of the beam-plasma instability into a beam-dissipative one is demonstrated.

The considered cylindrical beam-plasma structure is as described in Sect. 7.1 (see Fig. 7.1). The electron beam is described as a flow of particles with the same transversal momentum and zero axial momentum as done in Sect. 7.1. To obtain the SFXM dispersion relation (7.12) and (7.13) one can apply the impedance boundary conditions as done in Sect. 7.1.

The SFXM with azimuthal mode index $m = 1$ is efficiently excited in plasmas with $Z = \Omega_e/|\omega_e| = 3.5$ due to interaction with the first cyclotron harmonic of the electron beam ($N = 1$) with the relative beam particle density $\alpha = n_b/n_e = 0.01$, where n_b is the beam particle density and n_e is the plasma particle density. The dependence of the growth rate on the effective wavenumber k_{ef} reaches its maximum for $k_{ef} \equiv mc/(\Omega_e R_1) = 0.24$. Figure 7.34 shows the results of numerical studying the dispersion relation for these conditions with an effective plasma particle collision frequency $\nu = 0.1|\omega_e|$. The solid curve corresponds to the dependence of the growth rate normalized by the electron cyclotron frequency $\mathrm{Im}(\omega/|\omega_e|)$ on α (in logarithmic scale for the abscissa axis). The parameter α varies from 6×10^{-8} to 0.01. The lower the beam density is the lower is the growth rate. This dependence is qualitatively clear and has been expected. The problem is to distinguish whether that region or the other range of $\alpha = n_b/n_e$ corresponds to beam-plasma or beam-dissipative instability. To solve this problem it was proposed in [44] to use the following dependence of the maximum growth rate on beam particle density n_b and collision frequency ν:

$$\mathrm{Im}(\omega) \propto n_b^x \nu^y. \tag{7.43}$$

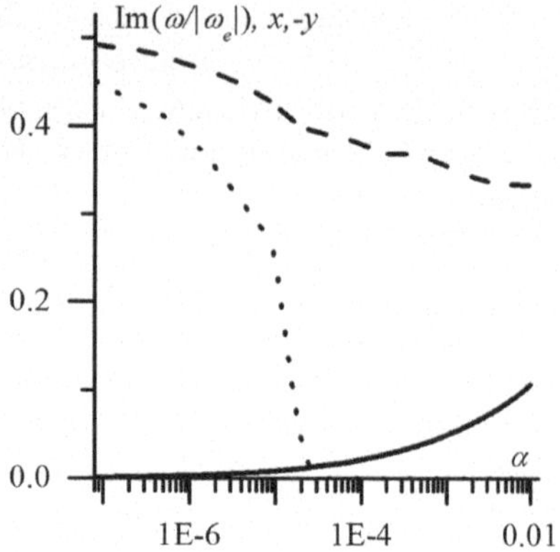

Fig. 7.34 Dependence of normalized growth rate, $\mathrm{Im}(\omega/|\omega_e|)$ (*solid line*), and the exponents x and $-y$ (see Eq. (7.43) (*dashed and dotted lines*, respectively) on the relative beam particle density α, for $m = 1$, $Z = \Omega_e/|\omega_e| = 3.5$, $k_{ef} = 0.24$, $R_2/R_1 = 1.2$, and $\nu = 0.1|\omega_e|$

The beam-plasma instability is well-known to be characterized by $x = 1/3$ and $y = 0$ [49], while in the case of beam-dissipative instability, $x = 1/2$ and $y = -1/2$ [70].

The dashed curve in Fig. 7.34 corresponds to the exponent x. It smoothly decreases from 0.492 to 0.333. Since the exponent y is known to be negative, the absolute value, $-y$, is taken in Fig. 7.34. The corresponding dependence is presented by the dotted line. The exponent increases from -0.45 to zero. A rapid change of y is observed for the transition value $\alpha_t \approx 1.21 \times 10^{-5}$. It is concluded in [44] that just the exponent y, which characterizes the dependence of the growth rate (7.43) on the collision frequency, is the most appropriate parameter to indicate whether the conditions of SFXM excitation correspond to the development of a beam-plasma or a beam-dissipative instability. On the other hand, for the SFXM with $m = 2$ the dependence of the growth rate on k_{ef} reaches its maximum for $k_{ef} = 0.53$ and the transition value appears to be higher, $\alpha_t \approx 2.08 \times 10^{-5}$.

Figure 7.35 shows the dependence of the SFXM growth rate $\mathrm{Im}(\omega/|\omega_e|)$ on the collision frequency normalized by the electron cyclotron frequency ($\nu/|\omega_e|$) for relatively high beam particle density, $\alpha = 0.01 >> \alpha_t$, for which Fig. 7.34 indicates that

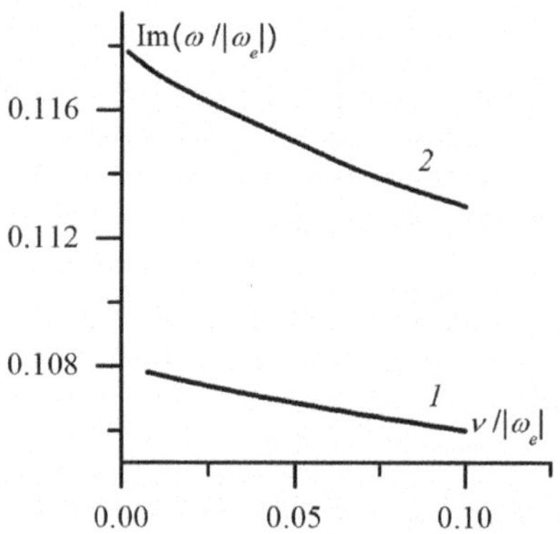

Fig. 7.35 Dependence of normalized growth rate $\mathrm{Im}(\omega/|\omega_e|)$ on the normalized collision frequency $\nu/|\omega_e|$ for $Z = \Omega_e/|\omega_e| = 3.5$, $R_2/R_1 = 1.2$, $\alpha = 0.01$, $m = 1$ ($k_{ef} = 0.24$) and $m = 2$ ($k_{ef} = 0.53$)

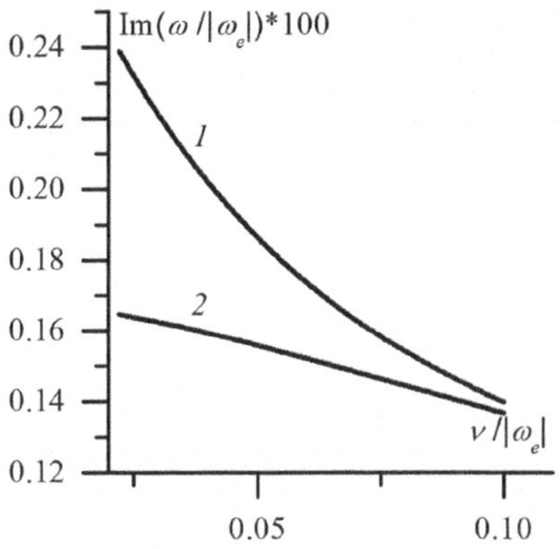

Fig. 7.36 Dependence of normalized growth rate $\mathrm{Im}(\omega/|\omega_e|)$ on the normalized collision frequency $\nu/|\omega_e|$, for $Z = 3.5$, $R_2/R_1 = 1.2$, $\alpha = 2 \times 10^{-7}$, $m = 1$ ($k_{ef} = 0.24$) and $m = 2$ ($k_{ef} = 0.53$)

the conditions of beam plasma instability are satisfied. Then for this α, the lower limit of the range for the collision frequency ν can be assumed as zero without violating the conditions of beam-plasma instability. The upper limit of the ν range is chosen to be $0.1|\omega_e|$ as it is in Fig. 7.34. The given numbers near the curves in Fig. 7.35 indicate the azimuthal mode index. Figure 7.35 evidently demonstrates a weak dependence of the SFXM growth rate on the collision frequency. Within the complete range $\nu << \omega$, the growth rate of the SFXM with $m = 1$ decreases only by less than 2%; and that of the SFXM with $m = 2$ by less than 4.4%.

The dependence of the SFXM growth rate $\mathrm{Im}(\omega/|\omega_e|)$ on $\nu\omega_e|$ for relatively weak electron beam particle density, $\alpha = 2 \times 10^{-7}$, for which Fig. 7.34 indicates realization of the conditions of beam-dissipative instability, is presented in Fig. 7.36. This beam density is 100 times lower than α_t for the SFXM with $m = 2$. This means that to ensure the conditions of beam-dissipative instability one should study the influence of dissipation on SFXM excitation within the range of collision frequencies for which the lower limit is $\sqrt[3]{100}$ times less than that in Fig. 7.34, which means $0.022|\omega_e| < \nu < 0.1|\omega_e|$. For collision frequencies $\nu < 0.022|\omega_e|$, the condition of beam-dissipative instability is violated. The growth rates of beam-dissipative instabilities are by two orders of magnitude lower than those of beam-plasma instabilities. The growth rate of a beam-dissipative instability decreases by a factor of 1.7 within the considered range of collision frequency for the SFXM with $m = 1$ and by a factor of 1.2 for the SFXM with $m = 2$. The dependence of the growth rate of the beam-dissipative instability on the collision frequency is stronger for the SFW with $m = 1$ than that for the SFW with $m = 2$. This is associated with the following feature: -0.5 is the minimum value of the exponent y. For the chosen beam-plasma parameters, $y = -0.429$ for the SFW with $m = 1$, and $y = -0.296$ for the SFW with $m = 2$.

The resonance nature of the instability manifests itself in the following feature of the dependence of the SFXM frequency $\mathrm{Re}(\omega)$ on the beam particle density. The difference between $\mathrm{Re}(\omega)$ and $N|\omega_e|/\gamma$, which is approximately the resonant denominator in the permittivity tensor components $\varepsilon_{ik}^{(b)}$ (7.18)–(7.20), decreases from $4.7 \times 10^8\ \mathrm{s}^{-1}$ for a relatively dense beam with $n_b = 2 \times 10^9\ \mathrm{cm}^{-3}$, to $7.0 \times 10^6\ \mathrm{s}^{-1}$ for a low density beam with $n_b = 4 \times 10^4\ \mathrm{cm}^{-3}$.

Table 7.3 SFXM frequency $\mathrm{Re}(\omega)$ and growth rate $\mathrm{Im}(\omega)$ in the case of a laboratory plasma experimental device for different values of the beam particle density which correspond to different regimes of instability

n_b (cm^{-3})	$\mathrm{Re}(\omega)$ (s^{-1})	$\mathrm{Im}(\omega)$ (s^{-1})	Type of instability
2×10^9	4.7×10^9	7.5×10^8	Beam-plasma
4×10^4	4.3×10^9	1.3×10^7	Beam-dissipative

In the following, the parameters of SFW beam excitation in a typical table-size University laboratory experimental device are estimated. Let the plasma column radius be $R_1 = 5$ cm. Then the radius of metal chamber should be $R_2 = 6$ cm. The maximum growth rate of the SFW with azimuthal mode index $m = 1$ can be reached for the effective wavenumber $k_{ef} = 0.24$, which means that the plasma frequency should be $\Omega_e = 2.5 \times 10^{10}$ s^{-1}. This corresponds to the plasma particle density $n_e = 2 \times 10^{11}$cm^{-3}. Since then $Z = \Omega_e/|\omega_e| = 3.5$, the electron cyclotron frequency is $|\omega_e| = 7 \times 10^9$ s^{-1} which corresponds to the magnetic field $B_0 = 0.04$ kG and a beam velocity $\upsilon_{\perp 0} = 2.4 \times 10^{10}$ cm/s. Let the collision frequency be $\nu = 0.05|\omega_e| = 3.5 \times 10^8$ s^{-1}. Then two cases which correspond to the different parts of the line $y(\alpha)$ in the Fig. 7.34 are considered: beam-dissipative instability ($\alpha << \alpha_t$) and beam-plasma instability ($\alpha >> \alpha_t$). The values of the corresponding beam particle density n_b, $\mathrm{Re}(\omega)$ and $\mathrm{Im}(\omega)$ are given in the Table 7.3.

This Section presents the results of analyzing the influence of dissipation on electron beam-wave interaction for the case of SFXMs in a cylindrical metal chamber partially filled by uniform plasma placed into an external axial static magnetic field [44]. Based on the classical theoretical basis [49, 70] two regimes of beam-wave interaction are distinguished: beam-plasma and beam-dissipative instabilities. The exponent y in the expression for their growth rate (see Eq. 7.43) which determines the dependence of the growth rate on the collision frequency is found to be a more appropriate indicator of whether beam-plasma or beam-dissipative instability takes place than the exponent x which characterizes the dependence of the growth rate on the beam particle density.

The growth rate of a beam-dissipative instability is shown to be much lower than that of a beam-plasma instability. In addition, it is proven that the collision frequency has more influence on the growth rate of a beam-dissipative instability than on that of a beam-plasma instability.

7.7 Initial Stage of Interaction Between Gyrating Relativistic Electron Beams and Long-Wavelength Electromagnetic Surface Waves

Surface flute waves (SFWs) are electromagnetic waves which propagate perpendicular to the axis of a cylindrical metallic waveguide both without any magnetic field and in the presence of an axial external static magnetic field B_0 [57]. Since there is no axial wavenumber there is the possibility to split the Maxwell equations for electromagnetic waves into two independent subsets. One describes SFWs with ordinary polarization (SFOMs). They propagate with the components E_z, H_r, H_φ of the electromagnetic field, and their dispersion properties do not depend on the presence of the axial external static magnetic field B_0. The other subset of the Maxwell equations describes SFWs with extraordinary polarization (SFXMs). They propagate with the components H_z, E_r, E_φ of the electromagnetic field, independently of the SFOMs.

If the plasma particle density is sufficiently large and the B_0 is weak enough so that $\Omega_e^2 > \omega_e^2$, SFXMs propagate in two frequency ranges (1.13) and (1.14) (Ω_e is the plasma (Langmuir) frequency of electrons, and ω_e is the electron cyclotron frequency). One of the ranges, being referred as low frequency (LF) range (1.13), includes the electron cyclotron frequency and is under consideration in this Section. The other range lies higher—above the upper hybrid frequency. It is out of scope of the present study, although SFXMs can be also excited there (see Sect. 10.1 for details) [36].

The assumption that the Langmuir frequency is higher than the electron cyclotron frequency cannot be considered as a strong limitation of applicability of the results obtained in the present Section. For the conditions of a modest University laboratory, a plasma particle density of n_e ~10^{12} cm^{-3} is not exotic. To provide the inequality $\Omega_e^2 > \omega_e^2$ for this n_e one should apply an external static magnetic field $B_0 \lesssim 3.2 \times 10^{-3}\ \sqrt{n_e}\ \sim\ 3$ kG. Even stronger B_0 can be used if semiconductor plasmas are considered [82] for which the density n_e ~10^{19} cm^{-3} of n-type conductivity electrons is achievable.

In the opposite case, $\Omega_e^2 < \omega_e^2$, the possibility of SFXM excitation in the ion cyclotron frequency range by a gyrating beam of light ions in cylindrical metallic waveguides partially filled with plasma is demonstrated below in Sect. 10.4 [83].

For the most beam-plasma parameters, LF SFXMs with azimuthal mode index m use to be excited due to interaction with the beam cyclotron harmonic $N = m$. This is true also for $m > 1$. However, the neighboring harmonics $N = m \pm 1$, $m \pm 2$ can also contribute to the LF SFXM excitation. Two physics mechanisms of resonance overlapping in excitation of SFXMs by rotating relativistic electron beams were demonstrated in Sect. 7.3 [59].

One of the most important drawbacks of all the previous numerous studies of SFXM beam excitation consists in neglecting the possible non-zero axial wavenumber k_z of the wave. E.g. in large-orbit gyrotrons [84, 85], the electromagnetic cavity mode is close to cutoff (very small k_z). Account for $k_z \neq 0$ can be carried out on the base of the perturbation theory. This approach was demonstrated to be successful for studying the dispersion properties of long-wavelength electromagnetic surface waves with arbitrary magnitude of the azimuthal mode index m propagating along the plasma-metal boundary in an axial external static magnetic field in Sect. 2.6 [86] and in a cylindrical waveguide partially filled by isotropic plasma in Ref. [87]. In presence of non-zero k_z the Maxwell equations do not split into independent subsets for SFOMs and SFXMs—these two modes appear to be coupled. However, for long-wavelength waves (with small magnitudes of the k_z) the coupling can be considered as weak, and k_z is the small parameter of the problem.

No doubt, numerical methods are powerful tools for solving the problems of plasma electrodynamics. For instance, the results of numerical study of the dispersion properties, wave field structure and energy characteristics of travelling electromagnetic non-symmetric surface waves with azimuthal mode indices $m = \pm 2, \pm 3$ in an axial external static magnetic field were considered in Ref. [88]. Influences of the effective collision frequency, the axial current, the dimensions of the coaxial waveguide system and the plasma particle density radial profile on dispersion properties and attenuation of symmetric ($m = 0$) electromagnetic waves were numerically studied in Ref. [89]. The solution of the dispersion equation in the case of dipolar waves ($m = \pm 1$) was reported there to be a more complicated problem because of the presence of the set of eigensolutions of the dispersion relation.

Electromagnetic waves propagating with the axial wavenumber k_z, azimuthal mode index m and frequency ω are under consideration in the present Section. Their wave fields can be written as: E_z and $H_z = f(r)\exp[i(k_z z + m\varphi - \omega t)]$. These waves are excited in a metallic cylindrical waveguide with inner radius R_2. The waveguide is partially filled by a cylindrical uniform plasma column of radius $R_1 < R_2$ (see Fig. 7.1). The electrodynamic properties of the plasma placed into a uniform axial external static magnetic field $\vec{B}_0 || \vec{z}$ are described by the permittivity tensor of a cold collisionless plasma ε_{ik} (1.2) [1, 21, 41].

One can derive the set (2.36) and (2.37) of two second order differential equations for the axial components of the electric E_z and magnetic B_z wave fields from the Maxwell equations. The right-hand sides of Eqs. (2.36) and (2.37) have the form (2.38) and (2.39) if small values of the first order in k_z are taken into account. The set of coupled Eqs. (2.36) and (2.37) for the axial components of the electromagnetic wave field E_z and B_z for arbitrary values of the axial wavenumber k_z and for a homogenous plasma in uniform axial static magnetic field was derived in [90].

In a plasma with the uniform particle density n_e, the depth of H-wave (ordinary polarized) $k_O{}^{-1}$ and E-wave (extraordinary polarized) $k_X{}^{-1}$ penetration into the plasma are determined through the components of the permittivity tensor ε_{ik} by Eq. (2.40).

The condition, that the real part of $k_X{}^2$ should be positive, determines the frequency ranges where the waves are of surface nature in the plasma, which means that the wave amplitudes decrease with going away from the plasma boundary into the plasma depth. The LF range is given by:

$$|\omega_e|\sqrt{(\Omega_i^2 + c^2k_z^2)/(\Omega_e^2 + \omega_e^2)} < \omega < |\omega_e|, \quad |\omega_e| < \omega < \omega_1, \tag{7.44}$$

where ω_1 is the cut-off frequency for bulk modes,

$$\omega_1 \approx -\frac{|\omega_e|}{2} + \sqrt{\frac{\omega_e^2}{4} + \frac{\Omega_e^2}{\varepsilon_r} + \frac{c^2\Omega_e^2 k_z^2}{\left(\sqrt{\Omega_e^2\varepsilon_r + 0.25\varepsilon_r^2\omega_e^2} - 0.5\varepsilon_r|\omega_e|\right)^2}}. \quad (7.45)$$

The range (7.44) coincides with that obtained Sect. 1.4.1 for SFXMs in the limiting case $k_z \to 0$ [57]. Note that $\mathrm{Re}(k_O{}^2) > 0$ in the range (7.44) as well.

The spatial distribution of the wave fields within the plasma column, $0 < r < R_1$, can be found by variation of constants,

$$B_z = \left(A_2 - A_1 \int\limits_{R_1}^{r} \frac{\tilde{\psi} k_X^2 \hat{K}\varphi dr}{W\left(\psi, \tilde{\psi}\right)}\right)\psi + A_1\tilde{\psi}\int\limits_{0}^{r} \frac{\psi k_X^2 \hat{K}\varphi dr}{W\left(\psi, \tilde{\psi}\right)}, \quad (7.46)$$

$$E_z = \left(A_1 - A_2 \int\limits_{R_1}^{r} \frac{\tilde{\varphi}\hat{M}\psi dr}{W(\varphi, \tilde{\varphi})}\right)\varphi + A_2\tilde{\varphi}\int\limits_{0}^{r} \frac{\varphi\hat{M}\psi dr}{W(\varphi, \tilde{\varphi})}. \quad (7.47)$$

In (7.46) and (7.47), the functions $\psi(r)$ and $\varphi(r)$ are solutions of the Eqs. (2.36) and (2.37) with zero right hand sides, respectively, which are finite at the cylinder axis, $r = 0$. The functions $\tilde{\psi}(r)$ and $\tilde{\varphi}(r)$ are the solutions of the same equations which are linearly independent of the functions ψ and φ. The solutions (7.46) and (7.47) include two constants of integration A_1 and A_2, since the other two constants of integration are already chosen to satisfy the boundary conditions that the wave fields E_z and B_z should be finite at the waveguide axis, where the functions $\tilde{\varphi}$ and $\tilde{\psi}$ are singular. The Wronskian $W(\psi,\tilde{\psi})$ of the functions $\psi(r)$ and $\tilde{\psi}(r)$ is equal to

$$W\left(\psi, \tilde{\psi}\right) = \psi\frac{\partial\tilde{\psi}}{\partial r} - \tilde{\psi}\frac{\partial\psi}{\partial r} \propto \frac{k_X^2}{r}. \quad (7.48)$$

The Wronskian of the functions $\varphi(r)$ and $\tilde{\varphi}(r)$ is inversely proportional to the radial coordinate r,

$$W(\varphi, \tilde{\varphi}) = \varphi\frac{\partial\tilde{\varphi}}{\partial r} - \tilde{\varphi}\frac{\partial\varphi}{\partial r} \propto \frac{1}{r}. \quad (7.49)$$

In the considered case of uniform plasma particle density and external static magnetic field, $\psi(r) = I_m(k_X r)$ is a modified Bessel function, $\tilde{\psi}(r) = K_m(k_x r)$ is a MacDonald function, $\varphi(r) = I_m(k_O r)$, and $\tilde{\varphi}(r) = K_m(k_O r)$ [30]. Their Wronskians (7.48) and (7.49) are equal to each other:

$$W(I_m(k_X r), K_m(k_X r)) = W(I_m(k_O r), K_m(k_O r)) = -1/r. \quad (7.50)$$

To derive the dispersion relation one needs also the expression for the tangential (azimuthal) electric wave field E_φ:

$$E_\varphi(r) = \frac{ic}{\omega N_x^2}\frac{dB_z}{dr} + \frac{i\mu}{N_x^2}N_\varphi B_z + \frac{N_z N_\varphi}{N_x^2}E_z + \frac{c\mu N_z}{\omega N_x^2}\frac{dE_z}{dr}. \quad (7.51)$$

Hereinafter, $N_\varphi = cm/(\omega r)$.

The electrodynamic properties of the space $R_1 < r < R_2$, where the electron flow gyrates, is described by the permittivity tensor $\varepsilon_{ik}{}^{(b)}$ given in Sect. 7.2 by Eqs. (7.18)–(7.20) [21, 34]. The electron flow particle density is assumed to be small, $\alpha \equiv n_b/n_e << 1$. This results in violation of the inequality $\Omega_e{}^2 > \omega_e{}^2$ assumed to be valid for the plasma region. In the beam region ($R_1 < r < R_2$), the plasma frequency is assumed to be sufficiently small, so that $\Omega_b{}^2 < \omega_e{}^2$, and the waves under study are of bulk nature. The electromagnetic wave fields are found from the Maxwell equations (2.36) and (2.37) with the following replacements:

$$\begin{gathered} k_O^2 \to k_O^{(b)2} = (\omega/c)^2\varepsilon_{33}^{(b)} - k_z^2,\, k_X^2 \to k_X^{(b)2} = (\omega/c)^2 N_X^{(b)2}, \\ N_X^2 \to N_X^{(b)2} = \varepsilon_{22}^{(b)} - N_z^2 + \left(\varepsilon_{12}^{(b)}\right)^2\left(\varepsilon_{11}^{(b)} - N_z^2\right)^{-1}, \\ \mu \to \mu^{(b)} = -i\varepsilon_{12}^{(b)}/\left(\varepsilon_{11}^{(b)} - N_z^2\right). \end{gathered} \quad (7.52)$$

Smallness of the electron beam particle density makes it possible to neglect the right-hand sides of Eqs. (2.36) and (2.37), and write down the solutions to the Maxwell equations in the beam region in the following form:

$$B_z^{(b)} = A_3\left[N_m'\left(k_X^{(b)}b\right)J_m\left(k_X^{(b)}r\right) - J_m'\left(k_X^{(b)}b\right)N_m\left(k_X^{(b)}r\right)\right], \quad (7.53)$$

$$E_z^{(b)} = A_4\left[N_m\left(k_O^{(b)}b\right)J_m\left(k_O^{(b)}r\right) - J_m\left(k_O^{(b)}b\right)N_m\left(k_O^{(b)}r\right)\right]. \quad (7.54)$$

The expressions (7.53) and (7.54) for both axial components of the electromagnetic field $B_z{}^{(b)}$ and $E_z{}^{(b)}$ already satisfy the boundary conditions at the metallic wall ($E_\varphi{}^{(b)}(R_2) = 0$, and $E_z{}^{(b)}(R_2) = 0$):

$$E_\varphi^{(b)}(r) = \frac{ic}{\omega N_X^{(b)2}}\frac{dB_z^{(b)}}{dr} + \frac{N_z N_\varphi}{N_X^{(b)2}}E_z, \quad (7.55)$$

and that is why they contain only one constant of integration each (A_3 and A_4).

The four constants of integration (A_1 and A_2 in (7.46) and (7.47), A_3 in (7.53) and A_4 in (7.54)) can be determined as the result of application of the following four boundary conditions. These are: continuity of the tangential wave fields E_φ, E_z, B_φ, and B_z at the plasma column interface $r = R_1$. To satisfy these boundary conditions the determinant constructed from the coefficients nearby the constants A_i should be equal to zero, which gives the dispersion relation. After calculating the determinant and neglecting the small terms of higher than the second order, the dispersion relation reads as

$$D_o D_x = D_1. \tag{7.56}$$

The general form of (7.56) is typical for dispersion relations of two modes with weak coupling [91]. In (7.56), $D_o = 0$ is dispersion relation of ordinary SFMs:

$$D_o = a_{21}a_{33} - a_{31}a_{23}, \tag{7.57}$$

$$a_{21} = \frac{ck_O}{\omega} I'_m(k_O R_1), \tag{7.58}$$

$$a_{23} = \frac{\omega}{ck_O^{(b)}} \left[J_m\left(k_O^{(b)} R_2\right) N'_m\left(k_O^{(b)} R_1\right) - N_m\left(k_O^{(b)} R_2\right) J'_m\left(k_O^{(b)} R_1\right) \right], \tag{7.59}$$

$$a_{31} = I_m(k_O R_1), \tag{7.60}$$

$$a_{33} = [J_m(k_O R_2) N_m(k_O R_1) - N_m(k_O R_2) J_m(k_O R_1)], \tag{7.61}$$

and $D_x = 0$ is dispersion relation of extraordinary SFMs:

$$D_x = a_{12}a_{44} - a_{14}a_{42}, \tag{7.62}$$

$$a_{12} = \frac{1}{N_X} I'_m(k_X R_1) + \frac{\mu N_\varphi}{N_X^2} I_m(k_X R_1), \tag{7.63}$$

$$a_{14} = \frac{\omega}{ck_X^{(b)}} \left[N'_m\left(k_X^{(b)} R_2\right) J'_m\left(k_X^{(b)} R_1\right) - N'_m\left(k_X^{(b)} R_1\right) J'_m\left(k_X^{(b)} R_2\right) \right], \tag{7.64}$$

$$a_{42} = I_m(k_X R_1), \tag{7.65}$$

$$a_{44} = \left[N'_m\left(k_X^{(b)} R_2\right) J_m(k_X R_1) - J'_m\left(k_X^{(b)} R_2\right) N_m(k_X R_1) \right]. \tag{7.66}$$

In (7.56), D_1 is their coupling parameter which has a value of the second order in the small parameter of the problem k_z:

$$\begin{aligned} D_1 = {} & a_{24}a_{42}(a_{13}a_{31} - a_{11}a_{33}) + (a_{14}a_{41} - a_{11}a_{44})(a_{23}a_{32} - a_{33}a_{22}) + \\ & + a_{13}a_{44}(a_{21}a_{32} - a_{31}a_{22}) + a_{41}a_{24}a_{12}a_{33}. \end{aligned} \tag{7.67}$$

Those elements in Eq. (7.67), which are given below, are linear in k_z:

$$\begin{aligned} a_{11} = {} & \tfrac{\mu k_z k_O}{k_X^2} I'_m(k_O R_1) + \tfrac{m k_z}{a k_X^2} I_m(k_O R_1) \\ & - i \int_0^{R_1} r I_m(k_X r) k_X^2 \hat{K} I_m(k_O r) dr \left[\tfrac{1}{N_X} K'_m(k_X R_1) + K_m(k_X R_1) \tfrac{\mu N_\varphi}{N_X^2} \right], \end{aligned} \tag{7.68}$$

$$a_{13} = \frac{mk_z}{ak_O^{(b)2}}\left[J_m\left(k_O^{(b)}R_1\right)N_m\left(k_O^{(b)}R_2\right) - J_m\left(k_O^{(b)}R_2\right)N_m\left(k_O^{(b)}R_1\right)\right] \quad (7.69)$$

$$a_{22} = -\frac{\mu k_z}{k_X}I'_m(k_x R_1) - \frac{mk_z}{ak_X^2}I_m(k_x R_1) + i\frac{ck_O}{\omega}K'_m(k_O R_1)\int_0^{R_1} rI_m(k_O r)\hat{M}I_m(k_X r)dr, \quad (7.70)$$

$$a_{24} = \frac{mk_z}{ak_X^{(b)2}}\left[N_m\left(k_X^{(b)}R_1\right)J'_m\left(k_X^{(b)}R_2\right) \quad N'_m\left(k_X^{(b)}R_2\right)J_m\left(k_X^{(b)}R_1\right)\right], \quad (7.71)$$

$$a_{32} = -iK_m(k_O R_1)\int_0^{R_1} rI_m(k_O r)\hat{M}I_m(k_X r)dr, \quad (7.72)$$

$$a_{41} = iK_m(k_O r)\int_0^{R_1} rI_m(k_X r)k_X^2\hat{K}I_m(k_O r)dr. \quad (7.73)$$

The dispersion relation (7.56) is analyzed here under the resonance condition (7.14). To carry out this numerical analysis the following dimensionless values are used in Figs. 7.37, 7.38, 7.39 and 7.40. The utilized abscissa axis is as it was in [34, 36, 57]: $k_{ef} = mc/(R_1\Omega_e)$. This is explained as follows. The characteristic azimuthal wavenumber in the waveguide with a plasma column of the radius R_1 is determined by $k_\varphi = m/R_1$, and the effective azimuthal wavenumber k_{ef} is m/R_1 normalized by the skin-depth $\delta = c/\Omega_e$. Increasing k_{ef} means decreasing plasma radius R_1 and/or decreasing plasma particle density n_e. The ordinate axis is chosen as imaginary part of the frequency normalized by the electron cyclotron frequency, $\mathrm{Im}(\omega/|\omega_e|)$, except

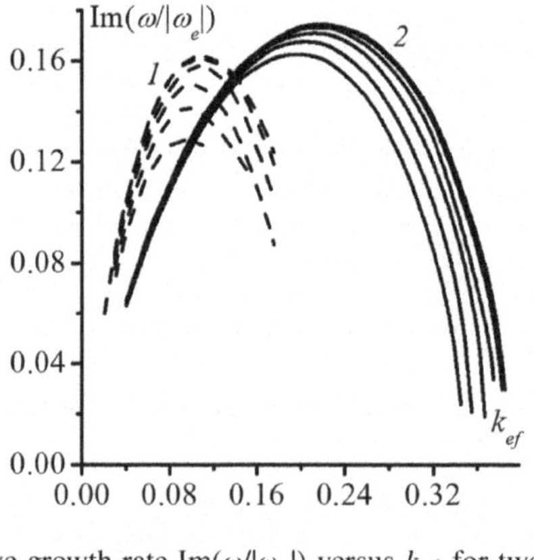

Fig. 7.37 Normalized wave growth rate $\mathrm{Im}(\omega/|\omega_e|)$ versus k_{ef} for two values of azimuthal wave indices: $m = 1$ (*dashed curves*), and $m = 2$ (*solid curves*). $Z = 7.5$, $\alpha = 0.01$, $R_2/R_1 = 1.2$. *The numbers* mark the azimuthal wave index. Each set of curves corresponds to six magnitudes of normalized axial wavenumber $k_zR_1 = 0.0$ (largest growth rate), 0.2, 0.4, 0.6, 0.8, 1

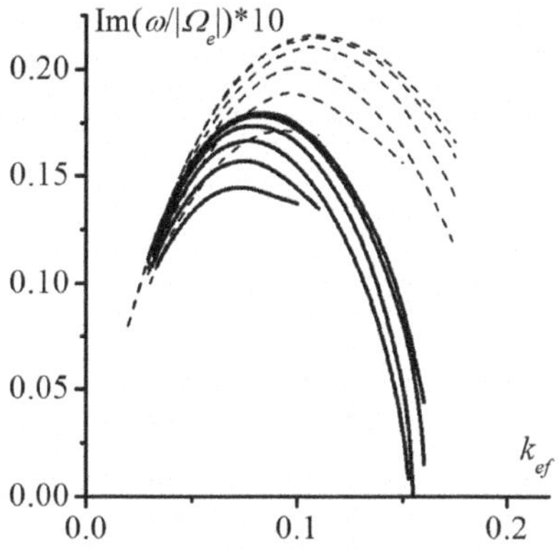

Fig. 7.38 Normalized wave growth rate $\mathrm{Im}(\omega/|\omega_e|)$ versus k_{ef} for two values of external static axial magnetic field: weaker ($Z = 10$) shown by *solid curves* and stronger ($Z = 7.5$) shown by *dashed curves*. $\alpha = 0.01$, $R_2/R_1 = 1.2$, and $m = 1$. Each set of curves corresponds to six values of normalized axial wavenumber $k_z R_1 = 0.0$ (largest growth rate), 0.2, 0.4, 0.6, 0.8, 1

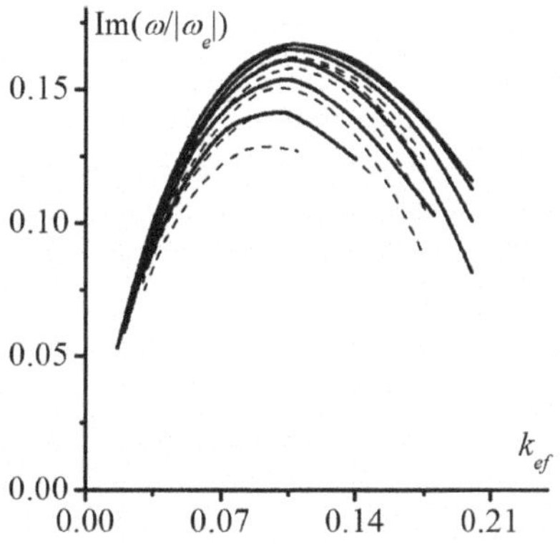

Fig. 7.39 Normalized wave growth rate $\mathrm{Im}(\omega/|\omega_e|)$ versus k_{ef} for two different values of the vacuum gap width: $R_2/R_1 = 1.1$ (*solid curves*) and $R_2/R_1 = 1.2$ (*dashed curves*). $m = 1$, $Z = 7.5$, $\alpha = 0.01$. Each set of curves corresponds to six magnitudes of normalized axial wavenumber $k_z R_1 = 0.0$ (largest growth rate), 0.2, 0.4, 0.6, 0.8, 1

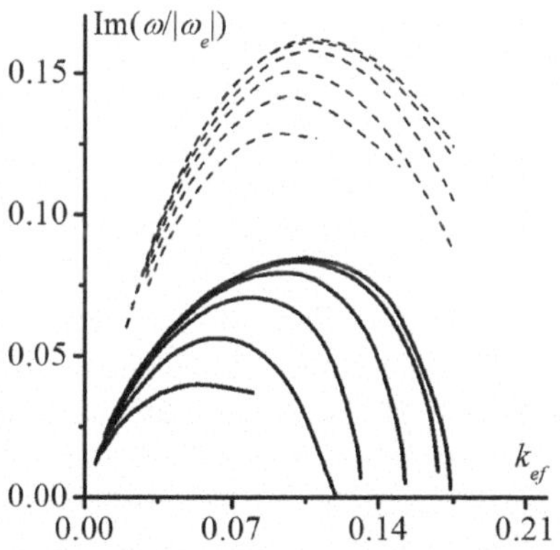

Fig. 7.40 Normalized wave growth rate $\mathrm{Im}(\omega/|\omega_e|)$ versus k_{ef} for two values of the beam particle density: $\alpha = 0.001$ (*solid curves*) and $\alpha = 0.01$ (*dashed curves*). $Z = 7.5$, $R_2/R_1 = 1.2$, and $m = 1$. Each set of curves corresponds to six magnitudes of normalized axial wavenumber $k_z R_1 = 0.0$ (largest growth rate), 0.2, 0.4, 0.6, 0.8, 1

of Fig. 7.38 in which the influence of the external static axial magnetic field B_0 is examined. Normalization by $|\omega_e|$ is improper for studying the effect of B_0 on excitation of long wavelength electromagnetic waves since $|\omega_e|$ is proportional to

B_0. That is why in Fig. 7.38 the imaginary part of the frequency normalized by the plasma frequency, $\mathrm{Im}(\omega/\Omega_e)$, is chosen as ordinate.

All Figs. 7.37, 7.38, 7.39 and 7.40 exhibit two sets of curves. One presents the wave growth rate $\mathrm{Im}(\omega)$ for $Z \equiv \Omega_e/|\omega_e| = 7.5$, $\alpha = 0.01$, $R_2/R_1 = 1.2$, and $m = 1$ (shown by dashed curves) which is chosen to serve for comparison. The other set contains six curves and depicts the reduction of the wave growth rate with increasing axial wavenumber from $k_z R_1 = 0$ to $k_z R_1 = 0.2, 0.4, 0.6, 0.8, 1$. Within this range of the dimensionless variable $k_z R_1$ the condition of applicability of the method of successive approximations is valid. For the case of long wavelength surface wave propagation in cylindrical metallic waveguide entirely filled by plasma, this condition was derived in [87] analytically as

$$(k_z R_1)^2 \ll 8m^2\left(1 + k_{ef}^2\right). \tag{7.74}$$

During carrying out the calculations in the present Section, specific attention is given to follow the change of the real part ($\mathrm{Re}(\omega/|\omega_e|)$) of the wave frequency. In the following, the maximum deviation of $\mathrm{Re}(\omega/|\omega_e|)$ calculated for $k_z R_1 = 1.0$ from that calculated for $k_z R_1 = 0$ is observed for the set of beam-plasma parameters which correspond to the conditions of the Fig. 7.40. In this case, $\mathrm{Re}(\omega/|\omega_e|)$ increases by 27% from $\mathrm{Re}(\omega/|\omega_e|) = 0.623$ to 0.793 with increasing $k_z R_1$ from $k_z R_1 = 0$ to $k_z R_1 = 1.0$. Increase of k_z from $k_z R_1 = 0$ to $k_z R_1 = 0.2$ causes a negligible decrease of $\mathrm{Im}(\omega/|\omega_e|)$, and the curves which correspond to $k_z R_1 = 0$ and $k_z R_1 = 0.2$ are almost indistinguishable in Figs. 7.37 and 7.39.

In Fig. 7.37, the wave growth rates are calculated for the following beam-plasma parameters: $Z = \Omega_e/|\omega_e| = 7.5$, $\alpha = 0.01$, and $R_2/R_1 = 1.2$. The two sets of curves correspond to two different values of the azimuthal mode index: $m = 1$ (dashed curves) and $m = 2$ (solid curves). The m-values are displayed nearby the sets. Increase of the axial wavenumber from $k_z = 0$ to $k_z R_1 = 1.0$ for the wave with $m = 1$ is accompanied by a 20% decrease of the maximum growth rate from $\mathrm{Im}(\omega/|\omega_e|) = 0.162$ which is observed for $k_{ef} = 0.105$ to $\mathrm{Im}(\omega/|\omega_e|) = 0.129$ which is achieved for smaller $k_{ef} = 0.09$. In other words, the magnitude of k_{ef} for which the growth rate reaches its maximum decreases by 14%. This tendency: decreasing of the k_{ef} value, for which the growth rate reaches the maximum magnitude, with increasing product $k_z R_1$, is a common feature of the curves in all Figs. 7.37, 7.38, 7.39 and 7.40.

To demonstrate how the increase of the azimuthal mode index m influences the dependence of the growth rate on the axial wavenumber, the growth rates are calculated also in Fig. 7.37 for the same beam-waveguide parameters as for the wave with $m = 1$ except of m which is increased to $m = 2$. In this case, the maximum growth rate decreases by 7% from $\mathrm{Im}(\omega/|\omega_e|) = 0.175$ which is observed for $k_{ef} = 0.22$ when $k_z = 0$ to $\mathrm{Im}(\omega/|\omega_e|) = 0.163$ which is achieved for $k_{ef} = 0.2$ when $k_z R_1 = 1.0$. The position of the growth rate maximum shifts only by 9%. This result (weaker effect of axial wavenumber on the beam wave excitation for larger azimuthal mode index) is predictable from the point of view of general physics since the wavenumber ratio k_z/k_φ is smaller for larger m.

To show the effect of the external static axial magnetic field B_0 on the beam wave excitation the growth rates are calculated in Fig. 7.38 for weaker B_0. All the beam-plasma waveguide parameters are the same as in Fig. 7.37 except of Z, which is increased to $Z = 10.0$. This is equivalent to the decrease of B_0 by 25%. In this case, the maximum growth rate decreases by 19% from $\mathrm{Im}(\omega/|\omega_e|) = 0.18$, which is observed for $k_{ef} = 0.08$ when $k_z = 0$, to $\mathrm{Im}(\omega/|\omega_e|) = 0.145$ which is achieved for $k_{ef} = 0.07$ when $k_z R_1 = 1.0$. So, the position of the growth rate maximum shifts only by 12%. The decrease of B_0 is accompanied by a weaker influence of the axial wavenumber on the wave growth rate.

The influence of the relative gap width $(R_2 - R_1)/R_1$ on the dependence of the growth rate on axial wavenumber is shown in Fig. 7.39. To calculate the growth rates in Fig. 7.39 the same beam-plasma parameters are used as in Fig. 7.37 except of the gap width which is smaller, $(R_2 - R_1)/R_1 = 0.1$ (solid curves), in this case. The maximum wave growth rate decreases by 15% from $\mathrm{Im}(\omega/|\omega_e|) = 0.167$ in Fig. 7.39, which is observed for $k_{ef} = 0.105$ when $k_z = 0$, to $\mathrm{Im}(\omega/|\omega_e|) = 0.142$ which is achieved for $k_{ef} = 0.1$ when $k_z R_1 = 1.0$. Therefore, the position of the growth rate maximum shifts only by 5%. The decrease of the gap width is also accompanied by a weaker influence of the axial wavenumber on the wave growth rate.

The wave beam-plasma instability growth rate is well-known to decrease with decreasing beam particle density. To demonstrate the effect of the parameter α on the dependence of the $\mathrm{Im}(\omega/|\omega_e|)$ versus the axial wavenumber the growth rates are calculated in Fig. 7.40 for the same beam-plasma waveguide parameters as in Fig. 7.37 except of α which is decreased by ten times to $\alpha = 0.001$ (solid curves). It should be noted that classical theory of the beam-plasma instability [49] predicts the decrease of the wave maximum growth rate caused by a decrease of the beam particle density by ten times by the factor ≈ 0.464 whereas the numerical analysis gives 0.525 which can be considered as a good qualitative agreement. The growth rate reaches its maximum for $k_z = 0$ in Fig. 7.40 for the same magnitude of $k_{ef} = 0.105$ as in Fig. 7.37. The maximum growth rate of the wave decreases by 53% from $\mathrm{Im}(\omega/|\omega_e|) = 0.085$ in Fig. 7.40 which is observed when $k_z = 0$ and to $\mathrm{Im}(\omega/|\omega_e|) = 0.04$ which is achieved for $k_{ef} = 0.05$ when $k_z R_1 = 1.0$. This means that the position of the growth rate maximum shifts by 52%. The decrease of the beam particle density enhances the influence of the axial wavenumber on the wave growth rate.

As a practical example to evaluate the possibility of experimental realization of long wavelength surface wave excitation, the following parameters of the beam-plasma system are chosen [92]. The plasma frequency is $\Omega_e \approx 1.261 \times 10^{11}\ \mathrm{s}^{-1}$ for the plasma particle density $n_e \approx 5 \times 10^{12}\ \mathrm{cm}^{-3}$. The effective wavenumber is $k_{ef} \approx 0.059$ for the first azimuthal mode index $m = 1$ and the plasma column radius $R_1 \approx 4$ cm. The electron cyclotron frequency is $|\omega_e| \approx 1.616 \times 10^{10}\ \mathrm{s}^{-1}$ for the external static axial magnetic field $B_0 \approx 920$ G. The wave radial phase shift $\Delta\varphi$ within the gap $R_1 < r < R_2 \approx 4.8$ cm, where the electron beam gyrates, appears to be small, $\Delta\varphi \approx 0.167$. The beam electrons move with the velocity $\upsilon \approx 0.923c$, which corresponds to the relativistic Lorentz factor $\gamma \approx 2.591$. With this parameters, long wavelength surface waves with $k_z = 0.09$ cm^{-1} are excited with the frequency $f \approx 1$ GHz and the growth rate $\mathrm{Im}(\omega) \approx 7.8 \times 10^7$ s^{-1}.

In the present Section, the initial stage of surface type electromagnetic wave excitation by a relativistic electron flow which gyrates in a narrow gap separating a dense plasma column from a metallic wall of a cylindrical waveguide is studied [93]. The consideration is limited to the case of long-wavelength waves with small magnitudes of axial wavenumber k_z, however the azimuthal mode index m can have an arbitrary value. The assumption of small axial wavenumber makes it possible to apply the theory of successive approximations with account for small values of the first order in k_z in the expressions for the electromagnetic wave fields. This results in deriving the dispersion relation (7.56) which has the form typical for coupled waves. In the present case, these coupled waves are ordinary and extraordinary polarized electromagnetic waves, which are known to propagate independently in the limit $k_z \to 0$, in which they are surface flute waves [57]. They are weakly coupled by the small axial wavenumber k_z.

Applicability of the theory of successive approximations for studying the long-wavelength wave beam excitation has been demonstrated [93]. The dispersion relation (7.56), which describes the initial (linear in respect to the wave amplitude) stage of the wave excitation, is derived. The excitation is caused by efficient beam-wave interaction. The model of the electron beam as a flow of oscillators is applied [21, 34]. The wave growth rate, which is the positive imaginary part of the complex frequency, arises due to existence of resonant denominators in the expressions (7.18)–(7.20) for the permittivity tensor. These resonances correspond to harmonics of the electron cyclotron frequency.

The dispersion relation is analyzed numerically. The influence of the beam-plasma waveguide parameters on the wave growth rate is studied. In general, increasing axial wavenumber results in decreasing maximum growth rate and increasing plasma particle density and plasma column radius for which the waves are excited most efficiently. The influence of the axial wavenumber on the wave growth rate increases with decreasing beam particle density and azimuthal mode index, and decreases with decreasing the external static axial magnetic field and gap width where the electron flow gyrates [93]. The results obtained in this Section are of interest for plasma electronics, in particular, for solid state electronics [82].

7.8 Conclusions

SFXMs are demonstrated theoretically to be efficiently excited by annular relativistic electron beams gyrating in an external static axial magnetic fields along large Larmor orbits around the plasma column in cylindrical metal waveguides partially filled by plasma. The wave-beam interaction is shown to be the most efficient under conditions when the wave eigenfrequency is close to a harmonic of the electron cyclotron frequency. The investigation is carried out via linear approximation in respect of the wave amplitude on the base of the oscillator flow model [21]. The expressions

(7.18)–(7.20) for the components of the dielectric permittivity tensor of the space where the beam rotates are carefully checked. Misprints are found in the formulas in [21] and corrected. The argument of the cylindrical functions in the expressions (7.18)–(7.20) has been simplified. Growth rates of SFXMs are demonstrated to reach their maximum for the values of the effective wavenumber in the order of unit. The excitation takes place for SFXMs with positive azimuthal mode indices only since this is the direction of the electron gyration in the external static axial magnetic field.

SFXM beam excitation is specifically studied in the two regimes: beam-plasma instability and beam-dissipative instability. Transition between both regimes is carefully investigated. It is concluded that just the exponent, which characterizes the dependence of the growth rate on the collision frequency, is the most appropriate parameter to distinguish whether the conditions of LF SFXM excitation correspond to the development of a beam-plasma or a beam-dissipative instability.

Most attention is paid to the case of a narrow gap between the plasma column and the metal wall where the electron beam gyrates. Just in this case, the plasma parameters have the strongest influence on the SFXM dispersion properties and growth rates. In this case, the zeroth ($l = 0$) radial modes of SFXMs propagate. In waveguides with a wide gap, higher radial modes of SFXMs can propagate. Their eigenfrequencies are higher than those of $l = 0$ modes, and in addition, experimentalist can get the advantage of dealing with higher frequency without any change in the waveguide design. However, their growth rates are significantly smaller than those of the zeroth modes. This issue restricts application of higher radial modes of SFXMs in plasma electronics.

Two mechanisms of resonance overlapping in the excitation of SFXMs by gyrating relativistic electron beams are discussed. One of them is well-known also for devices of plasma electronics with longitudinal electron beams. It is due to the enhancement of the beam particle density. The other is native to gyrating particle beams. It is caused by widening the gap where the beam gyrates, so that this layer is sufficiently wide for formation of two or more groups of electrons, which are in resonance with the wave.

The influences of some waveguide inhomogeneities on SFXM excitation have been studied. One of these non-uniformities is a non-circular cross-section of the plasma filling, which does not essentially affect the LF SFXW excitation by the annular electron beam. This can be explained by the fact that the effective wavenumber range (the range of the beam-plasma system parameters) wherein the excitation of LF SFXWs is efficient, differs from the range where the non-circularity of the plasma-vacuum interface essentially influences the dispersion properties of these waves. This fact can be useful for technological purposes: small deviations from the designed geometry of gas discharge tubes do not decrease the efficiency of the LF SFXW excitation by the electron beam. The problem of wave interaction with the annular electron beam differs technologically from that with a linear electron beam. Designing electronic devices based on linear beams requires usually extensive calculations to provide an exact phase relation between the beam and the wave. That is why a corrugated slow-wave interaction circuit often has a changing corrugation period (tapering) along the direction of electron beam propagation. Such method is not necessary to enhance the beam-wave interaction for annular electron beams since the beam rotating along the Larmor orbit comes back again to the same part of the device.

The other non-uniformity is observed in the case of SFXM propagation in toroidal devices. The toroidal non-uniformity of the external static magnetic field affects most essentially the dispersion properties of the LF SFXMs when their eigenfrequency is close to the absolute value of the electron cyclotron frequency. This spatial non-uniformity influences both the maximal magnitude of the SFXM beam instability growth rate and the range of the plasma waveguide parameters, for which SFXMs can be efficiently excited. However, the toroidal non-uniformity of the static magnetic field causes electron drift, which can restrict the efficiency of the power transfer from the electron beam to the excited waves.

The method of successive approximations has been applied to study beam excitation of surface electromagnetic waves with small axial wavenumber. Increasing axial wavenumber is shown to result in a decreasing maximum growth rate, an increased plasma particle density and a larger plasma column radius for which the waves are excited most efficiently. Despite the applicability of the obtained results is restricted to the range of small axial wavenumbers, they are true for arbitrary positive azimuthal mode index.

References

1. Kondratenko, A. N., & Kuklin, V. M. (1988). *Basis of plasma electronics*. Energoatomizdat. (in Russian).
2. Rukhadze, A. A., Bogdankevich, L. S., Rosinskii, S. Y., & Ruhlin, S. Y. (1980). *Physics of high current relativistic electron beams*. Atomizdat. (in Russian).
3. Kuzelev, M. V., & Rukhadze, A. A. (1990). *Electrodynamics of dense electron beams in plasmas*. Nauka. (in Russian).
4. Miller, R. B. (1982). *An introduction to the physics of intense charged particle beams*. PlenumPress.
5. Lawson, J. D. (1977). *The physics of charged—particle beams*. Clarendon Press.
6. Kalmykova, S. S., & Kurilko, V. I. (1988). Physical mechanisms for the hydrodynamic beam-plasma instability. *Soviet Physics Uspekhi, 31*(8), 750–762.
7. Davidson, R. C., & Tsang, KT. (1988). Analysis of magnetron instability for relativistic non-neutral electron flow in cylindrical high-voltage diode. *Laser and Particle Beams, 6*(4), 661–685.
8. Gisler, G. R. (1987). Particle in cell simulations of azimuthal instabilities of relativistic electron layers. *Physics of Fluids, 30*(7), 2199–2208.
9. Kainer, S., Gaffey, J. D., Price, C. P., et al. (1988). Nonlinear wave interactions and evolutions of a ring-beam distribution of energetic electrons. *Physics of Fluids, 31*(8), 2283–2284.
10. Lau, Y.Y. (1988). Radiation generated by rotating electron beams. In *Proceeding of Symposium Non-Neutral Plasma Physics* (pp. 210–223). USA.
11. Yeremka, V. D., Zhurakhovsky, V. A., & Shestopalov, V. P. (1989). Large-orbit peniotron profiled from the law of through gyroresonance. *Radiotehnika I Elektronika, 34*(9), 1900–1907. (in Russian).
12. Saito, H., & Wurtele, J. S. (1987). The linear theory of the circular free-electron laser. *Physics of Fluids, 30*(7), 2209–2220.

13. Tsang, K. T., & Hafisi, B. (1987). Magnetic drift induced ion-cyclotron mass instability. *Physics of Fluids, 30*(3), 804–806.
14. Danly, B. G., & Temkin, R. J. (1986). Generalized nonlinear harmonic gyrotron theory. *Physics of Fluids, 29*(2), 561–567.
15. Kho, T. H., Lin, A. T., & Chen, L. (1988). Gyrophase—coherent electron cyclotron maser. *Physics of Fluids, 31*(10), 3120–3126.
16. Chen, K. W., Kim, S. H., & McKinley, M. C. (1987). Low-frequency azimuthally propagating (diocotron) waves in a non-neutral electron beam column. *Physics of Fluids, 33*(10), 3306–3308.
17. Brandenburg, J. E., & Lee, E. P. (1988). A model of horse instabilities in rotating electron beams. *Physics of Fluids, 31*(11), 3403–3411.
18. Davies, J. A. (1989). Conditions for absolute instability in the cyclotron resonance maser. *Physics of Fluids B, 1*(3), 663–669.
19. Lawson, W., & Striffler, C. D. (1985). A general linear growth rate formula for large orbit, annular electron beams. *Physics of Fluids, 28*(9), 2868–2977.
20. Kitsenko, A. B., Pankratov, I. M., & Stepanov, K. N. (1974). Nonlinear stage of excitation of monochromatic surface waves by a flow of charged particles in a magnetic field. *Zhurnal Tehnicheskoi Fiziki, 44*, 2430–2432. (in Russian).
21. Alexandrov, A. F., Bogdankevich, L. S., & Rukhadze, A. A. (1984). *Principles of plasma electrodynamics*. Springer.
22. Panchenko, I. P., & Kuklin, V. M. (1987). Dissipative modes of beam instability under the conditions of abnormal Doppler effect. *Ukrainskii Fizicheskii Zhurnal, 32*(12), 1827–1832. (in Russian).
23. Carr, W., Jones, R., Boyd, D., et al. (1974). Saturation of beam-plasma instability in a lossy plasma. *Physics of Fluids, 17*(8), 1638–1643.
24. Dumbrajs, O., & Nusinovich, G. S. (2004). Coaxial gyrotrons: past, present, and future. IEEE *Transactions on Plasma Science, 32*(3), 934–946.
25. Nusinovich, G. S., Read, M. E., Dumbrajs, O., et al. (1994). Theory of gyrotrons with coaxial resonators. *IEEE Transactions on Electron Devices, 41*(3), 433–438.
26. Correa, R. A., & Barosso, J. J. (1993). Space charge effects of gyrotron electron beams in coaxial cavities. *International Journal of Electronics, 74*(1), 131–136.
27. Iatrou, C. T., Kern, S., & Pavelyev, A. B. (1996). Coaxial cavities with corrugated inner conductor for gyrotrons. *IEEE Transactions on Microwave Theory and Techniques, 44*(1), 56–64.
28. Rzesnicki, T., Piosczyk, B., Kern, S., et al. (2010). 2.2-MW record power of the 170-GHz European preprototype coaxial-cavity gyrotron for ITER. *IEEE Transactions on Plasma Science, 38*(6), 1141–1149.
29. Aers, G. C., & Boardman, A. D. (1978). The theory of semiconductor magnetoplasmon-polariton surface modes: Voigt geometry. *Journal of Physics C: Solid State Physics, 11*(5), 945–959.
30. Abramowitz, M., & Stegun, I. A., (Eds.) (1964). *Handbook of mathematical functions*. New-York: National Bureau of Standards, Applied Mathematics, Series 55.
31. Girka, V. A., Girka, I. A., Olefir, V. P., et al. (1991). Generation of electromagnetic waves with an annular relativistic electron beam. *Soviet Technical Physics Letters, 17*(1), 35–36.
32. Beletski, N. N., Svetlichnyj, V. M., Khalamejda, D. D., & Iakovenko, V. M. (1991). *Electromagnetic phenomena in HF range in inhomogeneous semiconductor structures*. Kyiv: Naukova Dumka. (in Russian).
33. Levitskii, S. M., & Koshevaya, S. V. (1986). *Vacuum and solid state electronics of super high frequencies*. Kyiv: Vyshcha shkola. (in Russian).
34. Girka, V. O., Girka, I. O., Morgal, Ya. I., et al. (2011). Excitation of azimuthal surface modes by annular electron beams in the electron cyclotron frequency range. *Physica Scripta, 84*(2), 025505
35. Girka, V. A., Girka, I. A., Kondratenko, A. N., et al. (1988). Azimuthal surface waves of magnetoactive plasma waveguides. *Soviet Journal of Communications Technology and Electronics, 33*(8), 37–41.

36. Girka, V. O., Girka, I. O., & Pavlenko, I. V. (2011). Excitation of azimuthal surface modes by relativistic flows of electrons in high-frequency range. *Plasma Physics Reports, 37*(5), 447–454.
37. Girka, V. A., & Girka, I. A. (1991). Azimuthal surface waves in a nonuniform plasma cylinder. *Radiophysics and Quantum Electronics, 34*(4), 324–328.
38. Girka, V. A., Girka, I. A., & Tkachenko, V. I. (1996). Excitation of azimuthal surface modes in cylindrical semiconductor structures in the presence of drift motion of electrons' flow. *Soviet Technical Physics, 66*(4), 114–120.
39. Smith, R. A. (1978). *Semiconductors*. Cambridge University Press.
40. Raizer, Yu. P. (1991). *Gas discharge physics*. Springer.
41. Krall, N. A., & Trivelpiece, A. W. (1986). *Principles of Plasma Physics*. San Francisco Press.
42. Dorodnitsyn, A. A. (1952). Asymptotic laws of distribution of the characteristic values for certain special forms of differential equations of the second order. *Uspekhi Matematicheskikh Nauk (N.S.) 7*(6(52)), 3–96 (in Russian).
43. Girka, V. O., Girka, I. O., Kostenko, A. V., et al. (2014). Excitation of azimuthal surface waves in toroidal waveguide by rotating electron beam at the range of electron cyclotron resonance. *Progress in Electromagnetics Research B, 57*, 267–277.
44. Girka, I. O., & Thumm, M. (2016). Excitation of azimuthal surface waves in the electron cyclotron frequency range by a rotating electron beam in presence of dissipation. *Physics of Plasmas, 23*, 122124.
45. Thumm, M. (2020). State-of the-art of high-power gyro-devices and free electron masers. *Journal of Infrared, Millimeter, and Terahertz Waves, 41*(1), 1–140.
46. Li, F., He, W., Cross, A. W., et al. (2010). Design and simulation of a ~390 GHz seventh harmonic gyrotron using a large orbit electron beam. *Journal of Physics D: Applied Physics, 43*, 155204.
47. Li, X., Lang, J., Alfadhl, Y., et al. (2015). Study of an eighth-harmonic large-orbit gyrotron in the terahertz band. *IEEE Transactions on Plasma Science, 43*, 506–514.
48. Ogura, K., Shirai, A., Ogata, M., et al. (2016). Beam interactions with surface waves and higher order modes in oversized G-band slow-wave structure. *IEEE Transactions on Plasma Science, 44*(2), 201–210.
49. Akhiezer, A. I., & Fainberg, Ya. B. (1949). On the interaction of a charged particle beam with electron plasma. *Doklady AN SSSR, 69*, 555–556. (in Russian). (Akhiezer, A. I., & Fainberg, Ya. B. (2008). On the interaction of a charged particle beam with electron plasma. *Ukrainian Journal of Physics, 53*, 87–88. (in English, translated and reprinted from *Dokl. Akad.Nauk SSSR, 69*, 555–556 (1949))).
50. Destler, W. W., Chojnacki, E., Hoeberling, R. F., et al. (1988). High-power microwave generation from large-orbit devices. *IEEE Transactions on Plasma Science, 16*(2), 71–89.
51. Velazco, J. E. (2000). Continuous rotating-wave electron beam accelerator. U.S. Patent No. 6,060,833. *Washington, D.C.: U.S. Patent and Trademark Office*.
52. Girka, I., Pavlenko, I., & Thumm, M. (2018). Excitation of higher radial modes of azimuthal surface waves in the electron cyclotron frequency range by rotating relativistic flow of electrons in cylindrical waveguides partially filled by plasmas. *Physics of Plasmas, 25*(5), 052109.
53. Stix, T. H. (1992). *Waves in plasmas*. AIP.
54. Longinov, A. V., & Stepanov, K. N. (1992). Radio-frequency plasma heating in tokamaks in the ion-cyclotron frequency range. In A. G. Litvak (Ed.), (pp. 93–237), AIP.
55. Sagdeev, R. Z., Usikov, D. A., & Zaslavsky, G. M. (1988). *Nonlinear physics: From the pendulum to turbulence and chaos*. Harwood Academic Publishers.
56. Girka, I. O., Kondratenko, V. M., & Thumm, M. (2018). Higher radial modes of azimuthal surface waves in magnetoactive cylindrical plasma waveguides. *Journal of Plasma Physics, 84*, 905840603.
57. Girka, V., Girka, I., & Thumm, M. (2014). *Surface flute waves in plasmas. Theory and applications*. Springer.
58. Girka, V. O., & Puzyr'kov, SYu. (2002). Nonlinear interaction of an annular electron beam with azimuthal surface waves. *Plasma Physics Reports, 28*(4), 351–358.

59. Girka, I., Pavlenko, I., & Thumm, M. (2018). Two mechanisms of resonance overlapping in excitation of azimuthal surface waves by rotating relativistic electron beams. *Physics of Plasmas, 25*(5), 052111.
60. Chojnacki, R., & Destler, W. W. (1987). Microwave radiation from a low- energy rotating electron beam in an azimuthally periodic magnetic wiggler field. *IEEE Journal of Quantum Electronics, 23*, 1605–1609.
61. Girka, V. O., & Girka, I. O. (2002). Effect of toroidal magnetic field variations on the spectra of azimuthal surface waves in metal waveguides entirely filled with plasma. *Plasma Physics Reports, 28*(3), 190–195.
62. Girka, V. A., & Girka, I. A. (1992). Coupled azimuthal surface waves in a nonuniform current–carrying plasma cylinder. *Soviet Journal of Communications Technology and Electronics, 37*(4), 23–29.
63. Girka, V. O., Girka, I. O., Girka, A. V., et al. (2011). Theory of azimuthal surface waves propagating in nonuniform waveguides. *Journal of Plasma Physics, 77*(4), 493–519.
64. Karpov, SYu., & Stolyarov, S. N. (1993). Propagation and transformation of electromagnetic waves in one- dimensional periodic structures. *Physics- Uspekhi, 36*(1), 1–22.
65. Wu, J., Xiong, C., & Liu, S. (1995). Excitation of microwave by an annular electron beam in a plasma-filled dielectric lined waveguide. *International Journal of Infrared and Millimeter Waves, 16*(9), 1573–1581.
66. Norreys, P. A., Green, J. S., Davies, J. R., et al. (2006). Observation of annular electron beam transport in multi-TeraWatt laser-solid interactions. *Plasma Physics and Controlled Fusion, 48*(2), L11–L22.
67. Girka, O. I., Girka, V. O., Girka, I. O., et al. (2007). Effect of the shape of the cross section of a plasma-dielectric interface on the dispersion properties of azimuthal surface modes. *Plasma Physics Reports, 23*(2), 91–101.
68. Girka, O. I., Girka, V. O., Girka, I. O., et al. (2007). Resonant effect of the noncircular shape of the plasma surface on the dispersion properties of extraordinary azimuthal surface modes in magnetoactive waveguides. *Plasma Physics Reports, 23*(7), 543–552.
69. Girka, V. O., Girka, I. O., & Pavlenko, I. V. (2012). Influence of the plasma column cross-section non-circularity on the excitation of the azimuthal surface waves in electron cyclotron frequency range by annular electron beam. *Progress In Electromagnetics Research M (PIERM), 26*, 39–53.
70. Bludman, S. A., Watson, K. M., & Rosenbluth, M. N. (1960). Statistical mechanics of relativistic streams. *II. Physics of Fluids, 3*(5), 747–757.
71. Carlevaro, N., Fanelli, D., Garbet, X., et al. (2014). Beam–plasma instability and fast particles: the Lynden-Bell approach. *Plasma Physics and Controlled Fusion, 56*(3), 035013.
72. Dubinov, A. E., Petrik, A. G., Kurkin, S. A., et al. (2016). Beam-plasma instability in charged plasma in the absence of ions. *Physics of Plasmas, 23*(4), 042105.
73. Rostomian, E. V. (2000). Space-time evolution of beam-plasma instabilities. *Physics of Plasmas, 7*(5), 1595–1600.
74. Ballai, I., Oliver, R., & Alexandrou, M. (2015). Dissipative instability in partially ionised prominence plasmas. *Astronomy and Astrophysics, 577*, A82.
75. Anderson, T. (2011). *Plasma Antennas.* Artech House.
76. Borg, G. G., Harris, J. H., Miljak, D. G., et al. (1999). Application of plasma columns to radio-frequency antennas. *Applied Physics Letters, 74*(22), 3272–3274.
77. Cerri, G., De Leo, R., Primiani, V. M., et al. (2008). Measurement of the properties of a plasma column used as a radiated element. *IEEE Transactions on IMT, 57*(2), 242–247.
78. Rayner, J. P., Whichello, A. P., & Cheetham, A. D. (2004). Physical characteristics of plasma antennas. *IEEE Transactions on Plasma Science, 32*(1), 269–281.
79. Wu, M., Wen, B. Y., & Zhou, H. (2009). Ionospheric clutter suppression in HF surface wave radar. *Journal of Electromagnetic Waves and Applications, 23*(10), 1265–1272.
80. Ostrikov, K., Neyts, E. C., & Meyyappan, M. (2013). Plasma nanoscience: from nano-solids in plasmas to nano-plasmas in solids. *Advances in Physics, 62*(2), 113–224.

81. Girka, V. O., Girka, I. O., & Pavlenko, I. V. (2001). Electrodynamic model of the gas discharge sustained by azimuthal surface waves. *Contributions to Plasma Physics, 41*, 393–406.
82. Sah, C.-T. (1991). *Fundamentals of solid state electronics*. World Scientific Publishing.
83. Girka, V. O., Girka, I. O., & Pavlenko, I. V. (2011). Excitation of ion azimuthal surface modes in a magnetized plasma by annular flow of light ions. *Progress in Electromagnetics Research M (PIERM), 21*, 267–278.
84. Bratman, V. L, Kalynov, Yu. K., Manuilov, V. N., & Samsonov, S. V. (2005). Large orbit gyrotron at submillimeter waves. In *Conference of Digest 30th International Conference on Infrared and Millimeter Waves and 13th International Conference on Terahertz Electronics* (pp. 443–444). Williamsburg, VA, USA.
85. Bratman, V. L, Kalynov, Yu. K., Manuilov, V. N. & Samsonov, S. V. (2005). Large-orbit gyrotron operation at submillimeter waves. In A. G. Litvak (Ed.),*Proceweding 6th International Workshop on Strong Microwaves in Plasmas* (Vol. 1, pp. 150–155). Nizhny Novgorod, Institute of Applied Physics, Russian Academy of Sciences, Nizhny Novgorod.
86. Girka, V. O., & Girka, I. O. (2002). Asymmetric long-wavelength surface modes of isotropic plasma waveguides. *Plasma Physics Reports, 28*(8), 682–689.
87. Girka, V. O., & Girka, I. O. (2002). Asymmetric long-wavelength surface modes in magnetized plasma waveguides entirely filled with plasma. *Plasma Physics Reports, 28*(11), 916–924.
88. Azarenkov, N. A., Olefir, V. P., & Sporov, A. E. (2001). Quadrupole and octopole electromagnetic waves in slightly nonuniform magnetized plasma column. *Physica Scripta, 63*, 36–42.
89. Olefir, V. P., & Sporov, A. E. (2011). Symmetric and dipolar electromagnetic waves in coaxial structure filled by non-uniform dissipative plasma with azimuth magnetic field. *Problems of Atomic Science and Technology, No 1, Series: Plasma Physics, 17*, 47–49.
90. Aronov, B. A., Bogdankevich, L. S., & Rukhadze, A. A. (1973). Electromagnetic irradiation under excitation of surface waves in plasma by electron beam. *Technical Physics, 43*(12), 2493–2498. (in Russian).
91. Miller, S. E. (1954). Coupled wave theory and waveguide applications. *Bell System Technical Journal, 33*(3), 661–719.
92. Kolyada, Yu. E., Fainberg, Ya. B., Kornilov, E. A., Ognivenko, V. V, & Kiyashko, V. A. (1976). Interaction of a high-current electron beam with a plasma in open cavity in a mirror system. *Soviet Technical Physics Letters, 2*, 348–350.
93. Girka I. O., Girka O. I., & Thumm M. (2019). Initial stage of interaction between gyrating relativistic electron beam and long wavelength surface electromagnetic waves in cylindrical metallic wave-guides partially filled with plasma. *Physics of Plasmas, 26*(4), 042118.

Chapter 8
Excitation of Surface Flute Waves by Electron Beams Gyrating Along Large Larmor Orbits in Coaxial Waveguides

To enhance the efficiency of devices of super high-frequency electronics it is necessary to solve a number of problems. An important place among them is occupied by studying the spectra of the electromagnetic eigenwaves excited in these devices, and the problems of transportation of charged particle beams interacting with these waves (see, for example, [1–3] and references therein). Utilizing plasmas in such devices allows solving a number of these important problems, including enhancement of the limiting current flowing in these devices, expansion of the frequency spectrum of excited oscillations, and providing possibilities for control of the excitation process etc. Interaction of charged particle beams with plasma eigenwaves is already used in plasma electronics for generation and amplification of electromagnetic radiation for a long time [4–9]. The major attention is given to studying the processes of wave excitation in magneto-active plasma waveguides [10–14]. However, it is impossible to consider the theory of beam-plasma instabilities as completely developed since the process of charged particle beam interaction with eigenwaves of waveguides essentially depends on a considerable number of factors, including the dispersion properties of these waves, their polarization and spatial distribution of their fields, as well as the geometry and design features of the waveguide.

Studying the interaction of annular charged particle beams in plasma-filled waveguides is considered as very perspective [15–17]. In this case, one has to expect enhancing the efficiency of plasma electronics devices in comparison with the generators working on longitudinal particle beams. For example, the efficiency of RF generators with longitudinal beams is restricted in linear theory by the device length. In devices with annular beams, the electrons, rotating along Larmor orbits in the gap separating the chamber wall from the plasma column, transmit their energy to an electromagnetic wave until they reach the plasma surface as a result of their deceleration. Thus, electrons make total trajectories much longer than the interaction length of devices employing longitudinal beams. This, first, allows hoping for achievement of higher efficiencies in devices with annular beams in comparison to devices using longitudinal beams and, second, gives the chance to design devices which are compact in the longitudinal direction.

© The Author(s), under exclusive license to Springer Nature Switzerland AG 2022
I. Girka and M. Thumm, *Surface Flute Waves in Plasmas*, Springer Series on Atomic, Optical, and Plasma Physics 120, https://doi.org/10.1007/978-3-030-98210-2_8

Along with the generation of powerful electromagnetic radiation, relativistic electron beams are proposed to be used, for example, in plasma confinement experiments. In paper [18] an example is given of calculating a steady state compact toroidal beam consisting of an immobile ion background and a relativistic electron beam.

The generated electromagnetic waves can be surface type waves whose excitation by charged particles beams is characterized by interesting distinctive features [5, 9]. In the monograph [19], a comprehensive study of the properties of surface eigenwaves propagating perpendicular to the axis of symmetry in various cylindrical waveguides filled with magneto-active plasma is presented. The present Chapter continues studying the beam excitation of SFXMS started in the previous chapter. These waves were shown to propagate in two frequency ranges: the low-frequency one [near the electron cyclotron frequency—see Eq. (1.3)] and the high-frequency one [above the upper hybrid resonance—see Eq. (1.14)]. In Chap. 7, the possibility of SFXM excitation in the low-frequency range by relativistic electron beams in cylindrical metal waveguides partially filled with a plasma column is discussed. The surface nature of the studied waves makes their dispersion properties weakly dependent on the properties of the plasma core near the waveguide axis, which is sufficiently far from the plasma-vacuum interface. This makes it possible to replace this space by an inner metal conductor and transform the structure into a plasma-filled coaxial waveguide.

In the present Chapter, the beam excitation of both HF and LF SFXMs with positive azimuthal mode indices m in coaxial cylindrical waveguides is studied because the direction of their propagation coincides with the direction of electron rotation in the axial magnetic field [20]. For both frequency ranges, two possibilities of injecting the annular electron beam are considered. First, electrons rotate along large Larmor orbits around the plasma coaxial layer. This case is referred to as external beam. It is studied in Sects. 8.1 and 8.3. Second, the electron flow rotates between the inner conductor and plasma coaxial layer. This case is called as internal beam. It is studied in Sects. 8.2 and 8.4.

The wave-field components have the following dependence on the co-ordinates and time: $E_r, E_\varphi, H_z(r, \varphi, t) = f(r)\exp(im\varphi - i\omega t)$. One can take into account a non-zero but sufficiently small axial wavenumber k_z employing the method of successive approximations as it was shown in the Sect. 7.7.

The plasma particle density is assumed to be homogeneous. This assumption is particularly valid for the cases when the surface waves are used to produce and sustain gas discharges for technological applications. In gas discharges sustained by surface waves propagation, the plasma density is homogeneous in the edge layer because their amplitudes reach a maximum on the plasma surface unlike for the case of bulk (volume) waves [21]. The influence of a weak radial inhomogeneity of the plasma density on the dispersion properties of SFXMs was studied in Sect. 5.1 [22]. The case of a linear plasma density profile near the boundary was also studied in Sect. 5.1. In this case, the SFXM dispersion properties are similar to those in the case of uniform plasma with replacing the penetration depth of these waves into the plasma by a definite "effective" value of this parameter [23].

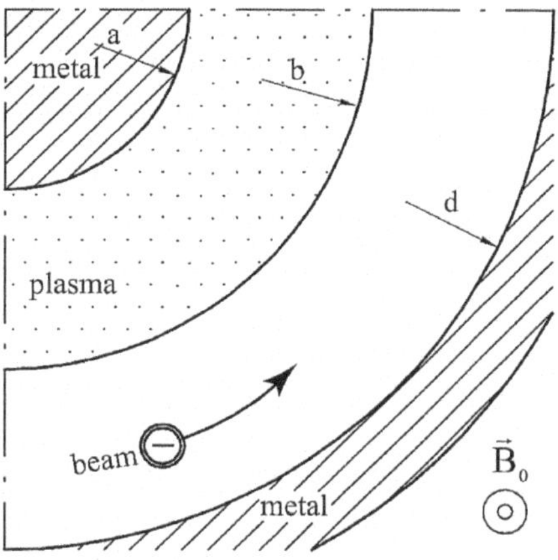

Fig. 8.1 Scheme of the coaxial beam-plasma-waveguide structure, where an electron beam gyrates around the outer surface of the plasma layer

In the present Chapter, the plasma particle density is assumed to be sufficiently high to satisfy the inequality $\Omega_e{}^2 > \varepsilon_r \omega_e{}^2$, where Ω_e and ω_e are the electron plasma and cyclotron frequencies, respectively, ε_r is the relative dielectric constant of a metamaterial, of a crystal or of the crystal lattice of a semiconductor ($\varepsilon_r > 1$). For gas plasmas one has $\varepsilon_r = 1$. This inequality, $\Omega_e{}^2 > \varepsilon_r \omega_e{}^2$, is surely realized in semiconductor plasmas but it can also be realized in laboratory gas plasmas when the external static magnetic field is sufficiently weak.

The electron beam is described by the model of oscillators with identical transversal momentum $p_{\perp 0}$ and zero axial momentum, $p_z = 0$, like in the Chap. 7. Doing that the considered beam-plasma system is assumed to be compensated both with respect to current and charge density. Such flow of electrons is described by the equilibrium distribution function (7.1) [24]. The electrodynamical properties of the waveguide space, which is occupied by the electron beam, is described by the dielectric permittivity tensor $\varepsilon_{ik}^{(b)}$ [24, 25]. The expressions for the three components of the $\varepsilon_{ik}^{(b)}$ tensor, which are used in the further calculations of the present Chapter are given by (7.18)–(7.20).

Solving the Maxwell equations in the space occupied by the electron beam for SFXM fields with application of the components of the $\varepsilon_{ik}^{(b)}$ tensor mentioned above, one can find a Bessel type equation for the magnetic component $H_z^{\sim}$ of the wave field. Within the plasma column, $H_z^{\sim}$ is described by a modified Bessel type equation. The azimuthal component of the SFXM electric wave field is expressed through the axial magnetic wave field via Eq. (1.10). To obtain the SFXM dispersion relation one can apply standard boundary conditions [26–29] (see also Sect. 1.4.4). The initial stage of the beam-plasma instability is studied via solving the dispersion relation assuming the validity of the resonance condition (7.14).

Excitation of electromagnetic waves by a gyrating electron beam possesses the following feature. The velocity of the beam is determined by the external magnetic field and the radii of the plasma column and the metal chamber $|\omega_e|b/\gamma < \upsilon < |\omega_e|d/\gamma$ (see Fig. 8.1). One can make the following conclusions from this circumstance. First, if the beam radius is close to the inner radius of the external metal waveguide wall, a larger part of the kinetic beam energy can be transformed into electromagnetic waves up to the moment when the beam reaches the plasma surface as the result of deceleration. On the other hand, if the beam rotates close to the metal wall it interacts with a weaker electric field, so that the wave-beam interaction is characterized by lower efficiency. Second, the role of the SFXM azimuthal wavenumber is played by the ratio of the azimuthal mode index to the outer radius of the plasma column, $k_\varphi = m/b$. Thus, the condition of resonant beam excitation, $\omega = |\omega_e|N/\gamma$ depends also on the azimuthal mode index.

It is well-known [7, 30–34], that the excitation of electromagnetic waves takes place in the most efficient way just nearby the eigenfrequencies of the plasma waveguide. Analysis of the dispersion relation (7.12) shows, that the influence of an electron beam with low particle density ($\alpha \ll 1$) can be sufficient in the case of the presence of resonance denominators in the expressions (7.18)–(7.20) for the components of the permittivity tensor $\varepsilon_{ik}^{(b)}$. Just nearby the intersection points of the SFXM dispersion curves, constructed in absence of the electron beam, $\alpha = 0$, with the curves $\omega = |\omega_e|N/\gamma$, along which the components of the tensor $\varepsilon_{ik}^{(b)}$ have resonance denominators, excitation of these waves takes place.

The goal of the present Chapter is to study the influence of the presence of an internal conductor located in a cylindrical waveguide on the excitation of SFXMs. The second objective is to find favorable conditions for excitation of these SFXMs in such type plasma filled coaxial waveguides.

The other reason to study SFXM excitation in coaxial plasmas by internal rotating electron beams is as follows. If the inner conductor is a hollow metal cylinder one has an empty space inside the waveguide structure ($r < a$, see Fig. 8.1) which can be used for placing a source producing the rotating electron beam there. In the present study, the problem of providing such type of gyrating electron flow is not investigated. However, the possibility to produce an electron beam gyrating in the inner gap that separates the plasma coaxial layer just from the internal metal wall is demonstrated in [35]. That is why the question: how one can produce an electron flow rotating in the narrow gap that separates the plasma coaxial layer, is out of scope in the present monograph.

8.1 Excitation of Surface Flute Waves in the Electron Cyclotron Frequency Range by an Electron Beam Gyrating Around a Coaxial Plasma Layer

A waveguide structure which consists of a cylindrical coaxial metal waveguide with a circular cross-section and magneto-active plasma filling is considered in [26]. The radius of the inner conductor is $r = a$, the radius of the inner wall of the outer waveguide is $r = d$ (Fig. 8.1). The coaxial line is partially filled with a plasma layer, $a < r < b$. The waveguide is made of a perfect conductor (metal with infinite conductivity). An external homogeneous static magnetic field is directed along the waveguide axis: $\vec{B}_0 \| \vec{z}$. The waveguide is assumed to be uniform along the cylinder axis: $\partial/\partial z = 0$. The plasma particle density is assumed to be homogeneous.

It is assumed that a rotating electron beam is injected into the narrow vacuum gap $b < r < d$ between the plasma layer and the external metal chamber. Numerical analysis shows that under the present conditions in order to realize efficient beam-plasma interaction the electron beam should be moderately relativistic. It should be mentioned here that moderately relativistic electron beams are utilized in experiments with Large Orbit Gyrotrons (LOGs) which use such axis encircling electron beams [36–39].

Applying linear boundary conditions allows one to obtain the dispersion relation that describes the excitation of SFXMs in the considered coaxial plasma filled waveguide [26]:

$$\begin{aligned}
&\left\{ J_m(\zeta_b)\left[N_m'(\zeta_d) + \frac{\mu_b m}{\zeta_d} N_m(\zeta_d)\right] - N_m(\zeta_b)\left[J_m'(\zeta_d) + \frac{\mu_b m}{\zeta_d} J_m(\zeta_d)\right]\right\} \\
&\times \left\{ I_m(\xi_b)\left[K_m'(\xi_a) + \frac{\mu m}{\xi_a} K_m(\xi_a)\right] - K_m(\xi_b)\left[I_m'(\xi_a) + \frac{\mu m}{\xi_a} I_m(\xi_a)\right]\right\}^{-1} \\
&= \frac{k_\perp}{k\sqrt{\psi_b}} \left\{ \left[J_m'(\zeta_b) + \frac{\mu_b m}{\zeta_b} J_m(\zeta_b)\right]\left[N_m'(\zeta_d) + \frac{\mu_b m}{\zeta_d} N_m(\zeta_d)\right]\right. \\
&\left. - \left[J_m'(\zeta_d) + \frac{\mu_b m}{\zeta_d} J_m(\zeta_d)\right]\left[N_m'(\zeta_b) + \frac{\mu_b m}{\zeta_b} N_m(\zeta_b)\right]\right\} \\
&\times \left\{ \left[I_m'(\xi_a) + \frac{\mu m}{\xi_a} I_m(\xi_a)\right]\left[K_m'(\xi_b) + \frac{\mu m}{\xi_b} K_m(\xi_b)\right]\right. \\
&\left. - \left[I_m'(\xi_b) + \frac{\mu m}{\xi_b} I_m(\xi_b)\right]\left[K_m'(\xi_a) + \frac{\mu m}{\xi_a} K_m(\xi_a)\right]\right\}^{-1}, \qquad (8.1)
\end{aligned}$$

where $\mu_b = -i\varepsilon_{12}^{(b)}/\varepsilon_{11}^{(b)}$, $\xi = k_\perp r$, $\zeta_b = \zeta(b)$, $\zeta_d = \zeta(d)$, $\xi_a = \xi(a)$, $\xi_b = \xi(b)$.

Equation (8.1) is analyzed here numerically under the resonance condition (7.14). Results of the numerical analysis of the dispersion relation (8.1) are plotted in Figs. 8.2, 8.3, 8.4, 8.5, 8.6 and 8.7. There the dashed and solid curves correspond to the cases of a narrow ($a/b = 0.1$) and a wide ($a/b = 0.9$) internal conductor, respectively. The ratio a/b characterizes mainly the relative size of the inner conductor. More complete and comprehensive information can be obtained from direct calculations of absolute values of the thickness of the plasma layer, which is located near the inner conductor of the coaxial waveguide.

Another parameter, which can be used for characterization of the plasma layer, is the ratio of the effective wavenumber $k_{ef} = mc/(b\Omega_e)$ and its critical value (cut-off wavenumber) k_{cr}. If the ratio $k_{ef}/k_{cr} < 1$ is valid, the penetration depth of the LF SFXM is lower than the thickness of the plasma layer, so the inequality $k_\perp(b-a) > 1$ is satisfied. In this limiting case the studied SFXM does not feel the interface of the inner conductor. Thus, the electrodynamical properties of such

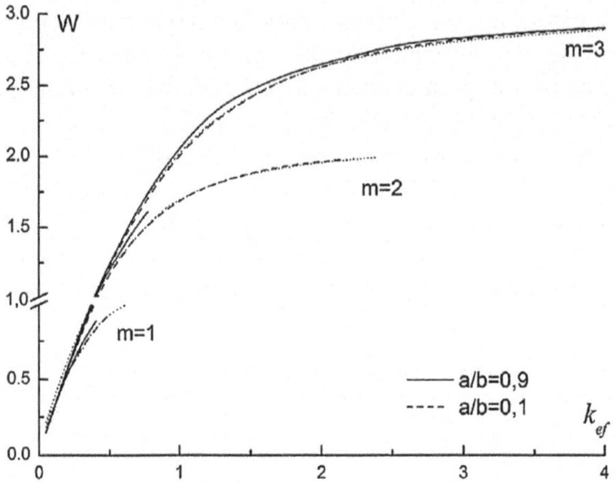

Fig. 8.2 Normalized eigenfrequencies of SFXMs ($W \equiv \omega/|\omega_e|$) versus effective wavenumber k_{ef} for different ratios of the plasma layer radii: $a/b = 0.9$ (*solid curves*), and $a/b = 0.1$ (*dashed curves*). $m = 1, 2, 3$ with $Z = 3$. The *dotted curves*, which practically coincide with the *dashed curves*, show the dependence of the corresponding cyclotron harmonics N/γ on k_{ef}, namely, $N = 1$ for the first azimuthal mode index, $N = 2$ for the second mode index and $N = 3$ for the azimuthal mode index $m = 3$

Fig. 8.3 Normalized LF SFXM growth rate $\beta \equiv \text{Im}(\omega/|\omega_e|)$ versus effective wavenumber for different azimuthal mode indices: $m = 1, 2, 3$. *Solid curves* ($a/b = 0.9$), *dashed curves* ($a/b = 0.1$). $Z = 3$

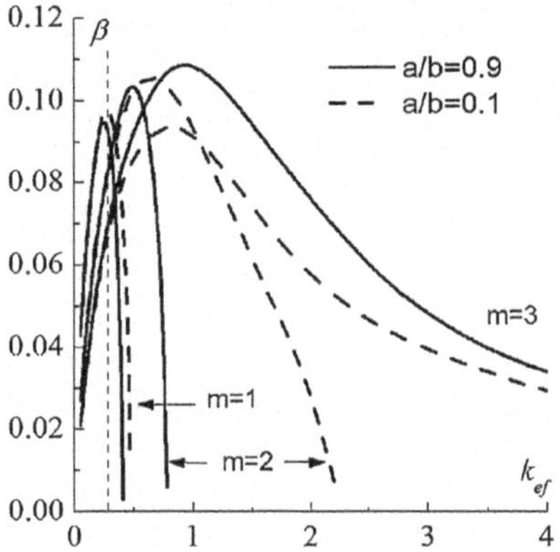

coaxial structures are the same as of a simple plasma filled cylindrical waveguide. Analyzing the Figs. 8.3, 8.4 and 8.7 in this limiting case, one can conclude that for a beam excitation of SFXMs there is almost no difference between coaxial waveguides with wide or narrow inner conductors. That is why the limiting case $k_{ef} < k_{cr}$ can be called as the case of a thick plasma layer and the limiting case $k_{ef} > k_{cr}$ the case of a thin plasma layer.

In the following, some beam-plasma parameters, which are needed for realization of the proposed plasma-waveguide structure, are estimated [26]. If the plasma density is $n_e = 10^{11}$ cm^{-3}, then the electron plasma (Langmuir) frequency is $\Omega_e \approx 2 \times$

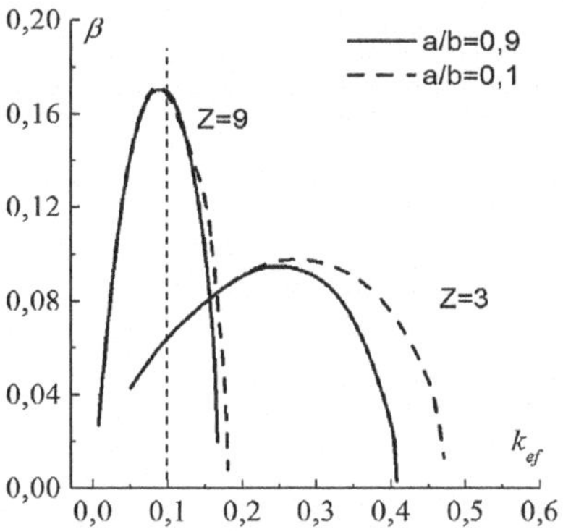

Fig. 8.4 Normalized LF SFXM growth rate $\beta \equiv \mathrm{Im}(\omega/|\omega_e|)$ versus effective wavenumber for two magnetic fields: stronger (Z = 3) and weaker (Z = 9). *Solid curves* ($a/b = 0.9$), *dashed curves* ($a/b = 0.1$). $m = 1$

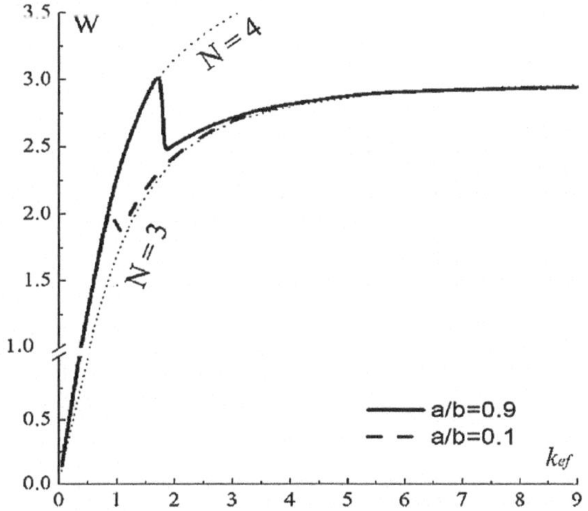

Fig. 8.5 Normalized eigenfrequencies of SFXMs versus effective wavenumber for $m = 4$ with $Z = 3$. *Solid curves* ($a/b = 0.9$), *dashed curves* ($a/b = 0.1$). *Dotted curves* show the dependences of the resonant *curves* N/γ on k_{ef}. *The numbers* indicate the cyclotron harmonics: $N = 3$ and 4

10^{10} s^{-1}, the skin depth is $c/\Omega_e = 1.5$ cm, and for the ratio between the Langmuir frequency and the electron cyclotron frequency of $Z = \Omega_e/|\omega_e|=3$, the axial magnetic field should be as high as $B_0 \approx 350$ G, which can easily be obtained experimentally. For the SFXM with azimuthal mode index $m = 2$ and effective wavenumber $k_{ef} \approx 0.5$ one can reach maximum values of the growth rate $\mathrm{Im}(\omega/|\omega_e|) \equiv \beta \approx 0.11$. In this case, the outer radius of the coaxial plasma layer should be $b = 6$ cm and the radius of the inner wall of the outer electrode of the studied coaxial waveguide is $d = 7.2$ cm. Thus, regimes of a narrow internal conductor $a/b = 0.1$ and a wide internal conductor $a/b = 0.9$ can be realized for the internal conductor radii $a = 0.6$ cm and $a = 5.4$ cm, respectively.

The analysis of the dispersion curves plotted in Fig. 8.2 proves that the value of the LF SFXM eigenfrequencies with $m = 2$ is $\mathrm{Re}(\omega) \approx 1.5|\omega_e|$ for a coaxial waveguide with a wide inner conductor and is a little bit less than $1.5|\omega_e|$ for the case of a narrow coaxial waveguide. Generally speaking, the LF SFXM eigenfrequencies for $m = 1$,

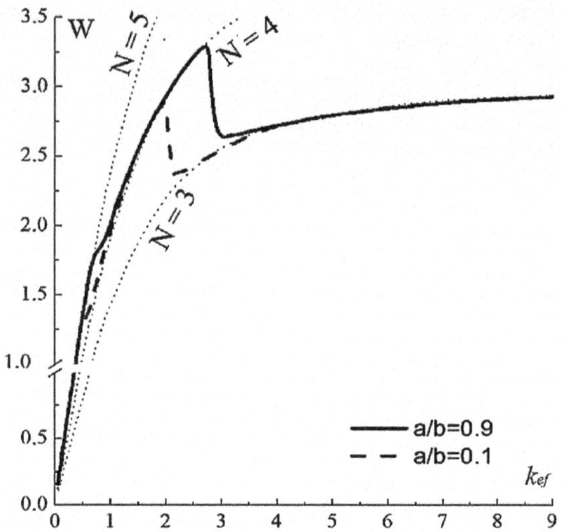

Fig. 8.6 Normalized eigenfrequencies of SFXMs versus effective wavenumber for $m = 5$ with $Z = 3$. *Solid curves* ($a/b = 0.9$), *dashed curves* ($a/b = 0.1$). *Dotted curves* show the dependences of the resonant *curves* N/γ on k_{ef}. *The numbers* indicate the cyclotron harmonics: $N = 3$, 4 and 5

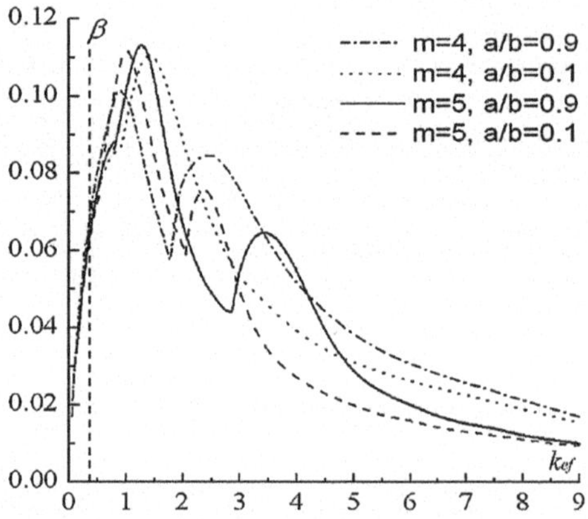

Fig. 8.7 Normalized LF SFXM growth rates versus effective wavenumber for $m = 4$ and 5 with $Z = 3$ and different radius ratios a/b

2 and 3 (see Fig. 8.2) and for $m = 4$ and 5 (see Figs. 8.5 and 8.6, respectively) are higher for the limiting case of a wide coaxial waveguide than for the case of a narrow coaxial waveguide. However, this difference is not essential, with exception of a narrow region of wavenumbers, where the phenomenon of mode competition is observed. Typical values of the eigenfrequencies for these waves excited under the condition of maximum growth rates with $m = 1$ and 3 are $\mathrm{Re}(\omega) \approx 0.7|\omega_e|$ and $\mathrm{Re}(\omega) \approx 2.0|\omega_e|$, respectively (as shown in Fig. 8.2).

Increasing the azimuthal mode index of the excited SFXM leads to an increase of the k_{ef} value for which the maximum $\mathrm{Im}(\omega/|\omega_e|)$ can be realized. Therefore, for the case $m = 3$ one needs a coaxial waveguide with smaller radial size, namely, for $d =$ 5.4 cm, $b = 4.5$ cm, the radius of the inner conductor can be either 0.45 or 4.05 cm for the regimes of narrow or wide conductors, respectively. The eigenfrequencies of the excited SFXM for any of the values of azimuthal mode index are lower than the upper limit of the LF range, which can be approximated as $\omega \approx \Omega_e$ (see Figs. 8.2, 8.5 and 8.6). It should be also underlined that changing the value of the azimuthal mode index does not lead to an essential increase of $\mathrm{Im}(\omega/|\omega_e|)$, its values are about 0.11 for $m = 2, 3, 4$ and 5 (see Figs. 8.3, 8.4, and 8.7).

The estimations presented above show that the thickness of the plasma layer in such coaxial waveguides can be $b - a \approx 0.5 \div 5$ cm. These are not extremely thin layers, however, production and maintaining of a relatively thin homogeneous plasma layer in such coaxial waveguides can be a difficult technical problem. Because of that, the most prospective way to solve this technical problem can be application of modern meta-materials (see e.g. [40–43]). Their application can allow one to solve another problem as well, namely, the production of high-density plasmas.

Figures 8.3 and 8.4 demonstrate the dependence of $\beta \equiv \mathrm{Im}(\omega/|\omega_e|) = f(k_{ef})$ for the waves with different azimuthal mode indices $m \leq Z \equiv \Omega_e/|\omega_e|$. First of all, one has to mention that the resonance condition for these cases is realized for only one single electron cyclotron harmonic N, which are depicted by dotted curves in Fig. 8.2. These curves overlap with the corresponding dispersion curves for SFXMs propagating in coaxial waveguides with a narrow inner conductor (dashed lines). Because of this the corresponding curves of $\mathrm{Im}(\omega/|\omega_e|)$ have a "single bell" shape. Increasing the azimuthal mode index leads to an increase of the effective wavenumber range where SFXMs can be efficiently excited [26].

If the beam-plasma parameters meet the condition $k_{ef} < k_{cr}$ (i.e. the penetration depth of the wave is shorter than the plasma layer thickness), the eigenmodes of the studied waveguide do not "feel" any influence of the inner conductor. This means that both growth rate and eigenfrequencies of the studied SFXM are almost the same for both tested limiting cases, namely for a narrow internal conductor with $a/b = 0.1$ and for a wide internal conductor with $a/b = 0.9$. Nevertheless, introduction of the inner conductor results in some enhancement of the magnitude of the LF SFXM eigenfrequency (see Figs. 8.2, 8.5 and 8.6). The influence of increasing m value on the maximum growth rate $\mathrm{Max}\{\mathrm{Im}(\omega/|\omega_e|)\}$ is not monotonous because it is described by Bessel functions of the first kind [26]. The plasma radius ratio a/b differently affects $\mathrm{Im}(\omega/|\omega_e|)$ for various values of m. If $Z = 3$, then for the LF SFXM with $m = 1$ (see Fig. 8.3) $\mathrm{Max}\{\mathrm{Im}(\omega/|\omega_e|)\} = 0.098$ for $k_{ef} = 0.28$ in the case of narrow inner conductor and $\mathrm{Max}\{\mathrm{Im}(\omega/|\omega_e|)\} = 0.096$ for $k_{ef} = 0.25$ in the opposite limiting case. For the LF SFXM with azimuthal mode index $m = 2$, $\mathrm{Max}\{\mathrm{Im}(\omega/|\omega_e|)\} = 0.106$ if $k_{ef} = 0.64$ in the case $a/b = 0.1$, and $\mathrm{Max}\{\mathrm{Im}(\omega/|\omega_e|\} = 0.105$ if $k_{ef} = 0.5$ for the coaxial structure with $a/b = 0.9$. The same dependence of $\mathrm{Max}\{\mathrm{Im}(\omega/|\omega_e|\} = f(a/b)$ is observed in the cases of $m = 4$, but opposite dependences are realized for the odd azimuthal mode indices $m = 3$ and $m = 5$ (see Figs. 8.3 and 8.7).

Unlike that, the influence of the applied magnetic field B_0 has a definite effect. The larger the parameter $Z = \Omega_e/|\omega_e|$ is (weaker magnetic field), the less pronounced is the influence of the inner conductor on the excitation of SFXMs. This result is illustrated in Fig. 8.4 for $m = 1$. The "bell" of the curve $\mathrm{Im}(\omega/|\omega_e|) = f(k_{ef})$ in the case of $a/b = 0.9$ (solid curve) is narrower than in the case of $a/b = 0.1$ (dashed curve). The range of effective wavenumbers where SFXM excitation is efficient (width of the "bell") decreases from 21% in the case of $Z = 3$ to 8% in the case of $Z = 9$. Finally, an increasing Z parameter leads to increasing maximum $\mathrm{Im}(\omega/|\omega_e|)$ value (see Fig. 8.4).

Mode competition is shown in Figs. 8.5, 8.6 and 8.7 for the excitation of LF SFXMs with $m = 4$ and $m = 5$. The curves $W \equiv \omega/|\omega_e| = f(k_{ef})$ for these waves and the related dotted curves N/γ [normalized cyclotron harmonics of the electron beam, which interact with these azimuthal modes under the condition (7.14)] are depicted in Figs. 8.5 and 8.6, respectively.

In Fig. 8.5 one can see that in the region $k_{ef} < 1$ the dispersion curve overlaps with the dotted curve $N = 4$. However, then in the region $k_{ef} > 1$, to satisfy the resonance condition (7.14), it has to "jump" to the dotted curve $N = 3$ in both of the limiting cases, namely: the wide ($a/b = 0.9$) and the narrow ($a/b = 0.1$) inner conductors. In the case of $a/b = 0.9$ transition from one cyclotron resonance line to another one takes place at $k_{ef} \approx 1.785$, while it occurs at $k_{ef} \approx 0.95$ for the case of $a/b = 0.1$ (see Fig. 8.5).

The two maxima of growth rates: $\mathrm{Max}\{\mathrm{Im}(\omega/|\omega_e|\} = 0.087$ for $k_{ef} \approx 0.75$ and $\mathrm{Max}\{\mathrm{Im}(\omega/|\omega_e|\} = 0.112$ for $k_{ef} \approx 1.41$, which exist for the SFXM with $m = 4$ propagating in the waveguide with a narrow internal conductor (dotted curve in Fig. 8.7), move to the new positions: $\mathrm{Max}\{\mathrm{Im}(\omega/|\omega_e|\} = 0.102$ for $k_{ef} = 0.88$ and $\mathrm{Max}\{\mathrm{Im}(\omega/|\omega_e|\} = 0.089$ for $k_{ef} = 2.35$ if one has the case of a waveguide with wide inner conductor (see dash-dotted curve in Fig. 8.7). Therefore, an increase of the inner conductor radius leads to a transition of the resonant beam instability of the LF SFXM with $m = 4$ from the regime of overlapping of neighboring cyclotron resonances to the regime of separate interactions of this wave with the fourth, $N = 4$, cyclotron harmonic and with the third, $N = 3$, cyclotron harmonic, which happen for two noticeably different values of k_{ef}, namely $k_{ef} \approx 0.88$ and $k_{ef} \approx 2.35$. In this case (SFXMs with $m = 4$) the right boundary of the range where these waves can be excited efficiently shifts toward the larger values of effective wavenumbers k_{ef}.

In the case of the SFXM with $m = 5$, excitation transfer from the regime of overlapping cyclotron resonances exists for very small values of k_{ef} (for details see Fig. 8.6), when this wave first interacts with the fifth cyclotron harmonic and then the dispersion curve "jumps" to the fourth ($N = 4$) cyclotron harmonic. That is why in the case of a narrow inner conductor (dashed curve) this regime is practically invisible in Fig. 8.7 and weakly manifests itself in the limiting case of a wide inner conductor (solid curve). A further increase of k_{ef} leads to the situation when resonance condition (7.14) can be satisfied for the SFXM with $m = 5$ only for the third cyclotron harmonic. Therefore, one can see "jumping" of the SFXM dispersion curves (Fig. 8.6) from the curve $N = 4$ to the curve $N = 3$ for k_{ef} approximately equal to 1.9 and 2.3 in the limiting cases of a narrow and wide inner conductor, respectively. That is why one can clearly see in Fig. 8.7 the regime of separate interaction of this wave with the fourth and third cyclotron harmonics of the applied electron beam. A widening of the range of k_{ef}, where the SFXM can be efficiently excited due to interaction with the rotating electron beam, happens in this case in the same manner as in the case $m = 4$.

Summarizing the obtained results concerning the resonant beam-plasma instability of LF SFXMs, propagating in a coaxial waveguide in the outer interface of the plasma filling, one can state the following features of this phenomenon. The higher the plasma density is, the smaller is the influence of the inner conductor on electron beam excitation. Generally speaking, the maximum value of the growth rate is not very sensitive to the radius of the internal conductor. The inner conductor exerts the main influence on the width of the effective wavenumbers range, where these eigenmodes can be excited with essentially large values of their growth rates [26]. Increase of the azimuthal mode index leads: first, to an increase of the $\mathrm{Max}\{\mathrm{Im}(\omega/|\omega_e|\}$ (but the rate of this increase quickly turns to zero); second, to the shift of the position of $\mathrm{Max}\{\mathrm{Im}(\omega/|\omega_e|\}$ to the range of larger values of k_{ef}, and third, to increasing width of the k_{ef} range, where SFXMs can be efficiently excited. Further increasing m value does not lead to an increase of $\mathrm{Max}\{\mathrm{Im}(\omega/|\omega_e|\}$, because fulfillment of the resonance condition (7.14) in this case is possible for higher values of N which in its turn leads to decreasing Bessel function values, which determine the value of the "beam" part of the dispersion relation (8.1).

Increase of the parameter Z leads: first, to an increase of the $\mathrm{Max}\{\mathrm{Im}(\omega/|\omega_e|\}$ value, second, to a shift of $\mathrm{Max}\{\mathrm{Im}(\omega/|\omega_e|\}$ position toward smaller values of k_{ef}, and third, to decreasing width of the k_{ef} range, where SFXMs can be efficiently excited. The similar influences of the m and Z values on the magnitude of $\mathrm{Max}\{\mathrm{Im}(\omega/|\omega_e|\}$ is explained by their analogous influence on the LF SFXM eigenfrequency. Opposite influences on the position of $\mathrm{Max}\{\mathrm{Im}(\omega/|\omega_e|\}$ and on the width of the k_{ef} range, where SFXMs can be efficiently excited, are connected with the choice of the normalizing factor. Since $k_{ef} = mc/(\Omega_e b)$, an increase of the plasma layer density leads to the decrease of k_{ef}. That is why one can see the corresponding shift of the $\mathrm{Max}\{\mathrm{Im}(\omega/|\omega_e|\}$ position and decrease of the width of the k_{ef} range, where SFXMs can be efficiently excited.

Beam excitation of LF SFXMs with negative azimuthal mode index is not studied here because the direction of their rotation is opposite to the direction of electron gyration in the axial magnetic field [1, 6, 7, 9, 20, 24, 44]. This is convenient from practical point of view as it provides a not so dense frequency spectrum and therefore allows one to avoid excitation of parasitic modes [26–30, 39]. Excitation of eigenmodes of a cylindrical waveguide by a rotating electron beam has good prospects to be applied for efficient, broadband amplification of microwave radiation [39]. It should be underlined that in many novel types of gyro-devices (see references given above) application of thin, specifically helical beams of electrons encircling the axes of the axial symmetrical magneto-active microwave systems is employed.

These LF SFXMs excited by rotating electron beams could be applied for sustaining of microwave gas discharges [21] and as operating modes in plasma-antenna and plasma-radar systems [45]. Practical results presented in [46–49] prove that surface waves have a good prospect for application in plasma-antenna systems. This relatively new class of antennas utilizes just a plasma column instead of metal wires as a highly conducting medium. The most important specific property of a gas discharge antenna is the possibility of fast switched-off which allows one to apply plasma-antenna systems for developing devices for wireless communication.

8.2 Excitation of Surface Flute Waves in the Electron Cyclotron Frequency Range by an Electron Beam Gyrating Around the Inner Conductor Inside a Coaxial Plasma Layer

In the previous section, SFXMs were shown to be efficiently excited by a flow of electrons rotating along Larmor orbits in the gap which separates the coaxial plasma layer from the external metal waveguide wall [26]. They are waves with positive azimuthal mode index *m*, which propagate in the same direction as electrons, which gyrate in the magnetic field. These LF SFXMs were shown in Sect. 2.2 to propagate with only positive azimuthal mode indices along the boundary of the coaxial plasma layer and the external metal wall [50]. That is why it was natural to expect in advance their excitation by the flow of electrons rotating along Larmor orbits in the gap that separates the coaxial plasma layer from the external metal wall. Since the previous Sect. 8.1 is devoted to the case of the coaxial waveguide where the rotating electron beam is located nearby the outer interface of the plasma layer, then, to complete the present investigation of beam-plasma interaction applied for excitation of SFWs in coaxial plasma waveguides, one should study the peculiarities of this interaction in the case when the beam electrons gyrate between the inner metallic conductor and the coaxial plasma layer.

A cylindrical coaxial metal waveguide, which contains a magnetized plasma layer, is considered. The radius of the inner metal conductor is $r = d$ and the radius of the outer wall is $r = a$ (see Fig. 8.8). The waveguide is made from a perfect conductor and is assumed to be uniform along the cylinder axis: $\partial/\partial z \equiv 0$. The coaxial line is partially filled with a plasma layer in the region $b < r < a$. An external homogeneous constant magnetic field is directed along the waveguide axis: $\vec{B}_0 \| \vec{z}$.

It is assumed that a gyrating electron beam is injected into the narrow gap between the coaxial plasma layer and the internal metal conductor of the waveguide ($d < r < b$). The beam is modeled as a flow of oscillators with the same transverse momentum

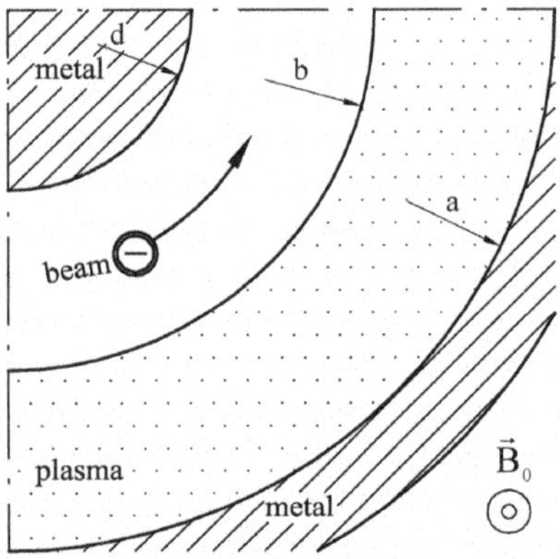

Fig. 8.8 Scheme of the coaxial beam-plasma-waveguide structure, where the electron beam gyrates inside the inner surface of the plasma layer

$p_{\perp 0}$ and zero axial momentum $p_z = 0$ as it is described in Sect. 8.1. The electron beam is assumed to be compensated in respect to electric current and charge. The ratio of the particle densities of beam and plasma electrons is small: $\alpha = n_b/n_e \ll 1$. Such an electron beam is described by the equilibrium distribution function (7.1). The electrodynamic properties of the waveguide part occupied by the electron beam are described by the dielectric permittivity tensor $\varepsilon_{ik}^{(b)}$ [24]. For solving the considered problem it is sufficient to apply the expressions for $\varepsilon_{ik}^{(b)}$ tensor given in the Sect. 7.2 [see expressions (7.18)–(7.20)].

The fields of SFXMs which propagate in the considered waveguide are described by special cylindrical functions. The main distinguishing feature of the present problem as compared with the case considered in the previous section is the difference in the applied boundary conditions [27]. This is explained by the difference in the position of the gyrating electron beam studied in these two cases. For the wave fields in the present case one can apply the following boundary conditions.

- The tangential component of the SFXM electric field is equal to zero on both metal surfaces: $E_\varphi(r = a) = E_\varphi(r = d) = 0$.
- The tangential components of the electric and magnetic fields of the SFXM are continuous on the interface between the plasma and vacuum region filled with the low density electron beam: $E_\varphi(r = b - 0) = E_\varphi(r = b + 0)$, $H_z(r = b - 0) = H_z(r = b + 0)$.

Applying these boundary conditions allows one to obtain the dispersion relation (8.1) which describes the SFXM excitation in the considered waveguide [27]. The dispersion relation is analyzed here numerically under the resonance condition (7.14) presented in the Sect. 7.1.

The results of the numerical analysis of dispersion relation (8.1) are plotted in Figs. 8.9, 8.10 and 8.11. In all these figures the dashed curves correspond to the case, when the parameters of the plasma filled coaxial waveguide are as follows: $d/b = 0.8$, $a/b = 1.1$, $Z = \Omega_e/|\omega_e| = 3$, and $\alpha = 0.01$. In all the figures the dependences of growth rates versus effective wavenumber $k_{ef} = mc/(b\Omega_e)$ are plotted.

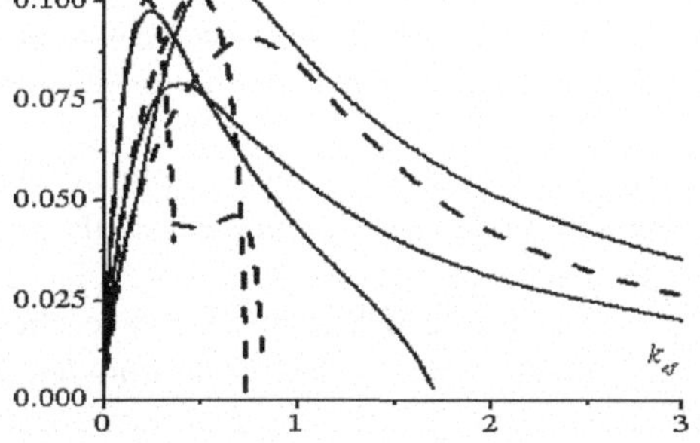

Fig. 8.9 Normalized LF SFXM growth rate $\mathrm{Im}(\omega/|\omega_e|)$ versus effective wavenumber; $m = 1, 2$ and 3, $Z = 3$, $d/b = 0.8$, $\alpha = 0.01$, *dashed curves* for $a/b = 1.1$, *solid curves* for $a/b = 5$. The case $a/b = 1.1$ and $m = 3$ shows overlapping resonances

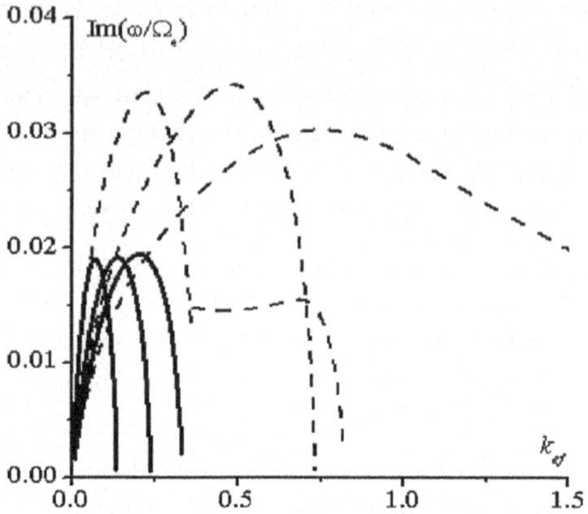

Fig. 8.10 Normalized LF SFXM growth rate $\mathrm{Im}(\omega/\Omega_e)$ versus effective wavenumber for different values of external static axial magnetic field: stronger field (Z = 3)—the dependence is shown by *dashed curves*; and weaker field (Z = 9)—the dependence is shown by *solid curves*. $m = 1$, 2 and 3, $d/b = 0.8$, $a/b = 1.1$, $\alpha = 0.01$. The case Z = 3 and $m = 3$ shows overlapping resonances

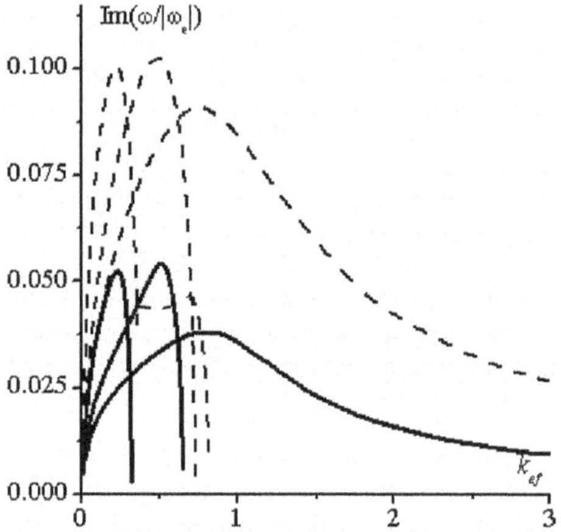

Fig. 8.11 Normalized LF SFXM growth rate $\mathrm{Im}(\omega/|\omega_e|)$ versus effective wavenumber for different values of the beam particle density: $\alpha = 0.01$ (*dashed curves*), and $\alpha = 0.00125$ (*solid curves*). $m = 1$, 2 and 3, Z = 3, $d/b = 0.8$, $a/b = 1.1$

In the Figs. 8.9 and 8.11 the growth rates are normalized to the absolute value of the electron cyclotron frequency, however, in Fig. 8.10 they are normalized to the electron plasma (Langmuir) frequency.

All the cases of LF SFXM excitation concern the following azimuthal mode indices: $m = +1, +2$ and $+3$. In all the cases, the higher the azimuthal mode index is, the larger is the effective wavenumber for which the dependence reaches its maximum. That is why the value of the azimuthal mode index is not shown in the plots.

The dashed curves, which are plotted in the Figs. 8.9, 8.10 and 8.11 for the first azimuthal mode $m = 1$, demonstrate the phenomenon of modes competition. This means that the curve consists of two "bells". In other words, these curves demonstrate that in this case the SFXM growth rate reaches its maximum twice. The first maximum of the dimensionless growth rate $\mathrm{Im}(\omega/|\omega_e|)$ is reached for $k_{ef} \approx 0.2$ (its value is $\mathrm{Max}\{\mathrm{Im}(\omega/|\omega_e|)\} \approx 0.101$), and the second maximum $\mathrm{Max}\{\mathrm{Im}(\omega/|\omega_e|)\} \approx 0.047$

Table 8.1 Comparison of maximum growth rate and corresponding effective wavenumbers calculated for coaxial waveguides where electrons gyrate in a narrow gap between the coaxial plasma layer and the inner rod of the waveguide ("internal beam"), or in a gap between the coaxial plasma layer and the inner wall of the outer waveguide ("external beam")

Azimuthal mode index	Internal electron beam		External electron beam	
	Max{Im(ω/|ω_e|)}	k_{ef}	Max{Im(ω/|ω_e|)}	k_{ef}
+1	0.101 0.047	0.2 0.7	0.096	0.25
+2	0.103	0.5	0.105	0.5
+3	0.091	0.75	0.113	0.8

can be seen for $k_{ef} \approx 0.7$. This is explained by the fact that the electron beam is considered in the present approximation as a flow of oscillators. If $k_{ef} < 0.36$, then the resonant interaction between LF SFXMs and the first cyclotron harmonic, $N = 1$, prevails. If $k_{ef} > 0.36$, then LF SFXM excitation is caused by their resonant interaction with the second cyclotron harmonic, $N = 2$. Excitation of LF SFXMs with the second and the third azimuthal mode indices (namely, for the cases $m = 2$ and $m = 3$) does not show any mode competition under the considered conditions. These SFXMs are excited due to their resonant interaction with a single cyclotron harmonic, namely, with $N = 2$ and $N = 3$, respectively.

The achieved results and their comparison with the results obtained in Sect. 8.1 are summarized below in the form of tables. The data in Table 8.1 convincingly demonstrates that both maximum values of SFXM growth rate and their position in the corresponding plot do not differ significantly for these two cases. The only qualitative difference consists in the fact of mode competition for SFXMs with $m = +1$ [27], mentioned above.

Figure 8.9 presents the results of studying the influence of the width of the coaxial plasma layer on the dependence of Im(ω/|ω_e|) for waves with different azimuthal mode indices $m \leq Z$. The solid curves are calculated for the case of a wide coaxial plasma layer, $a/b = 5$. The other beam-plasma parameters are the same as for the dashed curves, where $a/b = 1.1$ (thin plasma layer). The "bells" in the case of wide plasma layer are broader. This means that LF SFXM excitation can be observed for a broader set of waveguide parameters: for smaller waveguides and less dense plasmas. The data in Table 8.2 demonstrates that both maximum values of SFXM growth rate and their position in the plot also do not differ significantly for these two cases.

The dependence of growth rate on k_{ef} for SFXMs with $m = +3$ represents the phenomenon of overlapping resonances. For $k_{ef} < 0.175$ LF SFXMs are excited due to their interaction with the third cyclotron harmonic ($N = 3$). If $k_{ef} > 0.175$ then SFXM interaction with the second cyclotron harmonic ($N = 2$) prevails. Transition from the third to the second cyclotron harmonic is almost invisible on the curve unlike in the case for SFXM with $m = +1$ (dashed curve) [27].

The influence of the strength of the applied magnetic field $\vec{B}_0$ has a definite effect on LF SFXM-electron beam excitation. The higher the $\vec{B}_0$ value is, the larger are the SFXM growth rates and the broader is the range of k_{ef} for which efficient SFXM

Table 8.2 Comparison of maximum growth rate and corresponding effective wavenumbers calculated for a coaxial waveguide with inner thick ($a/b = 5$) and thin plasma layer ($a/b = 1.1$), respectively

Azimuthal mode index	Thin plasma layer		Thick plasma layer	
	$\mathrm{Max}\{\mathrm{Im}(\omega/|\omega_e|)\}$	k_{ef}	$\mathrm{Max}\{\mathrm{Im}(\omega/|\omega_e|)\}$	k_{ef}
+1	0.101 0.047	0.2 0.7	0.098	0.25
+2	0.103	0.5	0.079	0.4
+3	0.091	0.75	0.107	0.6

Table 8.3 Comparison of maximum growth rate and corresponding effective wavenumbers calculated for a waveguide immersed into a relatively strong ($Z = 3$) or weak ($Z = 9$) external axial static magnetic field

Azimuthal mode index	Stronger magnetic field		Weaker magnetic field	
	$\mathrm{Max}\{\mathrm{Im}(\omega/\Omega_e)\}$	k_{ef}	$\mathrm{Max}\{\mathrm{Im}(\omega/\Omega_e)\}$	k_{ef}
+1	0.03367 0.01567	0.2 0.7	0.0191	0.07
+2	0.03433	0.5	0.0192	0.14
+3	0.0303	0.75	0.0194	0.205

excitation is observed. This result is illustrated in Fig. 8.10, where the normalization factor is changed from $1/|\omega_e|$ to $1/\Omega_e$. This makes it possible to exclude the value of $\vec{B}_0$ from the magnitudes of both abscissa and ordinate. The solid curves are calculated for a weaker B_0 field, namely $Z = 9$. The maximum values of the growth rates are much smaller in this case as compared with the case of stronger B_0 field ($Z = 3$). The maxima of the growth rates for the case of weaker B_0 are reached for lower values of k_{ef} as compared with the case of stronger B_0 field, $Z = 3$ (see Table 8.3).

Figure 8.11 demonstrates the influence of the electron beam particle density on LF SFXM excitation. Decreasing ratio $\alpha = n_b/n_e$ results in decreasing values of growth rate and width of effective wavenumber k_{ef} ranges where LF SFXMs are efficiently excited by a gyrating electron beam. In particular, the solid curve corresponding to the azimuthal mode index $m = +1$ does not show mode competition since excitation of the SFXM with $m = +1$ by an electron beam with low density is caused by wave interaction with the first cyclotron harmonic of the beam ($N = 1$), unlike in the case of a relatively dense beam (dashed curve) [27]. Classical theory of beam-plasma interaction [1, 7, 9, 44, 47] predicts that the maximum value of the growth rate is proportional to the cubic root of the electron beam particle density. Then, decreasing α from $\alpha = 0.01$ to $\alpha = 0.00125$, that is by a factor of eight, should be accompanied by a reduction of the maximum value of the growth rate by a factor of two for the low beam density case. The results are summarized in Table 8.4.

The beam-plasma parameters which are needed for realization of the proposed waveguide design are estimated below. Using the data [35] one can obtain a gyrating

Table 8.4 Comparison of maximum growth rate and corresponding effective wavenumbers calculated for LF SFXM excitation by a relatively dense ($\alpha = 0.01$) electron beam and by a low density ($\alpha = 0.00125$) electron beam

Azimuthal mode index	High-density electron beam		Low-density electron beam	
	$\text{Max}\{\text{Im}(\omega/\Omega_e)\}$	k_{ef}	$\text{Max}\{\text{Im}(\omega/\Omega_e)\}$	k_{ef}
+1	0.101 0.047	0.2 0.7	0.053	0.25
+2	0.103	0.5	0.055	0.5
+3	0.091	0.75	0.038	0.8

electron beam with a radius of about 6.4 cm. If the density of the plasma layer in the proposed coaxial waveguide is $n_p \approx 8 \times 10^{10}$ cm^{-3}, then the Langmuir frequency is $\Omega_e \approx 8 \times 10^{10}\text{s}^{-1}$, and the skin-depth is $\delta \approx 1.6$ cm. Therefore, the ratio between Langmuir frequency and electron cyclotron frequency is $Z = 3$, if the axial magnetic is $B_0 \approx 315$ G, which can easily be obtained experimentally. For SFXMs with the azimuthal mode index $m = 2$ and effective wavenumber $k_{ef} \approx 0.5$ one can reach maximum values of the growth rate $Im(\omega/|\omega_e|)$. In this case the outer radius of the coaxial plasma layer should be $b \approx 6.5$ cm, the radius of the inner wall of the outer electrode of the studied coaxial waveguide is $a \approx 7.15$ *cm* (for the case $a/b = 1.1$), and the radius of the inner electrode is $d \approx 5.2$ *cm*. Analysis of this data shows that the thickness of the plasma layer in such coaxial waveguides is $a - b \approx 0.5$ cm.

However, one can see a lot of technological problems concerning practical realization of the proposed model. First, an increasing ratio a/b does not automatically lead to decreasing growth rates, and second, even $a - b \approx 0.5$ cm is not an extremely thin layer, nevertheless, production and maintenance of a relatively thin homogeneous plasma layer in such coaxial waveguides can be a difficult technical problem. The prospective way to solve this technological challenge can be to employ modern meta-materials (see, e.g. [40–43]). Their application also will allow one to solve another technical problem, which has to be solved in order to realize the proposed device, namely, the production of a high-density plasma.

Summarizing the results presented in this Section one can see that an initial stage of the resonant electron beam-plasma instability of SFXMs propagating in a coaxial waveguide, which is partially filled with cold plasma, caused by their interaction with the gyrating beam of electrons, has been studied in details. The maximum value of the growth rate $\text{Im}(\omega)$ is not very sensitive to the radius of the external waveguide wall. The main influence, which is caused by the external conductor, is that on the width of the effective wavenumber range, where these waves can be excited successfully. The increase of the azimuthal mode index leads to the shift of the maximum of the SFXM growth rates into the range of larger values of the effective wavenumber k_{ef} and to a broader width of the k_{ef} range, where SFXMs can be efficiently excited. Increasing the strength of external static magnetic field B_0 leads to the same consequences for $\text{Im}(\omega)$ as an increase of the azimuthal mode index m as well as to a significant increase of the $\text{Max}\{\text{Im}(\omega)\}$ value. The similar influence of the values of m and B_0

on the magnitude of Max{Im(ω)} is explained by their analogous influences on the LF SFXM eigenfrequency.

8.3 Excitation of Surface Flute Waves Above the Upper-Hybrid Frequency by an External Beam of Relativistic Electrons

Surface nature of SFXMs means that the amplitudes of the wave fields decrease when going from the plasma boundary to the axis in cylindrical metal waveguides partially filled by plasma. That is why their dispersion properties weakly depend on the presence of an internal conductor in the structure. This initiates the present study of beam excitation of SFXMs in the high-frequency (HF) range in coaxial waveguides. The initial stage of interaction between the gyrating beam of electrons which move along their large Larmor orbits in the narrow gap between the outer boundary of the coaxial plasma and the inner wall of the outer part of the coaxial metal waveguide and SFXMs in the electron cyclotron frequency range (low frequency range) has been studied in Sect. 8.1 [26]. It was demonstrated that the higher the plasma density is, the smaller is the influence of the internal conductor on the beam excitation and that the maximum value of the SFXM growth rate is not very sensitive to the radius of the internal conductor.

In coaxial metal waveguides entirely filled with plasma, SFXMs are known to be unidirectional waves [50]. In other words, SFXMs propagate along the plasma-metal boundary in definite directions, which are opposite in the two frequency regions: the electron cyclotron frequency range ((1.13), LF range) and the range higher than the upper-hybrid frequency ((1.14), HF range). Application of a dielectric (vacuum) layer strongly affects the dispersion properties of SFXMs even in the case of a thin layer [51]. In particular, the presence of a thin dielectric layer results in replacing the SFXM uni-directionality by their non-reciprocity.

A structure that consists of a cylindrical coaxial metal waveguide with a circular cross-section is considered in the present Section. The radius of the inner conductor is $r = a$ and the radius of the inner wall of the external waveguide structure is $r = d$ (Fig. 8.1). The coaxial line is partially filled with a plasma layer in the region $a < r < b$. The present study is restricted by the assumption of a thin vacuum layer, $(d - b) < b$. This assumption is applied to be sure that just the zeroth radial mode is under consideration. In this case one can expect the strongest influence of the plasma layer parameters on the dispersion properties of SFXMs. However, no definite value in the limit $(d - b) < b$ is taken into account in the present analytics.

The waveguide structure is made of a perfect conductor (metal with infinite electrical conductivity). An external homogeneous stationary magnetic field is directed along the waveguide axis: $\vec{B}_0 \| \vec{z}$. The waveguide is assumed to be uniform along the cylinder axis: $\partial/\partial z = 0$.

Extraordinary polarized electromagnetic waves (the non-zero components are E_r, E_φ and H_z) propagate inside the coaxial waveguide strictly perpendicularly to the external magnetic field $\vec{B}_0$. Ordinary polarized electromagnetic waves with the non-zero components H_r, H_φ and E_z are not considered here, since their dispersion properties do not depend on the strength of the external static magnetic field.

In the plasma region, the electric field displacement vector and the electric field vector are assumed to be connected by the permittivity tensor ε_j of a weakly collisional cold magnetized plasma [20, 24]. The magnetic component of a SFXM is described by a superposition of modified Bessel functions of the first and second kinds. Their electric components can be expressed through the magnetic wave field due to the Maxwell equations [see Eq. (1.10)].

The dispersion relation [28], which can be used for finding the growth rates of HF SFXMs in the coaxial plasma waveguide, has the form (8.1).

In Figs. 8.12, 8.13, 8.14 and 8.15, the results of the present numerical study of the HF SFXM beam excitation are plotted. Growth rates, normalized to the electron cyclotron frequency, Im$[\omega/|\omega_e|]$, and multiplied by 10^3, are taken along the ordinate axis. The effective wavenumber, $k_{ef} = |m|c/(b\Omega_e)$ is counted along the abscissa axis. The ratio $|m|/b$ plays the role of the wave vector in the case of wave propagation in azimuthal direction in the coaxial waveguide. The effective wavenumber is given by this wave vector normalized to the skin-depth $\delta = c/\Omega_e$. In all the figures presented here, only the largest growth rates are shown, which are related to the short wavelength limit. In the long wavelength limit, $k_{ef} \to 0$, the SFXM dispersion curves are crossed by an infinite number of curves $\omega/|\omega_e| = N/\gamma$ with the higher N values. However, the higher N is the lower is the value of the growth rate. This is explained by the fact that the growth rate is inversely proportional to the cubic root of the derivative, $[\partial D^{(0)}/\partial\omega]^{1/3}$ (where $D^{(0)} = 0$ is the dispersion relation in the absence of the electron beam), that turns to infinity in the long wavelength limit, $k_{ef} \to 0$.

Doing the calculations, the radial phase change is checked to ensure that just the zeroth radial SFXM mode is under consideration. The two indices m and N are

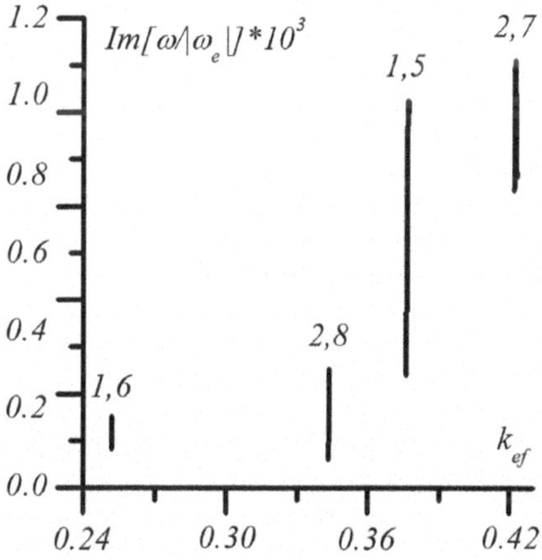

Fig. 8.12 Normalized growth rates of SFXMs with azimuthal indices $m = 1$ and 2 versus effective wavenumber for $Z = 3.5$; $d/b = 1.2$; $a/b = 0.75$; $\Omega_b{}^2 = 0.01\Omega_e{}^2$. The *number pairs* close to the curves indicate m,N where N is the cyclotron harmonic number

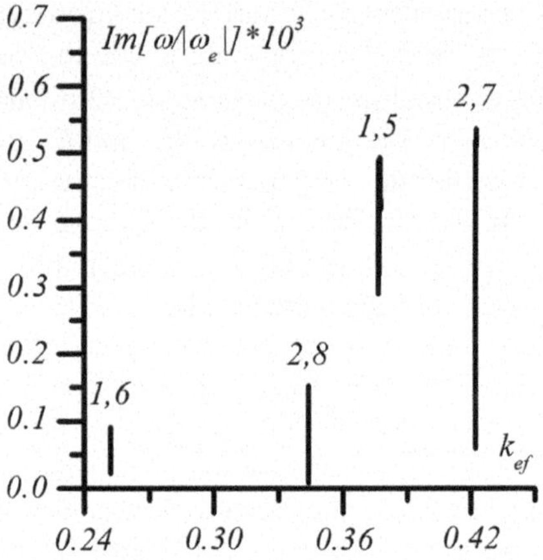

Fig. 8.13 Normalized growth rates of SFXMs with azimuthal indices $m = 1$ and 2 versus effective wavenumber for $Z = 3.5$; $d/b = 1.2$; $a/b = 0.75$; and lower beam particle density $\Omega_b{}^2 = 0.01\Omega_e{}^2/8$. The *number pairs* close to the *curves* indicate m,N where N is the cyclotron harmonic number

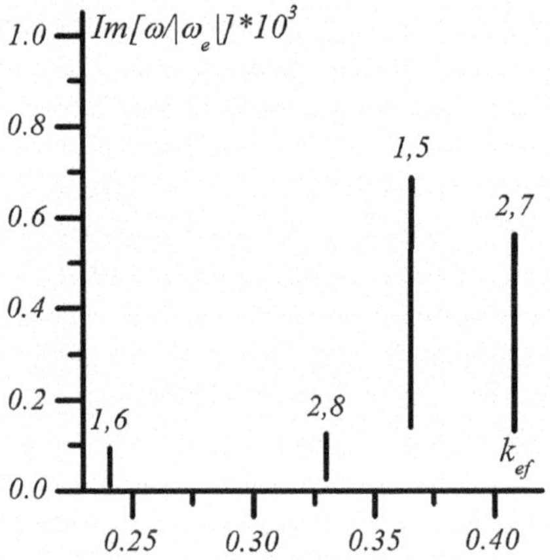

Fig. 8.14 Normalized growth rates of SFXMs with azimuthal indices $m = 1, 2$ versus effective wavenumber for $Z = 3.5$; $a/b = 0.75$; $\Omega_b{}^2 = 0.01\Omega_e{}^2$; and smaller vacuum gap $d/b = 1.1$. The *number pairs* close to the curves indicate m,N where N is the cyclotron harmonic number

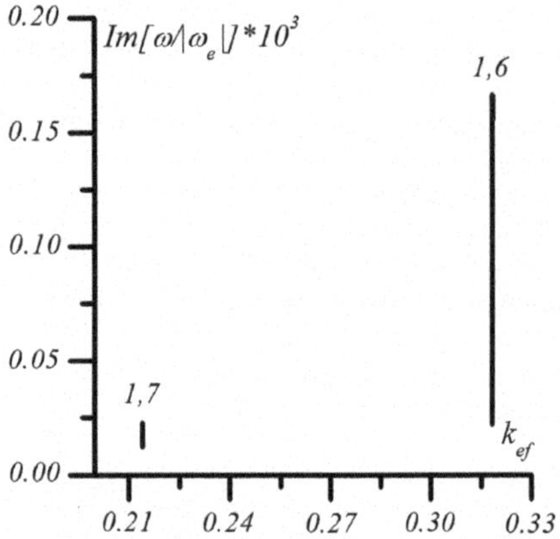

Fig. 8.15 Normalized growth rates of SFXMs with azimuthal indices $m = 1, 2$ versus effective wavenumber for $d/b = 1.2$; $a/b = 0.75$; $\Omega_b{}^2 = 0.01\Omega_e{}^2$; and weaker static magnetic field $Z = 4.5$. The *number pairs* close to the *lines* indicate m,N where N is the cyclotron harmonic number

assigned to each curve in Figs. 8.12, 8.13, 8.14 and 8.15. The first index denotes the azimuthal mode index of the wave and the second one the number of the beam cyclotron harmonic that causes the strongest impact on the wave-beam interaction.

The growth rates of the HF ASW beam instability are plotted in Fig. 8.12 for the azimuthal mode indices $m = 1$ and 2. The following values of the plasma waveguide parameters are chosen to make it possible to compare the results with those in Sect. 8.1: $Z = \Omega_e/|\omega_e| = 3.5$; $\Delta \equiv (d-b)/b = 0.2$, $a/b = 0.75$, and $\alpha = 0.01$. In this case the largest value of the effective wavenumber where the dispersion curve of the SFXM with the azimuthal mode index $m = 1$ is crossed by the curve $\omega/|\omega_e| = N/\gamma$, is $k_{ef} = 0.377$. The half-width of this maximum is very small, $\Delta k_{ef} \approx 6 \times 10^{-4}$, that is why the curve looks just like a vertical line. SFXM excitation is caused in this case by the wave interaction with the fifth cyclotron harmonic of the beam, $N = 5$. The dispersion curve for the SFXM with $m = 1$ is not crossed by any other curve $\omega/|\omega_e| = N/\gamma$ to the right of $k_{ef} = 0.377$. The maximum effective wavenumber for which the HF SFXM with $m = 1$ exists is $k_{ef} = 0.55$. At this k_{ef} the dispersion curve tends to the upper limit of the frequency range, $\omega = \omega_1$. Introducing the inner coaxial conductor results in changing the values of k_{ef} where the HF SFXM dispersion curve is crossed by the curve $\omega/|\omega_e| = N/\gamma$. For instance, this intersection point moves to the right for the case of the SFXM with $m = 1$ and the curve $N = 5$: from $k_{ef} = 0.356$ in the case of a cylindrical metal waveguide partially filled by plasma ($a = 0$) [52] to $k_{ef} = 0.377$ in the coaxial waveguide with $a/b = 0.75$.

This is accompanied by a relatively small decrease of the maximum value of the SFXM growth rate by 26%. The opposite situation takes place for the dispersion curve of the SFXM with $m = 2$ and the line $N = 7$—their intersection point moves to the left: from $k_{ef} = 0.566$ in the case of the cylindrical metal waveguide partially filled by plasma ($a = 0$) [52] to $k_{ef} = 0.422$ in the coaxial waveguide with $a/b = 0.75$. This is accompanied by a considerable decrease of the maximum value of the SFXM growth rate by the factor 4.25.

The words «beams of relativistic electrons» are applied in the title of the present Section. Now it is shown, that just relativistic electron beams really are required for excitation of HF SFXMs in coaxial waveguides. For example, for the SFXM with the azimuthal mode index $m = 1$ the most efficient excitation occurs at $k_{ef} \approx 0.377$. Under this condition the velocity of the electron beam should be $\upsilon \approx 0.64c$. Thus, the relativistic factor is equal to $\gamma \approx 1.3$. For the SFXM with $m = 2$ the requirements of excitation are even more rigid: $\upsilon \approx 0.83$c; therefore, the relativistic factor is equal to $\gamma \approx 1.79$, which corresponds to an acceleration voltage of approx. 400 kV.

As it is well-known from the very first discovery of the beam-plasma instability [53], the maximum value of the growth rates should be proportional to the cubic root of beam particle density. Results of studying the dependence of HF ASW growth rates under the condition of beam instability with azimuthal mode indices $m = 1$ and 2 on the effective wavenumber are presented in Fig. 8.13 for the same parameters as in Fig. 8.12, except of the beam particle density, which is eight times lower. According to theoretical predictions, eight times decreasing the beam particle density should be accompanied by a reduction of the growth rate by a factor of two. Numerical analyses show good agreement of the calculations and theoretical predictions.

Figures 8.12 and 8.13 look like the same with different scale regardless of being the result of calculations carried out for different beam particle densities. In the case $\Omega_b^2 = 0.01\Omega_e^2$ the maximum value of the growth rate of the SFXM with $m = 1$ is equal $\mathrm{Im}(\omega/|\omega_e|) = 0.999 \times 10^{-3}$ (line 1,5 in Fig. 8.12), whereas in the case $\Omega_b^2 = 0.01\Omega_e^2/8$ the maximum value of the growth rate is approx. two times lower, i.e. $\mathrm{Im}(\omega/|\omega_e|) = 0.496 \times 10^{-3}$ (line 1,5 in Fig. 8.13).

The dependence of HF SFXM growth rates on the thickness of the vacuum gap, where the electron beam moves, is demonstrated in Fig. 8.14. There, for comparison with the case $\Delta = 0.2$ considered in Fig. 8.12, the smaller value $\Delta = 0.1$ is chosen for these calculations. The growth rate for the SFXM with $m = 1$ reaches its maximum value for a lower value of k_{ef} ($k_{ef} = 0.366$) than in the case of a wide gap $\Delta = 0.2$ ($k_{ef} = 0.377$), shown in Fig. 8.12. This maximum value of the growth rate is lower in this case ($\mathrm{Im}(\omega/|\omega_e|) = 0.69 \times 10^{-3}$) as compared with the case of the wide gap ($\mathrm{Im}(\omega/|\omega_e|) = 0.99 \times 10^{-3}$).

Appreciable influence on the propagation of SFXMs is given by the strength of the external static magnetic field [51]. Therefore, this influence on HF SFXM excitation is analyzed in Fig. 8.15. To determine the effect of the magnetic field a weaker magnetic field ($Z = 4.5$) compared to the case shown in Fig. 8.14 ($Z = 3.5$) is chosen for calculations of the growth rate. One can see that a relatively small reduction of the strength of the external magnetic field essentially reduces the growth rate and thus the conditions of SFXM excitation.

The maximum of the growth rate for $m,N = 1,5$, that is situated in the Fig. 8.14 at $k_{ef} = 0.366$, is absent in Fig. 8.15. It moves with decreasing stationary magnetic field (increasing parameter Z) to the right and for $Z = 4.5$ leaves the range of k_{ef} where the SFXMs propagate.

The maximum growth rate for $m,N = 1,6$, that is located in Fig. 8.14 at $k_{ef} = 0.241$, shifts with decreasing stationary magnetic field to the right and for $Z = 4.5$ appears at $k_{ef} = 0.318$. The maximum value of the growth rate in this peak increases by the factor 1.77 from $\mathrm{Im}(\omega/|\omega_e|) = 9.44 \times 10^{-5}$ for $Z = 3.5$ to $\mathrm{Im}(\omega/|\omega_e|) = 1.67 \times 10^{-4}$ for $Z = 4.5$. This increase in growth rate, that accompanies the reduction of the stationary magnetic field, cannot be considered as a general rule. It is explained by the shift of the peak to the right along the axis of k_{ef} where the dispersion curve has a larger inclination.

The peak of the growth rate for $m,N = 2,7$, that is located at $k_{ef} = 0.408$ in Fig. 8.14, does not appear in Fig. 8.15 since the line $N = 7$ does not intersect the dispersion curve of the SFXM with $m = 2$ in this case. The intersection point of the line $N = 8$ with the dispersion curve of the SFXM with $m = 2$ moves from $k_{ef} = 0.33$ for stronger magnetic field, $Z = 3.5$, to the right ($k_{ef} = 0.344$) but the values of the growth rate are negligibly small in this case.

As noted in [52] for the case of no internal conductor, the dependence of SFXM beam instability growth rates on the strength of the external magnetic field is not monotonous. At gradual reduction of the magnetic field (that is accompanied by the increase of the parameter Z) the growth rate maxima move to the right, and those which correspond to SFXM interaction with cyclotron harmonics of the electron beam with smaller numbers N leave the range of k_{ef} where SFXMs exist. In addition, the maxima with higher N move to the right, have increasing values, and follow those with lower N going out of the range of k_{ef} where HF SFXMs propagate.

Now, the influence of the radius of the inner conductor on HF SFXM excitation by an electron beam is examined. Figure 8.16 shows the dependence of SFXM growth rates on the effective wavenumber for the same parameters of beam-coaxial structure as in Fig. 8.12, however, with a larger inner conductor. The width of the plasma layer is chosen to be a factor of two smaller than in the case given in Fig. 8.12 ($a/b = 0.875$). The peaks shown in Fig. 8.12 for the case of a wide plasma coaxial layer ($a/b = 0.75$) move to the left with narrowing the plasma layer width to $a/b = 0.875$ with negligible decrease of the value of the maximum growth rate. For instance, the peak for the mode $m,N = 1,5$ moves from the effective wavenumber $k_{ef} = 0.377$ in Fig. 8.12 to $k_{ef} = 0.357$ in Fig. 8.16. Herewith the maximum value of the growth rate slightly decreases from $\mathrm{Im}(\omega/|\omega_e|) = 9.99 \times 10^{-4}$ to $\mathrm{Im}(\omega/|\omega_e|) = 9.54 \times 10^{-4}$. The peak of the normalized growth rate for $m,N = 1,4$ remains out of the range of k_{ef} where this wave exists. However, the $m,N = 2,6$ mode appears at $k_{ef} = 0.532$ which has the highest peak in Fig. 8.16. Its normalized growth rate is $\mathrm{Im}(\omega/|\omega_e|) = 3.61 \times 10^{-3}$, with a half-width of $\Delta k_{ef} \approx 3 \times 10^{-3}$ leading to the highest and broadest peak compared to all the other modes shown in all the other Figs. 8.12, 8.13, 8.14 and 8.15.

In the following, the possibility of experimental verification of the phenomena studied in the present Section for excitation of electromagnetic radiation in the GHz range is estimated. The most favorable conditions for electron-beam excitation of HF SFXMs (the largest growth rate) among those presented in the figures above correspond to the highest peak in Fig. 8.16. It relates to interaction with the HF SFXM with $m = 2$ at the 6th cyclotron harmonic of the electron beam ($N = 6$). A coaxial line with the radius of plasma-vacuum boundary $b \sim 10$ cm is considered as an example. Then to meet the condition of Fig. 8.16 the radius of the internal conductor should be $a \sim 8.75$ cm and that of the outer conductor $d \sim 12$ cm. Since the maximum growth rate occurs at $k_{ef} = 0.532$, the electron plasma frequency

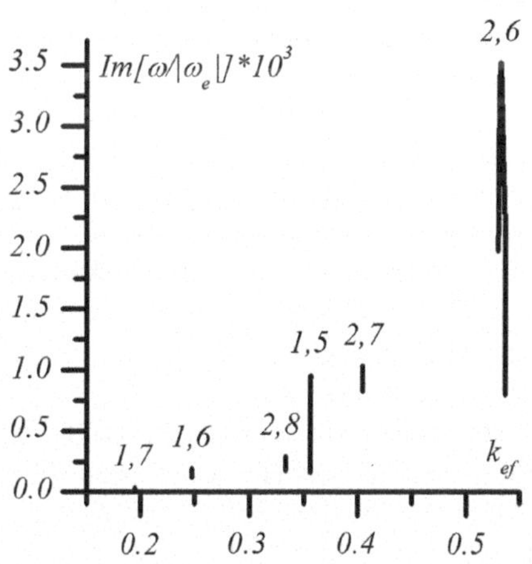

Fig. 8.16 Normalized growth rates of SFXMs with azimuthal indices $m = 1, 2$ versus effective wavenumber for $Z = 3.5$; $d/b = 1.2$; $\Omega_b^2 = 0.01\Omega_e^2$; and larger radius of the inner conductor $a/b = 0.875$. The *number pairs* close to the *curves* indicate m,N where N is the cyclotron harmonic number

should be equal to $\Omega_e \sim 1.13 \times 10^{10}$ s^{-1}, that corresponds to a plasma density of $n_e \sim 5 \times 10^{10}$ cm^{-3}. The plot in Fig. 8.16 is calculated for the parameter $Z = 3.5$. Therefore, the absolute value of electron cyclotron frequency should be equal $|\omega_e| \sim 3 \times 10^9$ s^{-1}, which corresponds to an external static magnetic field of $B_0 \sim 173$ G. In this case, the electron beam rotates with the velocity 0.763 c, which corresponds to relativistic factor $\gamma \sim 1.548$ (280 kV acceleration voltage). Then the considered HF SFXM propagates with the frequency $\omega \sim \Omega_e$ and it is excited with the growth rate $\mathrm{Im}(\omega) \sim 10^7$ s^{-1}.

In the present Section, in addition to the results obtained earlier in Sect. 8.1 the possibility to realize the conditions of electron-beam instability for SFXMs in the range above the upper-hybrid frequency in coaxial plasma-vacuum waveguides is proved [28]. The results of this study can be of interest for solving problems of plasma electronics.

Surface type nature of the electromagnetic waves considered here means that their amplitude decreases with going away from the plasma-vacuum boundary inside the plasma layer. That is why the electrodynamic properties of the medium placed nearby the waveguide axis affect the dispersion properties of these waves only weakly. This makes it possible to replace a portion of plasma column nearby the axis just by a metallic coaxial inner conductor. In an experimental device, this can reduce the costs of plasma production and device maintenance. Application of an internal conductor does not reduce HF SFXM beam excitation as compared to the case of its absence. Moreover, its application extends the range of plasma parameters for which SFXM excitation is efficient and even increases the peak values of SFXM growth rates [28].

Application of a wider gap Δ between plasma boundary and outer waveguide wall allows one to expect a higher efficiency of the generator of electromagnetic radiation and an increased growth rate of the HF SFXM beam-plasma instability [28].

The value of the external static magnetic field is characterized by the parameter $Z = \Omega_e/|\omega_e|$. A decrease of the magnetic field results in increasing Z. On one hand, this means that the HF range becomes wider, which facilitates the search of optimum conditions for SFXM excitation. On the other hand, this results in a decrease of the number of the cyclotron harmonics (interaction with them causes the SFXM excitation), and in a reduced maximum growth rate (with other experimental parameters being equal).

The character of the dependence of the efficiency of HF SFXM excitation by a relativistic electron beam on the plasma external radius b is explained by the dispersion properties of these SFXMs. In wide waveguides, $k_{ef} \to 0$, the SFXM frequency weakly depends on the plasma radius b, $\omega \propto k_{ef}^2$. However, in narrow waveguides, for $k_{ef} > 1/Z$, HF SFXMs do not propagate [54]. Therefore, it is desirable for HF SFXM excitation to choose a small radius of the plasma column for which the SFXM propagates in the generator with a frequency, which is close to upper bound of the HF range [28].

In comparison with the case of LF SFXM excitation in coaxial waveguides [26], the growth rates of HF SFXM beam-plasma instability are lower [28]. HF SFXMs are excited by rotating electron beams in narrow ranges of the coaxial plasma waveguide parameters, while for the same values of the beam particle density LF SFXMs are excited in the regime of resonance overlapping (LF SFXMs can be excited in the regime of separate resonances only for much lower beam densities than studied in the present Section). This is explained, first of all, by worsening the possibilities of realizing the resonance conditions (7.14) of beam-wave interaction. For HF SFXMs this requirement is demonstrated in the best case for the fifth cyclotron harmonic of the beam, $N = 5$, however, one should remind that the growth rate decreases with the number N, $\mathrm{Im}(\omega) \propto J_N^{2/3}(|m|)$.

8.4 Excitation of Surface Flute Waves Above the Upper-Hybrid Frequency by an Internal Gyrating Electron Beam

The study of the present Section expands the results obtained in Sect. 8.3 [28] to the case of an internal electron beam, as well as those obtained in Sect. 8.2 [27] for LF SFXM excitation by an internal electron beam to the higher frequency range above the upper-hybrid frequency.

The problem of producing a gyrating flow of electrons is out of scope of the present monograph. However, the possibility to produce an electron beam gyrating in the vacuum gap that separates the coaxial plasma from the internal metal wall is demonstrated in [35]. That is why the question arises whether HF SFXMs can be excited in a coaxial plasma waveguide by an internal electron beam. Thus, the aim of this Section is to find favorable conditions for HF SFXM excitation in the coaxial structure mentioned above. Here, only the initial (linear) stage of their resonant beam-plasma instability is studied.

The electron beam-plasma coaxial structure under consideration is shown in Fig. 8.8. This system is described in detail in Sect. 8.2 [29]. That is why one can formulate the problem here only briefly. In Fig. 8.8 the radius of internal metal conductor is $r = d$; the electron beam gyrates within the coaxial layer $d < r < b$; the coaxial plasma layer fills the space $b < r < a$ and has contact with the inner wall of the outer metal cylinder at $r = a$. The conductivity of the metal wall is assumed to be infinite. A stationary uniform magnetic field is oriented along the axis of the structure, $\vec{B}_0 \| \vec{z}$. The coaxial waveguide is assumed to be homogeneous along the axial co-ordinate. The model of uniform plasma density is applied, which is particularly suited for the case of surface type electromagnetic waves propagation unlike the case of bulk waves [21].

One should remind that electromagnetic waves of extraordinary polarization with the non-zero field components E_r, E_φ and H_z are studied, which propagate strictly perpendicularly to the external axial magnetic field $\vec{B}_0$. These wave fields have no dependence on the axial co-ordinate: $E_r, E_\varphi, H_z(r, \varphi, t) = f(r)\exp(im\varphi - i\omega t)$. Electron beam excitation of HF SFXMs is studied in the HF range (1.14).

Within the plasma layer the SFXM wave fields are described in terms of modified Bessel functions of the first kind, $I_m(\xi)$, and the second kind, $K_m(\xi)$, and their derivatives over the arguments. Here $\xi = k_{\perp} r$, where the value $1/k_{\perp}$ defines the penetration depth of the wave field into the plasma, $k_{\perp}^2 = k^2 \varepsilon_1 (\mu^2 - 1)$, $k = \omega/c$, and $\mu = \varepsilon_2/\varepsilon_1$, where $\varepsilon_{1,2}$ are the components of the plasma permittivity tensor [20, 24].

The model of the electron beam as a flow of oscillators is applied [24]. The electrodynamic properties of the spatial layer, $d < r < b$, where the electrons gyrate, can be described by the permittivity tensor $\varepsilon_{ik}^{(b)}$ [see (7.18)–(7.20)]. This tensor is represented in the form of series over the electron cyclotron harmonics. The electron beam is assumed to be compensated in respect to currents and charges; the beam particle density is small as compared to the plasma density, $\alpha \equiv n_b/n_e \ll 1$. This model is applicable for both ultra-relativistic and non-relativistic electron beams.

Within the space $d < r < b$, SFXM electromagnetic waves are described in terms of Bessel functions of the first kind, $J_m(\zeta)$, and the second kind, $N_m(\zeta)$, and their derivatives over the argument $\zeta = kr\sqrt{\Psi_b}$, where $\Psi_b = \varepsilon_{22}^{(b)} + (\varepsilon_{12}^{(b)})^2/\varepsilon_{11}^{(b)}$.

The well-known linear boundary conditions for the wave fields presented in Sect. 1.4.4 are applied [27]. Utilizing these boundary conditions allows one to obtain the dispersion relation (8.1) that describes HF SFXM excitation in the studied coaxial waveguide. This dispersion relation is analyzed numerically under the resonance condition (7.14).

To explore the parameter ranges of possible experimental devices, the results of the numerical studies on HF SFXM beam excitation are shown in Figs. 8.17, 8.18, 8.19, 8.20 and 8.21. Growth rates normalized by the electron cyclotron frequency, $\mathrm{Im}[\omega/|\omega_e|]$, and multiplied by 10^4 are plotted along the ordinate axis. The effective wavenumber, $k_{ef} = |m|c/(b\Omega_e)$, is counted along the abscissa axis.

As it is well-known (see e.g. [7, 24]), excitation of electromagnetic waves takes place in the most efficient way just nearby their eigenfrequencies. Analysis of the dispersion relation in the case of beam-plasma instability shows, that the influence of an electron beam with low particle density ($\alpha \ll 1$) is sufficient in the case of

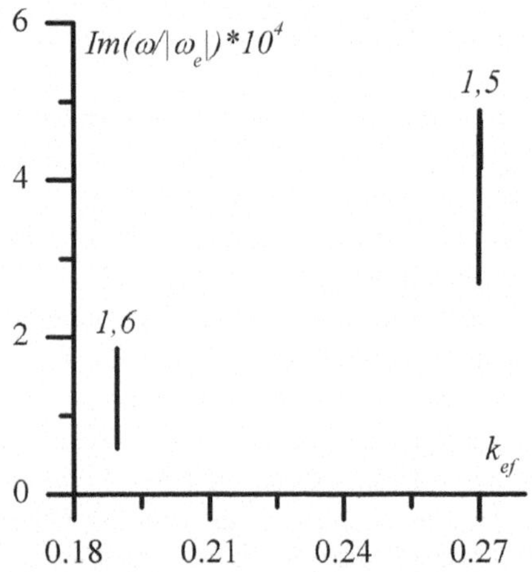

Fig. 8.17 Normalized HF SFXM growth rate $\mathrm{Im}(\omega/|\omega_e|)$ versus effective wavenumber for $m = 1$, $d/b = 0.8$, $\alpha = 0.01$, $a/b = 1.25$, and $Z = 3.5$. The *number pairs* close to the curves indicate *m,N* where *N* is the cyclotron harmonic number

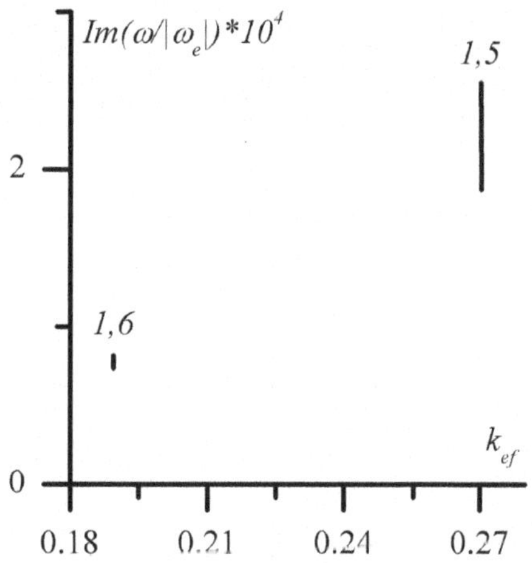

Fig. 8.18 Normalized HF SFXM growth rate Im($\omega/|\omega_e|$) versus effective wavenumber for $m = 1$, $d/b = 0.8$, $a/b = 1.25$, $Z = 3.5$, and lower beam particle density $\alpha = 0.00125$. The *number pairs* close to the lines indicate m,N where N is the cyclotron harmonic number

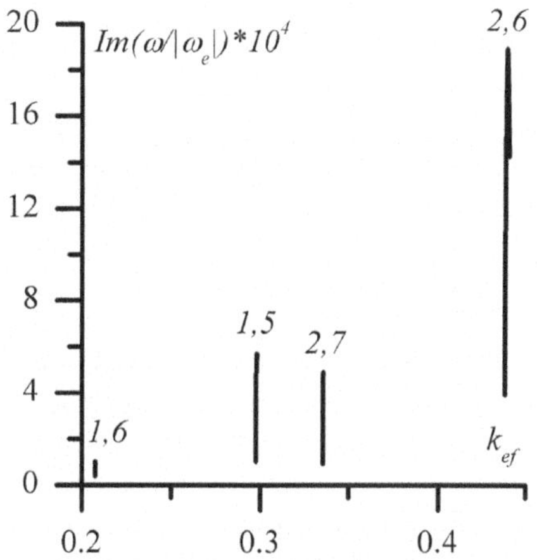

Fig. 8.19 Normalized HF SFXM growth rate Im($\omega/|\omega_e|$) versus effective wavenumber for $m = 1, 2$, $\alpha = 0.01$, $a/b = 1.25$, $Z = 3.5$, and a smaller value of the gap width $d/b = 0.9$. The *number pairs* close to the lines indicate m,N where N is the cyclotron harmonic number

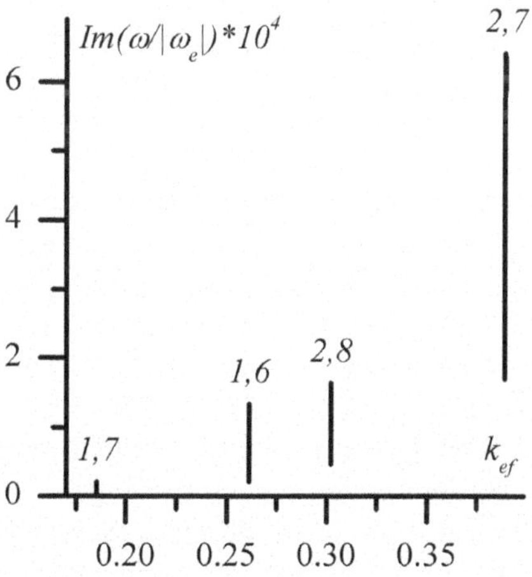

Fig. 8.20 Normalized HF SFXM growth rate Im($\omega/|\omega_e|$) versus effective wavenumber for $m = 1, 2$, $d/b = 0.8$, $\alpha = 0.01$, $a/b = 1.25$, and a weaker external static axial magnetic field $Z = 4.5$. The *number pairs* close to the lines indicate m,N where N is the cyclotron harmonic number

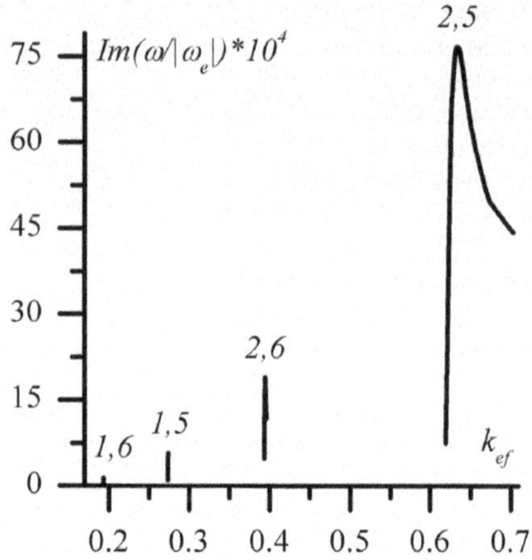

Fig. 8.21 Normalized HF SFXM growth rate $\mathrm{Im}(\omega/|\omega_e|)$ versus effective wavenumber for $m = 1, 2$, $d/b = 0.8$, $\alpha = 0.01$, $Z = 3.5$, and narrower plasma layer $a/b = 1.125$. The *number pairs* close to the curves indicate *m,N* where *N* is the cyclotron harmonic number

presence of resonance denominators in the expressions for $\varepsilon_{ik}^{(b)}$, i.e. in the case that condition (7.14) is valid.

The HF SFXM dispersion curve is known to start from $k_{ef} = 0$ where $\omega \approx \omega_{UH} + \beta k_{ef}{}^2$ (here $\beta > 0$ is a positive constant). For each set of beam-plasma parameters, there is a maximum value of effective wavenumber, $k_{\max}$. When $k_{ef} \to k_{\max}$ then the eigenfrequency approaches to the upper limit of the HF SFXM frequency range, $\omega \to \omega_2$.

The two indices *m,N* are assigned to each growth rate peak in Figs. 8.17, 8.18, 8.19, 8.20 and 8.21. The first index denotes the azimuthal mode index of the wave, and the second one the number of the electron beam cyclotron harmonic that causes the strongest contribution to the beam-wave interaction. In all the figures shown in the present Section, only the largest growth rates are plotted.

For a given azimuthal mode index *m*, there exists an infinite number ($N \to \infty$) of intersection points of the dispersion curve with the resonant lines $\omega = |\omega_e| N/\gamma$. For higher numbers *N* these points are located at smaller k_{ef}. The larger *N* is, the smaller is the maximum value of the growth rate peak. This is explained by two reasons. First, the growth rate is inversely proportional to the cubic root of the derivative, $[\mathrm{Im}(\omega/|\omega_e|)] \propto [\partial D^{(0)}/\partial\omega]^{-1/3}$ (where $D^{(0)} = 0$ is the dispersion relation when there is no electron beam). This derivative turns to infinity in the long-wavelength limit, $k_{ef} \to 0$, and has its minimum when $k_{ef} \to k_{\max}$. Second, the components of the permittivity tensor $\varepsilon_{ik}^{(b)}$ are proportional to Bessel functions of the order *N*, and Bessel functions are known to have lower values for higher orders and fixed argument.

For a given pair *m,N*, a change of the parameters of the coaxial beam-plasma waveguide causes a shift of the position $k_{ef} = k_{m,N}$ of the maximum of the corresponding growth rate peak. The closer $k_{m,N}$ is to $k_{\max}$, the larger is the peak value. This increase is explained by the fact that the dispersion curve has a steeper inclination when $k_{ef} \to k_{\max}$. If a change of a certain coaxial waveguide parameter causes the shift of $k_{m,N}$ to the right, then the peak *m,N* successively increases until $k_{m,N}$ reaches $k_{\max}$. With a further change of the parameter the peak leaves the range of k_{ef}

where the HF SFXM exists, and the cyclotron harmonic N stops to contribute to the HF SFXM electron beam excitation. Then the peak $m,N + 1$ starts to provide the most favorable conditions for HF SFXM beam excitation with the same azimuthal mode index. Such change of one of the coaxial waveguide parameters is referred as one that worsens the conditions for HF SFXM beam excitation.

Growth rates of HF SFXM beam instability are plotted in Fig. 8.17 for the following values of the system parameters: $Z = \Omega_e/|\omega_e| = 3.5$; $d/b = 0.8$, $a/b = 1.25$, and $\alpha = 0.01$. Just these values are chosen to make it possible to compare the results with those in the case of HF SFXM excitation by an external electron beam (see Sect. 8.3 [28]) with the following parameters: $Z = 3.5$; $d/b = 1.2$, $a/b = 0.75$, and $\alpha = 0.01$.

The parameter Z is chosen to have a moderate value for the following reason. For large Z (weak external magnetic field), the HF range (1.14) lies much higher than the electron cyclotron frequency. Then HF SFXMs can be excited due to interaction with the higher electron beam cyclotron harmonics N. As it is already explained above, the higher N is, the lower is the maximum value of the growth rate peak.

Table 8.5 summarizes the results of comparing the main characteristics of HF SFXM excitation in two types of coaxial waveguides: first—waveguides in which the electron flow gyrates in the narrow gap between the coaxial plasma layer and the inner conductor of the coaxial waveguide (internal electron beam studied in the present Section [29]), and second—waveguides in which the electrons gyrate in the gap between the coaxial plasma layer and the outer waveguide wall (external electron beam studied in the Sect. 8.3 [28]). The following parameters are presented to be compared: azimuthal mode index m, number N of cyclotron harmonic of the beam that causes beam-wave interaction, maximum value of the normalized growth rate in the corresponding peak, position of the maximum, $k_{m,N}$, and half-width of the peak, Δk_{ef}.

In the case of an internal beam, the dispersion curve of the HF SFXM with $m = 2$ does not intersect with the lines $\omega = |\omega_e| N/\gamma$ for $N < 11$. Intersection with the line $\omega = 11|\omega_e|/\gamma$ takes place for $k_{m,N} = 0.199$ that is not far from $k_{\max} = 0.210$. Nevertheless, the value of the corresponding growth rate is negligibly small.

Table 8.5 Comparison of the parameters of HF SFXM beam excitation by internal and external electron beams for the conditions of Fig. 8.17

	$\max[\mathrm{Im}(\omega/|\omega_e|)]$	$k_{m,N}$	$\Delta k_{ef} \times 10^4$
m,N	*Internal electron beam*		
1,5	5.08×10^{-4}	0.270	3.15
1,6	1.85×10^{-4}	0.190	0.479
m,N	*External electron beam*		
1,5	1.00×10^{-3}	0.377	10.0
1,6	1.83×10^{-4}	0.252	1.83
2,7	1.10×10^{-3}	0.422	11.0
2,8	3.04×10^{-4}	0.344	3.04

In the following, it is demonstrated that just relativistic electron beams are required for excitation of HF SFXMs in coaxial waveguides. For example, the most effective excitation occurs for HF SFXMs with the azimuthal mode index $m = 1$ at $k_{m,N} \approx 0.27$, where the velocity of the beam should be $\upsilon \approx 0.69\ c$ (for comparison: $\upsilon \approx 0.64\ c$ in the case of an external electron beam); thus the relativistic factor is equal to $\gamma \approx 1.38$ ($\gamma \approx 1.3$ in the case of an external beam).

As it is well-known from the very first discovery of electron beam-plasma instabilities [53], the maximum value of the growth rates is proportional to the cubic root of the beam particle density. Results of studying the dependence of HF SFXM growth rates under the condition of beam-plasma instability with azimuthal mode indices $m = 1$ and 2 on the effective wavenumber are shown in Fig. 8.18 for the same parameters as in Fig. 8.17, except of the beam particle density, which is eight times lower. According to the theoretical predictions, such a reduction of the electron beam particle density should be accompanied by a reduction of the growth rate by a factor of two. The numerical analysis shows good agreement of calculations and theoretical predictions. Figures 8.17 and 8.18 look like the same with different scale regardless of being the result of calculations carried out for different beam particle densities. In the case $\alpha = 0.01$ the maximum value of the normalized growth rate of the SFXM with $m = 1$ is equal to $\mathrm{Im}(\omega/|\omega_e|) = 5.08 \times 10^{-4}$ (line 1,5 in Fig. 8.17), whereas in the case $\alpha = 0.00125$ the maximum value of the growth rate is lower by a factor of 1.99, i.e. $\mathrm{Im}(\omega/|\omega_e|) = 2.55 \times 10^{-4}$ (line 1,5 in Fig. 8.18). This result clearly verifies the present model.

Table 8.6 allows to compare in detail the results of HF SFXM beam excitation under the conditions of a beam with lower density, $\alpha = 0.00125$, for the cases of internal and external electron beams.

The dependence of HF SFXM growth rates on the width of the vacuum gap, where the electron beam propagates, is demonstrated in Fig. 8.19. There, for comparison with the case $d/b = 0.8$ considered in Fig. 8.17, a smaller value of the gap width, $d/b = 0.9$, is taken for the calculations. In this case the growth rates of HF SFXMs with $m = 2$ become sufficiently larger. The intersection of the dispersion curve with the resonant line $\omega = 6|\omega_e|/\gamma$ becomes possible. This gives rise to identify gap broadening as a factor that worsens the HF SFXM beam excitation. Data which describes the

Table 8.6 Comparison of the parameters of HF SFXM beam excitation by internal and external electron beams for the conditions of Fig. 8.17, but with lower beam particle density

	$\max[\mathrm{Im}(\omega/|\omega_e|)]$	$k_{m,N}$	$\Delta k_{ef} \times 10^4$
m,N	*Internal electron beam*		
1,5	2.55×10^{-4}	0.270	1.17
1,6	8.17×10^{-5}	0.190	0.0557
m,N	*External electron beam*		
1,5	4.96×10^{-4}	0.377	4.96
1,6	9.11×10^{-5}	0.252	9.11
2,7	5.51×10^{-4}	0.422	5.51
2,8	1.53×10^{-4}	0.344	1.53

Table 8.7 Comparison of the parameters of HF SFXM beam excitation by internal and external electron beams for the conditions of Fig. 8.19, but with narrower vacuum gap

	max[Im($\omega/\lvert\omega_e\rvert$)]	$k_{m,N}$	$\Delta k_{ef} \times 10^4$
m,N	*Internal electron beam*		
1,5	5.66×10^{-4}	0.298	3.25
1,6	1.03×10^{-4}	0.208	0.170
2,6	1.91×10^{-3}	0.439	21.4
2,7	4.94×10^{-4}	0.336	2.22
m,N	*External electron beam*		
1,5	6.90×10^{-4}	0.366	6.90
1,6	9.44×10^{-5}	0.241	9.44
2,7	5.71×10^{-4}	0.408	5.71
2,8	1.26×10^{-4}	0.330	1.26

HF SFXM beam excitation in the coaxial waveguide with a narrow gap is given in Table 8.7. Because of the shift of $k_{1,5}$ in the case of internal beam to the right along the k_{ef} axis as compared to the case of wider gap, its maximum value is larger.

The strength of the external static magnetic field has appreciable influence on SFXM propagation [51], which is analyzed in Fig. 8.20. To determine the effect of the magnetic field a smaller strength, $Z = 4.5$, as compared to the case shown in Fig. 8.19, is chosen, the other parameters being the same. One can clearly see that the reduction of the external magnetic field essentially worsens the conditions of SFXM excitation. The growth rate peaks (1,5) and (2,6) which are present in Fig. 8.19 are absent in Fig. 8.20. They move with decreasing static magnetic field (increasing parameter Z) to the right and for $Z = 4.5$ leave the range of k_{ef}, where HF SFXMs propagate. This is accompanied by an increase of the maximum values of the kept peaks with (1,6) and (2,7) as compared to the case of stronger magnetic field, $Z = 3.5$. This increase is explained by the shift of the peak to the right along the k_{ef} axis where the dispersion curve has a steeper inclination. A detailed comparison of HF SFXM beam excitation in the weaker static magnetic field, $Z = 4.5$, in the cases of internal and external electron beams is given in Table 8.8.

Table 8.8 Comparison of the parameters of HF SFXM beam excitation by internal and external electron beams for the weaker external magnetic field (the conditions of Fig. 8.20)

	max[Im($\omega/\lvert\omega_e\rvert$)]	$k_{m,N}$	$\Delta k_{ef} \times 10^4$
m,N	*Internal electron beam*		
1,6	1.34×10^{-4}	0.261	0.705
1,7	2.10×10^{-5}	0.185	0.0196
2,7	6.43×10^{-4}	0.390	3.90
2,8	1.67×10^{-4}	0.302	0.567
m,N	*External electron beam*		
1,6	1.67×10^{-4}	0.318	1.67
1,7	2.29×10^{-5}	0.214	2.29

In the following, the influence of the width of the coaxial plasma layer on HF SFXM beam excitation is examined. Figure 8.21 shows the dependence of the HF SFXM growth rate on the effective wavenumber for the same parameters of the beam-coaxial plasma structure as in Fig. 8.17 except of the outer metal chamber radius a. The width of the plasma layer is chosen to be a factor of two smaller than in the case given in Fig. 8.17, $a/b = 1.125$.

The maxima shown in Fig. 8.17 for the case of a wider coaxial plasma layer, $a/b = 1.25$, move a little bit to the right when narrowing the plasma layer width to $a/b = 1.125$. The absolute value of the maximum magnitude of the growth rate changes only a little bit. The main impact of narrowing the plasma layer is the appearance of the possibility to excite HF SFXMs with azimuthal mode index $m = 2$ with essential growth rate and half-width which indicates the most appropriate conditions for HF SFXM beam excitation among those presented here. Then one can conclude that narrowing the width of the plasma coaxial layer strongly improves the conditions of HF SFXM beam excitation. Detailed description of the HF SFXM beam excitation parameters in a coaxial waveguide with narrow plasma layer in comparison with the case of external electron beam is given in Table 8.9.

The widening of the (2,5)-peak as compared to the (2,6)-maximum in Fig. 8.21 is explained by the HF SFXM dispersion properties. The dependence of the HF SFXM eigenfrequency normalized to the electron cyclotron frequency, $\omega/|\omega_e|$, versus the effective wavenumber k_{ef} is shown in Fig. 8.22 for the same plasma waveguide parameters as in Fig. 8.21: $m = 2$, $d/b = 0.8$, $a/b = 1.125$, and $Z = 3.5$. The resonance line $\omega/|\omega_e| = 5/\gamma$ crosses the dispersion curve with a smaller angle than the line $\omega/|\omega_e| = 6/\gamma$. In other words the line $\omega/|\omega_e| = 5/\gamma$ runs closer to the dispersion curve than the line $\omega/|\omega_e| = 6/\gamma$. Therefore, the resonance condition (7.14) for $N = 5$ remains to be valid within a wider range of k_{ef} than for $N = 6$.

Now, the possibility of experimental observation of the phenomena studied in the present Section for excitation of electromagnetic radiation in the GHz range is estimated. The most favorable conditions for HF SFXM beam excitation (the largest growth rate) among those presented in the figures above are related to the highest

Table 8.9 Comparison of the parameters of HF SFXM beam excitation by internal and external electron beams for the narrower plasma layer (the conditions of Fig. 8.21)

	$\max[\mathrm{Im}(\omega/|\omega_e|)]$	$k_{m,N}$	$\Delta k_{ef} \times 10^4$
m,N	*Internal electron beam*		
1,5	5.86×10^{-4}	0.274	3.31
1,6	1.46×10^{-4}	0.193	0.340
2,5	7.68×10^{-3}	0.633	830.0
2,6	1.92×10^{-3}	0.394	25.7
m,N	*External electron beam*		
1,5	9.54×10^{-4}	0.357	9.54
1,6	1.89×10^{-4}	0.247	1.89
2,6	3.61×10^{-3}	0.532	3.61
2,7	1.04×10^{-3}	0.405	1.04

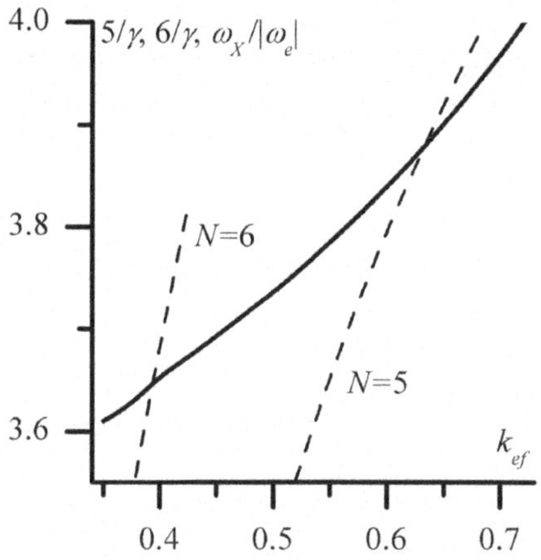

Fig. 8.22 Normalized HF SFXM eigenfrequency $\omega/|\omega_e|$ (*solid curve*), in absence of an electron beam, crosses the two resonance lines (*dashed lines*) $\omega/|\omega_e| = 6/\gamma$ and $\omega/|\omega_e| = 5/\gamma$ for $m = 2$, $d/b = 0.8$, $\alpha = 0.01$, $a/b = 1.125$, and $Z = 3.5$

peak in Fig. 8.21. It corresponds to interaction of the HF SFXM with $m = 2$ and the fifth cyclotron harmonic of the electron beam, $N = 5$. As an example, a coaxial line with $b \sim 10$ cm radius of plasma-vacuum boundary is considered. Then to meet the conditions of Fig. 8.21 the radius of the internal conductor should be $a \sim 8$ cm, and the inner radius of the external conductor $d \sim 11.25$ cm. Since the peak of the HF SFXM growth rate occurs at $k_{m,N} = 0.633$, the related Langmuir frequencyis $\Omega_e \sim 9.5 \times 10^9\ \mathrm{s}^{-1}$, which corresponds to a plasma particle density of $n_e \sim 2.8 \times 10^{10}\ \mathrm{cm}^{-3}$. The plot in Fig. 8.21 is calculated for the magnetic field parameter $Z = 3.5$, therefore the absolute value of the electron cyclotron frequency is $|\omega_e| \sim 2.7 \times 10^9\ \mathrm{s}^{-1}$, which corresponds to an induction of the magnetic field of $B_0 \sim 154$ G. In this case the electron beam rotates with the velocity $0.631\,c$ which corresponds to the relativistic factor $\gamma \sim 1.289$. The SFXM propagates with the frequency $\omega \sim 1.1\Omega_e$ and is excited with the growth rate $\mathrm{Im}(\omega) \sim 2 \times 10^7\ \mathrm{s}^{-1}$.

The example given above corresponds to a rather thick plasma layer with $b - a \approx 1.25$ cm. Plasma production and confinement can cause some engineering problems. To overcome these, one can apply modern metamaterials (see e.g. [40–43]). This provides also high plasma density, compared to the case under study here.

The following conclusions can be drawn from the studies carried out in the present Section. The initial stage of the resonant beam-plasma instability of HF SFXMs propagating along a plasma coaxial layer internal interface caused by their interaction with a gyrating beam of electrons, which move along large Larmor orbits in the narrow vacuum gap between the plasma layer and the internal conductor, is studied in this Section [29] and compared to the case considered in Sect. 8.3, when the electron beam propagates in the other vacuum gap between the plasma layer and the inner wall of the external waveguide [28].

HF SFXM excitation by an internal electron beam takes place in the regime of separated mode excitation: the distance between two resonances is much larger than their half-widths, $k_{m,N} - k_{m,N\pm 1} \gg \Delta k_{ef}$, while in the case of LF SFXM excitation resonance overlapping can already be reached by a beam with only moderate particle density [27]. Thus, the present study generalizes the possibility of SFXM beam excitation for the case of higher frequencies (upper-hybrid frequency range) as compared to the case of LF SFXMs, which propagate in the electron cyclotron frequency range. To get this advantage one has to accept restrictions for the range of parameters of the beam-plasma system where the excitation takes place.

Since the electromagnetic waves studied here are surface waves it is possible to get use of the internal part of the waveguide as compared to the case of HF SFXM excitation in the uniform plasma column studied in [52]. The feature that no plasma has to be generated in the internal part of the coaxial device makes the apparatus cheaper and leaves a space for installing an electron beam source like that described in [35].

Higher values of the HF SFXM growth rates are proven to be achievable in coaxial beam-plasma systems with higher azimuthal mode index m, beam particle density, thinner plasma layer, and gap where the electron beam gyrates.

The case of sufficiently dense plasmas is studied here, $\Omega_e^2 > \varepsilon_r \omega_e^2$, that provides higher SFXM frequencies in the range above the upper-hybrid frequency as compared to the previous study [27] when electron beam excitation of LF SFXMs in the electron cyclotron frequency range was investigated. One should remind that if the plasma particle density is small, so that $\Omega_e^2 < \varepsilon_r \omega_e^2$, then SFXMs can propagate in the range above the ion cyclotron frequency. In this case plasma ions play a prevailing role in electromagnetic wave oscillations [55]. The possibility to apply a gyrating ion beam for SFXM excitation in this case is demonstrated below in Sect. 10.4 [56, 57].

8.5 Conclusions

The prototype of a generator of electromagnetic radiation is presented in the previous chapter. It consists of a plasma column inside a metal waveguide. An electron beam gyrates within the vacuum gap, which separates the plasma column from the metal wall along large Larmor orbits in an external static axial magnetic field. In the present Chapter, two modifications of this prototype generator are presented. First, one gets advantage of the fact that the electromagnetic waves under consideration are of surface type within the plasma column. Their amplitudes decrease with going away from the plasma interface into the plasma depth. This peculiarity makes it possible to replace the internal part of the plasma column by a metal conductor. Such a change transforms the cylindrical plasma waveguide into a coaxial one. This modification is called as "external beam" since the electron beam gyrates around the coaxial plasma layer. The second modification differs from the first one by the placement of the space where the electron beam gyrates. In this case, the electron beam gyrates within a vacuum gap which separates the plasma coaxial layer from the internal conductor. This configuration is called as "internal beam". In this geometry, the inner conductor can be a hollow metal pipe, where the space inside the internal conductor can be utilized for installing an electron beam gun.

These two modifications possess the following two advantages as compared with the design discussed in Chap. 7. First, one does not need to produce and sustain a plasma in the space replaced by the internal conductor. This may make cheaper the construction and operation of the device. Second, introducing the internal conductor gives one tool more to a designer to control both the dispersion properties of SFXMs and their excitation.

In the case of a coaxial metal waveguide, which is entirely filled by plasma, surface flute X-modes are known to be unidirectional. In other words, SFXMs propagate in one definite direction in a definite frequency range. On the other hand, an electron beam can excite electromagnetic waves only along the direction of electron gyration in the external static axial magnetic field. These circumstances make a base to suggest in advance that an external (internal) electron beam can excite SFXMs in LF (HF) range only. However, the presence of the vacuum gap, where the electron beam gyrates, significantly changes the dispersion properties of SFXMs and in addition, excitation of SFXMs with positive azimuthal mode index m becomes possible in both frequency ranges for both modifications.

Application of an internal conductor extends the range of plasma parameters for which SFXM beam excitation is efficient and insignificantly increases the peak values of SFXM growth rates. The growth rates of electron beam excitation of SFXMs are larger in the case of external beam application.

References

1. Kuzelev, M. V., & Rukhadze, A. A. (1990). *Electrodynamics of intense electron beams in plasma*. Nauka. (in Russian).
2. Bogdankevich, L. S., Kuzelev, M. V., & Rukhadze, A. A. (1981). Plasma microwave electronics. *Soviet Physics Uspekhi, 24*(1), 1–16.
3. Azarenkov, N. A., Denisenko, I. B., & Ostrikov, K. N. (1995). A model of a large-area planar plasma producer based on surface wave propagation in a plasma-metal structure with a dielectric sheath. *Journal of Physics D: Applied Physics, 28*(12), 2465–2469.
4. Levitskij, S. M., & Koshevaja, S. V. (1986). *Vacuum and solid-state SHF electronics*. Vyshcha shkola. (in Russian).
5. Aronov, V. A., Bogdankevich, L. S., & Rukhadze, A. A. (1975). Bulk and surface electromagnetic waves in spatially bounded plasma and their excitation by relativistic electron beams. In Smirnov, B.M. (Ed.), *Plasma chemistry* (2, pp. 245–266). Atomizdat. (in Russian).
6. Lawson, J. D. (1977). *The physics of charged-particle beams*. Clarendon Press.
7. Miller, R. B. (1982). *An introduction to the physics of intense charged particle beams*. Plenum Press.
8. Andronov, A. A. (Ed.). (1986). *Semiconductor masers on a cyclotron resonance, Gorkiy, USSR: IAP AS of USSR*. (in Russian).
9. Kondratenko, A. N., & Kuklin, V. M. (1988). *Basics of plasma electronics*. Energoatomizdat. (in Russian).
10. Gisler, G. R. (1987). Particle-in-cell simulations of azimuthal instabilities in relativistic electron layers. *Physics of Fluids, 30*(7), 2199–2208.
11. Kainer, S., Gaffey, J. D., Price, C. P., Hu, X. W., & Zhou, G. C. (1988). Nonlinear wave interactions and evolution of a ring-beam distribution of energetic electrons. *Physics of Fluids, 31*(8), 2238–2248.

12. Lau, Y. Y. (1988). Radiation generated by rotating electron beams. In: *Proceedings of the symposium non-neutral plasma physics* (pp. 210–223). Washington, USA.
13. Barker, R. J., & Schamiloglu, E. (2001). *High-power microwaves sources and technologies*. IEEE Press/Wiley.
14. Kuzelev, M. V., Poezd, A. D., Rukhadze, A. A., Sveshnikov, A. G., & Yakunin, S. A. (1989). Application of conservative finite-difference schemes for modeling non-stationary processes in vircators. *Matematicheskoe Modelirovanie, 1*(10), 16–22. (in Russian).
15. Kho, T. H., Lin, A. T., & Chen, L. (1988). Gyrophase-coherent electron cyclotron maser. *Physics Fluids, 31*(10), 3120–3126.
16. Wu, J., Xiong, C., & Liu, S. (1995). Excitation of microwave by an annular electron beam in a plasma-filled dielectric lined waveguide. *International Journal of Infrared and Millimeter Waves, 16*(9), 1573–1581.
17. Norreys, P. A., Green, J. S., Davies, J. R., Tatarakis, M., Clark, E. L., Beg, F. N., et al. (2006). Observation of annular electron beam transport in multi-TeraWatt laser-solid interactions. *Plasma Physics and Controlled Fusion, 48*(2), L11–L22.
18. Kapitanov, A. N., Obrazkov, N. V., Sukhanov, L. A., Khlestkov, A. Yu., & Khlestkov, Yu. A. (2009). Solution of a set of Maxwell-Lorentz equations for a ring relativistic electron beam. *Plasma Physics Reports, 35*(6), 510–517.
19. Girka, V., Girka, I., & Thumm, M. (2014). *Surface flute waves in plasmas. Theory and applications*. Springer.
20. Krall, N. A., & Trivelpiece, A. W. (1986). *Principles of plasma physics*. San Francisco Press.
21. Aliev, Yu. M., Schlüter, H., & Shivarova, A. (2000). *Guided-wave produced plasmas*. Springer.
22. Girka, V. A., & Girka, I. A. (1990). Influence of plasma inhomogeneity on the spectra of azimuthal surface waves. *Radiophysics and Quantum Electronics, 33*(4), 516–517.
23. Girka, V. A., & Girka, I. A. (1991). Azimuthal surface waves in a nonuniform plasma cylinder. *Radiophysics and Quantum Electronics, 34*(4), 324–328.
24. Alexandrov, A. F., Bogdankevich, L. S., & Rukhadze, A. A. (1984). *Principles of plasma electrodynamics*. Springer.
25. Girka, V. O., Girka, I. O., Morgal, Ya. I., et al. (2011). Excitation of azimuthal surface modes by annular electron beams in the electron cyclotron frequency range. *Physica Scripta, 84*(2), 025505.
26. Blednov, O. S., Girka, I. O., Girka, V. O., & Pavlenko, I. V. (2014). Excitation of the azimuthal surface waves in electron cyclotron frequency range by rotating electron beam in a coaxial waveguide. *IEEE Transactions on Plasma Science, 42*(3), 735–741.
27. Blednov, O., Girka, I., Girka, V., et al. (2014). Excitation of the surface flute waves in electron cyclotron frequency range by internal rotating electron beam in a coaxial waveguide. *Physica Scripta, 89*(12), 125605.
28. Girka, I. O., & Thumm, M. (2016). Excitation of azimuthal surface modes above the upper-hybrid frequency by external relativistic flows of electrons in coaxial plasma-vacuum waveguide. *IEEE Transactions on Plasma Science, 44*(11), 2859–2866.
29. Girka, I. O., & Thumm, M. (2017). Excitation of electromagnetic waves above the upper-hybrid frequency by internal gyrating electron beam in a coaxial waveguide. *IEEE Transactions on Plasma Science, 45*(4), 623–630.
30. Humphries, S. (1990). *Charged particle beams*. Wiley.
31. Kiyoshi, Y. (1989). Industrial applications of pulse power and particle beams. *Laser Particle Beams, 7*(4), 733–741.
32. Saito, H., & Wurtele, J. S. (1987). The linear theory of the circular free-electron laser. *Physics of Fluids, 30*(7), 2209–2220.
33. Manheimer, W. M., Fernsler, R. F., & Gitlin, M. S. (1998). High power, fast, microwave components based on beam generated plasmas. *IEEE Transactions on Plasma Science, 26*(5), 1543–1555.
34. Murphy, D. P., Fernsler, R. F., Pechacek, R. E., & Meger, R. A. (2002). Microwave emission from plasmas produced by magnetically confined electron beams. *IEEE Transactions on Plasma Science, 30*(1), 436–441.

35. Velazco, J. E. (2000). *Continuous rotating-wave electron beam accelerator*. U.S. Patent No. 6,060,833. U.S. Patent and Trademark Office.
36. Li, F., He, W., Cross, A. W., et al. (2010). Design and simulation of a ~ 390 GHz seventh harmonic gyrotron using a large orbit electron beam. *Journal of Physics D: Applied Physics, 43*, 155204.
37. Li, X., Lang, J., Alfadhl, Y., et al. (2015). Study of an eighth-harmonic large-orbit gyrotron in the terahertz band. *IEEE Transactions on Plasma Science, 43*, 506–514.
38. Destler, W. W., Chojnacki, E., Hoeberling, R. F., et al. (1988). High-power microwave generation from large-orbit devices. *IEEE Transactions on Plasma Science, 16*(2), 71–89.
39. Bratman, V. L., Denisov, G. G., Kalynov, Yu. K., et al. (1999). Novel types of cyclotron resonance devices. In *Proceedings of the 4th International Workshop Strong Microwaves in Plasmas* (Vol. 2, pp. 683–702). Russian Federation.
40. Ostrikov, K., Neyts, E. C., & Meyyappan, M. (2013). Plasma nanoscience: from nano-solids in plasmas to nano-plasmas in solids. *Advances in Physics, 62*(2), 113-224.
41. Ilin, N. V., Kondratiev, I. G., & Smirnov, A. I. (2009). Planar waveguiding systems based on resonant meta materials. In *Proceedings of the 7th International Workshop Strong Microwaves in Plasmas* (Vol. 2, pp. 550–554). Institute of Applied Physics, RAS, Nizhny Novgorad, Russian Federation.
42. Ederra, I., Iriarte, J. C., Gonzalo, R., et al. (2011). Surface waves of finite size electromagnetic band gap woodpile structures. *Progress in Electromagnetics Research B, 28*, 19–34.
43. Sozuer, H. S., & Dowling, J. P. (1994). Photonic band calculations for woodpile structure. *Journal of Modern Optics, 41*(2), 231–239.
44. Rukhadze, A. A., Bogdankevich, L. S., Rosinskii, S. Y., & Ruhlin, S. Y. (1980). *Physics of high current relativistic electron beams*. Atomizdat. (in Russian).
45. Anderson, T. (2011). Plasma Antennas. Artech House.
46. Borg, G. G., Harris, J. H., Miljak, D. G., et al. (1999). Application of plasma columns to radio-frequency antennas. *Applied Physics Letters, 74*(22), 3272–3274.
47. Cerri, G., De Leo, R., Primiani, V. M., et al. (2008). Measurement of the properties of a plasma column used as a radiated element. *IEEE Transactions on IMT, 57*(2), 242–247.
48. Rayner, J. P., Whichello, A. P., & Cheetham, A. D. (2004). Physical characteristics of plasma antennas. *IEEE Transactions on Plasma Science, 32*(1), 269–281.
49. Wu, M., Wen, B. Y., & Zhou, H. (2009). Ionospheric clutter suppression in HF surface wave radar. *Journal of Electromagnetic Waves and Applications, 23*(10), 1265–1272.
50. Azarenkov, N. A., Ostrikov, K. N., & Shcherbinina, I. B. (1991). Azimuthal surface waves in a coaxial semiconductor structure with metal walls. *Soviet Journal of Communication Technology and Electronics, 36*(2), 68–72.
51. Girka, V. A., Girka, I. A., Kondratenko, A. N., & Tkachenko, V. I. (1988). Azimuthal surface waves of magnetoactive plasma wavequides. *Soviet Journal of Communications Technology and Electronics, 33*(8), 37–41.
52. Girka, V. O., Girka, I. O., & Pavlenko, I. V. (2011). Excitation of azimuthal surface modes by relativistic flows of electrons in high-frequency range. *Plasma Physics Reports, 37*(5), 447–454.
53. Akhiezer, A. I., & Fainberg, Ya. B. (1949). On the interaction of a charged particle beam with electron plasma. Doklady AN SSSR, 69, 555–556. (in Russian). (Akhiezer, A. I., & Fainberg, Ya. B. (2008). On the interaction of a charged particle beam with electron plasma. *Ukrainian Journal of Physics, 53*, 87–88. (in English, translated and reprinted from Dokl. Akad.Nauk SSSR, 69, 555–556).
54. Girka, I. O., Girka, O. I., Girka, V. O., & Pavlenko, I. V. (2006). Coupled HF azimuthal waves in magnetoactive waveguide partially filled by current-carrying plasma. *Problems of Atomic Science and Technology, Series: Plasma Electronics and New Methods of Acceleration*, No. 5, 28–33.
55. Girka, I. O., & Kovtun, P. K. (1998). Azimuthal surface waves in a magnetized plasma. *Technical Physics, 43*(12), 1424–1427.

56. Girka, V. O., Girka, I. O., & Pavlenko, I. V. (2011). Excitation of ion azimuthal surface modes in a magnetized plasma by annular flow of light ions. *Progress in Electromagnetics Research M (PIERM), 21*, 267–278.
57. Girka, V. O., Girka, I. O., & Pavlenko, I. V. (2012). Possibility of excitation of azimuthal surface waves in a magnetized plasma by annular ion beams. *Technical Physics Letters, 38*(2), 178–180.

Chapter 9
Nonlinear Theory of Surface Flute Wave Excitation in Electron Cyclotron Frequency Range

Studies of plasma–beam interaction are attracting continuous interest primarily because of the major practical value of the expected results, which are finding extensive fields of application, ranging from particle beam heating of plasmas in controlled nuclear fusion devices and performing geophysical experiments in space to solving problems in plasma electronics. The theoretical study, carried out in the present Chapter, is devoted to solving the problem relevant to one line of plasma electronics research [1], which involves studying the interaction between charged particle beams and eigenmodes of plasma waveguides. Studies of electromagnetic wave generation processes governed by the parameters of the particle beam and the plasma waveguide are aimed at manufacturing radio frequency (RF) oscillators or amplifiers with the highest possible efficiency, permitting continuous frequency tuning over a wide range, and having comparatively small dimensions, the highest possible radiation power, and so on. Obviously, all these requirements cannot be satisfied simultaneously by a single device. The study in the present Chapter is an attempt to construct a nonlinear theory for a plasma electronics device, which could satisfy two criteria to a certain extent, i.e., it should offer continuous tuning of the RF in a fairly compact device.

At present, the conditions for the onset of beam-plasma instabilities and nonlinear interaction between charged particle beams and plasmas in strong external magnetic fields with convenient magnetic field strength have been studied in fairly great detail. The influence of the spectrum of generated waves on the nature of the beam–plasma interaction has also been analyzed without taking the influence of the plasma boundaries into account [2–4]. The bounded dimensions of the plasma not only lead to changes in the dynamics of the excitation of bulk waves but also creates conditions for the excitation of surface waves [5]. Surface waves have various distinguishing features compared with bulk waves: the energy of surface waves is transferred along the plasma boundary and can be extracted comparatively easily from the interaction zone [1].

© The Author(s), under exclusive license to Springer Nature Switzerland AG 2022

I. Girka and M. Thumm, *Surface Flute Waves in Plasmas*, Springer Series on Atomic, Optical, and Plasma Physics 120, https://doi.org/10.1007/978-3-030-98210-2_9

9.1 Nonlinear Theory of Surface Flute Wave Excitation by Electron Beams Gyrating Around a Plasma Column Under Condition of Resonant Beam-Plasma Instability

To get the set of differential equations which describes the nonlinear stage of SFXM excitation by a gyrating electron beam one has to apply the hydrodynamic plasma model, the Maxwell equations for describing the SFXM fields, and the equations of motion for describing the beam particles in the gap $R_1 < r < R_2$ between the plasma cylinder and inner wall of the metallic waveguide (see Fig. 7.1). Since the beam particle density is assumed to be small compared to the plasma density, $\alpha \equiv n_b/n_e << 1$, one can neglect the influence of the beam eigenfield on the electromagnetic field in which the electron beam moves [2, 6, 7]. In the plasma region, the SFXM magnetic wave field H_z can be written in terms of the modified Bessel functions $I_m(\xi)$, and the electric wave fields are described by the formulas (1.10).

The charged particle beam is modeled by the method of plasma simulation using particles [8]. In the region, which is occupied by the electron beam, the SFXM fields are described by the following set of differential equations,

$$E_r^B = -4\pi i j_r m H_z^B/\omega\varsigma, \tag{9.1}$$

$$E_\varphi^B = -i4\pi\ j_\varphi/\omega - i\partial H_z^B/\partial\varsigma, \tag{9.2}$$

$$\frac{\partial^2 H_z^B}{\partial\varsigma^2} + \frac{1}{\varsigma}\frac{\partial H_z^B}{\partial\varsigma} + \left(1 - \frac{m^2}{\varsigma^2}\right)H_z^B = F_b, \tag{9.3}$$

where the following notations are applied: $F_b = -4\pi\left[\partial\left(j_\varphi\varsigma\right)/\partial\varsigma - imj_r\right]/(\omega\varsigma)$, radial current density $j_r = -\ |e|\sum_{i=1}^{N}\delta(r - r_i)\ \delta(\varphi - \varphi_i)\ \partial r/\partial t$, azimuthal current density $j_\varphi = -\ |e|\sum_{i=1}^{N} r\delta(r - r_i)\ \delta(\varphi - \varphi_i)\ \partial\varphi/\partial t$, and $\varsigma\ = kr$; where $\delta(x - x_0)$ is the Dirac delta function. The set of Eqs. (9.1–9.3) can be solved by the method of variation of constants. Then the expressions for the SFXM wave field components have the following form in the region of the electron beam:

$$\begin{aligned} H_z^B &= C_1\ J_m(\varsigma) + C_2 N_m(\varsigma) \\ &\quad - \frac{\pi}{2} J_m(\varsigma)\int\limits_{\varsigma_1}^{\varsigma} \varsigma' N_m(\varsigma')F_b(\varsigma')d\varsigma' \\ &\quad + \frac{\pi}{2}\ N_m(\varsigma)\int\limits_{\varsigma_1}^{\varsigma} \varsigma' J_m(\varsigma')F_b(\varsigma')d\varsigma', \end{aligned} \tag{9.4}$$

$$E_r^B = -\frac{m}{\varsigma}\left[C_1 J_m(\varsigma) + C_2 N_m(\varsigma) - \frac{\pi}{2} J_m(\varsigma)\int_{\varsigma_1}^{\varsigma} \varsigma' N_m(\varsigma') F_b(\varsigma') d\varsigma'\right.$$

$$\left. + \frac{\pi}{2} N_m(\varsigma)\int_{\varsigma_1}^{\varsigma} \varsigma' J_m(\varsigma') F_b(\varsigma') d\varsigma'\right] + \frac{4\pi}{i\omega} j_r, \tag{9.5}$$

$$E_\varphi^B = -iC_1 J_m'(\varsigma) - iC_2 N_m'(\varsigma)$$

$$+ \frac{i\pi}{2} J_m' \int_{\varsigma_1}^{\varsigma} \varsigma' J_m(\varsigma') F_b(\varsigma') d\varsigma'$$

$$- \frac{i\pi}{2} N_m'(\varsigma)\int_{\varsigma_1}^{\varsigma} \varsigma' J_m(\varsigma') F_b(\varsigma') d\varsigma' + \frac{4\pi}{i\omega} j_\varphi, \tag{9.6}$$

where $\varsigma_1 = kR_1$. The upper index "B" means that the SFXM field is calculated for the space, which is occupied by the beam.

To derive the set of equations which describes the behavior of the envelope amplitude and that of the wave phase one can apply the linear boundary conditions at the axis, $r = 0$, and at the metal wall, $r = R_2$. However, at the plasma interface, $r = R_1$, along with the linear boundary condition for the axial magnetic wave field to be continuous, the following nonlinear boundary condition for the azimuthal electric wave field is applied:

$$E_\varphi^B(R_1 + 0) = E_\varphi(R_1 - 0). \tag{9.7}$$

The nonlinearity in Eq. (9.7) is caused by the fact that the expression for the field E_φ^B contains the electric beam current, which in turn is determined by the SFXM wave field. Applying the standard procedures of averaging [1, 4], one can find the following equations for the envelope amplitude and the wave phase in the single-mode approach with neglecting dissipative phenomena in the plasma:

$$\frac{dE}{dt} = -\frac{\alpha D_p}{NzPL}\sum_{i=1}^{N}\left[\frac{m}{w}\frac{d\rho_i}{dt} L_1(\varsigma_i) sin(\phi_i) + \rho_i^2 \frac{d\varphi_i}{dt} L_2(\varsigma_i) cos(\phi_i)\right], \tag{9.8}$$

$$\frac{d\theta}{dt} = \frac{\alpha D_p}{ENzPL}\sum_{i=1}^{N}\left[\rho_i^2 \frac{d\varphi_i}{dt} L_2(\varsigma_i)\sin(\phi_i) - \frac{m}{w}\frac{d\rho_i}{dt} L_1(\varsigma_i)\cos(\phi_i)\right]. \tag{9.9}$$

In (9.8) and (9.9), $E = E_y/B_0$ is the dimensionless envelope amplitude of the wave, θ is the wave phase, $\phi_i = m\varphi_i + \theta - \omega t$, $z = |\omega_e|\Omega_e^{-1}$, $\varsigma_2 = kR_2$, $w = \omega\Omega_e^{-1}$, $L = J_m(\varsigma_1)\, N'_m(\varsigma_2) - J'_m(\varsigma_2)\, N_m(\varsigma_1)$, $L_1(\varsigma_i) = J_m(\varsigma_i)\, N'_m(\varsigma_2) -$

$J'_m(\varsigma_2)\,N_m(\varsigma_i)$, $L_2(\varsigma_i) = J'_m(\varsigma_i)\,N'_m(\varsigma_2) - J'_m(\varsigma_2)\,N'_m(\varsigma_i)$, $D_p = \frac{I'_m(\psi\varsigma_1)}{\psi\, I_m(\psi\varsigma_1)} + \frac{m\varepsilon_2}{\varepsilon_1\psi^2\varsigma_1}$, $D = D_p + \frac{J'_m(\varsigma_1)N'_m(\varsigma_2) - J'_m(\varsigma_2)N'_m(\varsigma_1)}{J_m(\varsigma_1)N'_m(\varsigma_2) - N_m(\varsigma_1)J'_m(\varsigma_2)}$, $P = \frac{dD}{d\,w}$, and $\rho_i = r_i\,\Omega_e c^{-1}$. N is the number of beam particles.

It is convenient to express the equation of motion for the beam electrons via introducing the term of particle momentum $\vec{p} = \gamma\, m_e \vec{V}$ [7]:

$$\frac{d\,\vec{p}}{d\,t} = e\vec{E} + \frac{e}{c}\Big[\vec{V}, \left(\vec{H} + \vec{B}_0\right)\Big]. \tag{9.10}$$

Substituting the expression (9.10) into the equation for the SFXM field in the region $R_1 < r < R_2$ one can derive the following set of nonlinear equations, which describes the motion of the ith particle:

$$d\,\rho_i/d\,t = |\omega_e| v_i / z\,\gamma_i, \tag{9.11}$$

$$d\,\varphi_i/d\,t = |\omega_e|\,u_i/(z\,\gamma_i\,\rho_i), \tag{9.12}$$

$$\frac{d\,v_i}{d\,t} = E R_1\left(\frac{m}{R_1} - w\frac{u_i}{\gamma_i}\right)\left(\frac{1}{\varsigma_1 D_p} - \rho_i\right)\,\sin(\theta_i) - |\omega_e|\frac{u_i}{\gamma_i}\left(1 - \frac{u_i}{z\,\rho_i}\right), \tag{9.13}$$

$$\begin{aligned}\frac{d\,u_i}{d\,t} &= \frac{E w R_1}{\varsigma_1}\left[\rho_i - 1 + \frac{\rho_i}{D_p}\left(\frac{m^2}{\varsigma_1} - \varsigma_1\right)\right]\cos(\phi_i)\\ &+ |\omega_e|\,\frac{v_i}{\gamma_i}\left(1 - \frac{u_i}{z\rho_i}\right)\\ &+ \frac{E w R_1 v_i}{\gamma_i}\left(\frac{1}{\varsigma_1 D_p} - \rho_i\right)\,\sin(\phi_i),\end{aligned} \tag{9.14}$$

where $\rho_i = r_i\,R_1{}^{-1} - 1$, $v = p_r/(m_e c)$, and $u = p_\varphi/(m_e c)$. Here v and u are the radial and azimuthal dimensionless momentum components of the beam particles, respectively.

Equations (9.8, 9.9, 9.13) and (9.14) are derived under the basic assumption that the region where the beam propagates is relatively narrow, $R_2 - R_1 << R_1$. This assumption makes it possible to significantly simplify these equations by using the asymptotic expansions of the cylindrical Bessel functions (see, e.g., [9]) and, thus, to substantially reduce a large amount of computer time required for the direct calculation of all cylindrical functions. The nonlinear interaction of a SFXM with an annular electron flow is investigated by solving the set of Eqs. (9.8, 9.9, 9.11–9.14) numerically.

The joint solution of this set of equations can be found by the numerical Runge–Kutta method of the fourth order [7]. In modeling the beam electrons, the number of particles is chosen as $N = 500$. The model of mirror reflection is employed to describe the interaction of the beam particles with the plasma interface and the metallic waveguide wall.

The following starting values of the parameters of the electron beam and the plasma waveguide are used here: wave amplitude $E = 10^{-3}$, wave phase $\theta = 0$ rad, radial beam particle momentum $v_i = 0$, azimuthal beam particle momentum $u_i = z\,\rho_i$ ($\pm 2\%$), and ratio of electron beam density to plasma particle density $\alpha = 10^{-2}$. The starting distribution of beam particles along the angle φ is assumed to be approximately uniform with only small random deviations ($\Delta\varphi = \pm 1\%$), and the beam particles are randomly distributed along the radius within the gap $R_1 < r < R_2$. The other waveguide and plasma parameters are chosen as follows: $\Delta = R_2/R_1 - 1 = 0.1$ and $|\omega_e| = 0.099\Omega_e$.

In the Figs. 9.1, 9.3 and 9.5, the beam particle distribution in the coordinate space is shown for the SFXMs with azimuthal mode index $m = 2$ at increasing time, whereas in Figs. 9.2, 9.4 and 9.6, one can see the change in the beam particle distribution within the phase space in course of time.

Figures 9.7 and 9.9 give the beam particle distribution in the coordinate space at the moment when the process of bunch formation is already completed for $m = 3$ and $m = 4$, respectively. In the phase space, the beam particles form the structures similar to a spindle (the number of spindle arms coincides with the mode index m). In Figs. 9.8 and 9.10, the beam particle distribution in phase space for the SFXMs with the azimuthal mode indices $m = 3$ and $m = 4$ is plotted. The obtained results confirm the conclusions of papers [10, 11] that the maximum value of the growth rate is for the effective wavenumber $k_{ef} \approx 0.4$ and that the SFXM amplitude becomes lower if k_{ef} is chosen to be equal to 0.2, 0.3, 0.5 or 0.6. Doing that, the decrease of the SFXM amplitude is more pronounced for $k_{ef} < 0.4$ than for $k_{ef} > 0.4$. Changing the sign of the azimuthal mode index m that determines the direction of SFXM propagation causes the breakdown of the instability as result of failing the resonance conditions. Decreasing ratio $|\omega_e|\ \Omega_e^{-1}$ and increasing parameter α cause shortening the time during which the SFXW amplitude reaches the maximum value.

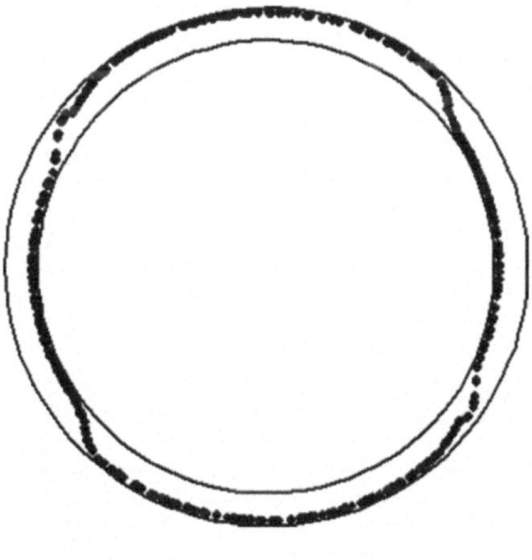

Fig. 9.1 Beam particle distribution in coordinate space for the SFXM with $m = 2$ at the normalized time moment $\tau = 0.02$

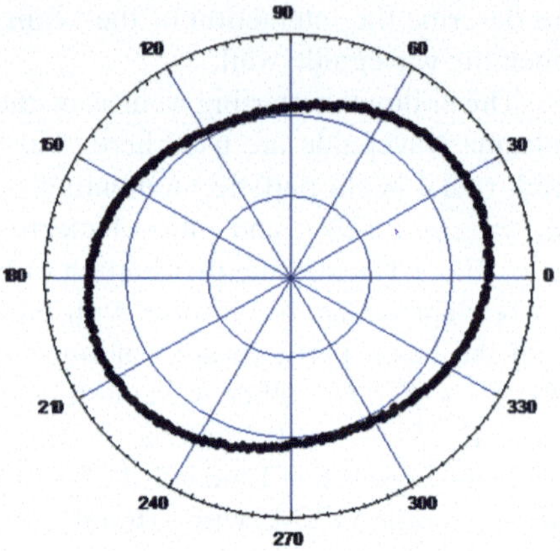

Fig. 9.2 Beam particle distribution in phase space for the SFXM with $m = 2$ at the normalized time moment $\tau = 0.02$

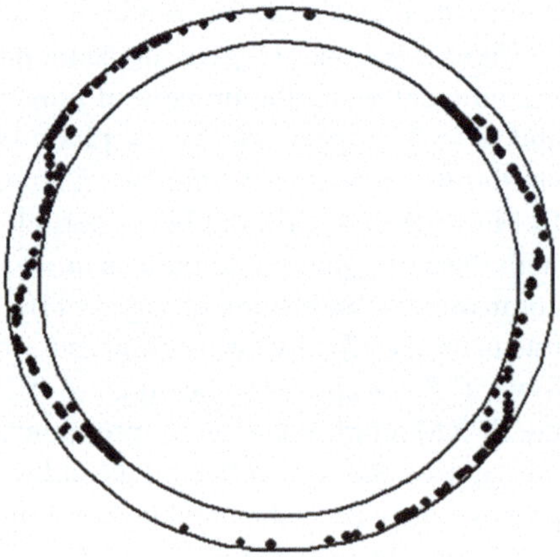

Fig. 9.3 Beam particle distribution in coordinate space for the SFXM with $m = 2$ at the normalized time moment $\tau = 0.06$

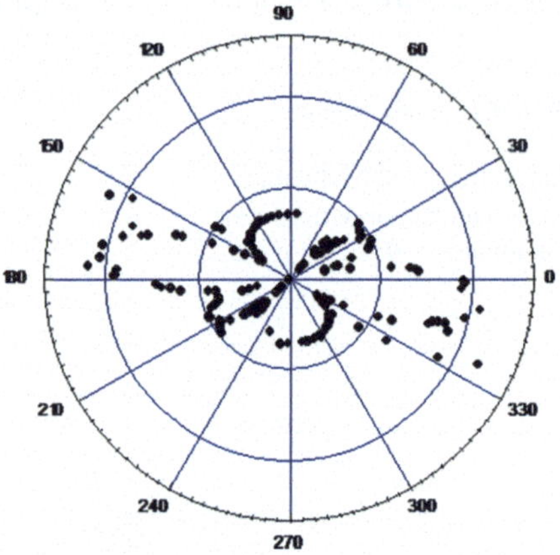

Fig. 9.4 Beam particle distribution in phase space for the SFXM with $m = 2$ at the normalized time moment $\tau = 0.06$

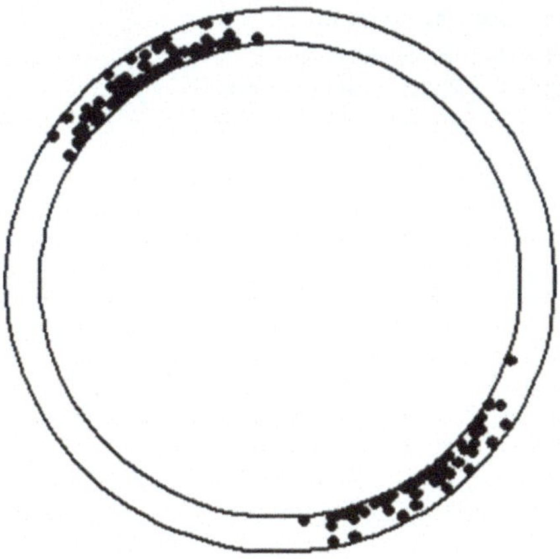

Fig. 9.5 Beam particles distribution in coordinate space for the SFXM with $m = 2$ at the normalized time moment $\tau = 0.3$

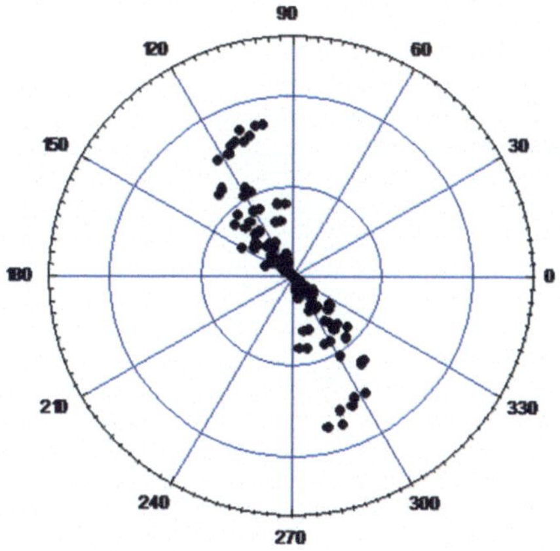

Fig. 9.6 Beam particle distribution in phase space for the SFXM with $m = 2$ at the normalized time moment $\tau= 0.3$

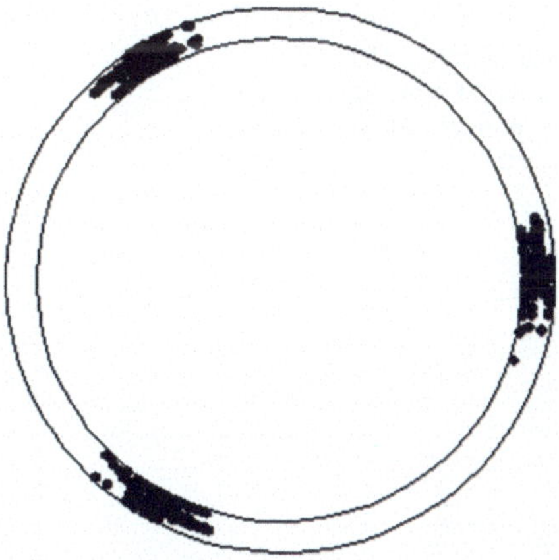

Fig. 9.7 Beam particle distribution in coordinate space for the SFXM with $m = 3$ at the normalized time moment $\tau = 0.3$

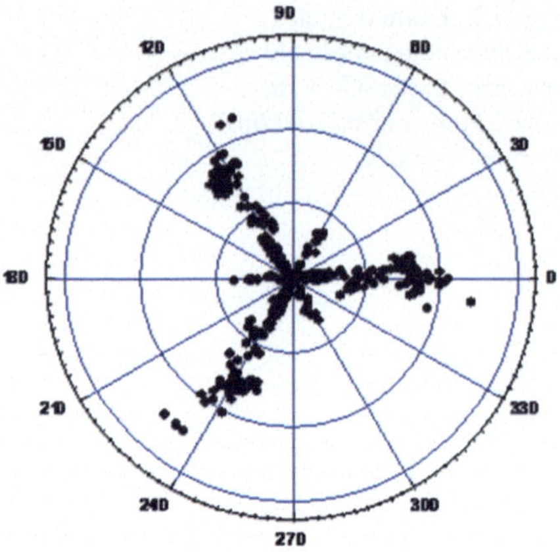

Fig. 9.8 Beam particle distribution in phase space for the SFXM with $m = 3$ at the normalized time moment $\tau = 0.3$

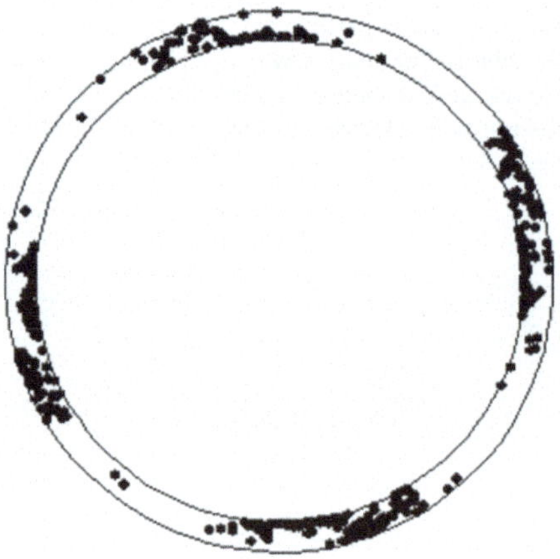

Fig. 9.9 Beam particle distribution in coordinate space for the SFXM with $m = 4$ at the normalized time moment $\tau = 0.3$

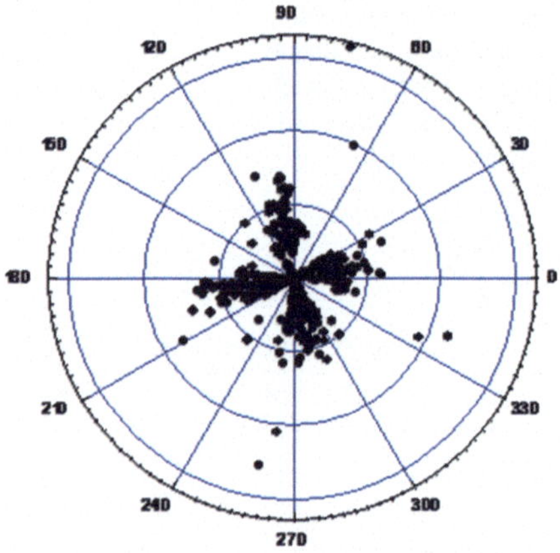

Fig. 9.10 Beam particle distribution in phase space for the SFXM with $m = 4$ at the normalized time moment $\tau = 0.3$

9.2 Nonlinear Theory of Surface Flute Wave Excitation by Electron Beams Gyrating Around a Plasma Column Under Condition of Beam-Dissipative Instability

The possibility to utilize the kinetic energy of a rotating beam of charged particles for excitation of microwave oscillations in cylindrical vacuum and plasma waveguides is being intensively studied. Doing that great attention is paid to magnetron devices [12, 13], which have much higher efficiency than that of free electron lasers [6] and electron cyclotron resonance masers [14–16].

In the previous section, the possibility of SFXM energy dissipation is neglected in analyzing the wave interaction with the charged beam particles. Under certain conditions, the collision frequency ν of the plasma particles can have a non-negligible value. In the present Section, the nonlinear stage of SFXM excitation caused by its interaction with a gyrating flow of electrons, which move over the plasma surface, is studied under the condition of a beam-dissipative resonance instability. This is also useful for understanding the complete physics picture of SFXM excitation. Development of beam-dissipative instabilities [17–22] has a number of specific features as compared to the case of beam-plasma instabilities. In a single-mode approach (if the width of the wave packet to be excited is assumed to be narrow) one can use the inverse Fourier transformation to get the equation for the amplitude of the surface wave [23]:

$$E_{\varphi}(R_1) = \mathrm{Re}D(\omega)E_{\varphi} + i\mathrm{Im}D(\omega)E_{\varphi} + i\frac{\partial\omega D}{\partial\omega}\frac{\partial E_{\varphi}}{\partial t} - E_{\varphi}\frac{\partial\omega D}{\partial\omega}\frac{\partial\theta}{\partial t}, \tag{9.15}$$

where E_{φ} is the amplitude and θ is the phase of the SFXW with azimuthal mode index m. One can derive the following approximate analytical expression for the imaginary part of $D(\omega)$:

$$\mathrm{Im}D(\omega) \approx \frac{\nu}{2\omega}\left[\frac{kI_{m+1}(\xi R_1)}{\xi I_m(\xi R_1)} + \frac{2|m|\omega^2}{\Omega_e^2 \varsigma_1}\right], \tag{9.16}$$

The presence of a non-zero imaginary part, $\mathrm{Im}D(\omega) \neq 0$, results in the change of the equation for the SFXM envelope amplitude as compared to the case of resonance beam-plasma instability (9.8). In the present case, the equation for the dimensionless envelope amplitude E is given by:

$$\begin{aligned}\frac{dE}{dt} &= -\frac{Ez}{P}\mathrm{Im}D(\omega)\\ &- \frac{\alpha D_p}{NzPL}\sum_{i=1}^{N}\left[\frac{mL_1(\varsigma_i)}{w}\frac{d\rho_i}{dt}\sin(\phi_i)\right.\\ &\left. + \rho_i^2\frac{d\varphi_i}{dt}L_2(\varsigma_i)\cos(\phi_i)\right],\end{aligned} \tag{9.17}$$

In Eq. (9.17), the same notations are used as in the previous Sect. 9.1. Equation (9.9) for the wave phase does not change under the condition $\mathrm{Im}\,D(\omega) \neq 0$. Thus, the nonlinear stage of a dissipative SFXM instability is described by Eq. (9.17) for the envelope amplitude, Eq. (9.9) for the wave phase and the set of Eqs. (9.11–9.14) for the coordinates and momenta of the beam particles [23].

The results of numerical studies on the nonlinear stage of beam-dissipative SFXM instability are plotted in Figs. 9.11 and 9.12 for the case: $m = 2$, $N = 500$, $R_1 = m\delta/0.39$, $z = 0.01$, and $\alpha = 0.05$. As one can see in Fig. 9.11, an increase in the value of ν results in enhancing the time interval during which the SFXM amplitude reaches its maximum while this maximum value itself becomes lower and the behavior of the amplitude oscillations at the stage of saturation becomes more irregular. These conclusions are in agreement with the general ones, which were obtained for SWs of other types [17]. Increase of the relative width of the gap, where the electron beam rotates, results in a decreasing linear growth rate. This is explained by the fact that the increase in the width causes the increase of the distance between the charged particles flow and the plasma surface where the SW amplitude has its maximum value. Increasing width also results in the increase of the value of the SFXM amplitude maximum and in the increase of the SFXM amplitude average value at the stage of saturation. Increasing SFXM azimuthal mode index m causes

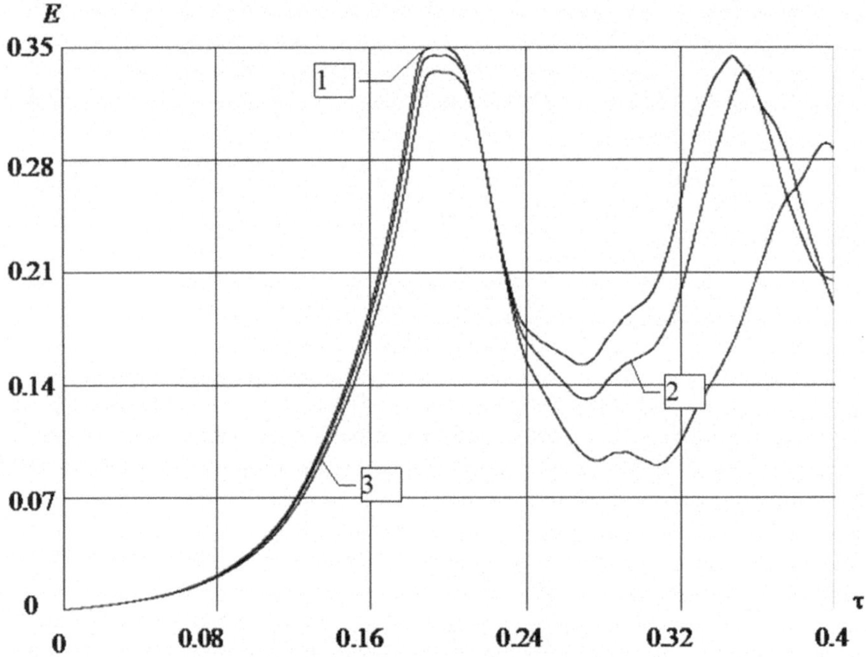

Fig. 9.11 Change of SFXM amplitude with time for $m = 2$. The *numbers* 1, 2 and 3 indicate the *curves* for $\nu = 0$; $0.1|\omega_e|$; $0.3|\omega_e|$, respectively

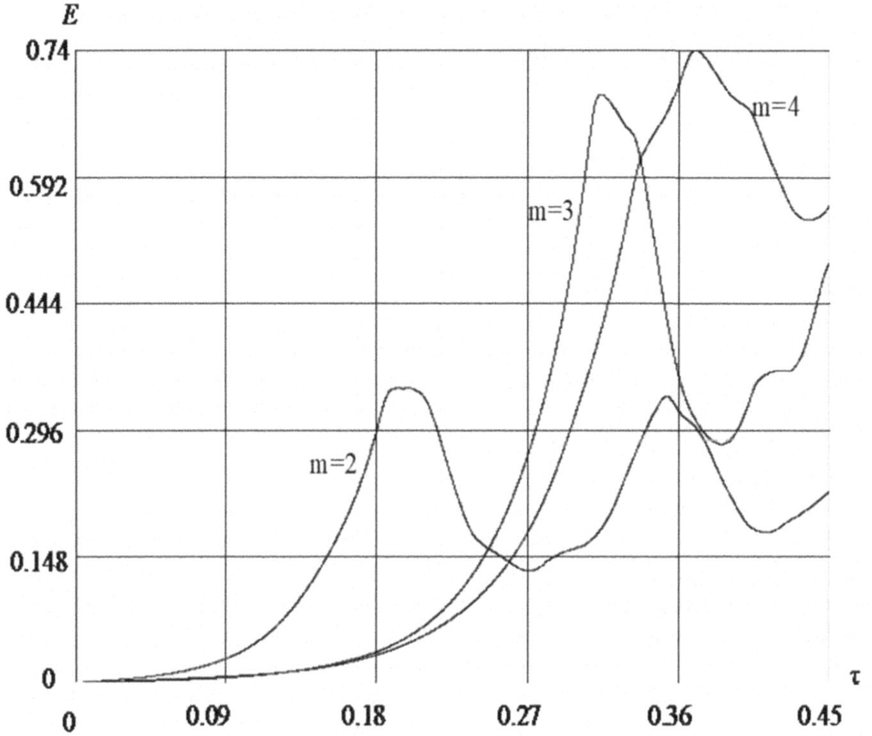

Fig. 9.12 Change of SFXM amplitude with time for $m = 2, 3, 4$ and $\nu = 0.1$, $\Delta = 0.1$

a decrease of the growth rate of the dissipative SFXM resonance instability along with an increase of the maximum of the envelope amplitude and average values at the stage of saturation. This is explained by the increase of the real part of the eigen-frequency, $Re(\omega)$, in this case, since just this parameter determines the amount of energy that can be transferred from the charged particles flow to wave excitation.

The proposed model of a microwave generator has the following advantages as compared to pure vacuum electron devices: smaller scale, the possibility to control the radiation spectrum via changing the external magnetic field and/or the density of internal semiconductor free electrons in solid state devices. In this case, one does not need to apply neither cryogenic technology nor super pure semiconductor materials. It looks like a convenient property of SFXMs that the wavenumber range, where SFXMs are excited efficiently, coincide with that, where it is easy to irradiate the wave energy through an axial slot antenna in the wall of the metal waveguide [24].

9.3 Nonlinear Theory of Surface Flute Wave Excitation by Modulated Annular Electron Beams

Active research on plasma filled structures during the last years is motivated first of all by the request to design electronic devices, which should be able to generate or amplify electromagnetic waves at high frequencies, be continuously tunable (over a broad frequency range) and be of a relatively small size. Various aspects of the problem of the interaction between electron beams and the eigenmodes of plasma filled waveguides are investigated e.g. in [1–4, 25–29]. There, the main attention is paid to studying the processes of wave excitation in magneto-active plasma waveguides. However, one cannot consider the theory of beam-plasma instabilities as completely developed, because the process of interaction between charged particle beams and the eigenwaves of plasma waveguides essentially depends on several factors, including the dispersion properties of these waves, their polarization, the spatial distribution of their fields, the geometrical peculiarities and the design features of the utilized waveguides.

To increase the efficiency of beam-plasma interaction, one can apply a preliminary spatial modulation (pre-bunching) of the charged particle beam [27, 29, 30]. One of the main branches of modulated beam application is the development of powerful super high frequency sources, see, e.g. [16], where a pre-bunched X-band (10 GHz) free electron maser amplifier with modulation input and power extraction output cavity was studied. The evolution of the velocity distribution function of a modulated beam during its motion in dense plasma was considered in [31]. There, the influence of background plasma inhomogeneity on the evolution of the modulated beam was studied numerically and the results obtained by simulations were compared with laboratory experiment data as well.

Since the nonlinear theory of SFXM excitation by a non-modulated charged particle beam due to beam-plasma instability was discussed in Sect. 9.1 and [7], and in order to further develop the theory of SFXM excitation, the present Section is devoted to the nonlinear theory of SFXM excitation by an annular electron beam, whose density is pre-modulated.

The schematic of the main features of the problem is described in Sect. 7.1 (see Fig. 7.1). The set of differential equations, which describes the nonlinear stage of the excitation of these modes by an electron beam, can be obtained from the hydrodynamic plasma equations, the Maxwell equations for the SFXM wave fields, and the momentum equation for the beam electrons. The beam is described by the method of plasma simulation using particles [8]. Since the beam and plasma particle densities are such that the inequality $n_b << n_e$ is valid, one can neglect both the effect of the beam on the properties of the plasma column and the effect of the self-field of the beam on the electromagnetic field in which the electron beam moves. The present consideration is restricted to the case of weak beam-plasma instability, where the density ratio $\alpha \equiv n_b/n_e << 1$ is the small parameter of the problem. Therefore, this theory describes the excitation of SFXMs, the increase of the wave envelop amplitude up to a maximum value, and the saturation of the instability due to the capturing of the beam particles into potential wells of the wave. However, such strong nonlinear effects as the oscillation of dense electron bunches in potential wells of the excited waves are out of scope of the present Section.

The tangential component of the SFXM magnetic wave field is described by an inhomogeneous Bessel equation [32], whose right-hand side $F_b(r)$ is determined by the electron beam, which moves around the plasma column. In the case when no flows of charged particles are present in the region $R_1 < r < R_2$ (which means $j_r = j_\varphi = 0$ and hence $F_b = 0$) the expressions for the SFXM wave fields coincide with those obtained in [33] for studying the dispersion properties of these modes in the case of a sufficiently dense plasma, where the electron plasma (Langmuir) frequency is larger than the electron cyclotron frequency. Thus, one can derive the expressions (9.4–9.6) for the components of the SFXM wave fields in the region occupied by the electron beam. The expressions (9.4) and (9.6) can be considered as Fourier-coefficients for the tangential components of the magnetic and electric wave fields in the space of time and azimuthal angle. They are functions of the radius r, the azimuthal mode index m, and the eigenfrequency ω.

In the plasma region ($r \leq R_1$) the spatial distribution of the SFXM axial magnetic wave field is described by the modified Bessel function [32] (see the definition (7.8) with the integration constant $C_4 = 0$). The radial and azimuthal electric wave fields can be found using this solution via Eq. (1.10).

Using standard procedures [7, 34] for the determination of the mean value and the separation of the amplitude and phase of the envelope of the excited waves, one can derive the Eqs. (9.8) and (9.9) for amplitude and phase of the envelope of the SFXM under the condition of resonant excitation of the eigenmode.

It is suitable to formulate the momentum Eq. (9.10), which describes the motion of the beam electrons, using terms of their momenta $\vec{p} = \gamma m_e \vec{V}$ (γ is the relativistic factor), because it allows one to take into account a weak relativism of the electron beam. The radial υ and azimuthal u dimensionless momenta of the beam particles and their time derivatives are determined by Eqs. (9.11–9.14).

Modulation of the electron beam means angular inhomogeneity of the beam particle density, $n_b = n_b(\varphi)$. Below, the dependence

$$n_b(\varphi) = (\pi/2) n_{b0} |cos(n\varphi)|, \tag{9.18}$$

is referred as weak modulation. The model (9.18) keeps the same overall number of beam particles regardless of the magnitude of n. The beam is called as strongly modulated if it consists of separate bunches:

$$\begin{aligned} n_b(\varphi) &= 2n_{b0}, \text{ if } 0 + 2\pi l/n < \varphi < (2l+1)\pi/n, \\ n_b(\varphi) &= 0, \text{ if } (2l+1)\pi/n < \varphi < 2(l+1)\pi/n, \end{aligned} \tag{9.19}$$

where $l = 0, 1, 2, \ldots$

The solution of the set of equations for amplitude (9.8) and phase (9.9) of the SFXM envelope, and the equation of motion for the beam electrons (9.11–9.14) is found by the fourth order Runge–Kutta method, which is widely used in studies of such problems [14, 27, 29, 35]. The intention to take a possible angular inhomogeneity of the beam particle density into account requires higher precision in numerical calculations as compared with the earlier studies in Sect. 9.1 for the case of initially angularly uniform beam. That is why the quantity of the beam macroparticles is chosen in the calculations presented in this Section to be $N = 2000$. To describe the interaction between the charged particle beam and the metal wall of the waveguide, the model of mirror reflection of the particles from the wall is applied as well (the particles do not disappear after such an interaction with the wall) [1, 7].

The results of the numerical analysis of the resonant beam-plasma instability of SFXMs are presented in Figs. 9.13, 9.14, 9.15, 9.16, 9.17 and 9.18. The condition for the resonant instability is given by (7.14) and means that the eigenfrequency ω of the SFXM is equal to the Nth harmonic of the electron cyclotron frequency ($\omega = N|\omega_e|/\gamma$).

One can conclude from the results of the linear theory of the interaction between annular electron beams and SFXMs [10, 11] that this interaction is most efficient if the azimuthal mode index m and the radius of the waveguide R_1 are chosen to satisfy the following approximate equation: $|m|\delta \approx 0.4R_1$. This condition is true for the calculations in the present Section.

Carrying out a numerical analysis of this resonant beam-plasma interaction, the following values of the parameters of the plasma filled waveguide and initial values of the electron beam parameters are applied: $E(\tau = 0) = 10^{-3}$ is the amplitude of the SFXM envelope, $\theta(\tau = 0) = 0$ is the phase of the envelope, $\upsilon_j(\tau = 0) = 0$ are

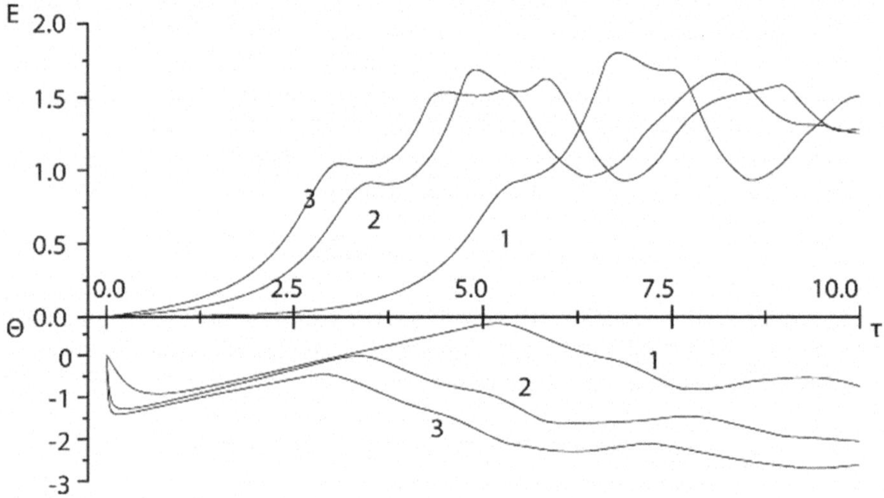

Fig. 9.13 Variation of the amplitude E of SFXM envelope (*in the upper half-space*) and its phase θ (*in the lower half-space*) with time τ in the cases: 1—angularly uniform beam; 2—weakly modulated beam; 3—strongly modulated beam. The azimuthal mode index is $m = 3$, and the gap width parameter $\Delta = 0.1$

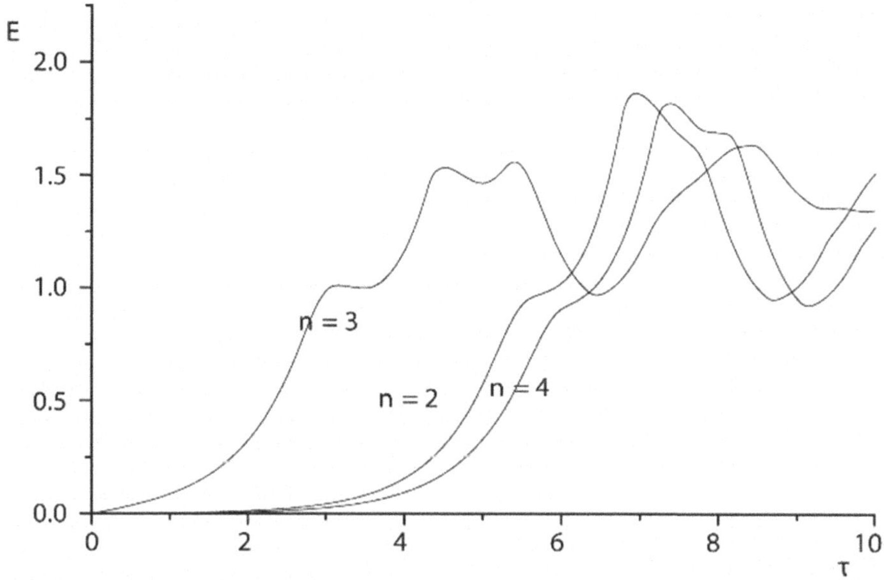

Fig. 9.14 Variation of the amplitude E of SFXM envelope with time τ for different magnitudes of the period of the initial spatial modulation of the beam with period: $n = 2$, 3 and 4. The azimuthal mode index is $m = 3$, and the gap width parameter $\Delta = 0.1$

the radial momenta of the beam particles, and $u_j(\tau = 0) \approx z\rho_j(\tau = 0)$ are the angular momenta of the particles. The latter approximate equation for the azimuthal momenta of the beam particles is valid within an accuracy of 2%. This means that at the initial moment of the electron beam-plasma interaction, the particles of the electron beam move with an azimuthal velocity, whose magnitude satisfies the condition $\upsilon_\varphi \approx |\omega_e|(R_1 + R_2)/(2\gamma)$.

Thus, initially the particles move in the gap between the plasma column and the waveguide wall without collisions with both the column and the wall. The normalized dimensionless particle density of the beam is $\alpha = 10^{-2}$, the SFXM azimuthal mode index is chosen to be $m = 3$. The initial location of the beam particles along the azimuthal angle φ is described by the sine-law (9.18). Along the radial co-ordinate the beam particles are distributed randomly within the space $R_1 + (R_2 - R_1)/3 < r < R_2 - (R_2 - R_1)/3$ of the gap between the plasma column and metal wall of the waveguide. The geometrical parameters of the waveguide are as follows: $\Delta = R_2/R_1 - 1 = 0.3$ (if another is not indicated), and the external static magnetic field and plasma particle density are chosen in such a manner that the following relation between electron cyclotron frequency and plasma frequency is valid: $Z \equiv \Omega_e/|\omega_e| = 10$. Here, the specific magnitude of the parameter $Z > 1$ is chosen arbitrarily. However, the condition for its large value was obtained in [33] in order to provide the possibility that the SFXM eigenfrequency belongs to the range of the electron cyclotron resonance. The calculations in the present Section are carried out for the cases when the azimuthal angular circumference 2π of the beam

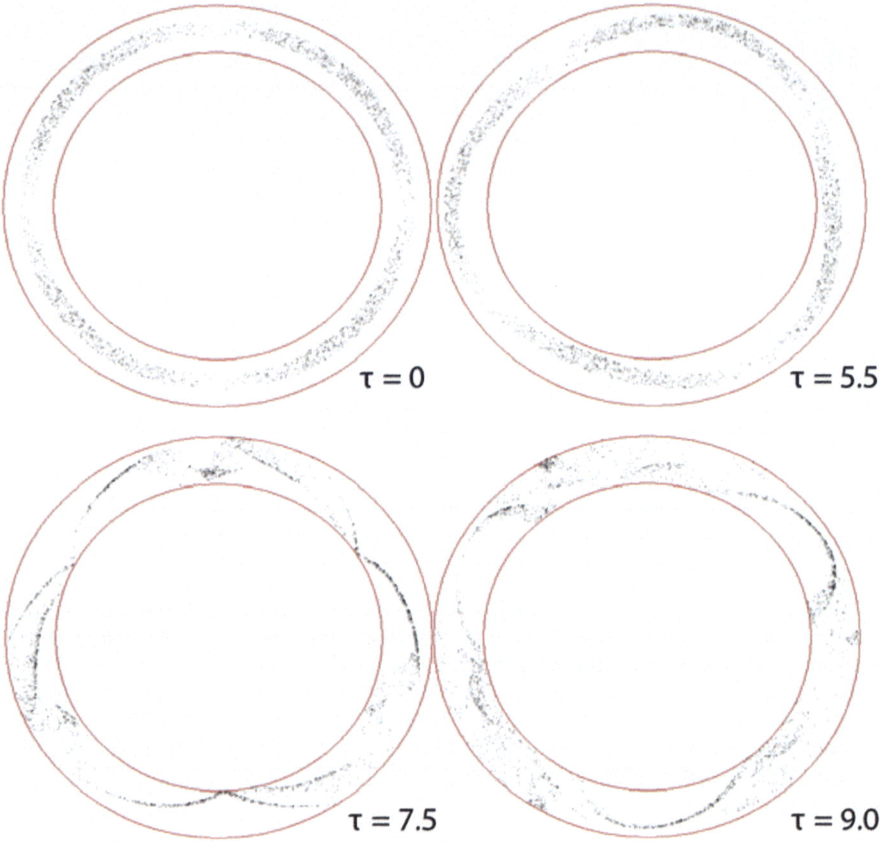

Fig. 9.15 Variation of spatial distribution of the beam particles with time for $m = 3$, $n = 4$, and $\Delta = 0.3$. The following moments of dimensionless time are chosen: $\tau = 0.0$, 5.5, 7.5 and 9.0

is equal to two, three and four periods ($n = 2$, 3 and 4) of the spatial modulation of the electron beam. Location of the macro-particles in both the co-ordinate space and phase space are indicated in the corresponding figures as dots. Hence, one can visually estimate, where the concentration of the beam particles is larger and where its magnitude is smaller.

One can see in Fig. 9.13, that the electron beam, whose density is spatially pre-modulated, interacts with the SFXM more efficiently than in the case of spatially uniform electron beam. The number 1 marks the curve obtained for the case of initially uniform angular distribution of the beam. Number 2 marks the curve calculated for the case of a continuous beam with weakly modulated density. The third curve marked by the number 3 presents the result obtained for the case of a strongly modulated beam, which initially is composed of spatially separated bunches. In the lower half-space of Fig. 9.13, the dependence of the SFXM envelope phase θ (in rad) on time $\tau \equiv |\omega_e|t$ is plotted. One can see that change of the phase θ for all

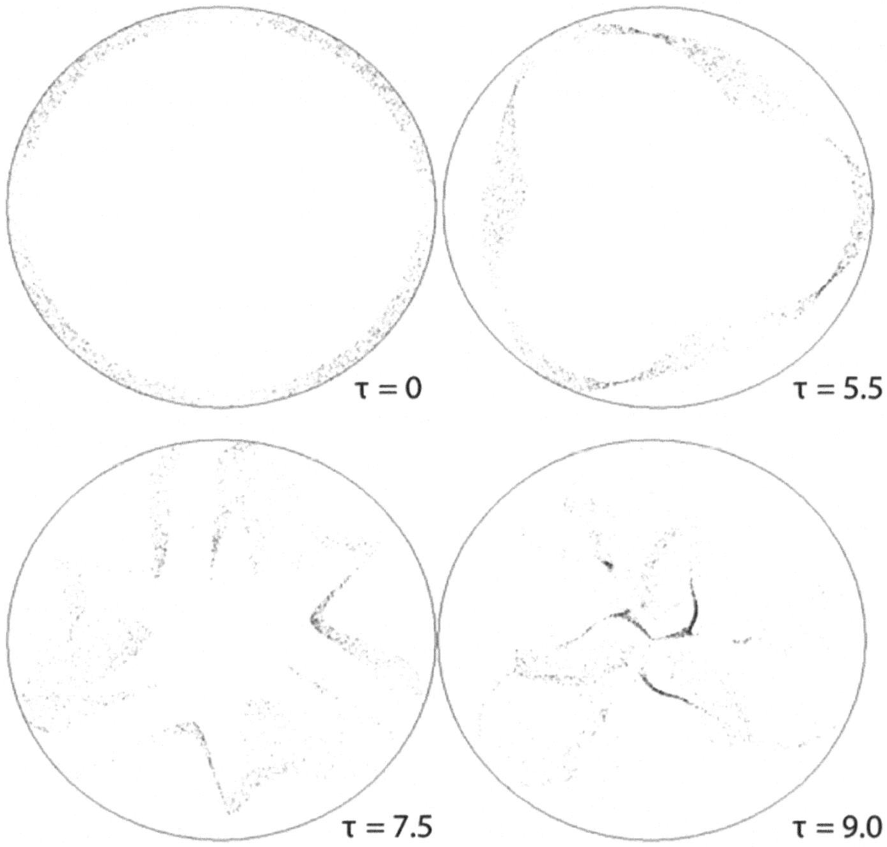

Fig. 9.16 Phase portraits of the beam particles: azimuthal momentum versus azimuthal angle for $m = 3$, $n = 4$, $\Delta = 0.3$. The external continuous circle represents the maximal magnitude of the azimuthal momentum $u = b/Z$. At the center of the circle $u = 0$

the calculated cases correlates with a change of the corresponding SFXM envelop amplitude with time.

The influence of the value of the spatial modulation period of the beam electrons on the development of the resonant beam-plasma instability of the SFXMs is also investigated. The results of this study are shown in Fig. 9.14. One can see that if the number n of periods of the spatial modulation of the annular electron beam at the initial moment of beam-plasma interaction coincides with the azimuthal mode index m, the resonant beam instability develops faster. If n is not equal to m, the resonant beam-plasma instability develops during the time interval, which is approximately equal to that interval, which is typical for the case of application of a spatially uniform (non-modulated) electron beam. The calculations are carried out until the time, when the SFXM envelop amplitude starts to oscillate nearby some value. This means that

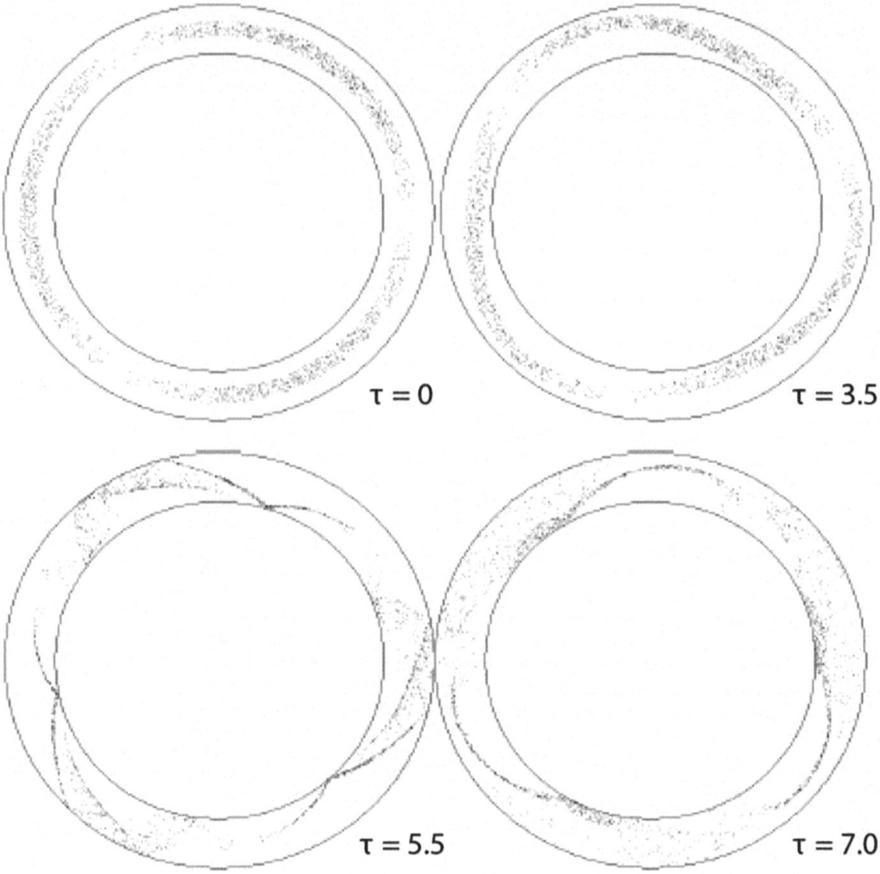

Fig. 9.17 The same as in Fig. 9.15, but for $n = 3$

the instability comes into the strongly nonlinear stage, where dense small-sized electron bunches interact with the wave, which has a finite value of the amplitude.

Figure 9.15 illustrates the dynamics of the spatial distribution of the beam particles in the case when the weakly modulated electron beam initially is characterized by the number of its modulation periods $n = 4$. One can see that the development of the resonant beam-plasma instability leads, first, to the formation of three groups of beam electrons and then capturing of them into potential wells of the SFXM with azimuthal mode index $m = 3$. Gradually the electrons, which initially were separated into four groups, start to roll down into the bottoms of three (since $m = 3$) potential wells of the SFXM. Then three groups of the captured beam particles start to create three dense bunches. To illustrate such development of the instability the following moments of time are chosen. The first is the initial moment. The second time moment corresponds to an intermediate magnitude of time between the starting moment and the moment, when the amplitude of the envelope reaches its first maximum. The third

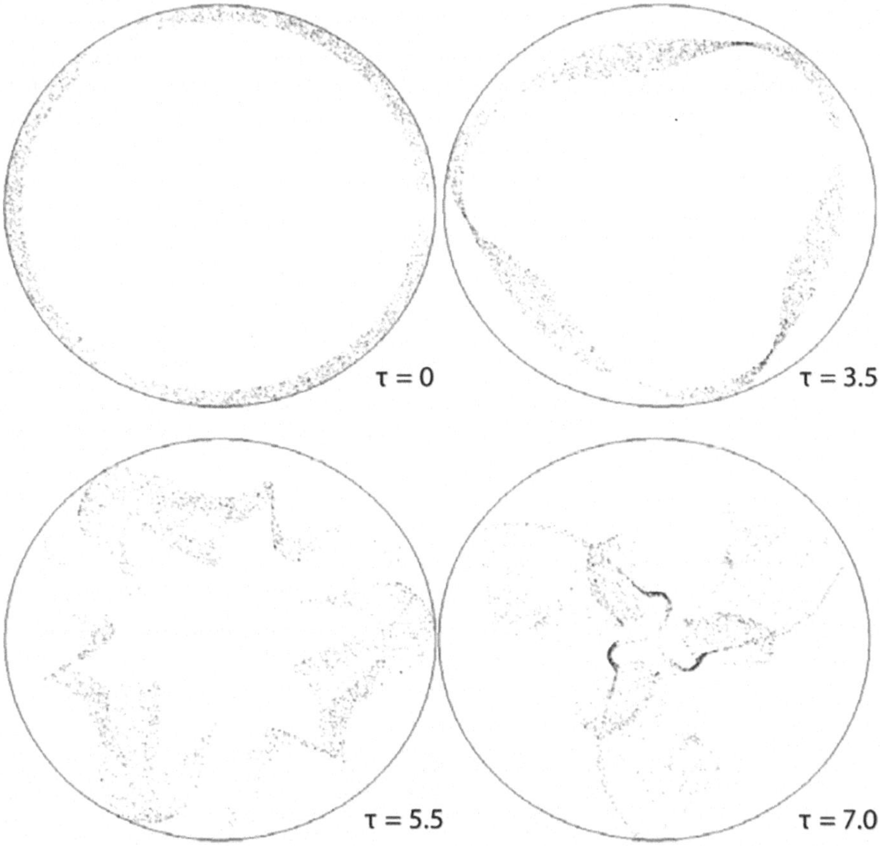

Fig. 9.18 The same as in Fig. 9.16, but for $n = 3$

moment corresponds to the first maximum of the amplitude of the SFXM envelope and the fourth time moment corresponds to its first minimum. In spite of the relatively small magnitude of the tangential component of the SFXM electric wave field [33] as compared with its magnetic wave field, one can see that finally just three bunches are formed, which correlates with the value of azimuthal mode index $m = 3$.

Phase portraits of the beam particles (azimuthal momenta versus azimuthal angles) calculated during the development of this instability are shown in Fig. 9.16 for the same moments of time, which are illustrated in Fig. 9.15. Here one can see the creation of three groups of the captured beam particles from initial four separated bunches of the beam electrons as well. The presented phase portraits illustrate losses of initial energy of the electron beam. One can see that gradually the azimuthal momenta of the electrons decrease (the dots, which represent the particles move to the center of the phase portrait), and the magnitudes of the electron momenta become closer to the bottoms of potential wells of the SFXM. One can also see that most of the beam electrons gather in a relatively narrow angular sectors, whose quantity is equal to the magnitude of the azimuthal mode index $m = 3$.

To compare the above discussed case of $n = 4$ with the case of $n = 3$, one has to study the results presented in Figs. 9.17 and 9.18. The distribution of the beam particles in coordinate space and phase space are plotted there, respectively. Once again, the coordinate distribution and phase portraits are given at the initial moment of time, an intermediate moment of time, the moment of the first maximum of the amplitude of the SFXM envelope and at the moment of its first minimum.

Comparison of the upper right phase portraits in Figs. 9.15 and 9.17 ($\tau = 5.5$) leads to the clear conclusion that the formation of the electron beam shape due to the influence of the SFXM with azimuthal mode index $m = 3$ happens faster in the case $n = 3$ as compared with the case $n = 4$. The process of gathering the beam electrons into three potential wells of the SFXM field also happens faster in the case $n = 3$ than in the case $n = 4$ (see the lower left phase portraits for $\tau = 7.5$ in Figs. 9.15 and 9.17). This means that the SFXM growth rate in the case $n = m = 3$ is larger than in the case $n = 4$. This conclusion is confirmed by the results presented in Fig. 9.14 (compare the curves marked by 3 and 4).

The second conclusion concerning the distinguishing features of the SFXM instability in the cases $n = 4$ and $n = 3$ can be made after comparing the corresponding phase portraits. This peculiarity is also seen in the plots with spatial distributions of the beam electrons illustrated as dots, but it is less pronounced as compared with the phase spaces filled with the illustrating dots. Thus, one can see that the distribution of the beam electrons in the SFXM field for the case $n = m = 3$ (Fig. 9.18) is practically symmetrical. In other words, the dots in all three groups are located in the same manner. However, in the case $n = 4$ (see the last three plots in Fig. 9.16), the location of the dots in these three groups is not symmetrical. This is connected with the fact that it is impossible to separate four initial groups of electrons into three identical groups at any time. Thus, in the case $n = 4$, one group of electrons consists of a larger quantity of particles as compared with the two other groups at the initial stage of the instability (it means until the beginning of the stage of saturation of the instability). Therefore, this dense bunch of electrons being in phase synchronism with the excited SFXM can transfer more quantity of its kinetic energy into wave energy as compared with the case $n = 3$. This is confirmed by results plotted in Fig. 9.14, where the maximum of the SFXM envelope amplitude in the case of $n = 3$ is smaller than that in the case of $n = 4$.

The resonant beam-plasma instability of surface flute modes is studied in the present Section numerically using the single-mode approximation. The nonlinear self-consistent set of differential equations is analyzed by the fourth order Runge–Kutta method. The obtained results allow one to make the following conclusions [36].

First, spatial pre-modulation of the electron beam leads to increasing growth rates of the resonant beam-plasma instability. The temporal interval necessary for the attainment of the saturation stage of this instability becomes shorter as compared with the case of uniform (non-modulated) electron beam. This is explained by the fact that the uniform electron beam is not separated into such a number of bunches, which could be able to move under the condition of phase synchronism with the excited SFXM.

Second, correlation between the number n of periods of the initial angular modulation of the beam electrons and the magnitude m of the azimuthal mode index of the studied SFXM influences strongly the development of its resonant beam-plasma instability. If the beam is initially composed of groups of electrons, whose quantity n is not equal to the azimuthal mode index m, the development of the SFXM instability has a qualitatively similar scenario as in the case of a non-modulated electron beam. This means, that the SFXM first of all separates the electrons into such a number of groups that is equal to m.

Third, strengthening of the degree of angular modulation of the electron beam and decreasing of the beam thickness lead to increasing SFXM instability growth rates. From the very beginning, the instability develops according to the prediction of the linear theory constructed for description of the interaction between an annular electron beam and SFXMs, which can propagate in different frequency ranges [11, 37]. At the nonlinear stage of the instability, changing of the SFXM envelop amplitude is similar to the results obtained in [7] for the case of a non-modulated beam.

Fourth, during the development of the resonant beam instability, the beam electrons lose their energy for excitation of the SFXM; the wave field captures the electrons into its potential wells. These electrons roll down into the bottom of the wells and gather there into groups of captured electrons with a small magnitude of azimuthal momentum. The quantity of these bunches is equal to the azimuthal mode index of the excited SFXM. Saturation of the SFXM instability is characterized by creation of high-density electron bunches by the captured electrons, which are localized in a space with narrow angular sizes.

The results obtained in the present Section can be useful for investigations in the branches of plasma electronics and gas discharges in plasmas, where SFXMs can be excited.

Higher order electron beam modulation is used e.g. in high harmonic gyrotrons [38]. The operation principle of gyrotrons is based on the generation of stimulated cyclotron radiation of the electrons (bremsstrahlung), which move along helical trajectories in a homogeneous magnetic field and interact resonantly in the cavity with a near-cutoff $TE_{m,p}$ mode ($k_z \approx 0$) at the Nth cyclotron harmonic, $\omega = N|\omega_e|/\gamma$, where ω is the wave frequency. For a fixed $TE_{m,p}$ cavity mode, the use of an axis-encircling electron beam with close-to-zero radii of the guiding centers of all particles provides a strong mode selection rule: the coupling factor between electron beam and $TE_{m,p}$ cavity mode differs from zero only when the azimuthal mode index m coincides with the resonance harmonic number N, $m = N$.

This means, that the spectrum of operating modes in so-called large orbit gyrotrons (LOGs) with axis-encircling electron beam is essentially rarer than in conventional gyrotrons with annular small-orbit helical electron beams. Correspondingly, even in the case of using the simplest cavity in the form of a cylindrical smoothly irregular waveguide, one can expect selective excitation of high electron cyclotron harmonics. As a result, it ensures the use of essentially simpler and cheaper sources of the magnetic field to obtain terahertz radiation. An important factor is that the LOG configuration allows operating at several harmonics while using the same electron gun, magnetic system, and tube collector. Thus, it is possible to work in a wide frequency range and change only the cavity and radiation output components [38].

9.4 Conclusions

Chapters 7 and 8 of the present book are devoted to elaborating the linear theory of SFXM interaction with an annular electron beam gyrating along large Larmor orbits around a cylindrical plasma column. The restriction to linear theory consists in consideration of the initial stage of beam-SFXM interaction when the wave amplitude is sufficiently small. The dependence of the positive imaginary part of the SFXM frequency on the parameters of both plasma waveguide and the beam is the main result of linear theory. This imaginary part is interpreted as the SFXM growth rate.

Obviously, the wave amplitude cannot increase exponentially for infinite time. Starting from some temporal moment, nonlinear phenomena in respect of the wave amplitude begin to influence the development of the beam-wave interaction. The present Chapter complements the theory of electron beam-SFXM interaction by description of this nonlinear stage, delivering new knowledge about the features of SFXMs. First, one can calculate how long does it take for SFXMs to reach saturation. Second, the wave amplitude at the stage of saturation can be predicted. Third, the beam-SFXM interaction at the nonlinear stage manifests itself through electron beam bunching both in the coordinate and phase spaces.

Studying nonlinear physics needs new mathematics as compared with that applied in Chaps. 7 and 8. A nonlinear boundary condition at the plasma interface (9.7) along with procedures of averaging over fast time variable processes are crucial points of this mathematics. The nonlinearity of the boundary condition is explained by the dependence of the SFXM wave field distribution within the space where the beam gyrates on the electric beam current, which in turn is determined by the SFXM wave field. The procedures of averaging are well-known tools in nonlinear physics. However, these procedures are applied for flute waves in the present monograph rather than for plane waves.

Preliminary modulation (pre-bunching) of the particle beam is a well-known method of enhancing the beam-wave interaction. In the present book, the preliminary modulation of the gyrating beam in azimuthal direction is considered to study its influence on the beam-SFXM interaction. The stronger the beam modulation is, the less time it takes for the SFXM amplitude to reach saturation. The main effect of the preliminary modulation of the beam can be observed for SFXMs with an angular period, which coincides with that of the beam modulation. In this case the time shortens almost by a factor of two.

References

1. Kondratenko, A. N., & Kuklin, V. M. (1988). *Basis of plasma electronics*. Energoatomizdat. (in Russian).
2. Rukhadze, A. A., Bogdankevich, L. S., Rosinskii, S. Y., & Ruhlin, S. Y. (1980). *Physics of high current relativistic electron beams*. Atomizdat. (in Russian).
3. Miller, R. B. (1982). *An introduction to the physics of intense charged particle beams*. Plenum Press.
4. Kuzelev, M. V., & Rukhadze, A. A. (1990). *Electrodynamics of dense electron beams in plasmas*. Nauka. (in Russian).
5. Kondratenko, A. N. (1985). *Surface and volume waves in a bounded plasma*, Energoatomizdat. (in Russian).
6. Saito, H., & Wurtele, J. S. (1987). The linear theory of the circular free-electron laser. *Physics of Fluids, 30*(7), 2209–2220.
7. Girka, V. O., & Puzyr'kov, S. Y. (2002). Nonlinear interaction of an annular electron beam with azimuthal surface waves. *Plasma Physics Reports, 28*(4), 351–358.
8. Birdsall, C. K., & Langdon, A. B. (2004). *Plasma physics via computer simulation, Series in plasma physics*. Taylor & Francis Group.
9. Abramowitz, M., & Stegun, I. A., (Eds). (1964). In *Handbook of mathematical functions*. National Bureau of Standards, Applied Mathematics, Series 55.
10. Girka, V. A., Girka, I. A., & Olefir, V. P., et al. (1991). Generation of electromagnetic waves with an annular relativistic electron beam. *Soviet Technical Physics Letters, 17*(1), 35–36.
11. Girka, V. O., Girka, I. O., & Morgal, Y. I., et al. (2011). Excitation of azimuthal surface modes by annular electron beams in the electron cyclotron frequency range. *Physica Scripta, 84*(2), 025505.
12. Davidson, R. C., & Tsang, K. T. (1988). Analysis of magnetron instability for relativistic non-neutral electron flow in cylindrical high-voltage diode. *Laser and Particle Beams, 6*(4), 661–685.
13. Riyopoulos, S. (1996). Magnetron theory. *Physics of Plasmas, 3*(3), 1137–1161.
14. Kho, T. H., Lin, A. T., & Chen, L. (1988). Gyrophase—coherent electron cyclotron maser. *Physics of Fluids, 31*(10), 3120–3126.
15. Davies, J. A. (1989). Conditions for absolute instability in the cyclotron resonance maser. *Physics of Fluids, B1*(3), 663–669.
16. Malek, F., Lucas, J., & Huang, Y. (2009). The engineering and construction of a pre-bunched free electron maser. *Progress in Electromagnetics Research, PIER, 95*, 19–38.
17. Panchenko, I. P., & Kuklin, V. M. (1987). Dissipative modes of beam instability under the conditions of abnormal Doppler effect. *Ukrainskii Fizicheskii Zhurnal, 32*(12), 1827–1832. (in Russian).
18. Girka, V. A., Girka, I. A., & Tkachenko, V. I. (1996). Excitation of azimuthal surface modes in cylindrical semiconductor structures in the presence of drift motion of electrons' flow. *Soviet Technical Physics, 66*(4), 114–120.
19. Bludman, S. A., Watson, K. M., & Rosenbluth, M. N. (1960). Statistical mechanics of relativistic streams II. *Physics of Fluids, 3*(5), 747–757.
20. Ballai, I., Oliver, R., & Alexandrou, M. (2015). Dissipative instability in partially ionised prominence plasmas. *Astronomy and Astrophysics, 577*, A82.
21. Ruderman, M. S., Verwichte, E., & Erdélyi, R., et al. (1996). Dissipative instability of the MHD tangential discontinuity in magnetized plasmas with anisotropic viscosity and thermal conductivity. *Journal of Plasma Physics, 56*(2), 285–306.
22. Tarasov, N., Perego, A. M., & Churkin, D. V., et al. (2016). Mode-locking via dissipative Faraday instability. *Nature Communications, 7*, 12441.
23. Girka, V. O., Kondratenko, A. M., & Sporov, A. E. (1999). Nonlinear theory of beam excitation of azimuthal surface modes in plasma-filled cylindrical metal waveguides. *Technical Physics, 44*(7), 814–818.

24. Girka, V. A., & Girka, I. A. (1992). Emission of azimuthal surface waves from narrow waveguide slot. *Soviet Journal of Communications Technology and Electronics, 37*(9), 32–35.
25. Lawson, J. D. (1977). *The physics of charged—particle beams*. Clarendon Press.
26. Kalmykova, S. S., & Kurilko, V. I. (1988). Physical mechanisms for the hydrodynamic beam-plasma instability. *Soviet Physics Uspekhi, 31*(8), 750–762.
27. Bogdankevich, L. S., Kuzelev, M. V., & Rukhadze, A. A. (1981). Plasma SHF electronics. *Physics-Uspekhi, 133*, 3–16.
28. Vlasov, A. N., Shkvarunets, A. G., & Rodgers, J. C., et al. (2000). Overmoded GW-class surface-wave microwave oscillator. *IEEE Transactions on Plasma Science, 28*(3), 505–560.
29. Humphries, S. (1990). *Charged particle beams*. Wiley and Sons Inc.
30. Barker, R. J., & Schamiloglu, E. (Eds.). (2001). *High-power microwave sources and technologies*. IEEE Press.
31. Anisimov, I. O., & Soloviova, M. J. (2011). The evolution of a modulated electron beam in a dense plasma barrier. *Plasma Physics Controlled Fusion, 53*(7), 074007.
32. Korenev, B. G. (2002). *Bessel functions and their applications*. Chapmen & Hall/CRC Press LLC.
33. Girka, V. O., Girka, I. O., Girka, A. V., & Pavlenko, I. V. (2010). Theory of azimuthal surface waves propagating in non-uniform waveguides. *Journal of Plasma Physics, 77*(4), 493–519.
34. Ederra, I., Iriarte, J. C., & Gonzalo, R., et al. (2011). Surface waves of finite size electromagnetic band gap woodpile structures. *Progress in Electromagnetics Research B, 28*, 19–34.
35. Kainer, S., Gaffey, J. D., & Price, C. P., et al. (1988). Nonlinear wave interactions and evolutions of a ring-beam distribution of energetic electrons. *Physics of Fluids, 31*(8), 2283–2284.
36. Girka, V. O., Puzyrkov, S. Y., & Nefodov, O. Y. (2013). Excitation of azimuthal eigen modes by modulated annular electron beam. *Progress in Electromagnetics Research B, 46*, 159–175.
37. Girka, V. O., Girka, I. O., & Pavlenko, I. V. (2011). Excitation of azimuthal surface modes by relativistic flows of electrons in high-frequency range. *Plasma Physics Reports, 37*(5), 447–454.
38. Kartikeyan, M. V., Borie, E., & Thumm, M. (2004). *Gyrotrons—high power microwave and millimeter wave technology*. Springer.

Chapter 10
Excitation of Surface Flute Waves with Frequencies Other Than Electron Cyclotron Frequency by Flow of Charged Particles Gyrating Along Large Larmor Orbits in Axial External Static Magnetic Field

Excitation of SFXMs by electron beams gyrating around a plasma column within the LF range (1.13) which includes the electron cyclotron frequency is under consideration in Chaps. 7–9. As it is mentioned in Sect. 1.4.1, SFXMs propagate within this range in the case of rather high plasma particle density and sufficiently low external stationary magnetic field, so that ${\Omega_e}^2 > {\omega_e}^2$. It is found there that in this case SFXMs can also propagate within the other frequency range (1.14), which is called as high-frequency (HF) range. It falls within the upper hybrid frequency ω_{UH} and the cut-off frequency for bulk waves: $\omega_2 = |\omega_e|/2 + \sqrt{\Omega_e^2 + \omega_e^2/4}$.

In the case of magnetized plasmas, ${\Omega_e}^2 < {\omega_e}^2$, the dispersion properties of SFXMs are studied in Sect. 2.3. There, SFXMs are shown to propagate within the frequency ranges (2.14) where SFXMs appear to be electromagnetic oscillations of the ion component in the plasma.

In the case of sufficiently dense plasmas, it is natural to expect SFXMs to be more efficiently excited by a gyrating electron beam just within the electron cyclotron frequency range as compared with the HF range. This is clearly demonstrated, in particular, in Chap. 8 via comparison of LF and HF SFXM excitation in coaxial waveguides. That is why just LF SFXM excitation is studied in more detail in the present monograph. However, SFXM excitation by gyrating electron beam is also possible within the HF range [1–3]. The initial stage of a resonant beam-plasma instability of HF SFXMs is investigated in the present Chapter in Sects. 10.1–10.3.

In addition, it is also possible to excite SFXMs in magnetized plasmas, by applying a flow of ions gyrating in the external static magnetic field [4, 5]. Linear theory of SFXM excitation by light ion beams is presented in the Sect. 10.4.

© The Author(s), under exclusive license to Springer Nature Switzerland AG 2022
I. Girka and M. Thumm, *Surface Flute Waves in Plasmas*, Springer Series on Atomic, Optical, and Plasma Physics 120, https://doi.org/10.1007/978-3-030-98210-2_10

10.1 Excitation of Surface Flute Waves Above Upper Hybrid Frequency by Electron Beams Gyrating Around a Plasma Column

In Ref. [6], the possibility of SFXW excitation by relativistic electron beams was shown. The peak value of the instability growth rate and its dependence on plasma particle density, azimuthal mode index, external static magnetic field and thickness of the vacuum gap between plasma and metallic waveguide wall was studied analytically. Because of restricted numerical resources which was typical at that time, one was lucky in Ref. [6] to find out the conditions for effective SFXW excitation in the case of sufficiently dense plasma filling ($\Omega_e = 10|\omega_e|$) in the LF range only. Note, that the HF range (1.14) looks very narrow, $1 + (2Z^2)^{-1} < \omega/\Omega_e < 1 + (2Z)^{-1}$, in units of the Langmuir frequencies for the case $Z = \Omega_e/|\omega_e| > 1$ under study. In 1991, this caused serious difficulties in the numerical study of the cumbersome dispersion relation. Thus, some questions, including those about possibility of HF SFXW excitation, influence of the ratio Z on SFXW interaction with electron beams, and presence of additional extremes in the curve of the dependence of the growth rate on the effective wavenumber, remained not considered. In the present Section, the study of the initial stage of SFXW excitation by charged particle beams, which started in Sect. 7.1, is extended to the HF range.

The schematic of the problem under consideration here is the same as in Sect. 7.1. A column of cold magneto-active plasma with radius R_1 is located concentrically inside a cylindrical metal waveguide with inner wall radius R_2 (see Fig. 7.1). An external static magnetic field is oriented along the system axis, $\vec{B}_0 || \vec{z}$. A relativistic annular electron beam is injected perpendicularly to the system axis into the gap $R_1 < r < R_2$ between the plasma column and the metal wall of the waveguide. The beam gyrating around the plasma is described by the model of oscillators with identical transversal momentum $p_{\perp 0}$ and zero axial momentum $p_z = 0$. Doing that, the considered beam-plasma system is assumed to be compensated both in respect to current and charge. Such flow of electrons is described by an equilibrium distribution function [7] (see Eq. (7.1)). The electrodynamic properties of the waveguide space, which is occupied by the beam, are described by the permittivity tensor $\varepsilon_{ik}^{(b)}$ [7] (see Eqs. (7.18)–(7.20)). To derive the SFXM dispersion relation one can apply dissipationless linear boundary conditions [8] (see also Sect. 1.4.4). This relation has the form (7.12).

To get the dependence of the SFXM growth rate on the beam-plasma parameters within the HF range (1.14) one assumes the resonance condition (7.14) to be valid. The ratio of the beam and plasma particle densities $\alpha = n_b/n_e$ is supposed to be small which allows to neglect the influence of the particle beam on the plasma properties.

As it is well-known [9–12], excitation of electromagnetic waves takes place in the most efficient way just nearby eigenfrequencies of plasma waveguides. Analysis of the dispersion relation (7.12) allows one to see, that the influence of a beam of low particle density ($\alpha \ll 1$) can already be sufficient in the case of occurrence of resonant denominators in the expressions for $\varepsilon_{ik}^{(b)}$. Excitation of SFXMs is possible just nearby the intersection points of these wave dispersion curves, constructed under the approximation $\alpha = 0$ in the HF range, with the lines $\omega = N|\omega_e|/\gamma$ along which the components of the permittivity tensor $\varepsilon_{ik}^{(b)}$ have resonant denominators.

As an example, the HF SFXM dispersion curves for the azimuthal mode indices $m = 1$ and 2 are plotted in Figs. 10.1 and 10.4, respectively. The effective wavenumber k_{ef} is counted along the abscissa axis, and the SFXM eigenfrequency normalized to the absolute value of the electron cyclotron frequency, $\omega/|\omega_e|$, is taken along the ordinate axis. The interval of ordinate values in these figures correspond to the boundaries of the HF range of SFXMs $2.236 < \omega/|\omega_e| < 2.562$. The following magnitudes of plasma and waveguide parameters are chosen: $Z = \Omega_e/|\omega_e| = 2$ and $\Delta = R_2/R_1 - 1 = 0.3$. In this case the dispersion curve for the $m = 1$ SFXM (bold solid curve) crosses with four resonance lines $\omega = N|\omega_e|/\gamma$, namely with the lines for $N = 7 \div 10$. For each value of the effective wavenumber (for each value of plasma column radius R_1), it is assumed that the relativistic electron beam rotates along a Larmor orbit which is precisely in the middle of the vacuum gap, which means with radius $r = R_1(1 + 0.5\Delta)$. This causes the dependence $\gamma = \gamma(k_{ef})$. The most effective excitation of SFXMs is to be expected near the values of the effective wavenumber for which the condition (7.14) is satisfied. There are four such values of k_{ef} for the specified azimuthal mode index $m = 1$ in Fig. 10.1: $k_{ef} = 0.134$ for $N = 10$, $k_{ef} = 0.154$ for $N = 9$, $k_{ef} = 0.184$ for $N = 8$, and $k_{ef} = 0.222$ for $N = 7$.

The behavior of the growth rates of SFXM beam excitation in Fig. 10.2 is completely in agreement with the requirement of fulfillment of the resonance conditions depicted in Fig. 10.1. The dependence of the growth rates normalized to the absolute value of the electron cyclotron frequency, $\mathrm{Im}(\omega/|\omega_e|)$, versus the effective wavenumber is plotted in Fig. 10.2 for the same waveguide parameters as in Fig. 10.1: $m = 1$, $Z = 2.0$, and $\Delta = 0.3$. To get visible magnitudes of the SFXM growth rate the beam particle density is chosen to be sufficiently large: $\alpha = 0.1$. Increasing number N of the beam cyclotron harmonic which causes the SFXM excitation is accompanied by decreasing effective wavenumber for which excitation takes place, decreasing peak magnitude of the growth rate, and decreasing width of the resonance in terms

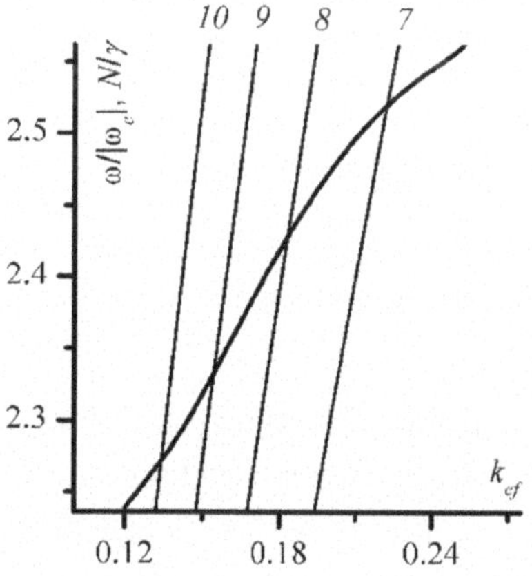

Fig. 10.1 Plot of the dispersion curve (*bold solid curve*) for the SFXM with azimuthal mode index $m = 1$ crossed by the resonance lines (*thin solid lines*) $N = \omega_\gamma/|\omega_e|$ for $Z = 2.0$ and $\Delta = 0.3$. The *numbers* close to the *thin lines* indicate the corresponding cyclotron harmonics: $N = 7, 8, 9$, and 10

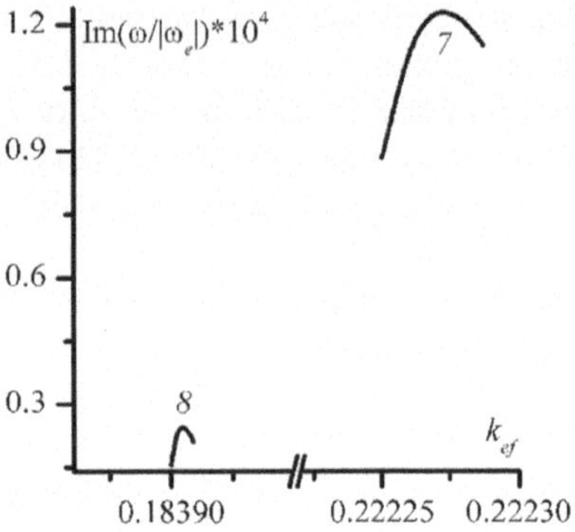

Fig. 10.2 Normalized growth rates of the SFXMs with azimuthal mode index $m = 1$ versus effective wavenumber for $Z = 2.0$, $\Delta = 0.3$, $\alpha = 0.1$. The *numbers* indicate the numbers of the cyclotron harmonics which provide SFXM excitation: $N = 7$ and 8

of k_{ef} (range of k_{ef} for which the growth rate is larger than half of its peak magnitude). For $N = 7$, the peak value of the growth rate $\mathrm{Im}(\omega/|\omega_e|) = 1.23 \times 10^{-4}$ is achieved for $k_{ef} = 0.22227$. Note that the resonance condition (7.14) is violated by 0.1% for this magnitude of k_{ef}. For $N = 8$, the maximum growth rate $\mathrm{Im}(\omega/|\omega_e|) = 2.45 \times 10^{-5}$ is much smaller than for $N = 7$. It is achieved for $k_{ef} = 0.184$ which is by 0.2% larger than that for which the resonance condition $\omega = 8|\omega_e|/\gamma$ is valid. The increase of the SFXM growth rates with rising k_{ef} is explained by the fact that the derivative $\partial D^{(0)}/\partial\omega$, to which the growth rate is inversely proportional, is much larger for wide ($k_{ef} \to 0$) than for narrow ($k_{ef} \sim 1$) waveguides.

One can also observe a decrease of the resonance width with increasing cyclotron number N from seven to eight. For $N = 8$, this width is about 10^{-5}, which is six times smaller than that for $N = 7$. Just this small magnitude of the resonance width (rather than that of the growth rate itself) causes most of the difficulties for numerical search of the growth rates for $N = 9$ and $N = 10$.

In general, the plot of the HF SFXM growth rate versus effective wavenumber in Fig. 10.2 looks like two separate resonances. In the case of LF SFXM beam excitation [6], the similar phenomenon was observed with the following difference. Namely, the regime of resonance overlapping was realized. Nevertheless, growth rates of LF SFXM excitation behave non-monotonously. In Ref. [6], this phenomenon was not explained. However, now it is clear, that with decreasing k_{ef} the resonance condition (7.14) for the beam-plasma instability is realized for higher order cyclotron harmonics N.

The dispersion relation (7.12) is derived with account for the relativistic factor γ. One has to show, that for excitation of HF SFXMs really just relativistic electron beams are required. For example, for the SFXM with azimuthal mode index $m = 1$ the peak magnitude of the growth rate occurs at $k_{ef} \approx 0.22227$. Under this condition, the velocity of the electron beam should be $\upsilon \approx 0.933c$, thus the relativistic factor is equal to $\gamma \approx 2.773$ (acceleration voltage $\approx$ 906 kV). Therefore, this condition for excitation of HF SFXMs differs from that studied in Sect. 7.1 for the LH SFXWs [6, 13] where the motion of the electron beam was characterized by the relativistic factor $\gamma = \sqrt{1 + p_{\perp 0}^2 m_e^{-2} c^{-2}}$, which did not exceed 1.7 (acceleration voltage $\approx$ 137 kV).

As it is well-known from analytical treatment of beam-plasma instabilities [14], the maximum value of the SFXM growth rate under the condition of beam-plasma instability is proportional to the cubic root of the beam particle density. Results of studying the dependence of SFXM growth rates under the condition of beam-plasma instability with azimuthal mode index $m = 1$ on the effective wavenumber are shown in Fig. 10.3 for different values of beam particle density. The dashed curves correspond to the beam particle density $\alpha = 0.1$. They are plotted in Fig. 10.3 for comparison with the case of twice larger density $\alpha = 0.2$ (solid curves). According to the theoretical predictions, the increase of the beam particle density by a factor of two should be accompanied by the increase of the growth rate by the factor of $\sqrt[3]{2} \approx 1.26$. The numerical analysis shows good agreement of calculations and analytical predictions. In the case of $N = 7$, the peak magnitude of the growth rate is $\text{Im}(\omega/|\omega_e|) = 1.23 \times 10^{-4}$ for $\alpha = 0.1$ as it was already mentioned, whereas for $\alpha = 0.2$ the peak value of the growth rate is 1.246 times larger, i.e. $\text{Im}(\omega/|\omega_e|) = 1.532 \times 10^{-4}$. In the case of $N = 8$, the peak magnitude of the growth rate is $\text{Im}(\omega/|\omega_e|) = 2.45 \times 10^{-5}$ for $\alpha = 0.1$, whereas for $\alpha = 0.2$, this peak value is 1.256 times larger, i.e. $\text{Im}(\omega/|\omega_e|) = 3.077 \times 10^{-5}$.

The positions of the maxima of the SFXM growth rates shift more to the right for larger values of α as compared with the values of k_{ef} where the dispersion curve of HF SFXM crosses the curves $\omega = N|\omega_e|/\gamma$. Doubling of α results in an increase of the shift Δk_{ef} by a factor of 2.167 (from $k_{ef} = 0.22227$ to $k_{ef} = 0.222585$) for $N = 7$, and by 2.105 for $N = 8$. The peak width naturally increases with increasing beam particle density α as well.

Increase of the azimuthal mode index m results in a significant increase in the number N of the beam cyclotron harmonics, which could contribute to HF SFXM excitation. As one can see in Fig. 10.4, these are $N = 19 \div 22$ for SFXM with

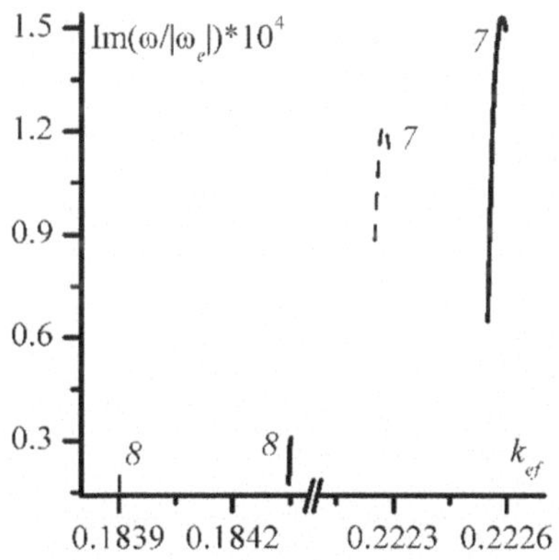

Fig. 10.3 Normalized growth rates for electron beam excitation of the SFXM with azimuthal mode index $m = 1$ versus effective wavenumber for different values of the beam particle density for $Z = 2.0$ and $\Delta = 0.3$. The *dashed curves* are plotted for the density $\alpha = 0.1$ whereas the *solid curves* show the case $\alpha = 0.2$. The *numbers* close to the *curves* indicate the corresponding cyclotron harmonics: $N = 7$ and 8

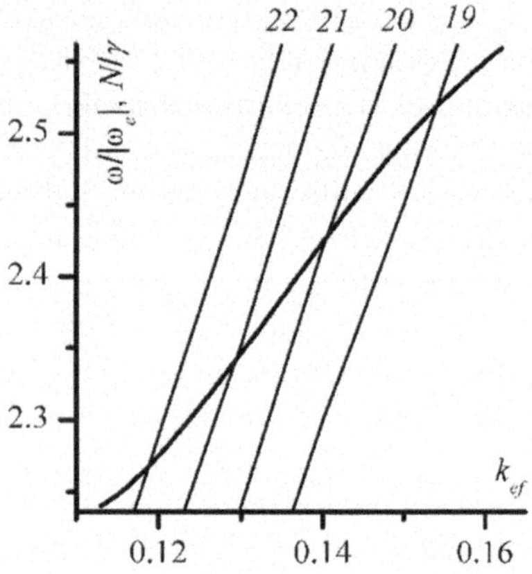

Fig. 10.4 Plot of the dispersion curve (*bold solid curve*) for the SFXM with azimuthal mode index $m = 2$ crossed by the resonance lines (*thin solid curves*) $N = \omega\gamma/|\omega_e|$ for $Z = 2.0$ and $\Delta = 0.3$. The *numbers* close to the *curves* indicate the corresponding cyclotron harmonics: $N = 19$, 20, 21 and 22

$m = 2$. As is known from analytical treatment and is convincingly demonstrated in Fig. 10.2, an increase of N is accompanied by a crucial decrease of both the growth rate and the peak width. One can expect that the maximum growth rate of the SFXM with $m = 2$ is by several orders smaller than for $m = 1$. This explains the failure in numerical studying the dispersion relation (7.12) in this case.

The dependencies of HF SFXM growth rates on the thickness of the vacuum gap, where the electron beam moves, are shown in Fig. 10.5. Thus, for comparison with the case $\Delta = 0.3$ considered above in Fig. 10.2 and presented in Fig. 10.5 by the short dashed curve, the growth rates for the larger value of the vacuum gap $\Delta = 0.35$ are shown by the solid curves. It should be underlined that the influence of this gap width on SFXM excitation growth rates consists not so much in an appreciable enhancement of the peak magnitude of the growth rate, as in increasing the number of

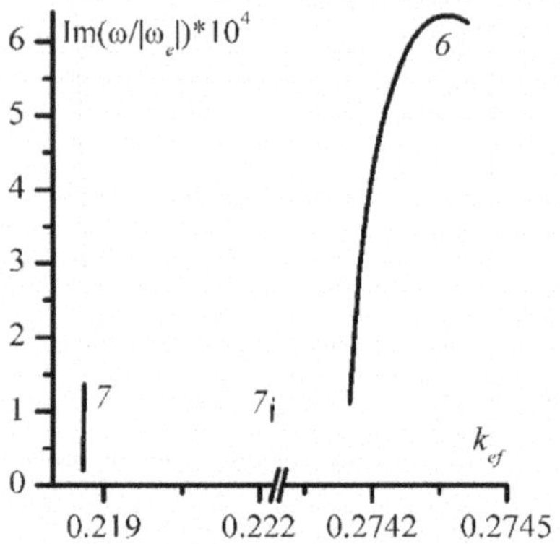

Fig. 10.5 Normalized growth rates of the SFXM with azimuthal mode index $m = 1$ versus effective wavenumber for different values of the vacuum gap width in the case $Z = 2.0$ and $\alpha = 0.1$. The short *dashed curve* corresponds to $\Delta = 0.3$ whereas the *solid curves* to the case $\Delta = 0.35$. The *numbers* close to the *curves* indicate the corresponding cyclotron harmonics: $N = 6$ and 7

intersection points of dispersion curves with resonance lines $\omega = N|\omega_e|/\gamma$. Thereby it results in an expansion of the ranges of k_{ef} values, where efficient excitation of SFXMs is possible. New (as compared with the case $\Delta = 0.3$) is the possibility for excitation of the SFXM with $m = 1$ which appears in a waveguide with $\Delta = 0.35$. It is introduced by the beam cyclotron harmonic $N = 6$. In this case the peak magnitude of the growth rate is $\text{Im}(\omega/|\omega_e|) = 6.368 \times 10^{-4}$. It is realized for $k_{ef} \approx 0.274$. This maximum is an order of magnitude wider than in the case $N = 7$.

Appreciable influence on SFXM propagation is exerted by the strength of the external static magnetic field [8]. The influence of the magnetic field on HF SFXM excitation is analyzed in Fig. 10.6. In this case normalization of the SFXW growth rates to the absolute value of electron cyclotron frequency applied for the plots presented in the previous figures becomes inconvenient. That is why in Fig. 10.6 the growth rates counted along the ordinate axis are normalized by the Langmuir frequency. Thus, both axes of the diagram are normalized to Ω_e. The solid curve corresponds to the case of stronger magnetic field, $Z = 1.8$, and for comparison the dashed curve presents the case of weaker magnetic field, $Z = 2$, shown in Fig. 10.2. One can see that reduction of the strength of the external static axial magnetic field by 10% worsens the conditions for SFXM excitation: the position of the maximum SFXM growth rate shifts by 15% to the right boundary of the k_{ef} range where HF SFXMs can exist.

In the following, the possibility of an experimental observation of the phenomenon studied here is estimated. As one can see from Fig. 10.5, the HF SFXM with azimuthal mode index $m = 1$ can be effectively excited at $k_{ef} \sim 0.274$. A laboratory device with the radius $R_1 \sim 10$ cm of the plasma column is considered as an example. For the plasma density $n_e \sim 15 \times 10^{10}\,\text{cm}^{-3}$ the Langmuir frequency is equal to $\Omega_e \sim 2.2 \times 10^{10}\,\text{s}^{-1}$. The plot in Fig. 10.5 is calculated for the parameter $Z = 2$. Therefore, the absolute value of the electron cyclotron frequency is $|\omega_e| \sim 1.1 \times 10^{10}\,\text{s}^{-1}$,

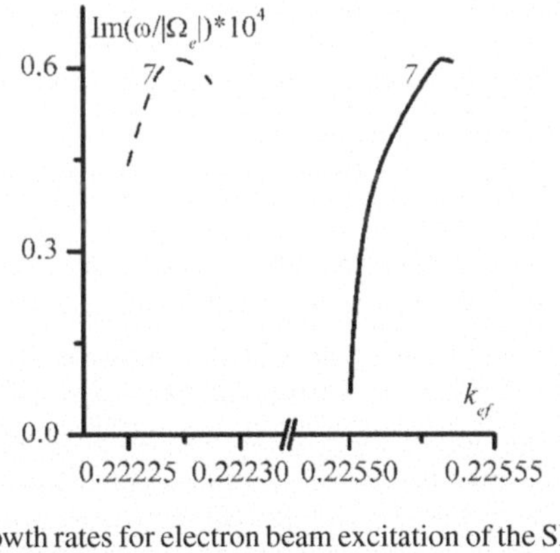

Fig. 10.6 Normalized growth rates for electron beam excitation of the SFXM with azimuthal mode index $m = 1$ versus effective wavenumber for various values of external constant magnetic field in the case $\alpha = 0.1$, $\Delta = 0.3$. The *dashed curve* shows the dependence in the case of $Z = 2.0$ and the *solid curve* for $Z = 1.8$. The *number* indicates the cyclotron harmonic $N = 7$ which causes interaction of the SFXM with the beam

which corresponds to the magnetic field induction $B_0 \sim 633$ G. Thus, the electron beam should rotate with the velocity 0.906*c*. In this case the SFXM propagates with a frequency of approx. $\omega \sim 2.8 \times 10^{10}\,\mathrm{s}^{-1}$, and is excited with the growth rate $\mathrm{Im}(\omega) \sim 7 \times 10^6\,\mathrm{s}^{-1}$.

In the present Section, in addition to results obtained in the Chap. 7, the possibility to realize the conditions of beam-plasma instabilities of SFXMs within the HF range is investigated. Results of numerical analysis qualitatively correspond to analytical predictions. It is shown that the maximum growth rates of resonant beam-plasma instabilities of HF SFXMs increase with decreasing external magnetic field, increasing thickness of the gap Δ between the plasma column and the metal wall of the waveguide, and increasing beam particle density [1]. In general, the dependence of the beam instability growth rate $\mathrm{Im}(\omega)$ on such plasma-waveguide parameters as plasma density, external axial static magnetic field and thickness of the vacuum gap has non-monotonous character. The latter is explained by the discrete arising or disappearing of the intersections of the HF SFXM dispersion curves with the resonance curves $\omega = N|\omega_e|/\gamma$.

The possibility to excite HF SFXMs with azimuthal mode index $m > 1$ looks to be very low.

Application of wider vacuum gaps Δ allows one to expect higher efficiencies of electromagnetic radiation generation. Moreover, increasing the thickness of the gap increases the quantity of cyclotron harmonics, which can interact with SFXMs, (or in other words, increases the quantity of ranges of the effective wavenumber, where the requirements for resonant beam instabilities of HF SFXMs can be realized) and hence leads to improving the conditions of SFXM excitation. Reduction of the gap Δ thickness leads to decreasing growth rates of the beam instability of HF SFXMs [1].

The character of the dependence of the efficiency of HF SFXM excitation by a relativistic electron beam on the plasma radius R_1 is explained by the dispersion properties of these SFXMs. In wide waveguides ($k_{ef} \to 0$), the HF SFXM frequency weakly depends on the radius R_1, since $\omega \propto k_{ef}{}^2$. However, in narrow waveguides, for $k_{ef} > 1/Z$, HF SFXMs do not propagate along the plasma-metal interface [15]. Therefore, for HF SFXM excitation it is desirable to choose a small radius of the plasma column for which the SFXM propagates at a frequency close to the upper bound of the HF range (1.14) [1].

In comparison with the case of LF SFXMs, growth rates of HF SFXM instability are lower. This is explained, first of all, by worsening the possibilities of realizing the resonance conditions (7.14) of beam-plasma interaction. For HF SFXMs with azimuthal mode index $m = 1$ this condition can be realized for the beam cyclotron harmonics $N \gg m$. Remind that the growth rate sharply decreases with the number of the cyclotron harmonic N, since $\mathrm{Im}(\omega) \propto J_N^{2/3}(|m|)$ [1]. Moreover, the resonances on the plots of HF SFXM growth rates versus effective wavenumber are very narrow. In addition, beam excitation of HF SFXMs takes place in the regime of separate resonances, even for rather large beam particle density $n_b = 0.1 n_e$.

10.2 Excitation of Surface Flute Waves Above Upper Hybrid Frequency by Annular Electron Beam in Waveguide with Non-circular Cross-Section of Plasma Column

Studying the excitation of electromagnetic waves in magnetized plasma waveguides is a matter of particular interest. For example, interaction of an electron beam in a plasma filled coaxial metal waveguide, whose inner wall is corrugated, was studied in [16, 17]. It was shown that for non-relativistic electron beams propagating in a plasma-filled slow-wave structure, the maximum value of instability growth rates is much higher than values of the growth rates obtained for vacuum structures. However, the development of the theory of beam-plasma instability cannot be considered as completed so far, since the interaction of a charged particle beam with the eigenwaves of different waveguides essentially depends on many factors, including the dispersion properties of these waves, their polarization, their spatial distribution of the fields, and the geometry and design features of these waveguides. From this point of view the study of the interaction of an annular charged particle beam with electromagnetic waves in plasma-filled metal waveguides is an important task of plasma physics [18–21].

Periodic spatial inhomogeneity of a medium along the direction of electromagnetic wave propagation can enhance the efficiency of plasma electronics devices. In Sect. 6.4 the results of a theoretical study of HF SFXM propagation in magneto-active waveguides, which are partially filled with a plasma column with non-circular cross-section, were presented [22]. The analysis was carried out there using the assumption that an external static magnetic field is directed along the axis of the metal waveguide, and the plasma column is separated from the circular waveguide wall with a thin dielectric gap between plasma and wall. It was shown that the frequency spectrum and the spectral composition of wave packets in this case can be controlled by choosing the shape of the cross section of the plasma column. Correction to the eigenfrequency of HF SFXMs was numerically computed as a function of the waveguide parameters [22]. The magnitude of the correction was shown to increase with decreasing magnetic field and increasing permittivity of the dielectric layer. Changing the width of the dielectric gap (taking into account the initial assumption that the gap is relatively narrow) has only a weak effect on the frequency correction $\Delta\omega_N$, but its absolute value is very sensitive to the magnitude of the angular corrugation period of the plasma-dielectric interface.

In Sect. 10.1, the interaction of an annular electron beam with HF SFXMs propagating near the interface of a concentric circular plasma cylinder is investigated [1]. This shape of the discharge gap cross section can be achieved in an experiment, for example, due to the use of a thin-walled glass tube or an appropriate confining magnetic field. The assumption of a circular cross-sectional shape of the plasma column and its location in the concentric metal waveguide can be intentionally or non-intentionally violated in the experiment. The cross section of the glass tube could be non-circular, intentionally, or non-intentionally because of wrong storage or transportation conditions, or the glass tube is non-concentrically located inside the metal waveguide.

In plasma traps with magnetic confinement (e.g. stellarators), the cross section of the plasma column is often non-circular one: elliptical, beanlike, triangular, etc.

The present Section is devoted to studying the possibility of increasing the growth rate of the HF SFXMs resonant beam-plasma instability at the initial stage by an appropriate choice of the cross-sectional shape of the plasma column.

Beam excitation of electromagnetic waves with extraordinary polarization (with components E_r, E_φ, H_z of the electromagnetic field) propagating in a perfectly conducting circular cylindrical metal chamber with the radius $r = R_2$ (see Fig. 10.7) is considered. A uniform external static magnetic field is directed along the axis of the waveguide, $\vec{B}_0 || \vec{z}$. The waveguide is considered to be uniform along the cylinder axis, $\partial/\partial z = 0$.

The plasma column with radius $R_1(\varphi)$ is enclosed within the metal chamber. Since the influence of the shape of the plasma-vacuum interface on the beam-wave interaction is investigated here only, the plasma particle density is assumed to be uniform. This assumption is valid, in particular, in the case of utilizing the surface waves for production and maintenance of a gas discharge. In this case, the radial uniformity of the plasma density is provided in the peripheral layer, whose thickness is of the order of the wave penetration depth into the plasma. The case of a solid-state plasma also corresponds to this assumption [7, 18].

Let the radial shape $R_1(\varphi)$ of the non-circular plasma column depend on the azimuthal angle by the following law:

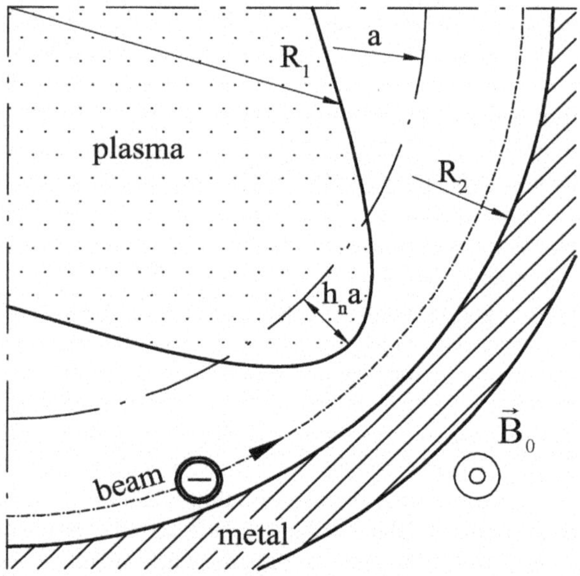

Fig. 10.7 Scheme of metal waveguide with circular cross section filled with a plasma column with non-circular cross section and electron beam gyrating in the dielectric layer between plasma and waveguide

$$R_1(\varphi) = a\left[1 + \sum_{n=1}^{\infty} h_n \cos(n\varphi + \varphi_n)\right], \qquad (10.1)$$

where a is the average radius of the plasma column. The parameters h_n describe the relative depth of the longitudinal corrugation of the plasma-vacuum interface and are considered as small parameters of the problem. Note that the expansion (10.1) of the radius of the plasma column cross section into Fourier series can model any shape. The different terms $\propto\cos(n\varphi + \varphi_n)$ in the sum (10.1), corresponding to different symmetries of corrugation, independently affect the dispersion properties of SFXMs, when considering only small terms, which are quadratic over h_n [22]. That is why the following analysis can be restricted to separate the treatment of the different small terms in (10.1) only:

$$R_1(\varphi) = a[1 + h_K \cos(K\varphi + \varphi_K)]. \qquad (10.2)$$

This restriction allows one to eliminate the influence of the other terms, but there are no fundamental problems to include a finite number of terms of the series (10.1) in the final dispersion equation to be studied. In the case $K = 1$, Eq. (10.2) describes a shift of the axes of waveguide and plasma column. The term with $K = 2$ describes an ellipticity of the plasma cross section, a triangularity is included by the term $K = 3$, etc.

The plasma is assumed to be cold and collisionless. Its electrodynamic properties are described by the plasma dielectric permittivity tensor (see (1.2) and, e.g., [7, 23]).

The annular electron beam is injected into the gap $R_1 < r < R_2$ between the plasma column and the metal wall of the waveguide. The electrodynamics properties of that part of the waveguide, which is occupied by the beam, are described via the dielectric permittivity tensor $\varepsilon_{ik}^{(b)}$ [7, 13]. Explicit expressions of the tensor components are given by Eqs. (7.18)–(7.20).

The dispersion properties of the hybrid waveguide structure considered in the present Section can be studied by the method of successive approximations. In the zeroth approximation, the plasma-vacuum interface is assumed to be circular and concentric with the metal chamber. The SFXM dispersion properties in this approximation can be described using the results presented in Sect. 2.1 [24]. If the radius of curvature of the interface between the plasma and the vacuum gap is constant along the azimuthal direction, SFXMs with different azimuthal mode indices propagate independently from each other. That is why in zeroth approximation over the small plasma radius perturbation parameter h_K one can assume that electromagnetic waves propagate with a single azimuthal mode number m. Due to the periodic spatial inhomogeneity of the plasma-vacuum interface (see Eq. (10.2)), SFXMs propagate in the form of wave packets. Each of these packets contains in addition to the fundamental harmonic, whose field varies proportionally to $\exp(im\varphi - i\omega t)$, an infinite number of satellite harmonic fields, which vary proportionally to $\exp[i(m \pm jK)\varphi - i\omega t]$ ($j = 1, 2, 3 \ldots$). According to the theory [25] of wave propagation in a medium, whose properties vary periodically along a given direction, the amplitudes of the satellite harmonics are small quantities of the order $(h_K)^j$ compared with the amplitude of the fundamental harmonic. In determining the dependence of the SFXM dispersion properties on the shape of the non-circular plasma-vacuum interface, the wave packet with the two nearest satellite harmonics, whose fields changed proportionally $\exp[i(m \pm K)\varphi - i\omega t]$, was considered in Sect. 6.5 [22]. The reason for this approach is that higher satellite harmonics make contributions only of third or higher order of smallness to the correction of the SFXMs eigenfrequency.

Taking into account the small terms of the first order over h_K does not change the amplitude of the fundamental harmonic. However, in the second approximation over the small parameter h_K the amplitude of the fundamental harmonic obtains a correction of the second order of smallness, which along with the amplitudes of the first satellite harmonics makes the contribution (6.39) of the second order of smallness to the dispersion relation, $D^{(2)}(\omega, m, K\ldots) \propto {h_K}^2$. As a result, instead of the dispersion relation (7.12) obtained in zeroth approximation, $D^{(0)}(\omega, m, \ldots) = 0$, the following dispersion equation: $D^{(0)}(\omega, m, \ldots) + D^{(2)}(\omega, m, K \ldots) = 0$ is derived in the second order approximation [2]. Doing that, the following boundary conditions, which are different from those in Sect. 1.4.4, are applied:

- the tangential component of the SFXM electric field is equal to zero at the metal surface:

$$E_\varphi(r = R_2) = 0; \tag{10.3}$$

- the tangential components of the SFXM electric and magnetic wave fields are continuous at the non-circular plasma-vacuum interface:

$$E_\tau(r = R_1 + 0) = E_\tau(r = R_1 - 0),\ H_z(r = R_1 + 0) = H_z(r = R_1 - 0); \tag{10.4}$$

- the SFXM wave fields are of finite value in the entire volume of the waveguide, in particular, at the axis of the waveguide:

$$H_z(r = 0) < \infty. \tag{10.5}$$

Application of these boundary conditions yields the following dispersion relation, which will be investigated in the present Section:

$$\frac{I'_m(k_\perp a)k}{k_\perp I_m(k_\perp a)} + \frac{\mu mka}{k_\perp^2 a^2} + D^{(2)}(\omega, m, K \ldots) = \frac{im\varepsilon_{12}^{(b)}}{\varepsilon_{11}^{(b)}\zeta_1\sqrt{\psi_b}} - \frac{J'_m(\zeta_1) - \Phi N'_m(\zeta_1)}{\sqrt{\psi_b}[J_m(\zeta_1) - \Phi N_m(\zeta_1)]}, \tag{10.6}$$

where the same notations as above, and in particular as in Sect. 7.1 are used. To analyze Eq. (10.6) one assumes that the HF SFXM frequency quite accurately corresponds to the resonance condition (7.14).

Results of numerical analysis of Eq. (10.6), which describes the excitation of HF SFXMs by an electron beam in the case of non-circular cross-section of the plasma column, are shown in Figs. 10.8, 10.9 and 10.10. The dependences of the resonant beam-plasma instability growth rate, normalized by the absolute value of the electron cyclotron frequency, on the effective wavenumber $k_{ef} = |m|c/(\Omega_e a)$ for the SFXM with azimuthal mode index $m = 1$ are shown in Fig. 10.8. Here, the following values of the plasma-beam system parameters are taken: $Z \equiv \Omega_e/|\omega_e| = 2$, $R_2 - a = 0.35a$, and $n_b = 0.1 n_e$. Just for these values of the parameters, the possibility of the most efficient excitation of HF SFXWs is shown in Sect. 10.1 for the case of circular shape of the plasma-vacuum interface. The numbers located nearby the solid curves indicate the values of the plasma shape parameter $K = 1, 2, 4$, which determines the

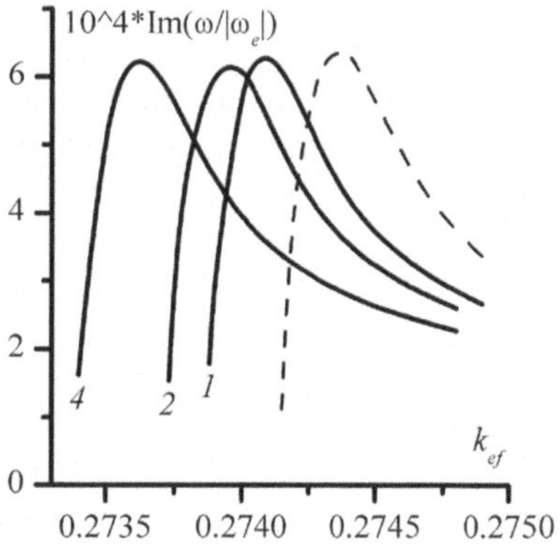

Fig. 10.8 Normalized growth rates of the HF SFXM beam-plasma instability versus the effective wavenumber k_{ef} for $m = 1, Z = 2.0, R_2 - a = 0.35a, \alpha = 0.1$, and $h_K = 0.1$ (*solid curves*). The numbers located nearby the *curves* indicate the value of the plasma shape parameter K. The dashed curve is for a concentric circular plasma

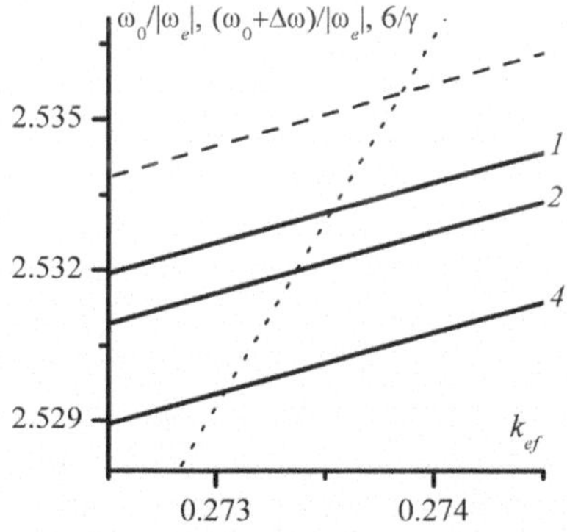

Fig. 10.9 Normalized HF SFXM eigenfrequencies in waveguides with circular ($h_K = 0.0$, *dashed line*) and non-circular ($h_K = 0.1$, *solid lines*) plasma columns for $m = 1, Z = 2.0, R_2 - a = 0.35a$, and $\alpha = 0.0$. The *numbers* located nearby the *solid lines* indicate the value of the plasma shape parameter K. The *dotted line* shows the resonance line $\omega = 6|\omega_e|/\gamma$

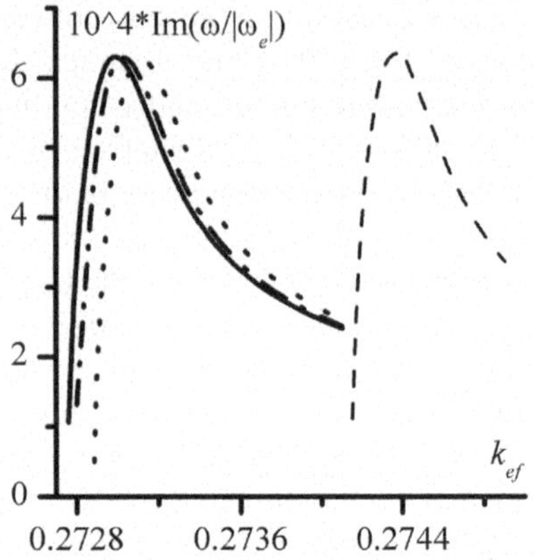

Fig. 10.10 Normalized growth rates of the HF SFXM beam-plasma instability versus the effective wavenumber k_{ef} for $m = 1, Z = 2.0, R_2 - a = 0.35a, \alpha = 0.1$, and $h_K = 0.1$ (*solid curve*) for the case of a squared-shape cross-section of the plasma column. The different curves are explained in the text

number of corrugations on the plasma-vacuum interface. The corrugation depth in this case is chosen to be $h_K = 0.1$. The *dashed curve* shows the dependence of the SFXM growth rate in the case of circular concentric plasma column. In Sect. 10.1 it is shown, that the main increase of the growth rates is observed near the right limit of the range of the effective wavenumber k_{ef}, where the HF SFXMs can exist, or in other words, it occurs near the top of the HF range (1.14), where these waves can exist. Thus, in this case HF SFXMs penetrate deeply into the plasma, $k_{\perp}a \ll 1$, because the equality $\varepsilon_2 \approx \varepsilon_1$ is satisfied near the top of the HF range. However, despite the increase of the plasma shape parameter K causes a slight down-shift of the position of maximum growth rate (see Fig. 10.8), it is not accompanied by a significant decrease of the growth rate. In the waveguide with a coaxial circular plasma column (dashed curve), the maximum magnitude of the growth rate $\mathrm{Im}(\omega/|\omega_e|) \approx 6.368 \times 10^{-4}$ is achieved for $k_{ef} \approx 0.27437$, whereas in the waveguide with non-circular plasma column cross-section with $K = 4$, the maximum magnitude of the growth rate is $\mathrm{Im}(\omega/|\omega_e|) \approx 6.38 \times 10^{-4}$ for the slightly reduced $k_{ef} \approx 0.2736$.

The shift of the position where the growth rate has its maximum to the range of smaller k_{ef} is explained by the behavior of the dispersion curves in the Fig. 10.9. The dashed curve corresponds to the case of circular cross-section concentric plasma column. The dotted line shows the resonance line $\omega/|\omega_e| = 6/\gamma$. The solid lines are the dispersion curves for the non-circular shape of the plasma column. The numbers located nearby the curves correspond to the magnitude of the plasma shape parameter K. One can see that the value of k_{ef}, for which the resonance curve crosses the dispersion curve in the case of the coaxial circular plasma column, is situated on the right of other crossing points. This corresponds to the right position of the maximum growth rate in this case in Fig. 10.8. The positions of the crossing points of the resonance curve with the dispersion curves, which correspond to the cases $K = 1, 2, 4$, successively shift to the left. This is in full agreement with the order of the positions of the maxima growth rates in Fig. 10.8. It should be emphasized that the method of successive approximations is still applicable in this case [22]. The correction to the HF SFXM frequency in this case amounts to: $\Delta\omega_4/|\omega_e| = 0.005$ for

$k_{ef} = 0.2736$, which corresponds to the maximum growth rate in the case of $K = 4$ in Fig. 10.8. Thus, the relative magnitude of the correction to the frequency is smaller than 0.2%.

As one can see from Fig. 10.8, increasing the magnitude of the corrugation parameter K leads to a larger shift of the position of the growth rate maximum to the left. Successive account for all the corrections to the shape of the plasma column cross-section could cause complete suppression of the beam-plasma instability. However, in the simulation of arbitrary shape of the cross-section one can neglect the contribution of the terms with large values of n in expression (10.1) due to a strong decrease of the corrugation depth h_n with increasing n, which corresponds to the conditions of applicability of the method of successive approximations. For example, when modelling a squared cross-section of the plasma column in the form of (10.1) the magnitude of the corrugation depth h_n decreases as the inverse square of the summation index:

$$h_n = 4\sqrt{2}(-1)^{\frac{n}{4}} / \left[n^2 \ln(3 + 2\sqrt{2}) \right]. \tag{10.7}$$

The plots in Fig. 10.10 show the convergence of the numerical results in the case of accounting ever-increasing quantity of summands in expression (10.1) for the case of the a squared cross-section of the plasma column. The dashed curve corresponds to the HF SFXM in a waveguide with concentric circular plasma cross-section. The dotted curve shows the SFXM growth rate in the case that in Eq. (10.1) only the first small summand is considered, which is proportional to $\cos(4\varphi)$. The maximum of the dotted curve is shifted as compared to the dashed curve by $\Delta k_{ef} = -1.3 \times 10^{-3}$. The dashed-dotted curve corresponds to considering two small terms in the sum (10.1), which are proportional to $\cos(4\varphi)$ and $\cos(8\varphi)$. In this case, the position of the maximum is shifted by only $\Delta k_{ef} = -1.0 \times 10^{-4}$ as compared to the dotted curve. Further refinement of the shape of the plasma cross-section contour by taking into account the next small summand, which is proportional to $\cos(12\varphi)$, gives a negligibly small correction to the position of the growth rate maximum, which is much smaller than $\Delta k_{ef} = -1.0 \times 10^{-4}$. This refinement changes the value of the growth rate by less than 0.07%. This is shown by the solid curve in Fig. 10.10.

The following conclusions result from the study carried out in the present Section. The appropriate choice of the shape of the plasma column cross-section can expand the range of effective wavenumbers (i.e., expand the range of values of the beam-plasma system parameters), where efficient excitation of HF SFXMs is possible, toward smaller plasma densities and smaller radii of the plasma column.

The numerical study demonstrates that the applicability of the method of successive approximations for the investigated problem of HF SFXMs excitation is closely connected with their dispersion properties. One should remind, that in the case of sufficiently dense plasma, HF SFXMs exist in a relatively narrow frequency range. In the case of wide waveguides ($a/\delta > |m|Z$) detuning of their frequency from the lower limit of the frequency band is a quadratic function over the effective wavenumber k_{ef}: $\omega - \omega_{UH} \propto {k_{ef}}^2$. These waves cannot propagate in narrow waveguides (if $a \ll \delta$).

Therefore, one would expect (and numerical analysis has confirmed this expectation) that there is an optimal magnitude of k_{ef} to excite these waves. This optimum frequency is close to the upper limit of the HF band of the SFXMs, $\omega \approx \omega_1$.

It should be noted, that the problem of interaction between an annular electron beam and SFXMs qualitatively differs from the problem of interaction between an electromagnetic wave and a longitudinal electron beam. In designing electronic devices, which apply longitudinal beams, as a rule, one performs special calculations in order to phase the beam and the wave. In this case, a corrugated line could have a variable period of corrugation (corrugation taper) along the direction of beam propagation. Such method of enhancing the interaction between the beam and the wave is not applicable in the case of an annular beam, because it unavoidably comes back to the same azimuthal position after gyrating along the Larmor orbit.

An influence of a corrugation of the metal waveguide on the rate of HF SEXM beam excitation is not investigated in the present monograph because it is well-known that a corrugation of the plasma cross section influences the dispersion properties of these waves much stronger than a deviation of the inner shape of the metal chamber from the circular one. This peculiarity of SFXMs dispersion properties is connected with the feature, that their power is located mainly nearby the plasma interface.

10.3 Transition Between Beam-Plasma and Beam-Dissipative Instability Regimes in Interaction of Relativistic Large Larmor Orbit Electron Beams and Surface Flute Waves Above Upper-Hybrid Frequency in Coaxial Plasma Waveguides

As it is well-known, the subset of the Maxwell equations for the extraordinary polarized electromagnetic waves can be reduced to one differential equation of second order, e.g. for the axial magnetic wave field $H_z^{(m)}$. In the collision-less limit (if the plasma is considered as a medium without collisions between the particles), this equation has the form of a Bessel equation within the plasma layer. In uniform plasmas, surface type solutions of this equation can be represented in terms of modified Bessel functions of the first and second kind, $I_m(k_\perp r)$ and $K_m(k_\perp r)$ (see Eq. (1.12)). The constants of integration $A_{1,2}$ both are non-zero in the case of coaxial plasma waveguides unlike in the case of cylindrical plasma columns when $A_2 = 0$. The condition of $k_\perp^2$ to be positive, $k_\perp^2 > 0$, indicates two frequency ranges ((1.13), LF range) and ((1.14), HF range), where the considered electromagnetic waves have surface nature: they do not oscillate along the radial coordinate and their amplitude decreases with going away from the plasma layer interface into the plasma volume.

Note that SFXMs are electromagnetic waves of surface nature in a plasma layer and in the mentioned two frequency ranges (1.13) and (1.14) only. Even within these frequency regions, but in the dielectric or vacuum regions between plasma and metal wall, SFXMs are volumetric waves: their amplitude oscillates with the radial coordinate.

Excitation of both LF and HF SFXMs in coaxial plasma waveguides by a flow of electrons rotating around the coaxial plasma layer is studied in detail in Chap. 8, where dissipation is neglected: the collision frequency is taken equal to zero, $\nu = 0$, and the regime of beam-plasma instability is considered.

On one hand, dissipation can be neglected in many plasmas. On the other hand, the collision frequency of plasma particles can be of moderate or even large magnitude as compared to the operating frequency. The possibility to excite LF SFXMs in the regime of beam-dissipative instability in a two-component semiconductor cylindrical structure was demonstrated in Sect. 7.1. The nonlinear theory of LF SFXW excitation by a rotating electron beam in the regime of beam-dissipative instability was developed in Sect. 9.2. The transition between beam-plasma and beam-dissipative instability regimes in the interaction of relativistic large Larmor orbit electron beams and LF SFXWs is studied in Sect. 7.6 [26].

HF SFXMs are preferable waves for practical applications due to their higher frequency and shorter wavelengths. Their excitation by an electron beam rotating around the coaxial plasma layer (external beam) in the regime of beam-plasma instability was studied in Sect. 8.3 [27]; and in the case of electron beam rotating inside the coaxial plasma layer (internal beam) in Sect. 8.4 [28].

To investigate HF SFXM excitation in the regime of beam-dissipative instability it is not sufficient only to take simply account for the collision frequency in the components of the plasma permittivity tensor during solving the dispersion relation. To have the right to declare that just the regime of beam-dissipative instability takes place one has to get convinced of the fact that those terms which are responsible for dissipation and beam motion prevail in the dispersion relation. In Sect. 7.6 the following different criteria for identifying whether a beam-plasma or a beam-dissipative instability takes place is applied [26]. Namely, it is proposed to use the classical analytical results [14, 29] in which analytical expressions for growth rates of beam-plasma and beam-dissipative instabilities were presented for the first time. The dependence of the growth rates on the beam particle density n_b and on the collision frequency ν was given there as follows:

$$\mathrm{Im}(\omega) \propto n_b^x \nu^y. \tag{10.8}$$

The beam-plasma instability is well-known to be characterized by $x = 1/3$ and $y = 0$ [14], while the case of beam-dissipative instability is described by $x = 1/2$ and $y = -1/2$ [29]. The character of the dependence of the growth rate on the collision frequency was found in Sect. 7.6 to be a more appropriate indicator of whether the beam-plasma or the beam-dissipative instability takes place than the character of the growth rate dependence on the beam particle density [26]. The dependence of $\mathrm{Im}(\omega)$ on ν was demonstrated in Sect. 7.6 to have a step junction between the one range where $y \ll 1$ (this was treated as the case of the beam-plasma instability) and the other range where $y \approx -1/2$ (this was identified as the case of the beam-dissipative instability) [26]. In the present Section, the influence of dissipation on HF SFXM beam excitation and the transition of the regime of beam-plasma instability to that of beam-dissipative instability with consecutive account for dissipation are investigated.

Electromagnetic waves of surface type in coaxial plasma-vacuum waveguides of two types are considered. In the first case (Fig. 8.1), the flow of electrons rotates along Larmor radii (large orbit configuration) in the gap, which separates the coaxial plasma layer from the inner wall of the outer cylindrical loss-free (infinite conductivity) metal chamber. The inner conductor is also assumed to be ideal (conductivity is also infinite). This case is referred as case of an external electron beam. In the second case (Fig. 8.8), the electrons rotate in the gap between the plasma layer and the internal conductor with infinite conductivity. This case is referred as case of an internal electron beam.

An external stationary magnetic field is oriented along the axis of the system, $\vec{B}_0 || \vec{z}$. The electrodynamic properties of the plasma layer in presence of dissipation are described by the permittivity tensor [30]:

$$\varepsilon_1 = 1 - \frac{\Omega_e^2(\omega + i\nu)}{\omega\left[(\omega + i\nu)^2 - \omega_e^2\right]}, \varepsilon_2 = -\frac{\Omega_e^2 \omega_e}{\omega\left[(\omega + i\nu)^2 - \omega_e^2\right]}. \tag{10.9}$$

The radial distribution of the SFXM axial magnetic field within the coaxial plasma layer can be written in the form (1.12) with two non-zero constants of integration $A_{1,2}$.

The electron beam is described in Sect. 7.1 as a flow of particles with zero axial momentum and having the same transversal momentum [7]. The radial distribution of the SFXM axial magnetic wave field within the beam layer is governed by the Bessel equation. Its solution contains two more constants of integration.

To derive the dispersion relation, one needs four boundary conditions. They are as follows. The tangential components of the SFXM electric wave field should be equal to zero at the metal interfaces, $E_\varphi(r = a, d) = 0$. The tangential components of the SFXM wave fields should be continuous at the plasma-beam interface, $r = b$. The dispersion relation, which is the subject of the present Section, has the form (8.1) with utilizing the permittivity tensor components (10.9).

The HF SFXM with azimuthal mode index $m = 1$ was shown in [27] to be efficiently excited in the coaxial plasma layer with an external beam for the case: $Z = \Omega_e/|\omega_e| = 3.75$, $d/b = 1.2$, $a/b = 0.75$, due to interaction with the fifth cyclotron harmonic of the electron beam ($N = 5$) with the relative beam particle density $\alpha = n_b/n_e = 0.0125 \div 0.1$, where n_b is the beam particle density and n_e is plasma particle density. To solve the dispersion relation the Python 3.5.1 software [31] has been used.

The HF SFXM growth rate reaches its maximum for the effective wavenumber $k_{ef} \equiv mc/(\Omega_e b) = 0.437$. The lower the beam particle density is, the smaller is the growth rate. This dependency is qualitatively clear and predictable. The problem is to distinguish whether that or the other range of α corresponds to beam-plasma or dissipative instability. To solve this problem the dependence of the growth rate on the beam particle density n_b and on the collision frequency ν (Eq. 10.8) is used [3].

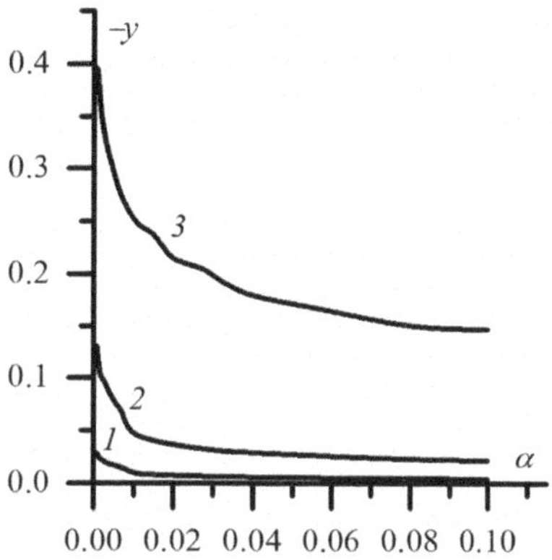

Fig. 10.11 Dependence of the exponent $-y$ (see Eq. (10.8)) on the relative beam particle density α, for $m = 1, Z = \Omega_e/|\omega_e| = 3.75$, $k_{ef} = 0.437$, $d/b = 1.2$, and $a/b = 0.75$. The three curves are for $\nu = 0.0001|\omega_e|$ (*curve 1*), $\nu = 0.0005|\omega_e|$ (*curve 2*), and $\nu = 0.0025|\omega_e|$ (*curve 3*)

Figures 10.11 and 10.13 present results of numerical studies of the dispersion relation for the coaxial beam-plasma system described above with an effective plasma particle collision frequency $\nu = 0.0001|\omega_e|$ (curve 1), $\nu = 0.0005|\omega_e|$ (curve 2), and $\nu = 0.0025|\omega_e|$ (curve 3), respectively. The exponents x and y appear to depend on both variables, α and ν. That is why the exponents are calculated numerically as follows:

$$x = \frac{\alpha}{\mathrm{Im}(\omega)} \frac{\Delta \mathrm{Im}(\omega)}{\Delta \alpha}, \quad y = \frac{\nu}{\mathrm{Im}(\omega)} \frac{\Delta \mathrm{Im}(\omega)}{\Delta \nu}. \tag{10.10}$$

Figure 10.12 shows a smooth decrease of the exponent x from 0.461 to 0.335, almost according to the theoretical predictions. The following important feature distinguishes this decrease for HF SFXMs from that observed in [26] for LF SFXMs. The decrease of x for HF SFXMs takes place in the same ν-range as for LF SFXMs, but to cover the range one has to recruit several magnitudes of ν in the case of HF SFXMs, whereas in the case of LF SFXMs, the exponent x changes smoothly

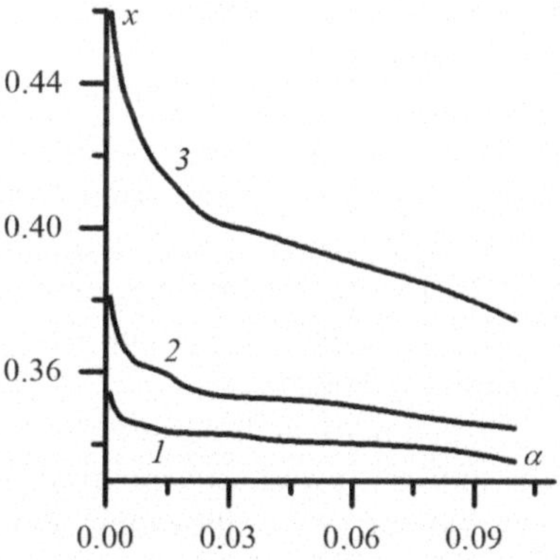

Fig. 10.12 Dependence of the exponent x (see Eq. (10.8)) on the relative beam particle density α, for $m = 1, Z = \Omega_e/|\omega_e| = 3.75$, $k_{ef} = 0.437$, $d/b = 1.2$, and $a/b = 0.75$. The three curves are for $\nu = 0.0001|\omega_e|$ (*curve 1*), $\nu = 0.0005|\omega_e|$ (*curve 2*), and $\nu = 0.0025|\omega_e|$ (*curve 3*)

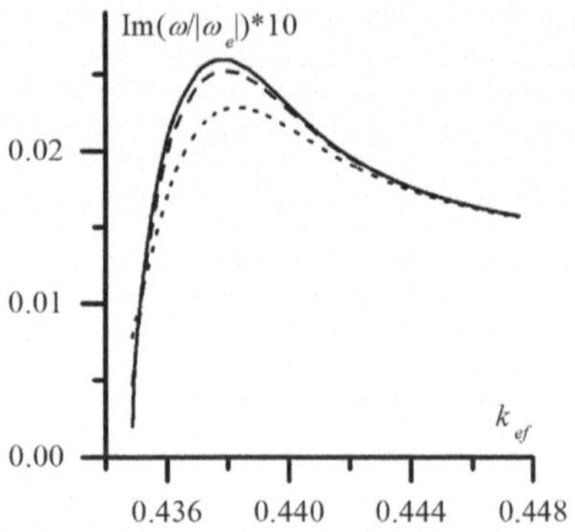

Fig. 10.13 Dependence of the normalized growth rate, $\mathrm{Im}(\omega/|\omega_e|)$, on the effective wavenumber k_{ef}, for $m = 1$, $\alpha = 0.1$, $Z = \Omega_e/|\omega_e| = 3.75$, $d/b = 1.2$, and $a/b = 0.75$ (external beam). The three curves are for $\nu = 0.0|\omega_e|$ (*solid curve*), $\nu = 0.0005|\omega_e|$ (*dashed curve*), $\nu = 0.0025|\omega_e|$ (*dotted curve*)

from the magnitude typical for a beam-dissipative instability to that typical for a beam-plasma instability.

As far as the exponent y is concerned (see Fig. 10.11), its dependence on α in the case of HF SFXMs has two features, which distinguish it from that in the case of LF SFXMs [26]. The first one is the same as that for the exponent x: in the case of HF SFXMs, the increase of the exponent y from -0.4 to zero can be shown to happen when ν changes by several orders of magnitude, while in the case of LF SFXMs, the exponent y changes its magnitude smoothly from that typical for a beam-dissipative instability to that typical for a beam-plasma instability. The second one is a smooth increase of the exponent y with α in the case of HF SFXMs in contrary to the case of LF SFXMs for which the curve looks like two almost horizontal segments joined by a rapid rise. This behavior of the dependence of the exponent y on α makes it possible to indicate a precise transition from one mode of instability to the other in the case of LF SFXMs [26]. Such transition is not so clear for HF SFXMs, since the exponents x and y vary smoothly.

As already mentioned earlier, in the present monograph, the problem of generating the electron beam in the studied beam-plasma coaxial structure is not discussed. This can be done e.g. in the way proposed in [32]. In this case one can get advantage of utilizing the space inside the inner conductor of the coaxial structure for placing the beam source there.

In Figs. 10.13 and 10.14, the dependence of the HF SFXM growth rate on the effective wavenumber k_{ef} is plotted for the cases of external and internal electron beams, respectively. The solid curves show the dependence without dissipation, $\nu = 0$. The curves, which are calculated for $\nu = 0.0001|\omega_e|$, are not shown in Figs. 10.13 and 10.14 since they are indistinguishable from those calculated for $\nu = 0$. The dashed curves correspond to the case of $\nu = 0.0005|\omega_e|$, and the dotted ones to $\nu = 0.0025|\omega_e|$.

The geometrical dimensions of the structures in Figs. 10.13 and 10.14 are chosen to be similar to provide the possibility of comparison. The width of the coaxial plasma layer is about 25% of the radius of the plasma-vacuum interface, $r = b$ and the width of the vacuum layer, where the electron beam is propagating, covers about 20% of the radius b.

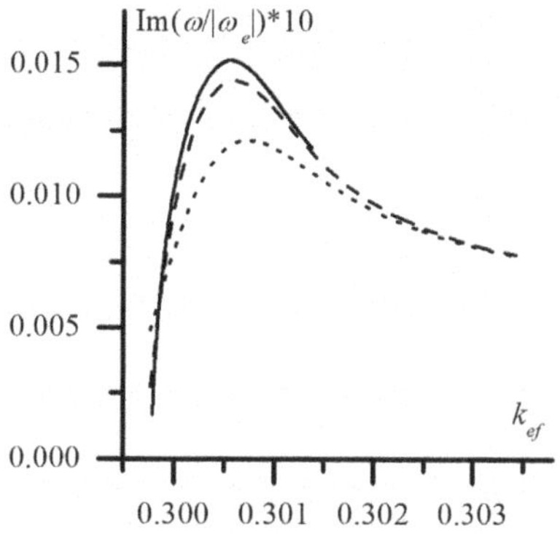

Fig. 10.14 Dependence of the normalized growth rate, Im(ω/|ω_e|), on the effective wavenumber k_{ef}, for $m = 1$, $\alpha = 0.1$, $Z = \Omega_e/|\omega_e|$ $= 3.75$, $d/b = 0.8$, and $a/b = 1.25$ (internal beam). The three curves are for $\nu = 0.0|\omega_e|$ (*solid curve*), $\nu = 0.0005|\omega_e|$ (*dashed curve*), $\nu = 0.0025|\omega_e|$ (*dotted curve*)

The dependence of the eigenfrequency of HF SFXMs on the effective wavenumber is shown in Sect. 2.1 to be almost quadratic, $\omega \approx \omega_{UH} + \beta\, k_{ef}^2$ (here β is a constant, ω_{UH} is the upper hybrid frequency which is the lower limit of the frequency range for HF SFXMs), in the range of small k_{ef} [8]. HF SFXMs do not propagate in the case of large effective wavenumbers. One can estimate analytically the maximum value of k_{ef}, for which HF SFXMs can propagate in the absence of a vacuum layer using the following inequality: $k_{ef} < Z^{-1}$. When approaching the right limit of the k_{ef} range where HF SFXMs propagate, the dependence the eigenfrequency of HF SFXMs on k_{ef} becomes more pronounced. That is why the maximum values of HF SFXM growth rates are known to decrease with moving of the position of the maximum on the abscissa axis to the left. Since the dependence of the growth rate on k_{ef} for the case of an internal beam (Fig. 10.14) takes place for smaller values of k_{ef} ($k_{ef} \approx 0.3006$) than for the case of an external beam ($k_{ef} \approx 0.438$), its maximum value is lower, max[Im(ω/|ω_e|)] $\approx 1.52 \times 10^{-3}$, in the absence of collisions, than in the case of an external beam, where max[Im(ω/|ω_e|)] $\approx 2.59 \times 10^{-3}$. Dissipation, $\nu = 0.0025|\omega_e|$, causes a stronger decrease of the maximum values of HF SFXM growth rates for the case of an internal electron beam (max[Im(ω/|ω_e|)] $\approx 1.18 \times 10^{-3}$) than for an external electron beam (max[Im(ω/|ω_e|)] $\approx 2.29 \times 10^{-3}$).

The definition "relativistic" is used in the title of the present Section to characterize the electron beam. Below, it is tested whether just relativistic electron flows are necessary to excite HF SFXMs in coaxial plasma structures or not. The maximum value of the HF SFXM growth rate is realized in the beam-plasma coaxial structure with external beam for the electron beam velocity $\upsilon = 0.556c$. In the case of the structure with internal beam, the Larmor radius should be smaller, that is why the beam velocity should be higher, $\upsilon = 0.624c$. Nevertheless, in both cases the beam velocity is not highly relativistic.

The following conclusions can be made from the study in the present Section. The initial stage of SFXM excitation in coaxial plasmas above the upper-hybrid frequency (HF range) caused by their interaction with electron flows rotating along large Larmor orbits is studied here taking dissipation into account [3]. Dissipation cannot be neglected in many practical applications like plasmas in metals and semiconductors. In addition, also, in gaseous plasmas its influence can be important. The frequency range studied here is higher than that of the electron cyclotron frequency region investigated earlier in [26], which explains the actuality of the present study since practical applications need electromagnetic waves with shorter wavelengths.

The lower the maximum value of the HF SFXM growth rate is, the stronger is its decrease caused by dissipation. No qualitative difference is found in the impact of dissipation on HF SFXM excitation in coaxial plasma-filled waveguides with an internal electron beam as compared to that with an external electron beam. No super relativistic velocities of the electron beams are needed to excite HF SFXMs in the both structures considered here.

The influence of dissipation on HF SFXM excitation significantly differs from that in the case of LF SFXMs studied in Sect. 7.6 [26]. The character of the dependence of the HF SFXM growth rate on such physical properties of the beam-plasma coaxial structure as beam particle density and plasma collision frequency does not demonstrate any rapid transition from the regime of beam-plasma instability to that of beam-dissipative instability. The exponents x and y (see Eq. (10.8)) which characterize the dependences of the HF SFXM growth rate on the beam particle density and collision frequency vary smoothly. That is why in the general case one cannot speak about a rapid transition from a pure beam-plasma to a pure beam-dissipative instability in electron beam excitation of HF SFXMs.

10.4 Excitation of Surface Flute Waves in Ion Cyclotron Frequency Range by Light Ion Beams Gyrating Around a Plasma Column in Strongly Magnetized Plasma Waveguides

The previous Chaps. 7 and 9 are devoted to studying SFXM excitation in the range of the electron cyclotron frequency (1.13). As it was mentioned in Sect. 1.4.4, SFXMs propagate in this range if the plasma density is sufficiently large and the stationary magnetic field is weak enough to provide the inequality $\Omega_e > |\omega_e|$. Along with this case, it was shown in Sect. 2.3 that the electromagnetic oscillations related to the ion component of the plasma can propagate with negative azimuthal mode index m in a cylindrical metal waveguide partially filled with a cold strongly magnetized plasma, $\Omega_e < |\omega_e|$, [33]. It was found that the minimum eigenfrequency of these oscillations is several times the ion cyclotron frequency. That is why, at first glance, the beam excitation of these waves looks to be possible only due to their interaction with beam ions oscillating at very high harmonics of the ion cyclotron frequency. Therefore, this process would be expected to be inefficient. However, in contrast to the oscillations at frequencies in the electron cyclotron resonance range, the plasma oscillations of some ion species can be excited by an ion beam consisting of another ion species. If the beam ions are lighter than the ions of the basic plasma, the cyclotron frequency of the beam ions can be approximately equal to the eigenfrequency of the SFXM. Therefore, one should expect that the excitation of SFXMs by such a light ion beam can be much more efficient than that with an ion beam with same ion species as the plasma. In the present Section, the initial stage of beam excitation of ion SFXMs in strongly magnetized plasma is studied in details.

The schematic of the problem is the same as that one considered in Sect. 7.1. The device is modeled by a metallic cylindrical waveguide with inner radius R_2. A cold magneto-active plasma column with radius R_1 is located concentrically inside the waveguide. An external static magnetic field is directed along the axis of the system, $\vec{B}_0 || \vec{z}$. The annular ion beam, which is modeled as a flow of oscillators with the same transverse momentum $p_{\perp 0}$ and zero axial momentum $p_z = 0$, is assumed to be injected into the gap between the plasma column and the waveguide metal wall, $R_1 < r < R_2$. The plasma-beam system is also assumed to be compensated in respect to both the electric currents and charges. The ion beam is characterized by the equilibrium distribution function (7.1). The electrodynamic properties of the waveguide space where the beam rotates are described by the permittivity tensor $\varepsilon_{ik}^{(b)}$ [4, 5]. The expressions for the three components of the $\varepsilon_{ik}^{(b)}$ tensor, which are used in the further calculations, are given by Eqs. (7.18)–(7.20).

The dispersion relation describing the wave excitation by the ion beam has the form (7.12). It can be solved analytically under assumption of the resonance condition

$$\omega = \omega_X + \delta\omega = N\omega_b\gamma^{-1} + \delta\omega, \tag{10.11}$$

where N is a positive integer, ω_X is the eigenfrequency of the XSFM in the absence of the beam, ω_b is the cyclotron frequency of the beam particles, and $\delta\omega$ is the frequency correction caused by the beam-plasma interaction, $|\delta\omega| \ll \omega_0$. Under the condition of resonant beam-plasma instability (10.11), the expression for the maximum value of the SFXM growth rate can be represented in the general form (7.15) [6]. The ratio of the beam and plasma particle densities $\alpha = n_b/n_e$ is assumed to be small, $\alpha \ll 1$.

Excitation of SFXWs with azimuthal mode indices $m = -2, -3, -4$ in a neon plasma by an alpha-particle beam is studied numerically [4]. The growth rates of beam instabilities normalized to the ion plasma frequency, $\mathrm{Im}(\omega)/\Omega_i$, are counted along the ordinate axis in Figs. 10.15, 10.16 and 10.17. The dimensionless abscissa is chosen in the form $\theta = \omega_i/\Omega_i$, since the analytical expression for the SFXM eigenfrequencies obtained in [33] contains this combination as a principal parameter. Thus, one can say that the plots show the dependence of growth rates on the external static magnetic field. The plasma density is chosen to be sufficiently large, $c/(\Omega_i R_1) = 5/6$. This allows enhancing the role of the term in the expression for the tensor component $\varepsilon_{22}^{(b)}$ (7.20), which is proportional to $\upsilon_{\perp 0}{}^2$ and that makes the main contribution to the growth rate. For the calculation of the plots in Figs. 10.15 and 10.17, the ion beam particle density is assumed to be sufficiently small, $\alpha = 10^{-6}$.

The conditions for SFXM instability in dependence on the azimuthal mode index m are compared in Fig. 10.15. The dependencies of normalized growth rates are shown for the SFXMs with azimuthal mode index $m = -2$ (Fig. 10.15a), $m = -3$

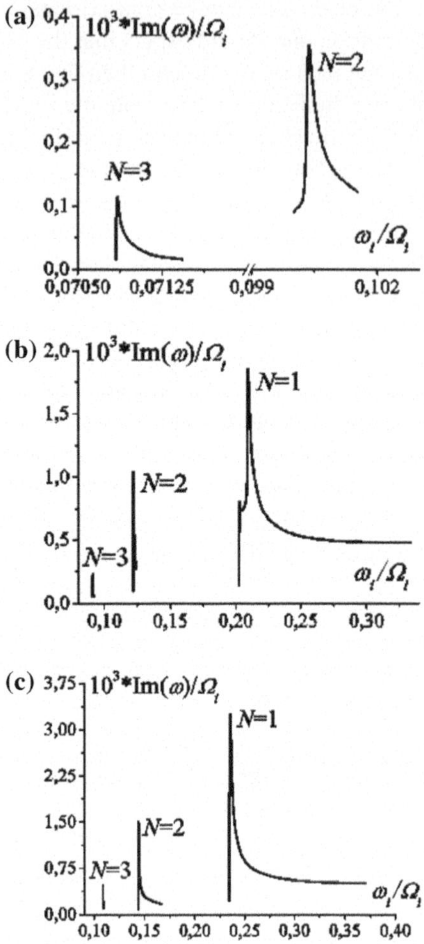

Fig. 10.15 **a** SFXM growth rates normalized to the plasma frequency of the basic plasma, $\mathrm{Im}(\omega)/\Omega_i$, versus the normalized external axial static magnetic field $\theta = \omega_i/\Omega_i$ for $\alpha = 10^{-6}$, $\Delta = 0.1$ and $m = -2$. N is cyclotron harmonic number of the ion beam. **b** The same as in Fig. 10.15a, but for $m = -3$, **c** The same as in Fig. 10.15a, but for $m = -4$

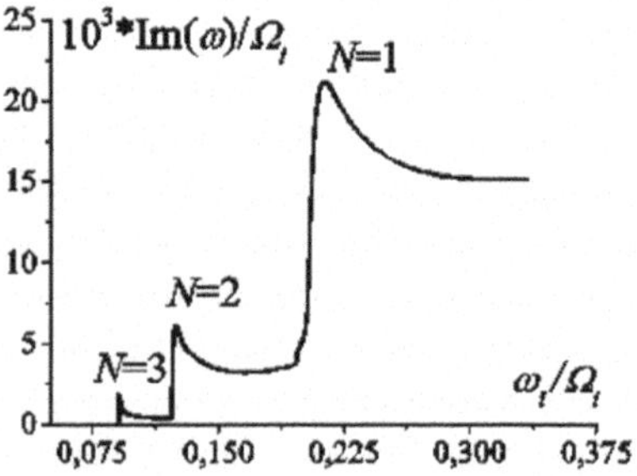

Fig. 10.16 The same as in Fig. 10.15b, but for higher alpha beam density, $\alpha = 10^{-3}$

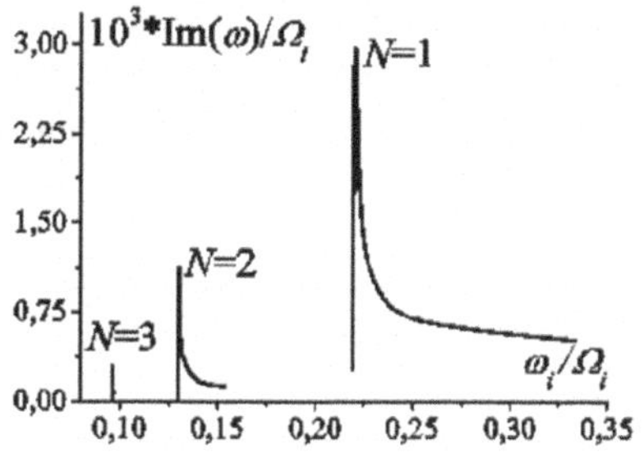

Fig. 10.17 The same as in Fig. 10.15b, but for a wider beam gap, $\Delta = 0.135$

(Fig. 10.15b), and $m = -4$ (Fig. 10.15c). The relative width of the gap between the plasma surface and the metal chamber is assumed to be $\Delta = R_2/R_1 - 1 = 0.1$.

In the following, the curves in Fig. 10.15b will be discussed in details. The highest peak on the plot is a result of SFXW interaction with alpha particles oscillating at their first cyclotron harmonic. Therefore, it is designated as $N = 1$. The peak value of the growth rate $\mathrm{Im}(\omega) \approx 1.9 \times 10^{-3}\Omega_i$ is obtained for $\theta \approx 0.21$. In this case the real part of the wave frequency is close to the fifth ion cyclotron harmonic of neon, $\mathrm{Re}(\omega) \approx 4.9\omega_i$. The velocity of the alpha particles is $\upsilon_{\perp 0} \approx 0.255c$ in this case and the relativistic factor is given by $\gamma \approx 1.034$, indicating a very weak relativism of the alpha particle beam. The second peak $\mathrm{Im}(\omega) \approx 1.05 \times 10^{-3}\Omega_i$, is reached at $\theta \approx 0.12$. This excitation is caused by SFXM interaction with the beam alpha particles oscillating at their second cyclotron harmonic, which is concluded from the real part of the frequency value, $\mathrm{Re}(\omega) \approx 9.8\omega_i$. The lower growth rate in comparison with the first peak is explained, first of all, by a decrease of the external magnetic field and as a sequence the beam velocity, which is $\upsilon_{\perp 0} \approx 0.152c$ ($\gamma \approx 1.012$) in this case. Finally, the third very small peak shown in Fig. 10.15b, $\mathrm{Im}(\omega) \approx 0.25 \times 10^{-3}\Omega_i$, is caused by the interaction of the neon plasma with the third cyclotron harmonic of the alpha particle beam, $\mathrm{Re}(\omega) \approx 15.0\omega_i$ (compare with the second peak). In this case $\theta \approx 0.09$ and $\upsilon_{\perp 0} \approx 0.114c$ ($\gamma \approx 1.007$).

Just as in the case of electron SFXM excitation [6, 13], oscillations of the ion plasma component with higher magnitude of the azimuthal mode index are excited with higher growth rates. The maxima of the peaks in Fig. 10.15c for $m = -4$ are higher and achieved for stronger magnetic fields in comparison with the mode $m = -3$: at the $(\mathrm{Im}(\omega)/\Omega_i)$, (ω_i/Ω_i))-points (0.235, 3.27) for $N = 1$, (0.144, 1.51) for $N = 3$, and (0.11, 0.48) for $N = 3$, respectively. The maxima of the peaks in Fig. 10.15a are lower and achieved for weaker magnetic fields than for the mode $m = -3$: $\mathrm{Im}(\omega) \approx 0.356 \times 10^{-3}\Omega_i$ as a result of the SFXM interaction with beam alpha particles, which oscillate at their second cyclotron harmonic, in this case $\theta \approx 0.1$ and $\mathrm{Im}(\omega) \approx 0.114 \times 10^{-3}\Omega_i$ for SFXM interaction with beam alpha particles, which oscillate at the third cyclotron harmonic, while $\theta \approx 0.07$. It should be noted that the possibility of beam excitation for the SFXM with $m = -2$ when $N = 1$ is not shown. This is explained by the fact that the SFXM frequency increases with azimuthal mode index $|m|$, but even the SFXM frequency for $m = -3$ exceeds slightly the lower boundary of the frequency range where the waves propagate. Therefore, the propagation of the SFXM with $m = -2$ is impossible for the considered values of the ion beam-plasma parameters.

The plot in Fig. 10.16 is calculated for the same values of the parameters of the ion beam-plasma system, as in Fig. 10.15b, except the alpha particle beam density, which corresponds to $\alpha = 10^{-3}$ in this case. Undoubtedly, the change of the beam particle density does not affect the locations of the peaks which become wider and in addition, overlapping of different resonances is observed with increasing alpha particle beam density. An increase in the beam particle density by three orders up to $\alpha = 10^{-3}$ leads in Fig. 10.16, as follows from the analytical solutions (Eq. 10.12), to an increase of the growth rate by an order of magnitude: the peak height for $N = 1$ is $\mathrm{Im}(\omega) \approx 21.2 \times 10^{-3}\Omega_i$.

A plot of the calculated growth rate dependence on θ is shown in Fig. 10.17 for the same values of the beam-plasma parameters, as in Fig. 10.15b, except the width of the vacuum gap for the beam that is assumed to be equal $\Delta = 0.135$ in this case. An increase of the gap width leads to a shift of the peaks to the range of stronger external static magnetic field, they are observed at $\theta \approx 0.096$; 0.13 and 0.22, respectively. The longitudinal (in respect to the beam velocity) electric wave field is zero at the metal wall, and the trajectories of the ion beam particles move away from the metal wall, if the vacuum gap broadens. That is why the efficiency of alpha particle interaction with the wave increases and the growth rates become larger in comparison with the case $\Delta = 0.1$.

Now, the possibility of experimental realization of the investigated phenomena is estimated. As an example, an operating point on the top of the first growth rate maximum in Fig. 10.15b is chosen. Let the radius of the laboratory plasma be 10 cm. Then the condition $c/(\Omega_i a R_1) = 5/6$ used in the calculations needs the neon plasma density $n_{\mathrm{Ne}} \approx 1.5 \times 10^{14}\,\mathrm{cm}^{-3}$ and the corresponding alpha particle beam density $n_{\mathrm{He}} \approx 1.5 \times 10^{8}\,\mathrm{cm}^{-3}$. The maximum of the first peak on the plot is realized for the extremely high external magnetic field $B_0 \approx 1600$ kG! A non-relativistic alpha particle beam with $\upsilon_{\perp 0} \approx 0.255c$ is required to excite this SFXM. For these conditions, the wave is excited with the growth rate $\mathrm{Im}(\omega) \approx 6.85 \times 10^{6}\,\mathrm{s}^{-1}$ at the frequency $\omega \approx 3.69 \times 10^{9}\,\mathrm{s}^{-1}$.

It is clear that the requirement of such high magnetic field B_0 is the weakest point of the proposal. Currently, the highest experimentally achieved static magnetic field is 455 kG. That is why one has to search for conditions of SFXM excitation when the magnetic field induction would be as small as possible. As an example, the excitation of the SFXM with azimuthal mode index $m = -3$ in a helium plasma by a proton beam can be considered. Let the radius of the laboratory plasma be again 10 cm. Then the condition $c/(\Omega_i a R_1) = 1.6$ gives the He plasma density $n_{\mathrm{He}} \approx 8 \times 10^{12}\,\mathrm{cm}^{-3}$. The particle density of the proton beam is assumed to be by six orders of magnitude lower, which means $n_p \approx 8 \times 10^{6}\,\mathrm{cm}^{-3}$. Effective interaction of the wave with the 3rd cyclotron harmonic is realized for $B_0 \approx 200$ kG (corresponding to 20 T). Although this is also a very strong magnetic field, the method of its production is well known [34] and it would be sufficient to produce it in a small volume of the $3 \times 10^{3}\,\mathrm{cm}^{3}$ order. One should also recall, that no relativistic proton beam is required for wave excitation under these experimental conditions: $\upsilon_{\perp 0} \approx 0.08c$. In other words, it is a beam of protons with $W = 0.5 \times 10^{-12}$ J energy, which carries an electric current $I \approx 30$ mA with the current density of 3.2 mA/cm^2. Such beams can be obtained, for example, in linear proton accelerators (3 meV, 300 mA) with transverse focusing by a 100 keV, 85 A electron beam [35] or in linear accelerators such as LEDA [36]. In this case, the SFXM wave is excited with the growth rate of $7 \times 10^5\ \mathrm{s}^{-1}$ at the frequency $6 \times 10^9\ \mathrm{s}^{-1}$.

The following conclusions can be drawn from the studies carried out in the present Section. The possibility of excitation of ion SFXMs with negative azimuthal mode index in strongly magnetized plasmas by annular beams of light ions is investigated. Numerical calculations confirm that the resonance interaction of the beam with this type of surface waves can be realized in sufficiently strong external static magnetic fields. Effective excitation is shown to be possible for SFXMs when the normalized frequency is close to multiples of the mass ratio of the main plasma ions and the beam ions. The results of the numerical analysis are in a good agreement with the analytical calculations. The growth rate of the SFXM resonant beam instability is shown to increase with increasing absolute value of azimuthal mode index, width of the gap between the plasma column and the metal waveguide wall and with the beam particle density.

Though no relativistic ion beam is needed for SFXM excitation, nevertheless, the demand for very high external magnetic fields appears to be a strong limitation for realization of the necessary excitation conditions. This is explained by the following reason: the third term in the expression (7.20) for $\varepsilon_{22}^{(b)}$ (it is proportional to $\upsilon_{\perp 0}^2$) is responsible for the excitation of the waves. In the case of a weak magnetic field the velocity of ion beam particles $\upsilon_{\perp 0}$ appears to be small so that the first two terms prevail in expression (7.20). Therefore, the excitation of SFXMs does not take place. The demand for very strong static magnetic fields explains why neither the linear theory of ion SFXM excitation in coaxial, toroidal waveguides and waveguides with non-circular cross-section nor the nonlinear theory of their beam-plasma and beam-dissipative excitation are studied in the present monograph.

Excitation of SFXMs by annular ion beams is characterized by the following feature. The beam rotates in a strong external magnetic field in a vacuum gap between the plasma column and the metal waveguide chamber. Therefore, the beam velocity is defined by the induction of the external magnetic field and the radii of the plasma column and the metal chamber, $\omega_i R_1/\gamma < \upsilon_{\perp 0} < \omega_i R_2/\gamma$. When the beam radius is closer to the chamber wall, more kinetic energy can be transferred into electromagnetic waves before the beam reaches the plasma surface due to its deceleration. On the other hand, in this case the electric field of the excited wave is weaker. Thus, the efficiency of beam-plasma interaction is lower.

Higher efficiencies can be expected for wider gaps Δ between the plasma column and the metal wall of the waveguide where the ion beam rotates. Increase in Δ leads also to substantial increase of the growth rate of ion SFXW excitation, which was observed also in the case of electron SFXWs [6, 13].

In the present study, the value of the magnetic induction B_0 of the external static magnetic field is characterized by the parameter θ. The dependence of $\mathrm{Im}(\omega)$ on B_0 is non-monotonous, since the most effective beam excitation of SFXMs occurs for those

B_0 values when the curve of the SFXM eigenfrequency crosses the lines $\omega = \omega_b N/\gamma$ (see condition (10.11)). In general, smaller B_0 requires higher harmonics of the beam ion cyclotron frequency to provide the resonant interaction between particle beam and SFXM, leading to the smaller growth rates.

The present results can be of interest to solve different tasks in the field of plasma electronics. The SFXM energy can be easily radiated from the waveguide. If one makes a narrow ($\varphi_0 \ll \pi$) axial slot antenna in the metal wall of the waveguide, part of the SFXM energy is radiated into the free space. As shown in Sect. 2.5 for the case of SFXMs in the range of the electron cyclotron frequency with different azimuthal mode index, the damping rates caused by their radiation are less than the growth rates of beam instabilities in the range of small effective wavenumbers [37]. Note that the flux of electromagnetic radiation can be controlled by varying the azimuthal size φ_0 of the slot, since the damping rate caused by radiation of energy is proportional to φ_0 in linear approximation over φ_0.

10.5 Conclusions

Electromagnetic wave excitation as the result of the interaction of a flow of electrons gyrating around a cylindrical plasma column immersed in an external static axial magnetic field is considered in Chaps. 7–9. The most efficient interaction of gyrating electrons with electromagnetic waves naturally takes place in the range (1.13) of the electron cyclotron frequency. The theory of SFXM excitation by electron beams gyrating along large Larmor radii in the electron cyclotron frequency range is presented in many details in Chaps. 7–9. However, SFXMs are shown in Chaps. 1 and 2 to propagate also in the high frequency (HF) range (1.14) above the upper hybrid frequency in sufficiently dense plasmas, and in the ion cyclotron frequency range (2.14) in strongly magnetized plasmas.

SFXM excitation above the upper hybrid frequency is favorable because electromagnetic waves generated in this case have shorter wavelengths than in the case of electron cyclotron radiation. Dependencies of the SFXM growth rate on plasma waveguide parameters are not monotonous in this case. The best conditions for SFXM excitation above the upper hybrid frequency are observed for those plasma waveguide parameters, which provide an SFXM eigenfrequency close to the upper limit of the HF range (1.14). Corrugation of the plasma column surface in azimuthal direction slows down HF SFXMs. This mechanism provides broadening of the range of plasma waveguide parameters, for which efficient HF SFXM beam excitation can take place, toward lower plasma particle density and smaller radii of the plasma column. Dissipations naturally worsen electromagnetic wave excitation. The gradual decrease of the SFXM growth rate with increasing collision frequency and decreasing beam particle density does not allow to find any distinct border between beam-plasma and beam-dissipative instability regimes in the interaction of relativistic large Larmor orbit electron beams and HF SFXMs in contrast to the case of LF SFXMs studied in Sect. 7.6. However, the growth rates of SFXMs appear to be much smaller than in the case of the electron cyclotron frequency range. Moreover, notable beam-wave interaction takes place in much more narrow ranges of the plasma waveguide parameters. These unfavorable circumstances sufficiently straiten the possibilities of SFXM practical applications in this frequency range.

Luckily, no relativistic ion beam is required for SFXM excitation in the ion cyclotron frequency range. Effective excitation of ion SFXMs is shown to be possible when the normalized frequency is close to multiples of the mass ratio of the main plasma ions and the beam ions. However, the process needs also strong external static magnetic fields. These physical restrictions explain the minor attention which is given to SFXM excitation in the ion cyclotron frequency range in the present monograph.

References

1. Girka, V. O., Girka, I. O., & Pavlenko, I. V. (2011). Excitation of azimuthal surface modes by relativistic flows of electrons in high-frequency range. *Plasma Physics Reports, 37*(5), 447–454.
2. Girka, V. O., Girka, I. O., & Pavlenko, I. V. (2013). Excitation of high frequency azimuthal surface waves by annular electron beam in a waveguide with a non-circular interface of a plasma column. *Plasma Physics Reports, 39*(5), 399–405.
3. Girka, I. O., & Thumm, M. (2017). Transition between beam-plasma and beam-dissipative instability regimes in the interaction of relativistic large Larmor orbit electron beams and azimuthal surface waves above the upper-hybrid frequency in coaxial plasma waveguides. *IEEE Transactions on Plasma Science, 45*(8), 2208–2214.
4. Girka, V. O., Girka, I. O., & Pavlenko, I. V. (2011). Excitation of ion azimuthal surface modes in a magnetized plasma by annular flow of light ions. *Progress in Electromagnetics Research M (PIERM), 21*, 267–278.
5. Girka, V. O., Girka, I. O., & Pavlenko, I. V. (2012). Possibility of excitation of azimuthal surface waves in a magnetized plasma by annular ion beams. *Technical Physics Letters, 38*(2), 178–180.
6. Girka, V. A., Girka, I. A., Olefir, V. P., et al. (1991). Generation of electromagnetic waves with an annular relativistic electron beam. *Soviet Technical Physics Letters, 17*(1), 35–36.
7. Alexandrov, A. F., Bogdankevich, L. S., & Rukhadze, A. A. (1984). *Principles of plasma electrodynamics*. Springer.
8. Girka, V. A., Girka, I. A., Kondratenko, A. N., et al. (1988). Azimuthal surface waves of magnetoactive plasma waveguides. *Soviet Journal of Communications Technology and Electronics, 33*(8), 37–41.
9. Kuzelev, M. V., & Rukhadze, A. A. (1990). *Electrodynamics of dense electron beams in plasmas*. Nauka. (in Russian).
10. Lawson, J. D. (1977). *The physics of charged—particle beams*. Clarendon Press.
11. Levitskii, S. M., & Koshevaya, S. V. (1986). *Vacuum and solid state electronics of super high frequencies*. Vyshcha Shkola (in Russian).
12. Aronov, V. A., Bogdankevich, L. S., & Rukhadze, A. A. (1975). Bulk and surface electromagnetic waves in spatially bounded plasma and their excitation by relativistic electron beams. In Smirnov, B. M. (Ed.), *Plasma chemistry* (2nd Ed., pp. 245–266). Atomizdat (in Russian).
13. Girka, V. O., Girka, I. O., Morgal, Ya. I., et al. (2011). Excitation of azimuthal surface modes by annular electron beams in the range of electron cyclotron frequency. *Physica Scripta, 84*, 025505, 7 p.

14. Akhiezer, A. I., & Fainberg, Ya. B. (1949). On the interaction of a charged particle beam with electron plasma. *Doklady Akademii Nauk SSSR, 69*, 555–556 (in Russian). (Akhiezer, A. I., & Fainberg, Ya. B. (2008). On the interaction of a charged particle beam with electron plasma. *Ukrainian Journal Physics, 53*, 87–88 (in English, Translated and reprinted from *Doklady Akademii Nauk SSSR, 69*, 555–556 (1949)).
15. Girka, O. I., Girka, I. O., Girka, V. O., et al. (2006). Coupled HF azimuthal waves in magnetoactive waveguide partially filled by current-carrying plasma. *Problems of Atomic Science and Technology, NSC "Kharkov Institute of Physics & Technology", No. 5, Series: Plasma Electronics and New Methods of Acceleration* (5), 28–33.
16. Bogdankevich, L. S., Kuzelev, M. V., & Rukhadze, A. A. (1980). Theory of excitation of plasma-filled rippled-boundary resonators by relativistic electron beams. *Soviet Physics Technical Physics, 25*(2), 143–147.
17. Onishchenko, I. N., Sidorenko, D. Yu., & Sotnikov, G. V. (1995). Electron beam interaction with a plasma-filled rippled coaxial waveguide. *Plasma Physics Reports, 21*(8), 670–676.
18. Vukovic, S. (Ed.). (1986). *Surface waves in plasma and solids*. World Scientific.
19. Sprangle, P. (1976). Excitation of electromagnetic waves from a rotating annular relativistic e-beam. *Journal of Applied Physics, 47*(7), 2935–2940.
20. Chojnacki, R., & Destler, W. W. (1987). Microwave radiation from a low-energy rotating electron beam in an azimuthally periodic magnetic wiggler field. *IEEE Journal of Quantum Electronics, 23*(9), 1605–1609.
21. Destler, W. W., Chojnacki, E., Hoeberling, R. F., et al. (1988). High-power microwave generation from large-orbit devices. *IEEE Transactions on Plasma Science, 16*(2), 71–89.
22. Girka, V. O., Girka, I. O., & Pavlenko, I. V. (2012). Influence of the shape of the cross section of a plasma-dielectric interface on the dispersion properties of high frequency azimuthal surface modes. *Plasma Physics Reports, 38*(2), 126–137.
23. Krall, N. A., & Trivelpiece, A. W. (1986). *Principles of plasma physics*. San Francisco Press.
24. Girka, V. O., Girka, I. O., Girka, A. V., et al. (2011). Theory of azimuthal surface waves propagating in non-uniform waveguides. *Journal of Plasma Physics, 77*(4), 493–519.
25. Karpov, S. Yu., & Stolyarov, S. N. (1993). Propagation and transformation of electromagnetic waves in one-dimensional periodic structures. *Physics-Uspekhi, 36*, 1–22.
26. Girka, I., & Thumm, M. (2016). Excitation of azimuthal surface waves in the electron cyclotron frequency range by a rotating electron beam in presence of dissipation. *Physics of Plasmas, 23*, 122124.
27. Girka, I. O., & Thumm, M. (2016). Excitation of azimuthal surface modes above the upper-hybrid frequency by external relativistic flows of electrons in coaxial plasma-vacuum waveguide. *IEEE Transactions on Plasma Science, 44*(11), 2859–2866.
28. Girka, I. O., & Thumm, M. (2017). Excitation of electromagnetic waves above the upper-hybrid frequency by internal gyrating electron beam in a coaxial waveguide. *IEEE Transactions on Plasma Science, 45*(4), 623–630.
29. Bludman, S. A., Watson, K. M., & Rosenbluth, M. N. (1960). Statistical mechanisms of relativistic streams II. *Physics of Fluids, 3*(5), 747–757.
30. Brambilla, M. (1998). *Kinetic theory of plasma waves: Homogeneous plasmas. International Series on Monographs on Physics 96*. Clarendon Press.
31. Python 3.6.1, release date is 2017.03.21. https://www.python.org/
32. Velazco, J. E. (2000). *Continuous rotating-wave electron beam accelerator*. U.S. Patent No. 6,060,833. U.S. Patent and Trademark Office.
33. Girka, I. O., & Kovtun, P. K. (1998). Azimuthal surface waves in a magnetized plasma. *Technical Physics, 43*(12), 1424–1427.
34. Herlah, F. (Ed.). (1985). *Strong and ultrastrong magnetic fields and their applications*. Springer.
35. Ivanov, B. I., & Shulika, N. G. (2004). Influence of longitudinal focusing on changing the transversal emittance in accelerating the high-current proton beam. *Problems of Atomic Science and Technology, No. 4, Series: Plasma Electronics and New Methods of Acceleration* (4), 133–138 (in Russian).

36. Dominic Chan, K. C., Lawrence, G. P., & Schneider, J. D. (1998). Development of RF linac for high-current applications. *Nuclear Instruments and Methods in Physics Research, Section B: Beam Interactions with Materials and Atoms, 139*(1–4), 394–400.
37. Girka, V. O., & Girka, I. O. (1992). Emission of azimuthal surface waves from narrow waveguide slot. *Soviet Journal of Communications Technology and Electronics, 37*(9), 32–35.

Chapter 11
Applications of Surface Wave Propagation

This Chapter is devoted to considerations on possible applications of the surface waves (SWs) studied in the previous chapters. For convenience, in the present Chapter, the SI system of units is used for the measurements of observables. SWs have a wide field of applications, but here only the following areas are considered: plasma electronics, including devices utilized for generation of super high frequency radiation and plasma-antenna systems; plasma production, namely the plasma sources based on SWs propagation; surface plasmon-polaritons propagating in THz frequency bands; plasmonics devices, including biosensors, photovoltaic cells; and plasma nanoscience.

11.1 Surface Wave Applications for Plasma Electronics

Plasma electronics studies the generation and amplification of radio frequency (RF) and microwaves due to interaction between charged particle beams and plasma eigenwaves. To create a plasma electronics device one can often introduce a plasma into an already established vacuum electronic device. This allows one to enhance the performance of the device due to: operation at higher injected beam currents and improvement of the control on the wave-beam interaction region. Therefore, higher output power, higher efficiency and ability of broadband frequency tuning can be achieved. These effects have been observed [1] in the following electronic devices: plasma-filled relativistic gyrotron, plasma-filled smooth waveguide amplifier and relativistic plasma-filled backward wave oscillator (BWO). It should be mentioned that in a plasma-filled smooth wall waveguide, excitation of slow waves by electron beam injection becomes possible. In paper [1] the authors studied the case of a relativistic BWO, applying a strong and a weak guiding magnetic field 0.2–1 T, when the peak plasma density was higher than 10^{12} cm^{-3}. Their experimental work demonstrated a significant increase of the operating efficiency and a tunable upshift

© The Author(s), under exclusive license to Springer Nature Switzerland AG 2022

I. Girka and M. Thumm, *Surface Flute Waves in Plasmas*, Springer Series on Atomic, Optical, and Plasma Physics 120, https://doi.org/10.1007/978-3-030-98210-2_11

of the operating frequency of the plasma-filled device as compared with the case of a conventional vacuum device.

In the paper [2], results of theoretical and experimental studies of a large diameter microwave oscillator, operated in the Giga-Watt (GW) power range, were presented. The device consisted of a large cross-section, slow-wave structure with a unique profile of wall radius specifically designed to support SWs and to provide a strong electron beam-wave coupling at moderate acceleration voltage (500 kV), an internal adjustable microwave reflector, a coaxial microwave extraction section, and a coaxial magnetically insulated field emission electron gun. The authors reported on the results of preliminary experiments carried out at 8.3 GHz, where the power level exceeded 0.5 GW, and an efficiency of 15% was measured [2]. SWs were found to be supported by specific wall structures (slow wave structures) in large diameter waveguides in order to be synchronous with the electron beam. The SWs were reported to be crucial to avoid mode competition. The experiments showed that a large coupling impedance is important to obtain high efficiency interaction. A periodic structure, which satisfies the requested conditions, was found. One of the main conclusions of the report was, that the experiments showed, that increasing the beam current to 16.8 kA had a small effect on the generated power, but resulted in a downshift of the operating frequency by 500 MHz. The authors summarized that single-mode operation of a large-diameter SW oscillator had been successfully demonstrated, that the application of a moderate voltage electron beam could generate GW-level radiation, and that additional optimization of the extraction section (coaxial and output window regions) was required in order to increase the radiated power.

The development of powerful, frequency tunable and small sized devices generating high frequency (HF) electromagnetic waves was outlined in [3] to be one of the main purposes of plasma electronics. The problems of elaboration of different schemes of microwave generators for super-radiation of extended electron bunches, which move above corrugated metal plates, were studied in [3]. Excitation of SWs by the electron beam in a corrugated structure was outlined to be one of these mechanisms. Electromagnetic wave pulses with the power ~5 MW and duration ~40 ps were estimated to be obtained for the electron current density 1200 A/cm, corrugation period and depth 15×10^{-3} cm and 5×10^{-3} cm, respectively. Thus, one of the main conclusions of the report was that THz band super-radiance can be achieved by utilization of the Cherenkov-type beam-surface wave interaction in corrugated waveguide interaction circuits.

The same team of authors proposes in [4] to apply a planar relativistic SW oscillator with two dimensional periodic gratings for additional transverse synchronization of the electromagnetic emission from a wide rectilinear sheet electron beam in order to develop a Cherenkov millimeter-wave oscillator with GW output power. This paper was one in the branch of plasma-electronic research, which studied different mechanisms of Cherenkov emission of a sheet electron beam propagating in the vicinity of a periodically corrugated surface. In [4], a quasi-optical approach proposed by the authors in their earlier papers for describing the formation of a SW located nearby a shallow corrugated metal surface, was applied for the case of two-dimensional feedback.

Cherenkov devices based on interaction of electron beams with electromagnetic SWs in corrugated structures were pointed out in [5] to provide both longitudinal and transverse mode selection in oversized electrodynamic systems. A periodically corrugated section of a metal plate as a surface-wave resonator was considered. An optimum value of the corrugation depth providing the maximum Q factor of the fundamental mode was found. Moreover, the authors of [4–7] also discussed the possibility to realize a cylindrical version of a two-dimensional SW oscillator driven by an annular electron beam.

A Ka-band free-electron maser (FEM) based on two-dimensional feedback was experimentally studied in coaxial geometry at the University of Strathclyde (Glasgow, UK) [8]. In the W-band, a FEM with planar undulator was experimentally investigated at the Budker INP RAS (Novosibirsk, Russian Federation) [9]. Both, theoretical and experimental studies of Cherenkov masers with two-dimensional distributed feedback were reported in [10] to be in progress. Among relativistic masers of such type, surface wave oscillators were stressed to be beneficial due to their large electron-wave coupling impedance. In addition, formation of a SW was found to ensure a regular field distribution along the coordinate directed perpendicularly to the corrugated surface. The latter issue was found to be the way to solve the problem of mode selection over this coordinate. Two-dimensional periodic structures were exploited in SW oscillators both as a slow-wave structure and as a fast-wave high selective Bragg resonator, which provided effective mode control over the azimuthal coordinate [11]. Theoretical and experimental studies on oversized SW oscillators with two-dimensional periodically corrugated cylindrical interaction structure were reported in [10]. Simulations on nonlinear dynamics of the oscillator were performed and experimental research was carried out based on the thermionic cathode accelerator "Saturn".

This idea of a two-dimensional periodically corrugated cylindrical structure is very interesting in the context with the present book, because the materials presented in the previous chapters are devoted just to SWs propagating in waveguides with cylindrical geometry. Thus, experimental studying of their interaction with an annual electron beam seems to be useful for demonstration of practical applications of SFWs.

The paper [12] reviewed accomplishments in the theoretical explanations and experimental performances of plasma–assisted slow-wave oscillators (pasotrons). This type of plasma electronic devices distinguishes from other type devices by two special features, namely: it has a plasma cathode electron gun to generate a high-perveance electron beam and it has no guiding magnetic field. Its slow-wave structure can be either a corrugated waveguide or a helix, where the electron beam interacts with both, axial and radial electric wave fields. The electromagnetic eigenwaves observed there had their maximum of amplitudes nearby the walls of the pasotron; this is a specific peculiarity of SWs. That is why two-dimensional motion of the employed electron beam in this case allowed one to enhance the efficiency of their interaction unlike in the case of application of an axial external magnetic field, which makes the electron motion like a one-dimensional one. Experiments conducted at that time allowed the authors to report on obtaining 50% interaction efficiency and 1 MW of radiated power at an operating frequency ~1.2 GHz. The main efforts of the authors were directed to understanding the physics of ion focusing of the electron beam and the interaction between the beam and the eigenfields of the slow waves in order to achieve a microwave power of ~2 MW.

Cherenkov type interaction in a corrugated waveguide between slow waves and an electron beam was studied in [13]. Because of the spatial distribution of the studied SW field one had to direct the beam close to the waveguide wall. The authors applied a strong magnetic field $B_0 > 2$ T, which was created by a superconducting magnetic coil with cryogenic cooling. The performed experiments with high repetition rate of the pulses [13] showed that the output power was more than 1 GW, with an electronic efficiency close to 35%. The authors proposed the design of a backward wave oscillator (BWO) and an orotron with wavelengths: 3 and 0.8 cm, respectively, as prospective devices for microwave generation.

The paper [14] presents results of experimental investigation and operation of a plasma Cherenkov maser amplifier. The experimental apparatus consisted of four main parts: coaxial mode transformer, annual plasma column, plasma source, and transmission-line output. An external longitudinal pulsed magnetic field could be varied up to $B_0 = 2$ T. The microwave source was a magnetron with the following parameters: wavelength $\lambda = 2.42$ cm, pulse duration 10–100 μs, and output power $P = 50 - 150$ kW. It was found that the plasma column consisted of filaments with varying density, which were rotating azimuthally. The number of azimuthal plasma filaments was usually not larger than ten. The authors had decided to excite azimuthally symmetric coaxial plasma waveguide modes. The best experimental results concerning high-power microwave transmission down the coaxial waveguide of 30 cm length were obtained for the slow TM_{01} plasma mode. In this case $P \approx 150$ kW output power was obtained at $B_0 \approx 0.92$ T and $3 \cdot 10^{13}$ cm^{-3} plasma density, where the input frequency of the amplifier was four times less than the electron plasma frequency: $\omega_0 \cong \Omega_e/4$.

Significant experience in experimental studies of relativistic Cherenkov surface-wave oscillators (SWOs) was reported in [15]. Such sources employ synchronous interaction of a rectilinear electron beam with a slowed down fundamental harmonic of the electromagnetic wave in the waveguide with periodical corrugation of the side walls. The amplitude of this fundamental frequency wave decays with the distance from the corrugation, thus forming a surface wave. This peculiarity was found to open up the possibility of using oversized electrodynamic systems with much larger dimension-to-wavelength ratio, as compared to conventional BWOs [16], known to operate at the spatial harmonics of a low mode of a corrugated waveguide.

The propagation of azimuthally symmetric TM waves in periodically corrugated cylindrical waveguides and their excitation by relativistic electron beams was investigated in [17] within the quasi-optical approach. A criterion for the existence of SWs was derived via presenting the wave field in two parts, forward and backward wave. The spectrum of axial evanescent eigenmodes was estimated for a finite-length corrugated section. Based on this model, two oscillation regimes were recognized, namely, the π-mode excitation regime and the regime of backward surface wave oscillator. The viability of practical implementation of relativistic SW oscillators with power level of up to 140 MW in the sub-millimeter wavelength band was demonstrated.

The application of a plasma as an element of electronic devices was outlined in [18] to have a lot of advantages connected with the peculiarities of plasma—filled waveguides. However, enhancement of powerful HF oscillations excited by an electron beam in plasma-filled slow-wave structures was pointed out to be still far away from complete understanding. The authors of [18] presented results of numerical calculations of the influence of the electrodynamics characteristics with temporary evolution, caused by the plasma nonlinearity, on excitation of slow waves under the regimes of generation and amplification. The limitation of beam-wave energy exchange was stressed to be caused by trapping of the beam particles into the wave potential wells. Those inherent nonlinear mechanisms, which could lead to self-maintaining of the resonance between the particle bunch velocity and the slow-wave phase velocity, were studied. Increasing effective dissipation coefficient was found to lead to increasing energy, which could be lost by the trapped bunches on the one hand, but on the other hand to a decreasing quantity of trapped particles. Thus, competition of these two mechanisms was reported to determine an optimum value of the effective dissipative coefficient, which allowed one to realize maximum efficiency of slow-wave excitation. Some enhancement of the efficiency of slow-wave excitation was shown in [18] as well to be achieved under the regime of stochastic modulation of the plasma particle density.

The paper [19] presents results of experimental and theoretical investigations on a relativistic BWO (or carcinotron), which operated with the rotating TM_{11} eigenwave. Decompression and cyclotron selection permitted the authors to obtain stable generation of microwave pulses, which had no after-pulse. Many requested features of the carcinotron can be obtained by application of higher interacting eigenmodes, among them power output, pulse duration, and stability of microwave generation. Higher eigenmodes were employed by single-mode generation and reduction of unwanted conversion of the operating mode to parasitic modes on various waveguide non-uniformities to an admissible level. To do that the authors applied a circular waveguide with a weak single threaded helical corrugation. To remove the degeneration of the frequency spectrum of the eigenwaves propagating in mutually opposite azimuthal directions they used a specific direction of the focusing magnetic field which correlated with the waveguide corrugation. The carcinotron operated in the following regime: accelerating voltage $U \approx 0.5$ MV, beam current 2 kA, and pulse duration $\tau \approx 20$ ns. From analysis of the experimental data [19], the authors concluded that application of higher azimuthally rotating eigenmodes had allowed them to realize stable operation, which came close to the potential barrier, to improve the control on the regime of generation, to reduce the value of the focusing magnetic field, and to suppress generation of the after-pulse, which made utilization of carcinotrons prospective for radar systems.

The results of integrated experiments devoted to operation of cyclotron autoresonance masers (CARMs) and relativistic gyrotrons were reviewed in [20]. Operation of gyrotrons was outlined in [20] to be based on the relativistic effect of the electron mass on the cyclotron frequency. First, increase of the electron energy was noted to decrease their cyclotron frequency, accompanied by a decrease of the radiation frequency. Second, change of the radiation frequency of sub-relativistic electrons was pointed out to result in electrons outgoing from the cyclotron resonance with the excited wave, therefore transferring only a small part of their energy into the wave. Two obvious methods to compensate these defects of sub-relativistic gyrotrons were mentioned. The first possibility was connected with application of the Doppler effect if the electrons interact with a wave propagating not across an external magnetic field, but along it. In this case, as in free electron lasers (FELs), the radiation frequency was found to be significantly larger than the electron cyclotron frequency (Doppler upshift). The second possibility was connected with application of non-dipole interaction of the sub-relativistic electrons to excite waves at higher cyclotron harmonics.

The authors mentioned above elaborated a special magneto-optical system supplied with a high current direct-action accelerator and a special pumped kicker for the electron beam that allowed them to improve the quality of the axis-encircling helical beam for the CARM. The use of an electron beam with a small velocity spread had enhanced the CARM efficiency to the level of FELs. However, the further improvement of the beam quality was planned to be continued. Another serious problem, which was solved by the authors, was connected with excitation of parasitic modes, especially near-cut-off waves (gyrotron modes) and backward waves. To avoid this problem the authors of [20] had applied the lowest transverse mode of a cylindrical waveguide (TE_{11}). Thus, the maximum power at the wavelength of 7.9 ± 0.1 mm amounted to 13 ± 2 MW, at an efficiency of $26 \pm 5\%$.

As the first step to develop high-harmonic relativistic gyrotrons the authors conducted experiments with large orbit gyrotrons (LOGs). They designed two devices: LOG-1 and LOG-2. In their experiments, selective excitation of the modes from TE_{11} up to TE_{51} was observed. The maximum signal was registered at the third cyclotron harmonic with a wavelength of 6.1 mm at an output power of about 100 kW. The opportunity to increase the efficiency of both devices was discussed. Thus, according to the authors' experience [20] the development of high-efficiency CARMs and/or LOGs seemed to be easier than the development of relevant FELs.

Results of theoretical and experimental investigations of three types of gyro-devices, which applied thin helical electron beams encircling the axes of axially symmetrical magneto-active microwave systems, were presented in [21]. The possibility of efficient simultaneous generation of an auto-resonant mode at fundamental harmonic and a gyrotron mode at the second electron cyclotron harmonic was proved both analytically and experimentally there. The mentioned types of CARM unlike conventional gyro-devices applied mildly relativistic electron beams, which had essential values of both axial and azimuthal velocities.

The first device combined features of a CARM based on excitation of a traveling RF wave and a relativistic gyrotron with excitation of cyclotron harmonics near the cut-off frequency. High electronic efficiency was pointed out to be the main advantage of the CARM. On the other hand, operation of the CARM was found to be very sensitive to the distribution function of the applied gyrating electron beam, while production of a high-quality beam appeared to be a very difficult technical problem. Stable operation and selective excitation of a near-cutoff cyclotron mode in a simple open-ended cylindrical cavity with less sensitivity to velocity spread were reported to be the advantages of the second type of proposed devices, the relativistic gyrotron. However, its efficiency was found to decrease with increasing number of the electron cyclotron harmonic. The authors studied excitation of the modes TE_{11} and TE_{21} of a cylindrical cavity, which were excited by an axis-encircling electron beam at the fundamental (CARM) harmonic and second cyclotron (gyrotron) harmonic. The proposed design of the cyclotron resonance device was found to ensure a high efficiency of microwave generation due to excitation of the so-called "CARM" mode by bunches of electrons, which was provided by the second so-called "Gyrotron" mode. Just stable excitation of the "Gyrotron" mode allowed to fix the frequency of the generated "CARM" mode. The experiments [21] produced microwaves with 6 MW output power at 40 GHz with an electronic efficiency of 22% using a 460 keV electron beam with 60 A beam current. The axial magnetic field, which was applied in these experiments, was about 1 T.

The second design was a large orbit gyrotron (LOG). The advantage of its operation is connected with the fact that the excitation of high cyclotron harmonics is stronger at relatively high energy of the beam electrons. In this device the authors once more applied a thin mildly relativistic electron beam encircling the axis of the utilized electro-dynamic system [21]. The beam excited only co-rotating waves, which had azimuthal mode indices equal to the resonant harmonic numbers, respectively. Selective excitation of the first five cyclotron harmonics (TE_{11}—TE_{51} operating modes) was demonstrated in prototype experiments with a short-pulse LOG. The obtained results were applied to develop a long-pulse LOG with a thermionic electron gun. To produce an electron beam, which would be able to gyrate around the axis of the waveguide interaction circuit with a minimum displacement of its guiding center, with a perpendicular to parallel velocity ratio (pitch factor) of 1.2–1.5 in a resonant magnetic field 1.6–1.7 T, the authors designed a two-stage electron-optical system. In the first stage a thin electron beam was formed, and in the second one it acquired the requested transverse velocity passing through a kicker magnetic field. Measurements had shown that the efficiency of microwave generation was about 3.5% at an output power of 70–80 kW.

Interaction between a gyrating electron beam and eigenmodes of a cylindrical waveguide was emphasized in [21] to have good prospects to be applied for efficient, broadband amplification of microwave radiation. However, the limiting case of small axial wavenumbers was outlined to be characterized by the limited bandwidth resulting from the dispersion of the eigenwaves and by the danger to excite undesirable parasitic oscillations. On the other hand, in the limiting case of large axial wavenumbers, Doppler broadening of the resonance was noted to restrict the bandwidth. The authors proposed to use a waveguide with helical corrugation in contrast to conventional gyro-devices, which were found to have serious disadvantages connected with their high sensitivity to the quality of the electron beam (velocity spread) and excitation of spurious oscillations. Experiments at 9.4 GHz with this new type of 2nd harmonic gyro-TWT (traveling-wave-tube) showed a maximum output power of 1.1 MW at 29% efficiency. Such a broadband gyro-TWT with efficiency close to that of a comparable conventional gyrotron was concluded to be prospective as efficient high-power amplifier for applications in radar systems and broadcast telecommunications.

Substrate integrated waveguides (SIWs) have been proposed and are widely used in microwave and RF systems due to their low loss, low crosstalk, and compatibility with planar microwave circuits [22]. However, the large dimension of SIWs was pointed out to limit their application in compact microwave circuits [23]. To overcome this problem, several solutions were proposed, in particular, the ridged half-mode SIW (RHMSIW) [24]. However, the technologies suggested earlier could only miniaturize the transverse size without miniaturizing the longitudinal dimension. A spoof surface plasmon-polariton (SSPP) structure was proposed in [25] to be introduced into the metal ridge of a RHMSIW. The suggested structure had a compact lateral size because the ridge of the RHMSIW introduced a step capacitance in the waveguide. Due to the loading of the SSPP structure, the proposed RHMSIW-SSPP structure had the ability of strong electric field confinement. Compared with the RHMSIW structure, the RHMSIW-SSPP structure had low crosstalk. A prototype of the proposed RHMSIW-SSPP structure for the 4.5–14 GHz frequency band was designed and fabricated utilizing the printed circuit board process. Application of the SSPP structure made it possible to reduce the longitudinal size of the suggested RHMSIW-SSPP structure by over 50% and the lateral width by 40% as compared with the HMSIW (the longitudinal size was reduced by over 50% and the transverse width by 70% as compared with the SIW). Therefore, the proposed structure was concluded to have potential applications in miniaturized microwave and RF systems.

A cylindrical hollow plasma cavity was designed in [26] for a particle-driven plasma accelerator experiment for future linear electron–positron colliders. Just SWs instead of laser ionization were applied to generate a high-density, highly stable cylindrical hollow plasma. A Ø100 × 400 mm (length) × 30 mm (thickness) hollow plasma column with a density of 2.76×10^{15} cm^{-3} was obtained with 2 kW microwave power at 6 GHz. The fluctuations of the plasma density in longitudinal direction were less than 2%, which was satisfactory to provide a long and sufficiently stable plasma region for the wakefield acceleration experiment.

In [27, 28], SFXMs were demonstrated to be eigenwaves of plasma loaded, coaxial helix traveling-wave-tube-like waveguides. The cylindrical structure consisted of a central dielectric rod, placed inside a coaxial plasma layer with a metallic helix sheath on its outer interface, and a metal screen separated from the helix by another dielectric layer. Two cases were considered. First, the plasma particle density was sufficiently large, so that the electron plasma frequency was larger than the electron cyclotron frequency [27]. Second, the external axial static magnetic field was sufficiently large, so that the electron cyclotron frequency was higher than the electron plasma frequency [28]. The introduction of the internal dielectric rod (e.g. as helix support in a travelling-wave-tube (TWT)) was shown to cause the appearance of an additional interface dielectric-plasma, which in turn gave rise to the existence of one more solution of the dispersion relation for the SFMs. This could be of practical interest for designing plasma electronic coaxial TWTs for wide-band communication systems. The introduction of the cylindrical metallic helix sheath was shown to result in coupling of SFOMs and SFXMs. In this case, the zeroth radial modes of SFOMs play the role of driven waves. The coupling was shown to cause the appearance of one solution more of the dispersion relation of SFMs with positive azimuthal mode indices. Since SFMs do not transfer their energy in axial direction, their excitation should be considered as an undesired parasitic phenomenon in such coaxial TWTs. Therefore, the analysis provided in [27, 28] can be useful for avoiding parasitic energy losses and suppression of parasitic waves in plasma loaded coaxial TWTs, operating in the GHz-range of frequencies.

A surface-wave oscillator (SWO) was outlined to be one of the promising THz-wave sources based on excitation of a surface wave using an electron beam. Design and fabrication of a 200-GHz oversized SWO were reported in [29]. The surface wave was suggested to be formed by a periodic metal structure (corrugated cylindrical waveguide), and the frequency was determined by the geometry of the structure and the parameters of the electron beam. The field intensity of the surface wave rapidly decreased to zero with going away from the structured wall. Due to a manufacturing error of the corrugated waveguide, the frequency of the surface wave in the fabricated structure was reported to decrease to 189 GHz at the π-operating point in comparison with the 204 GHz design frequency. The effect of the error was evaluated by calculating the intensity distribution of the electric field of the surface wave and its growth rate due to the beam interaction. The SWO experiment was carried out using the fabricated waveguide. Intense radiation with a total output power of approximately 0.2 kW was generated with a 45-keV/60-A electron beam.

An efficient THz radiation scheme was reported in [30] which used a low-energy electron beam, including a single-electron bunch and a train of bunches, to excite cascaded gratings. The scheme combined the mechanisms of wakefield radiation and Smith-Purcell radiation by transforming a series of surface waves into coherent Smith-Purcell radiation. Compared with the conventional wakefield and Smith-Purcell radiation schemes, its radiation efficiency and power were shown to be remarkably higher. In addition, it changed the conventional incoherent Smith-Purcell radiation into coherent emission with continuously tunable frequency. Thus, the scheme offered a promising way of developing high power, coherent, and tunable terahertz radiation sources.

Rapid development of the electronics industry and wireless communications, application of electromagnetic (EM) waves in medical, communications, and military areas was outlined in [31] to cause EM pollution. Since excessive EM radiation threatens human health and causes damage to the central nervous system and immunity, the reduction of EM pollution was pointed out to become an urgent problem. The scattering of EM waves on target objects was suggested to be divided into specular and non-specular scattering. Specular scattering was noted to be effectively suppressed by shaping and coating technologies [31]. However, the contribution of non-specular scattering was found to become the primary concern. Non-specular scattering was reported to include edge diffraction, creeping wave scattering, and surface traveling wave (STW) scattering [32], with STW scattering being particularly dominant. Therefore, the suppression of STW scattering was stressed to become one of the essential ways to reduce non-specular scattering. Scattering and attenuation characteristics of STWs in an absorbing coating were reported in [33]. In particular, the angular location of the scattering peak and the EM loss of the STWs were calculated. The thickness of an absorbing coating was shown to be appropriate to achieve a large EM loss in a wide frequency bandwidth, and the magnetic loss was demonstrated to be dominant in the attenuation constant. Furthermore, the simulated results were verified by measurements with an uncertainty of less than 20%.

Uniaxial materials whose axial and tangential permittivities have opposite signs were introduced in [34] as those named as indefinite or hyperbolic media. In such materials, light propagation was found to be unusual leading to novel and often non-intuitive optical phenomena. As a natural hyperbolic material, hexagonal boron nitride (*hBN*) was shown to support hyperbolic phonon polaritons (HPPs) being potentially used to investigate exotic photonic properties including sub-diffractional focusing, spontaneous emission (SE), light steering, and thermal radiation. In [35], hyperbolic plasmon–phonon polaritons inside the two reststrahlen (RS) bands of *hBN* and surface plasmon–phonon polaritons outside the two RS bands were reported to be achieved, both of which could be actively tuned by adjusting the $Ge_2Sb_2Te_5$ (*GST*) phase state and graphene chemical potential. These characteristics might provide an available route to realize active tunability of both phonon dispersion and spontaneous emission (SE) enhancement of natural or artificial hyperbolic materials. The result was achieved due to application of multilayer graphene and *hBN* combined with the *GST* functional substrate. *GST* phase change was found to have a profound impact on hyperbolic phonon polaritons in the *hBN* layer, and an actively tunable SE was obtained.

On-chip manipulating and controlling the temporal and spatial evolution of light were outlined in [36] to be of crucial importance for information processing in future planar integrated nano-photonics. The spin and orbital angular momentum of light, which could be treated independently in classical macroscopic geometrical optics, appeared to be coupled on sub-wavelength scales. Spin–orbit interactions in a plasmonic achiral nano-coupler were used in [36] to unidirectionally excite surface plasmon polariton modes propagating in seamlessly integrated plasmonic slotted waveguides. The spin-dependent flow of light in the proposed nano-photonic circuit allowed on-chip electrical detection of the spin state of incident photons by integrating two *Ge*-based plasmonic-waveguide photodetectors. Consequently, the device served as a compact ($\sim 6 \times 18\ \mu m^2$) electrical sensor for photonic spin Hall dynamics.

The ability to manipulate light propagation at the nanoscale was outlined in [37] to be of vital importance for future integrated photonic circuits. The high contrast in the optical properties of the phase change material *GST* was exploited to control the propagation of surface plasmon polaritons at an Au/SiO_2 interface. Using grating couplers, normally incident light at $\lambda = 1.55\ \mu m$ was converted into propagating surface plasmons on an *Au* waveguide. Single laser pulses ($\lambda = 975$ nm) were applied to a thin film of *GST* placed on top of the device, which, by transition from its amorphous to crystalline structural phase, dramatically increased both its refractive index and absorption coefficient, thus inhibiting propagation of the plasmonic mode.

This phenomenon was investigated for different interaction lengths between the phase change material and the *Au* waveguide. Contrast values in the transmitted intensity up to several tens of percent were demonstrated.

Solar energy was outlined in [38] to be widely recognized as a clean and renewable energy source. Solar energy was pointed out to remain in the focus of extensive investigation due to the growing energy demand and environmental pollution problems. Thermionic converters were designed to generate electricity from thermal energy in a power cycle based on vacuum emission of electrons. On one hand this cycle was found to be thermodynamically efficient. On the other hand, its practical implementation was noted to be limited by the extreme temperatures required for electron emission (>1500 K). Recently, a plasmonic thermionic energy converter (PTEC) was suggested [38] to overcome the challenges concerning the high operation temperature. The cathode of the PTEC employed a metal nanostructure, which supported local surface plasmons and provided efficient coupling for incident light to the nanoscale surface. Subsequently, the nanoscale confinement of surface plasmons created highly energetic electrons through non-radiative decay and they were referred to as plasmon-induced non-thermalized hot electrons. In this case, the high temperatures required for efficient vacuum emission could be maintained in a subpopulation of photo-excited "hot" electrons during steady-state optical illumination, whereas the lattice temperature of the metal remained within the range of thermal stability, below 600 K. Thermionic devices constructed from optimized plasmonic absorbers showed performance that could outcompete other strategies of concentrated solar power conversion in terms of efficiency and thermal stability.

Superlens slabs were outlined in [39] to rely on the coherent superposition of multiply reflected passing waves to amplify and restore the fine details of an object at the image plane. Similar interactions with external components were found to introduce resonances outside of the superlens, if a superlens slab is placed in close vicinity to a source object and image detector. The driving mechanism for evanescent-wave amplification within a superlens slab was the resonant coupling of surface waves between slab interfaces. It was shown both analytically and numerically in [39] that the improved performance of the μ-negative lens was caused by the existence of a resonance within the free-space gap formed by the object-lens spacing. The coherent superposition of multiple reflections within this free-space cavity excited a surface plasmon along the incident lens interface that was a factor of n^2 stronger for the μ-negative lens than for the ε-negative lens (where n was the refraction index of the dielectric half-spaces) and provided a high μ-negative lens transmission. Account for all interactions within an object-lens-detector imaging system revealed a second mechanism for evanescent-wave amplification. This was explained by the existence of two independent cavity resonances that could excite surface plasmons on the single negative lens and enable passing-wave amplification: one in the free-space object-lens gap and one in the single-negative slab itself. The gap resonance was observed to be much stronger in the μ-negative imaging system than in the ε-negative imaging system, resulting in a transmission response that was less sensitive to lens thickness and spatial frequency and could, therefore, operate at larger imaging distances.

The interaction of an intense laser pulse with an over-dense plasma, possessing a sharp density gradient, was outlined in [40] to result in acceleration of charged particles to relativistic velocities. The irradiation of structured targets, such as periodic grooves (gratings) on a metal surface, by ultra-short laser pulses, was pointed out in [41] to be of particular interest for generating intense surface plasma waves (SPWs), which could store the laser energy and efficiently accelerate electrons. Ultra-short high-power lasers were found to deliver extreme light intensities ($\geq 10^{20}$ W/cm^2 and ≤ 30 fs) and drive large amplitude SPWs at over-dense plasma surfaces. The resulting current of energetic electrons was noted to be of a great interest for applications, potentially scaling with the laser amplitude, provided that the laser–plasma transfer to the accelerated particles mediated by SPWs at ultra-high intensity was still efficient. The best conditions for SPW excitation were identified in [42] by means of particle-in-cell (PIC) simulations. A strong correlation between the angle of optimum surface plasma wave excitation and the incidence angle of the laser beam, applied for optimizing the electron acceleration along the plasma surface, was shown. An adjustment of plasma density and plasma surface shape with the objective to push the limit of surface plasma wave excitation to higher laser intensity was also discussed. Those results opened the way to new experiments on forthcoming multi-petawatt laser systems.

Two possible mechanisms are considered in the present book, which can be applied for surface flute wave (SFW) excitation, namely by application of charged particle beams or external alternating electric fields. The references discussed in this Section are devoted, in particular, to excitation of bulk waveguide waves by gyrating electron beams. As it is indicated at the beginning of the present Section, plasma electronics is closely related to vacuum electronics. Thus, the papers [19–21], especially their parts concerned with application of electron beams gyrating along large Larmor orbits, which encircled the axis of the employed waveguide, can be considered as a definite confirmation of the possibility to utilize such type of electron beams for excitation of SFWs considered in the present book. The possibility to inject an electron beam into the gap between plasma column and metal wall of the waveguide (see, Fig. 2.1, where the region indicated as "dielectric" can be changed by "gyrating electron beam"), can be applied for excitation of SFWs.

One should remind that the radial distribution of the SFW field is described by Bessel functions of the first and second kinds (like any bulk wave) just in the vacuum region, and only out of this region (namely in the region of the plasma column) their field is described by modified Bessel functions which is typical for surface waves. Moreover, a gaseous plasma column can be exchanged by a semi-conducting cylinder, then (taking into account that the density of such a uniform and stable plasma is larger than 10^{14} cm^{-3}) one can generate electromagnetic waves in the THz frequency range. The given references prove the validity of the results presented here on beam excitation of SFWs.

11.2 Surface Wave Applications for Plasma-Antenna Systems

In recent years, surface waves (SWs) have obtained a new field of application, namely plasma-antenna systems. This new class of antennas is based on utilization of a plasma as medium with sufficiently high conductivity instead of metal wires and/or metal bodies. Taking into account the possibility to be quickly switched-off and/or switched-on, these antennas are considered first of all, for military application. However, they can be applied as well for civilian needs, mainly for development of technologies for wireless communication and for special selective shielding reflectors (see e.g. [43, 44]).

One of the first serious physical papers devoted to problems of plasma antenna design, which was published in a well-acknowledged journal, was article [45]. It contained a report on results of measurements of efficiencies up to 50% and radiation patterns for a single cylindrical plasma antenna. The authors proved that the distribution of electric current along the plasma antenna could be controlled by the plasma density, therefore, plasma columns could be used instead of metal elements in communication antennas. In their previous experiments, the plasma was produced by electrodes located at opposite ends of a dielectric tube, which contained a plasma column such as in a mains-driven fluorescent tube. This technique was reported to have some disadvantages, the main of them were the complexity of this construction and many possibilities to disturb the plasma antenna radiating properties, which could increase its radar cross-section.

In paper [46], it was proposed to drive a non-magnetized plasma, sustained in the discharge from one end of the discharge tube, by excitation of an axially symmetric SW. In this case there was no need to apply two electrodes, which was a great advantage of the proposed plasma antenna design. However, at the regime of low plasma density, the SW was found to become heavily damped and have a shorter wavelength than current oscillations on a metal wire antenna. Generally, the authors concluded that the efficiency of the proposed plasma antenna design was high sufficiently for RF communications.

Experiments performed by the authors of [47] confirmed that propagation of SWs, which sustained a plasma column, could replace a metal rod as the guiding element in RF antennas. Moreover, such plasma antennas would be of low radar detectability and would have negligible mutual coupling in the de-energized state. Their experiments confirmed that the efficiencies of plasma antennas and their noise level would be approximately the same as compared with the case of a corresponding conventional metal antenna. It was also shown that the relatively high efficiency of the SW driven plasma column could be predicted by a simple theoretical analysis.

In addition to such an important feature as a low radar cross-section, the proposed antenna system, which applied a plasma column as the conductor in an RF antenna, was reconfigurable with respect to shape, frequency, bandwidth and directivity. The authors concluded that cylindrical plasma discharge tubes could be applied as antenna array elements, whose weight, required space and cost were less than that for any metal construction. Using the idea of [46], the authors elaborated plasma antennas for the frequency range 3–300 MHz, which applied excited SWs. The technical advantages of this construction of the plasma antenna were as follows: no electrodes as sources of plasma contamination; simplification of the antenna design as compared with both dual electrode and laser driven sources; minimization of the antenna detectability by radar and complete elimination of the mutual coupling for the de-energized antenna. Plasma antennas with such construction were maintained by about one fifth of the power required for an equivalent fluorescent lighting tube. The authors indicated that the measured efficiency of the proposed plasma antenna construction was in good agreement with theoretical calculations. The SW-driven plasma antenna produced sufficiently low noise, which made it prospective for application in the field of broadcast communications.

It should be mentioned that the scientific team under the leadership of T. Anderson, President and I. Alexeff, Chief Scientist has performed an intensive investigation in the area of plasma antennas design, construction and application. The team reached indisputable achievements in this area of plasma science. In the paper [48], the authors demonstrated results of their experiments with various plasma antenna systems. Their activity was focused to the following main directions:

1. *Expansion of the working frequency range*. Their antenna systems could operate in a wide frequency range (500 MHz-20 GHz), which made them useful for many military purposes and wireless communication.
2. *Making the plasma antenna invisible.* It was shown that if the external electrical supply was switched-off then the discharge stopped and the plasma antenna reverted to a state of an ordinary dielectric tube, which only had a small radar scattering cross-section and was practically invisible for radars.
3. *Protection from electronic warfare*. The authors demonstrated that their antenna system could be so configured that electronic-warfare signals with high-frequency passed through the antenna without appreciable interaction, while it transmitted or received signals at a lower frequency.
4. *Operation in the regime of shielding.* Creation of high-density plasmas in the discharge tubes of the designed plasma antenna systems allowed one to apply them as reflector. When it was placed over a receiving horn and was switch-on, it prevented an unwanted electromagnetic signal from entering. When the plasma antenna system was switched-off, the signal passed through freely.
5. *Improvement of mechanical robustness of the antenna system.* The authors developed two kinds of robust plasma antennas. In the first design, glass tubes intended for plasma confinement were encapsulated in a dielectric block. In the second version, the plasma antennas were composed of flexible plastic tubes (but they could be cooled additionally).
6. *Reduction of electromagnetic noise of the antenna system.* It was found that the noise generated by a plasma antenna was not a difficult problem. In the present case the level of this noise was comparable to the noise of a copper wire antenna of the same configuration. However, the authors planned to realize some additional techniques to reduce the noise level in future.

7. *Alternative design of plasma antenna system.* The authors presented as well results concerning the elaboration of a coaxial plasma waveguide consisting of an outer metal shell and the plasma discharge tube, which played the role of the inner conductor. This device could be applied as a switch. The advantages of the proposed construction were that the metal shell of such a waveguide did not transmit electromagnetic signals if the plasma was energized, and this system had a small cross-section for radar signals.

As compared with the results published in the previous work the mentioned team succeeded enhancing of the plasma antenna efficiency for transmission and reception of electromagnetic signals. In [49] they concluded that plasma antennas could work as well as metal antennas. Their results concerning plasma antennas, which were operated in pulsed regimes, were as follows: demonstration of transmission and reception of signals in noncurrent-carrying plasmas; no nonlinear effects were observed in high-power regimes so far. Experiments devoted to the problem of noise in plasma antennas showed that in the most cases, noise was a few dB higher than that for metal antennas, which was always used for reference. Interesting results were obtained for solving the problem of the ignition of the plasma antenna influenced by a pulsed RF field. The authors concluded that the plasma might or might not ignite in an intense RF field, because first of all plasma start-up depended on the design of the plasma antenna.

Experiments performed in [50] were devoted to investigation of plasma antenna properties for the cases of different plasma antenna design. The main method of the paper was testing various configurations of the plasma column, which presented the main element of the antenna system. In the initial configuration, the plasma antenna consisted of a 30 cm long plasma column, which was excited by SWs. Two methods of increasing the efficiency of plasma antenna operation were considered: first, by increasing the antenna length, and second, by using antenna arrays. The requirement for huge power supplies, very long glass tubes and definite additional technical design intended on enhancing robustness of the antenna system to be taken into account for construction of very long plasma antennas was emphasized. Thus, application of antenna systems with long plasma columns was concluded to be a difficult engineering task. In this case, a phased array plasma antenna was suggested to be the best alternative to increase the antenna efficiency. Because of the necessity to calculate the complex geometry of the antenna elements and to sustain definite phase relations between these elements, conventional methods for the construction of an array plasma antenna were recognized, certainly, to be also a difficult task. However, for a linear array the authors [50] found a relatively simple way to avoid these difficulties based on the possibility to maintain the gas discharge volume as collection of volumes with spatially periodical plasma densities.

It was found that by variation of the supplying operating parameters, namely, type and operating pressure of the working gas, value of the input power and its frequency, radius and length of glass tube, a single linear plasma antenna, which consisted of single plasma column, could be transformed into multiple small plasma antenna elements. The authors also found a possibility to control the number, length, and separation between two plasma antenna elements. The performed experiments allowed them to investigate the efficiency of antenna operation, the radiation patterns as well as the directivity of the radiation. The obtained results confirmed the assumption that this configuration of plasma antenna could be treated as a phased array vertical linear plasma antenna, which produced a radiation pattern with higher directivity as compared with a single plasma antenna [50]. This study revealed the advantages of a plasma antenna over the conventional metal antenna due to its re-configurability and directivity of its electromagnetic radiation.

The paper [51] was devoted to consideration of a novel reconfigurable antenna concept based on the dynamic excitation of metal-like conductive plasma channels in high-resistivity silicon, which were activated by the injection of a DC current. These plasma channels were precisely determined and formed due to the high-resolution technology developed for fabrication of silicon wafers. The proposed reconfigurable plasma antennas enabled to realize regimes of frequency hopping, to perform beam shaping and to provide steering of the beams without utilization of a complex and high-cost external RF supply. The authors hoped that this concept of plasma antenna would allow to realize phased arrays at reduced cost as compared with conventional antenna systems. However, they considered challenges such as the nonlinear features of the P-I-N semiconductor junction and the needed increase of the antenna efficiency. Thus, further investigations were required.

According to the authors' opinion their novel approach, which applied P-I-N semiconductors junction, required much lower power levels, less expensive power sources and much more direct interface to the external feed supply as compared with optical techniques. The feasibility of the proposed antenna design depended on the possibility of creating sufficiently conductive patterns on semiconductor wafers. It was demonstrated that the carriers of electrical charge in semiconductors formed a plasma, which caused the semiconductor to behave like a metallic medium. The authors believed that future improvements of the proposed design of the plasma antenna would prove its feasibility, adaptivity and re-configurability.

Paper [52] contained results of experimental and theoretical studies, which were devoted to the examination of plasma antenna excitation applying an argon discharge sustained by SW propagation. The operation parameters of the device were as follows: working frequency 500 MHz, RF power up to 120 W and working gas pressures 0.03–0.5 mbar. The experimental observations showed that the length of the plasma column increased with the square root of the applied power and its density decreased approximately linearly from the wave launcher to the end of the plasma column. This confirmed the conclusions of a global theoretical model. The noise generated by the plasma discharge was measured for the operating frequencies 10–250 MHz and its excess level was found to be 17.2 ± 1.0 dB above 290 K, which corresponds to an Ohmic thermal noise at 1.4 ± 0.3 eV. Theoretical and experimental studies of the plasma antenna radiation pattern allowed the authors to conclude that the linear variation of the conductivity and the finite value of the column resistance led to an increase in the width of the main lobe.

According to the authors' opinion the proposed plasma antenna design had the following advantages over conventional metal antenna systems:

1. Taking into account that the plasma antenna was energized for a very short time when communication took place one could conclude that it was suitable for military HF communications. Since the tube operated by SW propagation under the regime of single-ended excitation it would be difficult to be detected by hostile radar.
2. Since plasma antenna arrays could be quickly reconfigured without excitation of perturbations from switched-off plasma elements, switched-off plasma antennas would not couple unwanted high-power signals into nearby receivers. Thus, they could be applied in cases, which required utilization of different antennas for several communication systems, located in close proximity. Hence it simplified, e.g. the layout of ship-borne maritime communication systems.
3. Since plasma elements of the designed antenna could be energized and de-energized in microseconds, fast transmission of data at a low-frequency carrier could be switched at the end of each bit, thus minimizing signal degradation by antenna ringing.
4. Since the effective length of the plasma antenna could be varied with the help of the applied RF power, this plasma antenna design allowed one to perform its reconfiguration for different transmitting frequencies.

In the case of completely plasma filled discharge tubes, SWs were experimentally shown to be reflected from the end of the tube. They deposited more energy into the top part of the plasma. Thus, one could obtain a more uniform distribution of the plasma along the column. High density of the discharge plasma was required in order to maximize the efficiency of the plasma antenna and to minimize the plasma noise. The authors found that at lower frequencies the noise of plasma antennas increased rapidly. They explained this fact by possible oscillations of plasma ions.

Ionospheric clutter was found in [53] to be one of the main clutters degrading the performance of HF surface wave radar (HFSWR). Thus, its suppression was noted to be an important problem for improvement of radar operation. A new orthogonal projection method, which allowed one to enhance the HFSWR capability for target detection sufficiently, was proposed in [53]. Thus, it became possible to detect vessels on the ocean surface confidently at distances up to a few hundred kilometers. Occurrence of ionospheric clutter was explained by the existence of some portion of vertically polarized HF electromagnetic waves, emitted by the HFSWR, which was found to propagate not along the ocean surface but upward, then being reflected from the ionospheric zone. Then electronic interference was reported to occur on the radar display as a consequence of Doppler shifting and widening of this signal. The proposed method for echo analysis allowed one to avoid installation of any auxiliary electronic element in the antenna system and to provide sufficient enhancement of the target detection capability even in the case of strong ionospheric clutter [53]. All the obtained conclusions were confirmed experimentally.

Book [44] was devoted to problems of practical utilization of SW propagation in gaseous plasmas, sustained in dielectric discharge tubes. The attention was paid mainly to the following possible applications of the SWs: plasma antennas and transmission lines, plasma reflectors and lenses, and plasma radiation. The fifth Chapter of the book dealt with operation of plasma filled microwave devices such as: filters and phase shifters. To obtain a plasma in a special dielectric tube the author proposed to ionize a neutral gas. As a rule, it was one of chemically inert gases like helium, neon, argon, xenon and krypton; mercury vapor could be applied as well. To perform the ionization process, one could either utilize a DC current flowing between two or more electrodes, which supplied a discharge tube, or apply SW propagation.

Using the simple model of a cold non-magnetized plasma the creation of a special "critical surface" was shown to be possible, where the plasma density had such a large value that the frequency of the studied signal was less than the electron plasma frequency. This allowed one to obtain a medium with negative dielectric permittivity, which would not permit this signal to pass through this surface. This physical phenomenon was the base for plasma reflector operation. So, it was clear that application of a weak degree of ionization would allow one to decrease the electron plasma frequency so that the signal would pass the plasma layer. In addition, the plasma shape and its dielectric properties could be designed in such way that it would act as a lens system as well.

The main concept of plasma antennas consists in the application of coupling between plasma eigenwaves and electromagnetic waves, which can be radiated from a tube filled by this plasma. In the case studied in the present book, the surface flute X-modes are leaking modes. Thus, there is no necessity to establish coupling between them and electromagnetic waves. At least flowing of a weak electric current allows one to realize coupling of the flute X-mode with the flute O-modes, which are of bulk type and therefore can be easily radiated from the waveguide.

The idea of a plasma filter was explained in [44] by the plasma features to reflect or transmit electromagnetic signals depending on the relation between the signal frequency and the cut-off frequency of the plasma. Therefore, changing either plasma density (in the case of non-magnetized plasma) or the strength of an external magnetic field (in the case of magnetized plasma) was mentioned as a method to tune the operating frequency and change the frequency band of the device. Thus, book [44] provided the possibility for readers to get acquainted with a wide spectrum of practical applications of plasma antennas, using the simple model of a cold non-magnetized plasma without deep consideration of physical mechanisms of the studied phenomena.

More complete information about the elaboration of plasma antennas was presented in book [43]. It gave much more information than reference [44]. This was achieved not only by the presentation of exact mathematical solutions for scattered and reflected electromagnetic fields, far-field radiation patterns and the detailed description of the performed experimental work. New (compared with reference [44]) chapters devoted to plasma antennas nesting, windowing and steering were presented there. Such designs allowed one to elaborate plasma antenna systems operating in multi-frequency bands and multi-radiation lobe regimes, to obtain zero mutual inductance for extinguished antennas, which could not be obtained by application of P-I-N diodes in the case of conventional metal antennas.

A special chapter of reference [43] was devoted to discussion of the results published by Borg et al. [45] on studying the problem of decreasing noise of plasma antennas. It was indicated there that to decrease the noise level one could apply a higher operating frequency in the GHz range and a regime of low pressure of the operating gas filling of the discharge tube unlike the experimental conditions, which were realized in the experiments of [45]. Two new chapters were devoted to the elaboration of smart plasma antennas and satellite plasma antennas, which benefited from low thermal noise. It was mentioned there, that in the past, smart satellite antennas were applied for military purposes. However, these designs of plasma antennas were noted to be utilized nowadays for establishing GPS—free positioning, for reconfiguration of signal beam width and numbers of lobes of a transmitting signal, for enhancing the useful part of the detected signal from satellite due to application of focusing and steering of the microwave beams, etc. Thus book [43] gave the readers not only the ability to understand the present advantages of plasma antennas over conventional metallic ones, but also to assess the prospects for further development of plasma antenna systems.

The case of a plasma antenna modeled by a cylindrical dielectric tube filled with non-magnetized plasma was under consideration in [54]. The study was carried out numerically using the solution of the Maxwell equations and the kinetic Boltzmann equation. It was concluded that for an electromagnetic signal with a frequency of about 500 MHz used for ionization of neutral gas with a density of 10^{16} cm^{-3} one could expect to obtain the best impedance matching of the plasma antenna. Increasing of the frequency led to increasing of the antenna radiation efficiency.

Considerable attention was emphasized in [55] to be attracted in the past decades to elaboration of different plasma-like meta-materials, which could be applied in plasma electronics. Extensive studies were performed to design novel high-quality antennas using various meta-materials. In [55], the authors considered the possibility to apply meta-materials in order to shrink the size of patch antennas. Namely, a two-dimensional cavity formed by a double positive-epsilon negative bi-layer waveguide was considered and studied to evaluate the radiation capability of its corresponding patch antenna. Based on the cavity model analysis of patch antennas, it was shown that good efficiency for broadside radiation of such a cavity-based rectangular patch antenna could be achieved when one layer of the cavity was partially shielded by perfect electrical conductivity boundaries.

Paper [56] was devoted to some theoretical problems related to the radiation of plasma antennas. The authors analyzed two kinds of mechanisms of plasma antenna radiation. The first one was the radiation of an electric current excited by SWs, which was similar to the radiation of a metal antenna. In the second case, electromagnetic radiation was connected with excitation of some transmission modes. This mechanism was similar to the radiation of a dielectric antenna; it could be realized because the electromagnetic waves partly cannot satisfy the total reflection condition in the plasma antenna for some specific frequency bands. The radiation properties of plasma antennas were studied numerically. The design of the antenna was as follows: a cylindrical dielectric tube was filled with non-magnetized uniform plasma. According to the conclusion of the authors the provided results could be important for further experimental investigation of the construction of effectively controlled plasma antennas and plasma antenna arrays.

Experimental studying an electric dipole surrounded with a magnetized plasma shell [57] proved that the input impedance of such an antenna could be matched with a feeder in a wide frequency range near the lower hybrid resonance. Radiated power of the studied plasma-antenna system increased by almost three orders of magnitude compared with the case of a conventional metal dipole antenna. An antenna immersed into non-magnetized plasma had resonant properties at the operating angular frequency ω_0, whose value was in the vicinity of the electron plasma frequency $\omega_0 \le \Omega_e$. This was connected with vanishing of the antenna input reactance. In this case a surface wave could propagate along the plasma surface, and the plasma reactance had an inductive feature. However, the collisional frequency was found to be often approximately that one for the case of non-magnetized plasmas because of the only possibility to realize low plasma density in this case. Unlike in the case of magnetized plasma a negative value of the plasma permittivity was suggested to be obtained for operating frequencies below the plasma lower resonance.

The authors of reference [57] reported on results of experimentally studying a magnetized plasma antenna system operating in the frequency range 30 kHz < ω_0 <2 MHz. A metal dipole with $2\,l = 80$ cm immersed into a magnetized plasma was located in a glass tube with a length of 100 cm and a diameter of 16 cm. The working gas was argon under 10^{-3} Torr pressure. The magnetic field of about 10^{-2} T was produced by two Helmholtz coils and the maximum plasma density was 10^{13} cm^{-3}. It relaxed with the characteristic time constant 500 μs. Experiments [57] proved that such a magnetized plasma antenna system had matching properties for frequencies less than the lower hybrid resonance and that application of the magnetized plasma shell allowed one to increase the radiated power by almost three orders of magnitude compared with a conventional metal dipole antenna without plasma shell.

The experimental characterization of a plasma dipole in the ultra-high frequency (UHF) band was presented in [58]. The antenna consisted on two customized argon plasma sources disposed as the arms of a conventional metallic dipole. The commercial electrodes used to produce the plasma acted also as signal coupler. The plasma parameters were measured to compute the electric permittivity. Subsequently, the reflection coefficient and the antenna gain were derived through computer simulations and measurements in a portion of the spectrum centered at 0.435 GHz. In [58], the antenna was characterized (both at plasma on and off) in terms of magnitude of the reflection coefficient and gain pattern. When the plasma was on, the resonance frequency of the antenna was 0.435 GHz, with a gain of –4.2 dBi. This value was appreciably lower compared to a conventional dipole due to higher losses within the plasma medium than in metal. When the plasma was off, the gain dropped to −12.3 dBi. This reduction was suggested to be increased by reducing the size of the electrodes, thus further enhancing the capability of the antenna to electrically disappear when the plasma was de-energized. A moderate level of cross polarization, namely −13 dBi, was induced by the cables used for the plasma generation. A good agreement between measurements and numerical results was attained. Although the proof of concept presented in [58] was promising, further development was required to enhance the antenna performance by increasing the plasma density. The electrodes were also suggested to be optimized in terms of shape and materials. Their size should be minimized while increasing the maximum sustainable current (e.g., by tuning figures of merit such as the secondary electron emission yield). Other gases were proposed to be employed for plasma production. Krypton and xenon were potential candidates due to their reduced ionization energy in respect to argon. The efficiency of the power processing unit was suggested to be further increased to minimize the electrical power budget required to operate the plasma dipole antenna.

A traditional tapered slot antenna, well known for its ultra-wide bandwidth, but low gain was considered in [59]. A novel design for the antenna was proposed in [60] where the radiation characteristics were enhanced by adding a parasitic patch that supported spoof surface plasmon polaritons (SSPPs). The energy in the flared slot could excite the odd-mode surface wave on the patch structure. SSPPs were found to be beneficial for the gain and directivity without affecting the bandwidth. The maximum gain improvement was about 5 dBi at 14 GHz. The measured results verified the enhanced performance. Thus, the paper provided an effective method to enhance the radiation performances of tapered slot antennas, which would be promising in future ultra-wideband systems.

A novel antenna approach was suggested in Ref. [61], which applied a high-speed plasma as a conducting channel. The behavior of an antenna with time-varying conductivity was presented. The conducting channel of the antenna was suggested to consist of a series of individually-controlled plasma cells. This antenna was assumed to be fed with pulse-amplitude modulated signals. The pulses in these signals would travel down the length of the antenna and would be attenuated near the antenna tip. This time-domain matching technique was suggested to make efficient matching possible at all frequencies. Plasma dispersion was included in the time-varying antenna scheme modelling utilizing a modified version of Telegrapher's equations. Theoretical plasma parameters were incorporated to develop requirements for plasma density and switching speed. The time-varying antenna scheme was shown to radiate more efficiently at low frequencies than existing electrically short antenna designs. It was noted that the behavior of a plasma transmission line could differ significantly from the behavior of a typical transmission line. These effects were especially noticeable for plasma columns with low electron density. A travelling surface wave supporting a propagating impulse was the primary mode examined in [61].

A band-pass filter (BPF) with a notch band was designed based on high-order mode of spoof surface plasmon polaritons (SSPPs) in [62]. The high-efficiency passband was developed from the first high-order mode (Mode 1) of SSPPs. Compared with the fundamental mode having low-pass response, the Mode 1 was reported to have the inherent band-pass response. Thus, the BPF generated by Mode 1 did not need to add an additional high-pass filtering structure and employed a gradient transition structure for excitation. The Mode 1 of SSPPs was suggested to be efficiently excited just with a simple trapezoidal transition converting the signals from a microstrip line to the SSPP waveguide. Moreover, by embedding a U-shaped resonator into the SSPP structure, a BPF with a notch band was constructed, where the location of the notch band was red-shifted when the length of resonator was increased. The measurement data of two fabricated prototypes agreed with the simulated ones, which verified the feasibility of the design.

The existence of an electromagnetic SW over a laminar ice-salt water structure was confirmed experimentally in [63]. The appearance of such an electromagnetic SW was found to influence on ice-sea radio paths [64]. Software for calculating the spatial and energy characteristics of Arctic radio systems was elaborated in [65]. Preliminary theoretical studies [66] of the propagation of radio waves over impedance media showed that electromagnetic fields over layered media with highly inductive surface impedance could achieve a large electrodynamic field increase relative to an infinitely conducting plane. The model of "ice-sea" layered medium representing a two-layer structure of the "dielectric on a conductor" type was applied to study surface wave propagation in the ice-covered water area of the Arctic Ocean. The novelty of the approach was in the substantiation of engineering methods for calculating the propagation of very low frequency, low frequency and medium frequency radio waves over a layered-inhomogeneous "ice-sea" medium using the Arctic Ocean ice cover maps when calculating Arctic radio lines.

Aiming at suppressing the scattering of electromagnetic SWs propagating on rough surface, the theoretical design and experimental realization of invisibility cloaks for SWs were presented in [67]. A non-magnetic transformation optics method was used. Near-unitary broadband transmission of electromagnetic SWs on rough surface was experimentally realized in the microwave regime.

An underwater portable radio antenna operating in the 50 MHz band and efficient for launching electromagnetic SWs at the seawater/air interface was presented in [68]. The antenna operation was based on field enhancement at the antenna tip and on an impedance-matching antenna enclosure, which was filled with deionized water. This enclosure made it possible to reduce the antenna dimensions and to improve the coupling of electromagnetic energy to the surrounding salt water medium. Due to the fact that the SW propagation length is significantly larger than the skin depth of conventional radio waves at the same frequency, this technique was suggested to be useful for broadband underwater wireless communication over a distance of several meters.

A versatile platform for manipulating the launching of SWs by engineering their phase profiles with phase discontinuities was developed in [69] by using orientation-controlled metallic apertures. Polarization-controllable anomalous SWs were experimentally launched and observed at THz frequencies. The proposed approach represented a significant step forward in developing SW-based plasmonic devices.

The experiments described in [57] can be considered as proves of the possibility to apply the generation of SFWs propagating around a cylindrical metal rod immersed into magnetized cold plasma for the development of plasma antenna systems.

The book by Ben Munk [70] provided a bright example of the crucial change of the properties of an object caused by the finiteness of its dimensions. The finiteness of the dimensions of an antenna array was reported to give rise to propagation of electromagnetic SWs on the array. The author outlined two ways in which this SWs could manifest itself. First, the antenna array was considered to be used as a frequency selective surface (FSS). In this case, the SW was found to lead to a significant increase in the bistatic scattering. In particular, a sizeable increase in the radar cross-section of objects was observed. Second, the structure was suggested to be used as a phased array. In this case, the presence of SWs was found to lead to dramatic variations of the terminal or scan impedance from column to column. Under these conditions, it would be very difficult to design a high-quality matching network, in particular, due to significant change of the maxima and minima of the scan impedance which move significantly with frequency and scan angle.

The finite antenna arrays considered in [70] did not operate with any plasma. In this respect this book is far from the contents of the present monograph. On the other hand, one can see direct analogy of the physical phenomena presented in [70] with that studied in the present book, whose main subject is electromagnetic SFWs. As it is already mentioned in Chap. 1, SFWs do not propagate in infinite uniform plasmas, as the surface wave studied in [70], which cannot be observed in infinite arrays. However, the presence of the plasma interface is an unavoidable feature of all the experimental and technological plasma devices, as real antenna arrays are always finite. Just the presence of the interface provides the conditions for SFW propagation. Finally, the present monograph convincingly confirms that its authors belong "*To the increasing number of my friends who realize that computer power is a supplement, not a substitute, for brain power*" (from epigraph to [70]).

11.3 Surface Waves Propagating in the Plasma Periphery of Magnetic Confinement Fusion Devices

The review [71] was devoted to a wide range of problems of ion cyclotron resonance heating (ICRH) and other applications of IC waves in magnetic confinement thermonuclear fusion plasma devices, those problems to be solved for future application of the IC frequency wave technologies in the International Thermonuclear Experimental Reactor (ITER) being built in Cadarache, France. RF power was mentioned to be lost during RF wave penetration through the plasma periphery, because of parasitic absorption like collisional damping, torus wall heating, and non-resonant damping, etc. The theory of so-called coaxial modes [72], known to propagate in the coaxial cavity formed by the metallic wall of the toroidal chamber and the plasma column with the dispersion relation: $k_{\|}^2 = \omega^2/c^2 - k_{\varphi}^2$, with k_{φ} as poloidal wavenumber, was applied in [72]. The author tried to answer the question: what is the minimum value of $k_{\|}$ (parallel wavenumber) requested for optimal ICRH of ITER plasmas. Assuming that all the power launched in coaxial modes did not contribute to bulk plasma heating, the lower limit of the parallel wavenumber $k_{\|} = 0.03\ \text{m}^{-1}$ was estimated for ITER. The importance of studying SW propagation during ICRH in fusion plasma devices was already emphasized earlier in [73–76] where the authors concluded that just propagation of different SWs is the most probable reason of such undesirable processes in fusion plasmas as plasma periphery heating and generation of impurity ions with high atomic numbers.

Slow waves excited in JET and other tokamaks during ICRH regimes were proved in [77] to produce larger plasma particle momentum than fast waves. This allowed the authors to predict that these waves could be able to drive plasma non-inductive plasma currents. On the other hand, there existed some contradictions between the theory, which indicated RF driven rotation, and the preliminary experimental results. Therefore, both experimental and theoretical work were continued.

The main results in solving problems of IC wave application in large-scale fusion devices like JET with superconducting coils applied for creation of the confining magnetic field were overviewed in paper [78], which could be considered as a milestone in this area of research. The problem of efficient conditioning technique for the walls of the fusion device chamber was reported there as a key one to be solved to support continuous periodical operation of ITER. This technique would be able to reduce the impurity content in the torus chamber, to control the hydrogen isotopic ratio on the inner surface of the chamber walls, to mitigate the long-term tritium inventory build-up in the chamber wall and even to clean the chamber walls after some disruptions. This conditioning was emphasized to be performed under the condition of presence of the residual external constant toroidal magnetic field produced by the superconducting coils. That is why glow discharge conditioning (GDC) of fusion devices, which was frequently utilized earlier, was defined to be useless in ITER. The authors of [78] proposed the IC wall conditioning (ICWC) technique as an alternative to the GDC method. The ICRH antennas, which are already installed for RF plasma heating in tokamaks and stellarators, working in different countries, were assumed to be applied for this technique. Preliminary results obtained at JET and TEXTOR were encouraging and the authors estimated the ICWC technique as one of the most promising to be able to support ITER with routine inter-pulse or/and overnight conditioning. They saw the following tasks for ICWC: first, to ignite safely and reliably an RF discharge by the ICRH hard ware, already installed in the fusion chamber; second, to couple the main part of the RF power to low density ($\sim 10^{10} - 10^{13}$ cm^{-3}) plasmas at the presence of large toroidal magnetic fields (~5 T); third, to develop RF power absorption schemes intended for obtaining a more uniform plasma discharge and for enhancing the conditioning effect.

In the opinion of the authors of the present book, the last task could be solved by application of discharges sustained by SWs. As it is known, see e.g. [46, 79], gas discharges sustained by various types of SWs have a lot of advantages as compared with discharges based on bulk wave propagation: SWs are slower than corresponding bulk waves, thus they can more effectively interact with plasma particles; the radial distribution of SW fields has the peculiarity that the maximum values are located nearby the plasma surface, therefore just the application of SWs will allow one to obtain a more uniform plasma discharge; excitation of SW eigenmodes is more efficient as compared with the case of bulk wave excitation.

Paper [80] was devoted to study *Ni* impurity generation in the JET plasma during operation of two different types of ICRH antennas. The authors stated that the quantity of metallic impurities in JET plasmas did not significantly affect the plasma heating in contrast to what was usually observed in fusion devices during additional ICRH. They considered two possibilities to decrease the impurity concentration in the fusion reactor without changing hard ware: to shift the plasma away from the ICRH antennas (which of course reduces the coupling of the IC waves into the plasma) or to increase the D_2 gas-puffing rate. Therefore, the most interesting result obtained there was the possibility to reduce the *Ni* concentration by changing the location of D_2 gas injection.

Paper [81] was devoted to studying a flute instability in the divertor region of a tandem mirror machine. The divertor was noted to be a compulsory element of large fusion devices intended for removing the helium ash produced by the reaction of fuel combustion in a thermonuclear tritium-deuterium reactor and the various impurities generated in the magnetically confined plasma due to active interaction between the plasma and the first wall of the reactor vessel. The GAMMA 10 A-divertor was proposed by the authors as a candidate for the upgrade of tandem mirror fusion device Gamma 10 in Tsukuba, Japan. A flute instability was reported to be the most dangerous instability in this open magnetic plasma system. It was found to manifest itself by appearance of a small perturbation of the plasma volume on the surface of the confined plasma column, which moved azimuthally around the plasma column. The shape of the plasma column cross-section deviated from a circular one. The quantity of such flutes located on the surface of the confined plasma column was described by the azimuthal mode index m (see e.g. [82]). Then this small perturbation (for the case $m = 1$) became larger and larger and finally touched the first wall of the reactor. This could lead to destroying the coating of the first wall and even to a failure of the plasma discharge.

The main purpose for the installation a divertor mirror cell was to carry on the simulation experiments on the divertor intended for application in a large torus like LHD, JET and ITER. Traditional tandem mirrors such as GAMMA 10 contained a space region with minimum value of an external magnetic field in order to stabilize the flute instability. However, existence of such a region with minimum magnetic field enhanced the neoclassical transport. Thus, elaboration of a new type of divertor was intended for solving this difficult problem, namely to stabilize the flute instability without a strong enhancement of the neoclassical transport. The reduced MHD simulation performed by the authors allowed one to find an answer to the question: what was the influence of the divertor region in their new design on the flute instability? It was found that the divertor region played a positive role in stabilizing bulk flute modes.

In their next paper [83], the authors continued investigations of the stability of bulk flute modes in the GAMMA 10A-divertor. They carried out a stability analysis by application of three-dimensional treatment of the basic equations describing the fluctuation of bulk flute modes. The stability boundary of these modes was found to essentially depend on the radial profiles of plasma density and temperature in the axisymmetric divertor mirror. The linear growth rate of the instability of the bulk flute mode with azimuthal wave index $m = 1$ was calculated for various radial profiles of plasma density and temperature in the GAMMA 10 A-divertor. The modes were found to be stabilized by a special form of the applied magnetic field profile in an anchor mirror cell, while this stabilization was performed due to sufficient plasma compressibility. Changing the definition of the specific plasma volume of a magnetic field line applied in the bulk flute mode stability criterion did not lead to changing results compared with the previous definition of this specific parameter. Thus, the results of paper [83] revealed that the linear phase of the flute instability was the same as that in the previous works on this topic.

The last two references are not devoted to the case of surface wave propagation. However, they concern the propagation of bulk flute modes and their stability. Thus, they also confirm the importance of studying surface flute eigenmodes, whose theory is presented in the previous chapters of book.

Moreover, not only an interface of two media but also a plasma filament appeared to produce conditions for propagation of electromagnetic waves of surface type. For example, density filaments (blobs) were reported in [84] to occur in experimental magnetic confinement devices. The presence of the filament interface provided the possibility for electromagnetic waves of surface type to propagate along the filaments. The waves were predicted to be common in NSTX and to carry RF power in the parallel direction away from the ICRF antenna. The waves could be excited by the ICRF waves incident upon the filament in front of the antenna and then continue propagating along the filament even far away from the antenna, where there were no incident RF waves. This explanation was consistent with the observed occurrence of this parallel transport on NSTX but not on other machines and was consistent with the experimentally observed increased power loss for low $k_{\parallel}$ of the launched ICRF waves. It also naturally explained the observed origin of the parallel RF power flow from the scrape-off layer in front of the antenna rather than from the antenna surface itself. This made the surface type waves a perfect candidate for the physical mechanism behind the experimentally observed power losses in the NSTX edge plasma [85, 86].

For numerical or technical reasons, a multilayer plasma particle density profile is often preferred instead of a continuous one in simulating fusion plasmas. In this case the number of layers has to be chosen as large as possible to avoid strong resonances and surface wave effects on the power spectrum and coupling strength (see e.g. [87]). Thus, in these studies SWs play a parasitic role, and have to be avoided. The method of modeling the plasma particle density with an inhomogeneous radial profile by a set of uniform plasma layers is successfully applied in Sect. 5.1 of the present monograph for numerical studying the dispersion properties of SFXMs.

Undesirable excitation of surface electromagnetic waves by RF antennas was reported in [88] to be expected in the edge of the plasma of DEMO and ITER. Application of the semi-analytic coupling code ANTITER II made it possible to predict SW propagation between the wall and the Alfvén resonance region using the designed antenna and a reference low coupling density profile of ITER in the modeling. The conditions were derived for which excitation of these modes could be strongly reduced. These conditions were satisfied by the triple-strap antenna of ASDEX Upgrade that allowed successful ICRH operation in a W-coated machine. They would also be achieved for the $(0\pi\pi0)$ phasing of the four columns of two poloidal triplets of radiating straps constituting the future ITER antenna array. Ways to further optimize the cancellation of these SWs were also investigated. However, the model applied was still in plane geometry.

Production of a poloidal component of the confining magnetic field by the Ohmic current in tokamaks is a principal feature of current-carrying fusion plasma models [89]. As a result, the magnetic field in tokamaks has two components (toroidal and poloidal) and become helically twisted (shear). Therefore, studying the influence of steady-state current on wave propagation and dissipation in current-carrying plasmas was emphasized to be important. Therefore, the spectra of fast and slow magnetosonic waves in a current-carrying plasma were determined. Doing that, the authors applied the expressions for the plasma permittivity tensor ε_{jk} with all nine non-zero components and successfully overcame the difficulties associated with this issue. However, another approach is carefully described in Chap. 4 of the present monograph [90, 91]. The latter consists in application of the local coordinate system associated with the external static magnetic field. This makes it possible to use the tensor $\varepsilon_{\mathrm{jk}}$ of the cold collisionless plasma with only five components in Ampère-Maxwell's circuital law, which significantly simplifies the algebra. From this point of view, the present monograph is of additional interest since it provides not only valuable physical results but also explains many research tools to be applied in studying electrodynamics of spatially inhomogeneous plasma. Worth to note is, that the influence of non-zero azimuthal components of the external static magnetic field on the spectra of fast magnetosonic waves in current-carrying cylindrical plasma waveguide was studied by Stepanov and Girka in 1990 [92] in the framework of the method of successive approximations. The investigations in [89] were carried out with neglecting the influence of toroidicity. The influence consists in the Shafranov shift of the magnetic axis and toroidal inhomogeneity of the confining magnetic field. The effect of both phenomena on the dispersion properties of SFXWs in a toroidal chamber is examined in Sects. 6.3 and 5.3 of the present monograph, respectively. Influence of the toroidicity and ellipticity of the plasma column on the spectra of fast magnetosonic and Alfvén waves was studied in [93]. Moreover, the approach with application of the local coordinate system is successfully used for studying the spectra of SFXWs in axially inhomogeneous external static magnetic field, which is typical for both tokamaks and stellarators, in Sect. 5.5. Results of investigating the dispersion properties of fast magnetosonic and Alfvén waves in cylindrical plasma waveguides with both weakly rippled and helical external static magnetic fields (with three non-zero components in cylindrical coordinates) were reported in [94, 95] and [96], respectively.

An important role of ICRH systems in fusion-relevant plasma machines was explained in [97] by their flexibility and relatively low cost. Compared with other auxiliary heating and current drive methods, the engineering challenges of ICRH systems were considered as relatively low. However, a significant physics challenge was emphasized to remain: understanding and mitigating edge plasma interactions which were frequently observed in high power ICRF experiments [97]. ICRH specific edge plasma interactions were found to be evidently associated with the development of RF sheaths. Formation of RF sheaths was observed there, where the plasma, strong RF wave fields and material surfaces coexisted. They resulted in enhanced plasma potentials relative to the wall, and therefore they increased the energy with which ions stroked the surface and sputtered impurities. RF sheaths were reported also to lead to enhanced plasma convection and surface power deposition (see e.g. [98]) which might be localized and cause material damage or enhanced first wall and antenna erosion. RF sheaths were suggested in [99] to be modeled by employing a sheath boundary condition in place of the more customary conducting wall boundary condition [99]. A method for post-processing results obtained with the conducting wall boundary condition was presented. Under certain assumptions, the post-processing method produced results that were equivalent to those obtained with the RF sheath boundary condition.

11.4 Surface Wave Applications in Nano-Technologies

Review [100] was devoted to a wide spectrum of technologies for metal plasma production, mainly to energetic condensation of metal plasmas, and to the formation of nanostructures using metal plasmas. The plasma production techniques were based on application of ionizing metal vapor obtained in high frequency and strong current discharges. Special attention was paid by the author to different forms of ionized metal vapor deposition, namely magnetron sputtering, metal self-sputtering, high-power impulse magnetron sputtering, cathode arcs and pulsed laser ablation. Metal plasmas produced by these methods were already utilized in several industrial techniques, such as the deposition of ultra-thin layers in the computer and data storage industry and the coating of special tools. Examples of industrial spotless arc deposition of titanium with rates up to 1 μm/s were realized.

Taking into account the possibility of surface flute X-mode propagation along the metal interface, the authors of the present book hope that SFXMs could be useful to develop some technologies of solids processing, namely polishing and etching. Then by the aid of advanced diagnostics, deposition and processing techniques, metal plasmas can be used for fabrication of photonic and biomedical devices.

Production of nanostructures by metal plasma was studied in [100] as well, including both the possibility to use the metal plasma for conformal coating or for completely filling trenches on a substrate surface with the nanostructures, which were primary prepared by lithography. The possibility to apply metal plasmas for fabrication of self-organized nanocomposites (i.e. production of a matrix of amorphous material filled with nanocrystals) and nanoporous materials also was discussed. The initial material could be decomposed and then the deposition of the prepared metal plasma could be performed at the meta-stable phase and as a result one could obtain highly porous nano-structural materials. Metal plasma techniques were demonstrated in [100] to allow one to obtain a trench filling and a conformal coating of a substrate surface down to the 100 nm size. Application of metal plasmas already allowed one to fabricate super hard (H > 40 GPa) and tough nanostructures. For instance, the possibility to obtain high hardness of multilayer composite films by cathode arc plasma deposition of titanium–aluminium and titanium–zirconium was known for more than thirty years. However, more interesting and prospective results were reported in [100] to be achieved by the incorporation of ions of such unique metals as yttrium or vanadium to stable phases of the both *TiN* and *AlN*. During their further thermal treatment, they were found to improve the oxidation resistance and also sufficiently reduce the coefficient of friction of the nanostructures (about twice in the high temperature range: 850–950 °C).

The term surface localized plasmon (SLP) was reminded in [101] to mean collective oscillations of conductivity electrons in some metal nanostructure excited by electromagnetic waves with wavelength within the visible light spectrum. Due to the light wavelength prevailing over the nanostructure scale, the conductivity electrons, affected by the light, were found to oscillate nearby the nanostructure. Surface localized plasmons were demonstrated in [102] to be used for enhancement of forward light scattering into a photoactive layer without over-shadowing by strong absorption of the light at the plasmon resonance wavelength of the *Ag* nanoparticles incorporated into solar cells. Such a solution of the problem supplemented the US Patent Application [103]. In this case, SLP resonances were shown to be shifted to the low-frequency edge of the light spectrum by varying the dielectric properties (therefore, both the refractive and reflective indices changed their values) of this solar cell.

Nanoparticles located at the front surface of solar cells were found in [102] to perform trapping of the light at wavelengths, which were above the surface plasmon resonance. Nanoparticles located on the rear of the cell provided trapping of the rest portion of the incident light. Using nanocomposite technique, namely introducing nano-crystallites of *Ag* into the commonly utilized silicon solar cells was shown to allow one to produce large-scale photovoltaic devices.

Plasmonic materials and phenomena were reported in [104] as those been widely studied and applied in multiple fields for a long time. The engineering of biosensor devices, offering label-free and real-time analysis of bio-molecular interactions with excellent performances was noted there as one of the most promising applications. R. W. Wood was mentioned there as the first who noticed between 1900 and 1912 an optical phenomenon which became the physical base for surface plasmon resonance (SPR) systems [105, 106]. He observed an unusual dark and light pattern in the reflected light of a metal-backed diffraction grating, and although he hypothesized about light–metal interactions, no clear answer was provided and the phenomenon was termed Wood's anomalies. Along the following years, theoretical studies and analyses were performed by Rayleigh [107], Wood [108], and Palmer [109, 110]. However, it was Fano in 1941 who concluded that these anomalies were related to SWs (surface plasmons) supported by the grating structure [111]. In the 1950s, further experimental research described the excitation of a surface plasma oscillation of the conducting electrons of the metal, which generated evanescent electromagnetic waves, and excitation of non-radiative surface plasma waves in silver by the method of frustrated total reflection [112–114]. Soon after, in the late 1960s, the controlled excitation of surface plasmons was achieved by Kretschmann and Raether [115] and Otto [116] by means of attenuated total reflection using a prism-coupler based systems. Its first application for sensing was demonstrated in 1983 [117]. On the other hand, Cullen et al. [118] were first in developing an SPR based on grating-coupler systems, demonstrating their sensing performance with immunoassays. Later, in the 1980s, the evanescent field SPR principle also found applications in spectroscopy, as for the interrogation of thin chemical and biological films, which was first demonstrated by Pockrand et al. [119].

In 1980, the first commercial SPR biosensor was launched (BIAcore. Pharmacia Biosensor AB, Sweden), and ten years later, it was first commercially released [120, 121]. Since then, the commercialization and use of SPR biosensor systems widely expanded and numerous manufacturers were selling different platforms with improved capabilities that were routinely employed in research laboratories or in pharmaceutical industry for the analysis of biochemical compounds and interactions. SPR sensors were sold as pure instrumentation, and thus, the application for specific biological or chemical assays had to be developed and optimized by the end user.

Review [122] was devoted to one more practical application of surface plasmon resonance (SPR). It was shown that the SPR as an optical technique could be applied for investigation of biological interactions. Unlike most of the known techniques the SPR was found to analyze the dynamics of surface events in real time and, therefore using it made it possible to determine and estimate rates of adsorption and desorption for a range of surface interactions. The authors of [122] concluded that utilization of the SPR to probe surface interactions had a lot of advantages as compared with other classical methods. In particular, it provided the possibility to rapidly monitor any dynamical process, such as adsorption or degradation. Besides, it made it possible to avoid the necessity to perform difficult complex preparation of biological and/or medical samples and relevant tested medical reagents and/or reagents for the respective investigations.

The SPR was shown in [122] to be a prospective technique for studying processes of drug release from biodegradable systems and be a fast, simple technique to examine the response of biological fluids to biomaterial surfaces. Moreover, the authors connected the future development of application of SPRs with the elaboration of biosensors (e.g. the production of a commercial biosensor (Pharmacia Biosensor, "BiaCore")) for the investigation of specific biological interactions including adsorption and desorption kinetics, antigen–antibody binding, and epitope mapping.

Paper [123] presented results of experimental investigations devoted to finding ways for widening the field of SPR applications as biosensors. The SPR, which was known to be exploited commercially in instruments such as the "BiaCore", was confirmed to be one of the most effective methods of detecting protein molecules (especially enzymes and antibodies) motion relative to a metal surface (as a rule this was a golden or silver film). The experiments were carried out to detect the effect of electric fields on these molecules. The application of an external electric field was found to lead to attraction or reflection of protein molecules and colloidal particles on the golden film. The direction of the motion depended on the charge state of the molecules.

Progress in elaboration of the first completely autonomous (electric power source was an ordinary 9 V battery) hand-held biosensor based on the SPR technique, which did not require a computer interface for its operation, was reported in [124]. Its responsivity was shown by testing experiments to be as small as 3.3×10^{-6} refractive index units. Thus, the sensor could detect such biological toxins as ricin at a concentration of 2×10^{-7} g/mL. The development of such type of sensors was connected with on-going needs of the security area to counteract the threat of bio-terrorism. A need for systems with a capability to quickly and correctly operate in situ to efficiently fight against the threat of applying bio-toxins and dangerous organisms such as ricin and anthrax was emphasized. Concerning other possible applications of the elaborated biosensor based on the SPR, the following fields were indicated: diagnostics of animal and plant pathogens in order to organize a quarantine zone, gene analysis, remote rapid testing of a malarial strain, and water purity analysis. At least it could be applied as detector of unexpected chemical processes in laboratory devices.

Article [125] was devoted to investigation of electromagnetic band gap (EBG) structures, also known as photonic band gap materials (photonic crystals). Such meta-materials were reported in [126] to represent periodic structures, which created forbidden frequency bands for propagation of definite electromagnetic waves. The periodicity of the structure was approximately one half-wavelength. The possibility of building up such materials with the requested electromagnetic properties made them attractive for applications such as antenna systems, elements of millimeter-wave electronic devices [48] and optical communication devices [127]. The woodpile structures were noted in [128] to be among the most popular EBG structures due to their advantages: large bandwidth, robustness and relatively simple fabrication.

In the papers [129, 130], the woodpile structure was shown to be used as antenna substrate, placing a planar antenna on top of its surface. The woodpile structure was reported to act as a reflector for the microwave antenna radiation due to non-transparency of its band-gap for electromagnetic waves. Moreover, utilization of woodpile structures for building up waveguides with increasing width near one of the edges of the waveguide allowed one to elaborate horn like antennas [131]. Additional surface modes were reported to appear in the case of various truncated woodpile structures. The eigenfrequencies of these surface modes were reported to be located in the band gaps of the structures, which would lead to their coupling and interfering with the usual working frequency range of the EBG structure.

Results of temporal characteristics measurements of surface-plasmon polaritons (SPPs) in the THz frequency band were reported in [132]. SPPs were generated by coupling THz radiation into dielectric films deposited on flat gold surfaces. This allowed the authors to measure the features of thin dielectric films and to demonstrate the capabilities of the SPP application for thin film spectroscopy.

The SPP technique confirmed its possibilities to measure optical and dielectric constants, dielectric film thickness, and uniformity of dielectric coatings of metal surfaces with high precision and to control light arrays over length scales, much smaller than the light wavelength. This allowed to apply SPPs in the area of plasmonics, e.g., for fabrication of nanoscale photonic circuits. Other branches for application of this technique were noted to be medical diagnostics and three-dimensional imaging systems for monitoring industrial processes. Biologically active substances like proteins, ribonucleic and deoxyribonucleic acids were found to have resonances in the THz range, thus SPP spectroscopy at THz frequencies could be applied for creation of new biosensors. Due to analysis of the obtained experimental results the authors of [132] pointed out the high sensitivity of this technique and the possibility to apply it for sensing of biomolecules with resonant transitions at THz frequencies.

The SPPs, propagating in the visible and infrared spectral ranges, were concluded by the authors of paper [133] to be studied well enough, because of their wide applications in spectroscopy and sensor development. However, the properties of SPPs in the THz range were evaluated to be far from complete understanding. Using a free electron laser with an average power of 10–50 W at wavelengths 130–140 μm the authors of [133] focused the radiation on the input mouth of a parallel-plate waveguide, which was formed by the surface of a sample and the lower face of a glass prism (both surfaces were covered with a 1 μm gold layer). The features of SPPs were found to strongly depend on the thickness of the dielectric layer deposited on the conductor surface. After reaching the end of the sample, SPPs radiated a strongly directional THz beam into the free space, which was suggested to be used for THz probing.

The paper [134] was devoted to studying the possibility to use modern meta-materials for an efficient guiding of surface electromagnetic waves in the GHz and THz frequency ranges. Two models were considered. First, surface wave propagation along the interface between a semi-bounded meta-material and vacuum was studied, and second, SWs guided by layers of meta-materials were investigated. SWs were found to propagate as symmetric and anti-symmetric modes in meta-material layer. Regimes of forward and backward propagation of the SWs were realized by varying the layer thickness and some parameters of the material. The authors outlined the possibility to control the slowing down of the SWs.

Recent editorials in the Journal of Applied Physics [135] provided comprehensive analysis of novel materials applied in plasmonics whose capabilities, in particular, ability to harvest and manipulate light at the nanoscale had turned the research on plasmonic nanomaterials into a booming field of study. Plasmonic materials were pointed out to offer new research directions in plasmonics, and be already applied in numerous technologies like photovoltaics, biochemical sensors, photocatalysis, and optical devices, as well as in the innovative fields of single-photon sources, all-optical switches, and quantum information processing [101]. Current capabilities of nanoscale fabrication enabled large amounts of electromagnetic energy to be concentrated in volumes much smaller than the wavelength of the radiation used to excite them. This was outlined to be of obvious interest not only for enhancing sensitivities but also for reducing sizes of logic elements. The range of applications of the novel plasmonic building blocks was expected by the authors of [135] to continuously grow in the coming years, opening exciting new opportunities in diverse areas as sensing, imaging, reconfigurable photonic displays, ultrafast photonics, quantum circuitry, and many more.

One of the recent perspectives in the Journal of Applied Physics [136] was devoted to the essential role of the plasma sheath in plasma–liquid interaction and its applications. Discussing the physical mechanisms, responsible for the ionization and electron emission, the authors suggested to pay more attention to excitation mechanisms characteristic to the condensed phase. In particular, they reminded the possible energy transfer between two neighboring electronically excited species or solvated electrons with following generation of an Auger or Auger-like electron emission event. Alternatively, the ensemble of such species at the interface might acquire excitations of a more collective nature. Collective excitations and their interactions were noted to be well-known in condensed matter physics, often expressed in terms of quasi-particles. Therefore, these collective excitations might play a crucial role at the plasma-liquid interface too. Electron emission was reported to occur in the case that the energy of surface plasmons exceeded the binding energy of an electron. The multi-plasma model was suggested by the authors as a useful toolbox to describe excitation and relaxation mechanisms at a plasma–liquid or plasma–solid interface in general.

The tutorials [137] in the Journal of Applied Physics have overviewed magneto-optical methods for magneto-plasmonics in noble metal nanostructures. The methods are based on the following features of surface-type waves. The interaction between photons and free electrons in a metal nanostructure can trigger their collective oscillation, which is responsible for the manifestation of fascinating optical properties. This collective motion can either propagate along the interface between the metal surface and a dielectric and is then called a surface plasmon polariton (SPP) or be confined in a nanostructure, in this case the motion is called as localized surface plasmons (LSPs). "When the energy and momentum of the incident electromagnetic field match the ones of the SPP (or LSP), the system is resonant, resulting in a large extinction of the incoming light together with a strong localization and enhancement of the electromagnetic field in a nano-sized region around the metal surface." These two features allow for light manipulation well below the diffraction limit, opening up a wide range of possible applications in the fields of nanophotonics [138], photocatalysis [139, 140], photovoltaics [141], and sensing [142–144]. Among the applications of plasmonics, an important one is represented by surface-enhanced spectroscopies, which exploit the localization and enhancement of the electromagnetic field to significantly increase the spectroscopic signal of a molecule placed close to the surface. This effect is at the origin of surface-enhanced Raman scattering (SERS) [145, 146] and surface-enhanced luminescence [147]. In particular, the typically weak signals in Raman spectroscopy can be improved up to a factor of 10^{14} by adsorbing the molecules on plasmonic nanostructures [148].

A comparison between sensing performance of traditional SPRs and magneto-optic SPRs (MOSPRs) transducing techniques was presented in [149]. MOSPR was outlined to come from an evolution of the traditional SPR platform aiming at modulating surface plasmon waves by the application of an external magnetic field in transverse configuration. The external magnetic field was pointed out to be well-known to induce a modification of the coupling of the incident light with the SPP during excitation of the plasmon resonance in these structures. Besides, these structures were found to lead to an enhancement in the magneto-optical (MO) activity during the SPP excitation. This phenomenon was exploited in [149] to demonstrate the possibility to use the enhanced MO signal as proper transducer signal for investigating biomolecular interactions in liquid phase. Higher sensing performance in terms of sensitivity and lower limit of detection of the MOSPR biosensor with respect to traditional SPR sensors was demonstrated.

The biosensors market is valued at 25.5 billion USD in 2021 and is projected to reach 36.7 billion USD by 2026 [150]. The technological progress experienced by the nanotechnology field was underlined in [151] to significantly contribute to this expansion of the biosensor market. The increasing demand for cost-effective, portable, and easy-to-use point of care (POC) platforms was pointed out to considerably raise the innovation in the design of biosensors as complete lab-on-a-chip (LOC) platforms. The platforms were noted to require low-sample consumption and to provide sensitivity and real-time response. Among biosensor devices, optical biosensors were found to show unquestionable advantages compared to other biosensor technologies. They demonstrated the capability to deliver label-free quantitative analysis and showed exceptional potential for multiplexing and miniaturization. Among them, those based on localized surface plasmon resonance (LSPR) were subjected to a great scientific interest as the novel counterpart of the well-established SPR sensor [151].

An LSPR-based POC system was implemented using an *Au* nanohole arrays in [152]. The detection configuration was based on surface plasmon-enhanced fluorescence spectroscopy rather than on detecting the LSPR wavelength shift, which was reported to significantly enhance the sensitivity. The POC design contemplated a handheld-sized version using the mobile phone camera to detect the fluorescence and a desktop version using a high-resolution microscopic camera for enhanced performance. The design incorporated disposable fluidics with either four or nine chambers controlled by micropumps to deliver the fluid. Although it required fluorescent labels and the sensitivity of the handheld version was limited, the design incorporated the necessary software to operate automatically and contemplated all the aspects to have an affordable portable POC to be used in remote areas. Another example, more elaborated from the biosensor point of view, implemented a label-free detection experiment for pesticide detection using *Ag* nano-slits and transmission measurements with a smartphone [153]. Although, in principle, it was a qualitative estimation based on naked-eye colorimetric detection, it was noted to be further optimized with image processing to get a more precise semi-quantitative analysis by developing spectrometric software for the smartphone and a POC design with integrated optics.

Sensing based on LSPR of plasmonic nanoparticles was reported in [154] to be realized through the transformation of the nano-plasmonic particle induced by the presence of a specific analyte at low concentrations or external stimulus to change the refractive index. The direct detection of molecules was carried out by the analysis of LSPR shifts, inelastic vibrational scattering via SERS and fluorescence emission via surface enhanced fluorescence (SEF). Plasmon coupling and plasmon hybridization, as well as antenna-mediated nano-plasmonic systems, offered possibilities to improve the sensitivity and to be used in various applications from sensing of catalytic reactions to tip-enhanced spectroscopies. Plasmon-mediated fluorescence enhancement was underlined to open new prospects for imaging due to the drastic fluorescence enhancement achievable for weak emitters. In addition, the coupling of spectroscopic methods with scanning-probe techniques was noticed to extend the purely optic information to spatially resolved optoelectronic information and thus to open a great potential in bio- and materials imaging. The current resolution was found to reach a few nm depending on the confinement technique of the local electromagnetic field, which allowed chemical resolution using plasmon-mediated tip-enhanced Raman scattering within the intramolecular range.

The first SEF sensor was reported in 1991 [155]. SEF was outlined to be also referred to as surface plasmon-enhanced fluorescence spectroscopy (SPFS), surface-enhanced fluorescence spectroscopy (SEFS), plasmon-enhanced fluorescence (PEF) or metal-enhanced fluorescence (MEF). So far, a standard naming convention for this type of sensor was noted not to exist. The SEF research was not active before 2007, but since this time, many nanostructures showed plasmonically enhanced fluorescence, coincident with the rise of controllable nanoparticle synthesis and lithography-based fabrication of nanostructured chips. The plasmon-induced fluorescence enhancement factor was reported in [156] to generally fall in the range of 10–100, with the possibility to be higher in optimized plasmonic nanostructures, even reaching up to 1340.

Plasmon-induced transparency (PIT) was pointed out in [157] to be an analog form of classical electromagnetic-induced transparency (EIT). It was found to be the destructive interference effect of the coupled beam in the quantum system that made the original opaque medium transparent. EIT was proved in different optical systems, including ultra-cold atomic gas, photonic crystal waveguide, electromagnetic meta-material, and sensors. H-shaped meta-material based on the semiconductor (*InSb*) was designed in [158]. By analyzing the magnetic field, surface current, and equivalent refractive index, the authors showed that the PIT effect was caused by the destructive interference of the so-called bright–bright mode. At the same time, the Lorentz equivalent model further explained the physical mechanism of PIT. The strong dispersion at the transparent peak was demonstrated to be of interest for designing slow light devices. The group delay was noted to be adjusted by changing the polarization angle. Besides, the dielectric constant of *InSb* was shown to be changed by controlling the temperature, making the PIT effect tunable. The possibility to use the phenomenon in temperature sensors and thermo-optical switches was suggested. A PIT modulation depth of 97.8% was obtained, which was by far better than in most other studies. Although the noted response speed of temperature regulation was slower than that based on photoelectric regulation, it provided new ideas and ways for PIT applications of semiconductor materials.

Perfect optical meta-material absorbers were noted in [159] to attract much attention due to crucial role of light absorption in many areas including optical sensors, photocatalysis, photoelectric technology, and utilization of energy. To design an impedance matching layer with the objective to minimize the reflection and improve absorption by using the loss of meta-materials was outlined to be one of the basic ideas of a perfect meta-material absorber. An anisotropic meta-surface was designed in [159] to realize high-Q perfect absorption at optical frequencies. The root cause of this phenomenon was revealed to be coupled excitation of the LSP mode and the SPP mode. This absorber could realize bi-functional switching and was expected to realize energy conversion of photons–SPP–photons. This research suggested to provide theoretical guidance for the further development of applications related to high-Q perfect absorption such as absorption filters with specific wavelengths and narrow band photon emitters.

The review [160] was devoted to theory and applications of the localized surface plasmon resonance (LSPR) in semiconductor nanocrystals (NCs). LSPR was reported to result in resonant absorption, scattering, and near field enhancement around the NC which could be tuned across a wide optical spectral range from visible to far-infrared by synthetically varying the doping level, as well as post synthetically. LSPR in doped semiconductor NCs was demonstrated to hold for a wide range of applications such as infrared spectroscopy, energy-saving technologies like smart windows and waste heat management, biomedical applications including therapy and imaging, and optical applications like two photon up-conversion, enhanced luminescence, and infrared metasurfaces. In particular, localization of an intense electromagnetic field inside and around the plasmonic NCs was emphasized to result in strong absorption of far-field illumination. The absorbed light was dissipated to the surroundings in the form of heat establishing NCs as local heat sources. The high absorption cross-section of plasmonic materials made them promising candidates for photoacoustic imaging, photothermal therapy and imaging, perfect infrared absorbers, waste heat recovery, and more.

An analog of electromagnetically induced transparency was revealed in meta-surfaces, composed from identical dielectric resonators of simple cylindrical shapes [161]. It was detected in numerical experiments at optical and microwave frequencies and confirmed by real experiments in the microwave range. The main specific of the observed phenomenon was its appearance at frequencies of electric dipolar resonances in meta-surface elements, when total reflection of incident waves instead of transmission was expected. Investigations of electric field distributions in meta-surfaces allowed for the conclusion that the observed types of distributions could be related to the formation of standing SWs.

The possibility to tune the wavelength of LSPR by nanoparticle composition, size, alloy proportion, etc. was noted in [162] to be a significant advantage of plasmonic nanostructures with sharp tips, narrow gaps, and aggregates. The nanostructures were reported to generally provide more enhanced Raman scattering as well.

Compact plasmonic switches consisting of a plasmonic coaxial waveguide side-coupled to a periodic array of two open-circuited coaxial stub resonators were designed in [163]. The greatly ($\sim$3.02 times) enhanced transmission cross-section of the double-slit coupler compared with one of the single-slit coupler was reported and explained by the excitation of surface plasmons at the interface between the slits and air by the incident light waves. Namely, the coupling of the surface plasmons into the slits led to an increase in the transmission cross-section of the structure.

The threefold role of plasmonics in nonlinear optics was outlined in the review [164]. First, by enhancing the effective nonlinearity, plasmonic nanostructures were noted to contribute to conventional photonics by allowing nonlinear effects to be applied with reduced optical power. Second, they were found to make it possible to scale down nonlinear components in size, which was important for the development of integrable photonic devices and, ultimately, fully functional nanophotonic circuitry. Third, the response time of plasmonic excitations was stressed to be ultrafast, allowing optical signals to be manipulated on femtosecond timescales. From the physical point of view, these applications were explained as follows. The possibility of light coupling to surface plasmons near the metal surface due to light interaction with metal nanostructures was introduced for the explanation. For extended metal surfaces, this was noted to give rise to surface plasmon polaritons (SPPs) known to be surface electromagnetic waves propagating at the metal–dielectric interface. For metal nanoparticles, the response was outlined to arise from localized surface plasmons (LSPs), whose resonances depended on the particle size and shape. This was emphasized to open opportunities for controlling light confinement on the nanoscale. In particular, periodically structured metal films and surfaces, the so-called plasmonic or surface plasmon polaritonic crystals (SPPCs), were noted to provide considerable versatility for tailoring the wavelength-dependent plasmonic resonances and the electromagnetic field enhancement in devices a few tens of wavelengths in size [165]. SPPCs were reported to have optical properties analogous to those of two-dimensional photonic crystals.

The linear theory of electromagnetic SW excitation in conductors was estimated in [101] to be well developed. The necessity to develop the nonlinear theory of SW excitation in conductors was demonstrated by the experiment reported in [166]. The possibility to excite SWs by a normally incident focused pulse was analyzed in application to the action of femtosecond pulses on conductors in [167]. The time-dependent ponderomotive force was shown to generate SWs in the THz frequency range. The possibility to significantly increase the total energy of the SWs when applying this method of excitation was demonstrated in [168] by using a cylindrical lens.

The theory of another nonlinear mechanism of SW excitation was reported in [169]. The generation of the drag current by a laser pulse focused by a cylindrical lens and incident not perpendicular to the surface of the conductor was announced to be the mechanism. This mechanism was found to be more efficient than the ponderomotive one due to the relatively high values of the effective frequencies of electron collisions for typical metals at room and higher temperatures.

Nanoscale devices harnessing strong-field laser-electron interactions in semiconductors showed promise for the next generation of optical computing devices [170]. Specifically, semiconductor ponderomotive electron acceleration allowed for increased switching speeds by moving free carriers via light instead of a conventional, externally applied, electrical bias. This, in turn, opened a new platform that revisited the operating principles of the original vacuum tubes, except operating at significantly increased speeds (THz), owing to the electrons being driven by surface plasmon fields.

Microellipsometric investigations of metal–insulator–metal (MIM) structures with regular square lattices of nanocylinders as one metallic layer were presented together with a theoretical description of all dipolar interactions in such a system in

[171]. The theoretical and experimental results were compared and clearly demonstrated the influence of surface plasmons on the optical properties of such MIM systems. The hybridization of LSPs and SPPs was revealed.

Regarding biological applications, second-harmonic generation (SHG) from plasmonic nanostructures was demonstrated in [172, 173] to be used to sensitively detect biological and chemical toxins. In an excellent review [172], details about the nonlinear optical response of metal nanoparticles and their applications in sensing biomolecules and other chemicals were discussed elaborately. SHG, as well as hyper-Rayleigh scattering (HRS), from individual plasmonic nanospheres was used to diagnose single-base-mismatch DNA hybridization [173], to detect Alzheimer's disease biomarkers [174], and to probe heavy metal ions [175]. HRS from gold nanorods was used to selectively detect and identify *Escherichia coli* bacteria [176]. HRS from oval-shaped nanoparticles also was used to detect breast cancer cells [177].

The atom was noted in [178] to set an ultimate scaling limit to Moore's law in the electronics industry. Photonics studies were outlined to deal with devices at the micrometer scale, while electronics research already explored atomic scales devices. However, photonic scaling, similar to electronics, was demonstrated in [178] also to be only limited by the atom. More precisely, an electrically controlled plasmonic switch operating at the atomic scale was introduced therein. The switch allowed for fast and reproducible switching by means of the relocation of an individual or, at most, a few atoms in a plasmonic cavity. Depending on the location of the atom either of two distinct plasmonic cavity resonance states was supported. Experimental results showed reversible digital optical switching with an extinction ratio of 9.2 dB and operation at room temperature up to MHz with femtojoule energy consumption for a single switch operation. This demonstration of an integrated quantum device allowing to control photons at the atomic level seemed to open intriguing perspectives for a fully integrated and highly scalable chip platform, a platform where optics, electronics, and memory might be controlled at the single-atom level.

Taking into account that surface flute waves can propagate either along a magnetized or a non-magnetized plasma interface the authors of the present book estimate their application for nanotechnologies as a very prospective.

11.5 Surface Wave Applications for Plasma Production

Plasmas obtained in different types of discharges are finding increasing use in materials processing, plasma chemistry, chemical analysis, and many other fields. One of the methods to sustain gas discharges was reported in [179] to be connected with excitation of electromagnetic surface waves (SWs). This paper reviewed the physical principles of operation and the design of SW plasma sources. Since the wave launcher is the central component of any plasma source, this review presented a description of several compact, efficient and easy to operate RF launchers specifically intended for plasma generation. Using them it was possible to sustain RF plasmas in the frequency range from 1 MHz up to 10 GHz, at working gas pressures from 10^{-5} Torr up to few times atmospheric pressure. However, in order to simplify the presentation, the authors restricted their considerations to the case of a low-pressure discharge sustained by SWs propagating along the axis of symmetry of a cylindrical dielectric tube. The length of the active zone of the discharge, where the RF power transferred to the plasma particles, was considered to be larger than the tube diameter.

SWs, as eigenmodes of the discharge tube, were proved experimentally to be excited by a very compact RF wave launcher, which took up only a small area of the discharge tube. The RF power was carried to the rest of the plasma column due to the discharge by the propagating wave. The obtained plasma was found to be stable and reproducible because a stable mode of a definite SW could propagate there. Thus, the level of electron density fluctuations was sufficiently low. The authors underlined an essential difference between the case, when the SW propagated along the interface of a plasma column created by some other methods and the case, when the eigen SW sustained the plasma column by itself. The fact that the discharge and the SW, which sustained it, depended upon each other allowed the authors to formulate an equation of the power balance between the power lost by the electrons and the power acquired from the SW field.

The papers [179, 180] studied specific features of the process of SWs excitation for the cases of azimuthally symmetric (azimuthal mode index $m = 0$) and azimuthally non-symmetric ($m \neq 0$) perturbations. Summarizing the results of the experimental investigations obtained for the cases $m = 0$ and $m = 1$ they concluded that the $m = 1$ non-symmetric mode could be launched to sustain a plasma column only if the product of the SW frequency f and tube radius R exceeded a definite critical value. In the case of paper [180] it was 2 GHz cm. However, the azimuthally symmetric $m = 0$ mode could propagate and sustain the plasma for each value of the SW frequency, which resulted in the product $f \cdot R$ <2 GHz m. This circumstance led the authors of review [179] to develop various $m = 0$ SW plasma sources. The main types of them are listed below.

The first of them is the LC Ro-box, which operates in the lower domain of the indicated frequency range (approximately 1–100 MHz). It consists of two separate units: the field applicator (or field shaping) module and a matchbox, whose impedance matching network contains LC elements. The second one is the stub Ro-box. This SW launcher uses the same field applicator as those of the LC Ro-box, but its power feed and impedance matching are different. The RF power is supplied via a coaxial feeder through a capacitive coupler. This impedance tuner is constructed of semi-rigid cable, which penetrates radially into the field applicator module. The third launcher is entitled as surfatron. It is the first SW launcher, which has found a wide range of applications as power plasma source intended for production of long plasma columns utilized for various physical studies, plasma chemistry, solids processing, laser engineering, etc. The surfatron is an integrated wave launcher, which performs both field shaping and impedance matching (the external stub used in the stub Ro-box launcher is now part of the surfatron body). The fourth type of SW launcher is the surfaguide. It has been elaborated because of restrictions in the power handling capability of the surfatron. Without doubt the surfaguide is the simplest of the surface wave launchers among those considered here. It consists of two parts:

RF field applicator, which is a section of rectangular waveguide of reduced height, tapered symmetrically with respect to the launching aperture and a movable plunger, which is located at one end of the applicator. The last type of SW launcher developed for sustaining gas discharges by propagation of azimuthally symmetric SWs is the waveguide-surfatron. It is a SW launcher consisting of both waveguide and coaxial line elements. Here the microwave power is supplied to a rectangular waveguide section by a generator and an air insulated coaxial line section is attached perpendicularly to the wide wall of the waveguide, so that its inner conductor extends into the waveguide.

The results of experiments reviewed by the authors enabled them to conclude that applications of SW sustained discharges employed a range of operating conditions and plasma parameters available with gas discharges. Namely, these types of discharges allowed one to obtain over-dense plasmas in relatively large discharge chambers. The plasma sustained by SW propagation in such discharges was characterized by a high level of reproducibility, convenience to utilize it for various technological needs and by low level of parasitic electromagnetic oscillations. This was explained by such useful features of the SWs as slow phase velocity, monomode regime of wave propagation and weak dependence of the wave impedance on plasma parameters [179]. The development of suitable wave launchers was concluded to be the main challenge of SW plasma applications, because the other elements of the plasma source could be elaborated by good established methods of electronic engineering.

Reflecting the broad utilization of RF gas discharges in the 1980–1990s, the review [181] concerned the theoretical description of these discharges sustained by SWs. As a rule, the discharges were confined in a cylindrical dielectric tube. The plasma column surrounded by its dielectric discharge tube was a self-supporting waveguide or an integral part of the waveguide structure when the dielectric tube was within a cylindrical metallic enclosure. These discharges appeared very convenient for studies since they were generally less expensive, easier to handle, more efficient, and more reliable sources of particles and radiation for physical studies and technical applications as compared with the various DC discharges. The review [181] of SW plasmas included models of plasmas sustained by SWs in various electromagnetic modes (azimuthally symmetric ($m = 0$) and dipolar ($m = 1$)) and considered the influence of an external static axial magnetic field. The basic results of these theoretical studies were reported to be in a good agreement with the available experimental data.

Paper [182] was one of the first reviewing articles devoted to the development of large-size, high-density $n_e \geq 10^{11}$ cm^{-3} microwave plasma production. It concerned plasma generation, which could be performed at low pressures (<20 mTorr) of the operating gas, without utilization of an external magnetic field. Deep understanding of the discharge physics allowed the authors to elaborate a microwave plasma production technique, which was characterized by a stable operating regime, generation of a large-diameter plasma column (the diameter was larger than the axial length of the column) sustained by azimuthally non-symmetric SWs, which were clearly observed in experiments using optical emission and microwave field measurements. The authors proved that just excitation of SWs gave them the possibility to reach such good results, thus after improvement of the antenna structure this technique was planned to be commercially available for processing of large scale flat panel displays and solar cells with diameter larger than one meter.

Comparative analysis of different types (including electron cyclotron resonance plasmas, helicon-wave excited plasmas and inductively coupled plasmas) of plasma production methods showed the advantages of this SW plasma source in comparison with other high-density sources. All that allowed the authors of [182] to conclude that in the regime of low gas pressure the slot-excited azimuthally non-symmetric SWs were attractive for sustaining a high density, large scale gas discharge, which produced non-magnetized plasma for solids processing technologies. Therefore, the theory of SFMs considered in the previous chapters of this book can be directly applied for explanation of the experimental results of reference [182].

Microwave production of plasmas is widely used in the fields of semiconductor processing and treatment of solids. It is characterized by the following features: higher electron density compared with RF capacitive discharges, possibility to operate at extremely reduced gas pressures, low energy of the ions striking the substrate, and a possible independent biasing of the substrate. That is why the authors of [183] proposed to concentrate the main efforts on theoretical and experimental studying of SW-produced plasmas influenced by an external magnetic field. It was found that in a magneto-active plasma column located in a dielectric tube of a finite thickness, fundamental surface hybrid waves $HE_{m,1}$ with azimuthal mode index m and radial mode number l (l-1 is the number of zeros of the axial electric wave field E_z along radial direction) could effectively sustain a microwave gas discharge. The produced plasma density increased with the power deposited into the discharge per unit axial length. Experiments proved that azimuthally non-symmetric SWs with azimuthal mode indices $m \pm 1$ had a larger attenuation coefficient than for the case of azimuthally symmetric surface waves. This fact could be considered as a witness on prospective application of SFWs for sustaining microwave gas discharges.

A step towards the definition of measurement techniques for the experimental characterization of surfaguide-fed plasma antenna was done in [184]. Two main problems were considered and solved. First, the presence of two RF signals at the same time on the same structure. Second, the design of a tool for the characterization of the plasma state. To solve the first problem a suitable measurement set-up was developed to evaluate and reduce the strong coupling between the pump and the radiated signals in order to prevent damage of the experimental arrangement and measurement errors. As far as the second problem was concerned, the measurement of the plasma state appeared to be complicated because the space, where the plasma was ignited (a glass tube), was not accessible. In this situation an indirect measurement procedure was carried out, involving a self-consistent technique that made it possible to retrieve the value of plasma conductivity using experimental data and simulations. Three important parameters were successfully measured: the efficiency of the plasma antenna, its length, and the column conductivity. The experimental analysis showed the surfaguide to be an effective device to excite plasma antennas and also suitable for array configurations. At the same time this study underlined the need not only to have a model to describe the interaction mechanism between a surface wave and the plasma but also to carry out a parametric investigation of the problem. For this purpose, a numerical tool was developed to help the optimization of all the parameters involved in plasma antenna design.

Plasma columns sustained by electromagnetic SWs under the free-fall regime, a still unexplored pressure domain for such plasmas, were experimentally investigated in [185]. Because of the relatively large values of the frequency (2.45 GHz) and plasma diameter (5.6 cm), the waves propagated either in the dipolar ($m = 1$) or quadrupolar ($m = 2$) mode rather than in the azimuthally symmetric mode ($m = 0$). The radial distributions of the electron density and temperature, measured by means of electrostatic probes, demonstrated the existence of a resonant absorption mechanism of the electric wave field by the electrons. The measured value of the power loss per electron was found to be four to ten times larger (!) than the value extrapolated from a diffusion model for the $m = 0$ mode. Such a discharge was suggested to supply plasma into an applicator operating with a multipolar magnetic confinement used for surface processing.

A high-density SW magnetized argon plasma operated in the very low-pressure regime together with a biased RF system was used in [186] to study physical etching characteristics of platinum thin films. The platinum etch rate was shown to strongly decrease as the operating pressure increased for a given DC self-bias voltage, which resulted from a decrease of the ion density at the sheath edge and from enhanced redeposition. Using a high-density plasma in the very low-pressure regime was found to yield high etching rates with a good selectivity over resist. Fence-free features were also achieved at bias voltages that, in contrast with reactive ion etching reactors, were only slightly above the platinum sputtering threshold.

Results of experimental approbation of a new method of generation of a stable surface microwave discharge in a boundary layer near a body streamlined of supersonic airflow were presented in [187]. Breakdown characteristics, spatial distribution of electric field, dynamics, space-temporary evolution of gas, and other characteristics of this type discharge were measured. This investigation was initiated by the needs to know, how supersonic aerodynamics influences the characteristics of gas flow near the surfaces of flying bodies. Since electrode discharges in such air flow were unstable, spatially non-uniform, and non-reproducible, the authors paid attention to a new version of the electrode-less microwave discharge. Their experiments showed that the most suitable type of discharge was a microwave discharge sustained by SWs. The SW can propagate in a gas as long as its energy is sufficient to create a plasma, whose density is no less than the critical density. Thus, there is a self-sustaining system. The experimental set up consisted of a vacuum chamber, magnetron with pulsed power $P \leq 300$ kW and wavelength $\lambda = 2.4$ cm, pulse duration $\tau = 1$–100 μs, system for microwave power input, system for creation of supersonic airflow with Mach number $M = 2$, and diagnostic system.

It was found in [187] that the minimum power, which was needed for generation of the discharge, decreased with increasing pressure in the air pressure range from 0.001 to 0.1 Torr. Then, the minimum power increased with increasing pressure in the higher pressure range 10 Torr $< p <$ 1000 Torr. The ionization frequency was an increasing function of the electron temperature, or (in other words) ratio of the plasma density divided by the electric field strength. The longitudinal size of the discharge was found to be a linear function of the duration of the microwave pulse. According to the obtained results of a comprehensive program of investigations, the authors concluded that microwave discharges sustained by SW propagation provided stable uniform luminous plasmas, which coated the whole surface of dielectric bodies immersed into the supersonic airflow and that the flow with the Mach number mentioned above did not influence the characteristics of this discharge.

The paper [188] presented results, which were a continuation of the previous work of these authors. They found that in the regime of low air pressure $p < 1$ Torr, where the electron-neutral molecule collisional frequency was much lower than the eigenfrequency of the applied SW, the discharge was very uniform and occupied a large volume, whose size increased with decreasing air pressure. The authors proposed to apply such type of discharge for treatment of solid surfaces, etching, deposition of films and special coatings. Under the regime of high pressure 30 Torr $< p$ the discharge had a complex form of thin plasma channels (filamentation), whose diameter was less than 0.1 cm. Formation of such discharges gave rise to shock waves. Thus, it could be applied for ignition of liquid hydrocarbon fuel.

The paper [189] was devoted to investigation of fundamental problems of microwave gas discharges: low threshold of the discharges at a metal–dielectric interface, thermal-ionization instability, and plasma resonance on its non-uniform density profile. Results of experiments on studying plasma generated by three new types of plasmatrons were discussed there.

The paper [190] presented a self-consistent model of an azimuthally symmetric TM SW discharge in hydrogen applying cylindrical geometry. The basic set of equations was derived from Maxwell equations in kinetic approximation. The performed simulation and analysis of experimental data allowed the authors to make the following conclusions. The electron density in this discharge sustained by SWs exceeded the critical value for the 3 cm wavelength band and reached 10^{13} cm^{-3}. The velocity of the SW front essentially depended on the plasma resonance at the front of the discharge. A significant role in the kinetic of the discharge was played by ion-molecular reactions, which led to ion conversion. In addition, the considered discharge was found to be a very efficient source of hydrogen.

It is known that neutron beam injection will be applied for plasma heating and non-inductive current drive in ITER. The basis of the injection method is neutralization of positively charged ions of hydrogen isotopes (D, T) accelerated up to MeV energies. The efficiency of neutralization determines the total efficiency of the neutral beam injection. Comparison of the gas neutralization method, which was earlier used and the plasma neutralization method, which was developed in [191] showed an undisputed advantage of the latter for the conditions of ITER, namely under the conditions of a large volume of confined plasma and a large strength of the applied external magnetic field. The concept of the device elaborated by the authors was based on application of a multipole magnetic trap with the following properties: closed drift of electrons connected with the gradient of an external magnetic field, and large magnetic field and large volume of the plasma neutralizer. The estimations made by the authors allowed them to affirm that the proposed design of plasma neutralizer would be more efficient compared with the usual gas neutralizer.

Taking into account that microwave discharges are intensively used in different applications, the authors [192] presented two examples of their applications. The first one was the plasma neutralizer, which could be considered as an attractive upgrade for Neutral Beam Injection Systems to be utilized in ITER. The second one was the plasma separator for treatment of spent nuclear fuel and radioactive wastes.

The authors obtained a sufficient increase of the neutralizer efficiency when changing the gas reagent by a plasma produced in a large magnetic trap under the regime of an electron cyclotron resonance (ECR) microwave gas discharge. This paper developed ideas proposed in [183] on advantages of ECR microwave discharges over other types of discharges and was a sequel of the previous paper [191]. Utilization of a plasma neutralizer was reported to allow one to increase the neutralization efficiency up to 80% in comparison to 60% achieved in the case of a gas neutralizer.

Application of plasmas produced by ECR plasma sources for ionization of spent nuclear fuel components was considered as the first step into a new technology developed by the authors. The separated elements were selectively heated in an ion cyclotron resonance chamber and finally selected in a toroidal magnetic field. The first experimental results allowed the authors to estimate the proposed design of a plasma separator for treatment of spent nuclear fuel and radioactive wastes as very good prospective [192].

Discharges sustained by SWs are under consideration of many groups of scientists because of their technological applications and also due to their utilization for fundamental investigations in various branches of physics. Concerning just the last mentioned aspect the steady state of SW produced discharges especially at low pressures (less than 5 Torr) and relatively low frequencies (less than 5 GHz) is studied actively for a long time. However, on the other hand, there are many applications where a long uniform plasma column can be generated for a short lifetime. The maximum frequency, at which one could obtain a short time discharge, was evaluated in [193] not to exceed 10 GHz at that time. In this frequency band the SW was usually launched into the discharge system by a surfaguide [179]. However, its operating power was limited by gas breakdown in the region, where the SW was excited.

That is why the authors of the paper [193] developed two novel types of launchers, which allowed them to apply much higher microwave power for generation of SW-sustained discharges under the pulsed regime. The first launcher was called as E-plane junction. It allowed to generate a long plasma column in a quartz tube placed in a single-mode waveguide. The tube had an inside diameter of 0.4 cm, an outside diameter of 0.6 cm, its length was 50 cm, and it was filled by various gases: air, argon etc., under the pressure 1–60 Torr. The experiments were conducted using microwaves with 3 cm wavelength, $\tau = 1$–$15\ \mu$s pulse duration, 30–150 kW pulse power, and 0.1–1 Hz repetition frequency. A high degree of absorption of the input power (its typical value was 70–90%) was observed over a wide range of gas pressures. That is why this type of microwave SW launcher was concluded to allow one to extend the operating power level.

The second type of launcher applied the circular waveguide TM_{01} mode for excitation of the operating SW. The discharges were sustained in quartz tubes with 0.6–0.9 cm outside diameter, and 30 cm length. These experiments were carried out for microwaves with 0.8 cm wavelength, $\tau = 0.5$–$10\ \mu s$ pulse duration, 30–60 kW pulse power, and 10–10^3 Hz repetition frequency. The authors of [193] performed detailed measurements of the spatial distribution of the excited SWs, and obtained a sufficiently large value of electron density in this discharge: $n_e \sim 10^{14}\ cm^{-3}$. This allowed them to conclude that this launcher for SW sustained high power microwave discharge had high efficiency.

The authors of [194] reported on the elaboration of a microwave discharge, which enabled the production of high-density (plasma electron density $n_e \geq 10^{11}\ cm^{-3}$), large-area (square $S > 5\ m^2$) flat plasma, owing to its efficient absorption of SWs. In order to supply uniform radiation of microwaves over an entire plasma surface, an array of slotted waveguide antennas and a novel compact power divider were developed. The plasma uniformity was verified with two-dimensional Langmuir probe measurements. Moreover, the authors proposed general guidelines for designing microwave slot antenna arrays for requested plasma sizes.

The authors of [195] presented results of their experiments on measurements of a dense plasma created in a microwave resonator. This was an example of specific application of a plasma discharge. An electron density up to $10^{15}\ cm^{-3}$ was obtained in a 9.4 GHz resonator cavity, which was equipped with a movable plunger at the end where the rectangular feeding waveguide was located. The second end was created by the reflecting over-dense plasma column. The microwave energy for this device was provided by a pulsed magnetron with the following parameters: 120 kW power and 1 μs pulse duration. This energy dissipated in the capillary discharge operated with 20–100 Torr helium. The discharge was sustained by propagation of microwave SWs along the protruding parts of the quartz capillarity. Analyzing the results of their experiments the authors were sure that the excitation of the SWs would provide suitable starting conditions for the detailed analysis of afterglow measurements been useful for studies of regimes of high plasma densities.

The authors of paper [196] described a technique for the measurement of the absolute electron density in low-pressure plasmas using propagation of electromagnetic SWs along a dielectric cylinder immersed into plasma. The transmission spectrum was measured between two antennas located at the ends of the cylinder and connected to a network analyzer. The plasma transmission probe (PTP) elaborated by the authors was promising for the measurement of densities larger than $10^9\ cm^{-3}$ at relatively high gas pressure (until 1 Torr). Experimental spectra measured in capacitively coupled RF discharge in argon at various plasma densities and pressures (40–750 mTorr) were presented. Plasma densities derived from the transmission spectra were compared with those obtained with a Langmuir probe and with a microwave 1/4-wavelength resonator at low pressure (4–45 mTorr). The densities determined by the PTP were found to be about two times lower compared to those obtained with other methods. The authors explained it by the influence of the density gradient of the pre-sheath plasma, as the PTP determined the sheath edge electron density, but not the bulk value.

In recent years, microwave discharges at atmospheric pressure have attracted an increasing interest in view of their technological applications. They are efficient sources of ions, excited atoms and molecules and light radiation for plasma chemistry, spectro-chemical analysis, lamp technology, plasma processing and laser excitation [197, 198]. The axial distributions of the electrons and 3p54s and 3p54p excited atoms in SW-sustained argon plasma columns were studied both experimentally and theoretically in the gas pressure range of 0.2–2.8 Torr in [199]. The plasma was produced by surface type waves in a Pyrex cylindrical tube with external radius of 0.6 cm by coupling the 2.45 GHz microwave energy via a surfatron-type launcher to produce the azimuthal ($m = 0$) mode. The tube was surrounded by a metal screen of radius 2.0 cm. Various diagnostic methods were used in the data acquisition: SW interferometry as well as emission and absorption spectroscopy. The agreement between experimental data and theoretical results was very good. It was confirmed that with increasing pressure the electron density increased while the populations of all the 4 s and 4p levels decreased. The latter was owing to the decrease of the average electron energy with pressure enhancement. It was established that the 4p populations (and the electron density) decreased from the wave launcher to the plasma column end, while the populations of the 4 s levels increased. The same device was used in [200] to obtain the electron temperature in an argon plasma column. The electron energy distribution function was verified to be nearly Maxwellian in a 0.8–2.8 Torr intermediate pressure range. The microwave power was coupled to the plasma by a device that produced a SW plasma.

The aerosol deposition (AD) method is a novel coating technique for preparing thick ceramic films at room temperature. The method is based on the impact adhesion of fine particles for the formation and micro-patterning of thick ceramic layers on substrates. In [201], a microwave plasma source (MPS) with a novel structure for plasma aerosol deposition (PAD) was developed as a plasma generator using ignition in a SW discharge in a low vacuum region, which enabled long-term (>4 h) operation with stable plasma conditions by introducing an effective cooling structure. This process was called "microwave plasma assisted aerosol deposition (μ-PAD)". Under the optimized conditions, Y_2O_3 coatings deposited on an aluminum substrate were investigated using a scanning electron microscope. After applying the μ-PAD for 10 min, a coating layer thickness of ~50 μm with porosity <10% was observed. By comparison with AD results with $\leq$10 μm thickness, the data published in [201] showed that the μ-PAD with the developed MPS was a highly feasible alternative to AD, without limitations, such as low adhesion for the coating layer in case of thick films.

An analytical expression to predict the axial distribution of electron density along the SW sustained plasma column was derived in [202]. The model was based on square root approximation of the dispersion relation giving a linear roll-off of the electron density distribution in the plasma column. The point was to use a more tractable analytical approximation of the dispersion relation of SWs. An expression for the attenuation coefficient, which included the collisional loss, was applied in studying the structure of the SW produced plasma column. The derived attenuation model predicted a linear axial attenuation of the electron density, as well as a nonlinear relation between the frequency dependent glass permittivity and the wavenumber. The nonlinearity arose from non-negligible thickness of the dielectric. This result was consistent with early Aliev's results of the electron density profile, which corresponded to the volume collisional absorption regime of SW sustained plasmas. The combined effects of SW amplitude attenuation due to collisions and frequency dependence of the group velocity caused an axial decrease of the SW axial power flux as the wave propagated away from the launcher. A short analysis of the sensitivity of the normalized frequency to the permittivity of the glass based on the quasi-static approximation of the dispersion relation of SWs was given.

Discharge characteristics of microwave plasmas excited with a large-area planar microwave launcher installed internally in a 60 cm-diameter cylindrical vacuum chamber were studied in [203]. With less than roughly 400 W microwave power at a *He* pressure of 14–27 Pa, the large volume-wave plasma (VWP) filled the entire chamber. Above 400 W, the plasma discharge made a sudden transition to higher-density, uniform surface-wave plasma (SWP) having a spatial uniformity of $\pm$ 3.5% over 30 cm in diameter. Electron energy probability functions in the downstream region were studied using Langmuir probe measurements with the Druyvesteyn method in both the SWP and VWP discharges.

The measurement of electron density is a key issue in understanding and controlling plasma applications. To date, plasma density in electric thrusters has been mainly evaluated with electrostatic techniques, such as the Langmuir probe, which could be quite invasive. In [204], the application of a microwave resonant probe, the curling probe, was proposed for diagnostics of an electrodeless plasma thruster. The three probes: CP700, CP1400, and CP2965, were applied, having a resonator length of 10.6, 4.4, and 1.8 cm, respectively. A limited analyzing range was observed for the CP700. When the plasma density was sufficiently high, i.e., when the plasma frequency was larger than the analyzing frequency, the relative plasma permittivity became negative ($\varepsilon_r < 0$) and the volume wave took the form of a decaying SW propagating at the interface between the plasma and the probe. This effect could be estimated from observing the measured reflectance spectra from the CP700 probe for different plasma densities. As the plasma density increased, the resonant frequency shifted to higher values and the reflectance spectrum flattened, meaning that less power was dissipated in the resonator. When the critical density value was reached, the SW abruptly decayed at the plasma–probe interface and was completely confined inside the structure under the antenna slot, where no plasma was present, resulting in a negative frequency shift.

Plasma-aided material processing is one of the key technologies in semiconductor industries. For example, fluorocarbon plasmas have widely been used in etching of such dielectric materials as SiO_2 and Si_3N_4. The recent trend towards ultra large-scale integration devices requires a high-density ($> 10^{11}$ cm^{-3}), low-pressure (<20 mTorr) and large diameter (>40 cm) plasma to achieve high-aspect-ratio etching under the 0.1 μm design rule. However, in such high-density plasmas the feed gas is strongly dissociated into a composition rich in F radicals and poor in CF_x radicals (x = 1–3), thus reducing the etching selectivity to the underlying Si layer. The correlation between the EEDF and plasma chemistry, mainly in the case of high-density fluorocarbon plasma excited by SWs was discussed in [205]. New diagnostic tools developed in the course of the study were described. Two types of electron heating modes were observed in a low-pressure planar microwave discharge: the low-density, high-temperature bulk heating mode ($T_e \sim 10$ eV), and the high-density, low-temperature surface heating mode ($T_e \sim 3$ eV). An abrupt transition between the two modes was reported to take place at the cut-off density. The surface heating was pointed out to lead to a bi-Maxwellian electron energy distribution function in the non-local kinetic regime. On the other hand, the bulk heating was noted to lead to an electron energy distribution function relatively close to a single Maxwellian distribution with electron temperature higher than the bulk temperature of the surface heating mode. Such differences in the electron energy distribution functions made the compositions of reactive species different between SW plasmas and inductively coupled plasmas. Namely, the radical measurements at the same electron density showed more F and fewer CF_x radicals in the inductively coupled plasma than in the SW plasma.

As it is shown in [206, 207] for simple models, surface flute waves are able to sustain gas discharges under the condition of utilization of an external magnetic field. Their eigenfrequencies are in the range of the electron cyclotron resonance. Therefore, they have good prospects to be applied for plasma production.

References

1. Shkvarunets, A. G., Kobayashi, S., Weaver, J., et al. (1996). Electromagnetic properties of corrugated and smooth waveguides filled with radially inhomogeneous plasma. *IEEE Transactions on Plasma Science, 24*(3), 905–917.
2. Vlasov, A. N., Shkvarunets, A. G., Rodgers, J. C., et al. (2000). Overmoded GW-class surface-wave microwave oscillator. *IEEE Transactions on Plasma Science, 28*(3), 505–560.
3. Ginzburg, N. S., Malkin, A. M., Sergeev, A. S., et al. (2001). Quasi-optical model of wave propagation near a shallow corrugated surface. *Applied Physics Letters, 99*, 121505.
4. Ginzburg, N. S., Malkin, A. M., Sergeev, A. S., et al. (2012). Powerful surface-wave oscillators with two-dimensional periodic structures. *Journal of Applied Physics, 100*, 143510.
5. Malkin, A. M., Ginzburg, N. S., Fil'chenkov, S. E., et al. (2017). Surface-wave Bragg resonators for terahertz frequency range. In *Proceedings of the 10th International Workshop "Strong Microwaves and Terahertz Waves: Sources and Applications"* (pp. 214–215). Nizhny Novgorod: Russian Federation.
6. Shin, Y. M., So, J. K., Jang, K. H., et al. (2007). Evanescent tunneling of an effective surface plasmon excited by convection electrons. *Physical Review Letters, 99*, 147402.
7. Bratman, V. L., Fedotov, A. E., Makhalov, P. B., et al. (2009). Smith-Purcell frequency multiplier with synchronization of radiation from a wide electron beam. *Applied Physics Letters, 94*, 061501.
8. Konoplev, I. V., Cross, A. W., Phelps, A. D. R., et al. (2007). Co-axial free-electron maser based on two-dimensional distributed feedback. *Physics Reviews E, 76*(5), 056406.

9. Arzhannikov, A. V., Ginzburg, N. S., Kalinin, P. V., et al. (2016). Using two-dimensional distributed feedback for synchronization of radiation from two parallel-sheet electron beams in a free-electron maser. *Physics Reviews Letter, 117*(11), 114801.
10. Ginzburg, N. S., Ilyakov, E. V., Kulagin, I. S., et al. (2017). Theoretical and experimental investigations of oversized Ka-band surface-wave oscillator based on 2D periodical corrugated structure. In *Proceedings of the 10th International Workshop "Strong Microwaves and Terahertz Waves: Sources and Applications* (pp. 253–254). Nizhny Novgorod: Russian Federation.
11. Ginzburg, N. S., Malkin, A. M., Sergeev, A. S., et al. (2013). Oversized co-axial and cylindrical surface-wave oscillators with two-dimensional periodic grating (quasi-optical model). *Journal of Applied Physics., 113*(10), 104504.
12. Nusinovich, G. S., Carmel, Y., Shkvarunets, A. G., et al. (2005). The pasotron: progress in the theory and experiments. *IEEE Transactions on Electron Devices, 52*(5), 845–856.
13. Bykov, V. M., Gubanov, V. P., Rostov, A. V., et al. (1991). Relativistic pulse-periodic microwave oscillators. In *Proceedings of the 1st International Workshop "Strong Microwaves in Plasmas"* (Vol. 2, pp. 866–871). Russian Federation.
14. Sclivanov, I. A., Fedotov, A. V., & Shkvarunets, A. G. (1991). The excitation of the fundamental mode of a coaxial plasma slow-wave structure of a PCM-amplifier. In *Proceedings of the 1st International Workshop "Strong Microwaves in Plasmas"* (Vol. 2, pp. 462–467). Russian Federation.
15. Li, S., Wang, J., & Wang, D. (2020). Relativistic surface wave oscillator in Y-band with large oversized structures modulated by dual reflectors. *Scientific Reports, 10*, 336.
16. Cao, Y., Song, Z., Wu, P., et al. (2017). Effective suppression of pulse shortening in a relativistic backward wave oscillator. *Physics of Plasmas, 24*(3), 033109.
17. Malkin, A. M., Zheleznov, I. V., Sergeev, A. S., et al. (2021). Quasi-optical theory of relativistic Cherenkov surface-wave oscillators with oversized cylindrical waveguides. *Physics of Plasmas, 28*(6), 063102.
18. Buts, V. A., Miroshnichenko, V. I., Ognivenko, V.V., et al. (1993). The excitation of a plasma-filled slow resonator by an electron beam. In N. Novgorod (Ed.), *Proceedings of the 2nd International Workshop "Strong Microwaves in Plasmas"* (Vol. 2, pp. 462–467). Russian Federation.
19. Kovalev, N. F., Kolganov, N. G., Palitsin, A. V., et al. (2000). Relativistic BWO with cyclotron selection of an asymmetric wave. In N. Novgorod (Ed.), *Proceedings of the 4th International Workshop "Strong Microwaves in Plasmas"* (Vol. 2, pp. 845–850). Russian Federation.
20. Bratman, V. L., Kalynov, Yu. K., Kolganov, N. G., et al. (1996). Cyclotron autoresonance masers and relativistic gyrotrons. In N. Novgorod (Ed.), *Proceedings of the 3rd International Workshop "Strong Microwaves in Plasmas"* (Vol. 2, pp. 745–761). Russian Federation.
21. Bratman, V. L., Denisov, G. G., Kalynov, Yu. K., et al. (2000). Novel types of cyclotron resonance devices. In: N. Novgorod (Ed.), *Proceedings of the 4th International Workshop "Strong Microwaves in Plasmas"* (Vol. 2, pp. 683–702), Russian Federation.
22. Deslandes, D., & Wu, K. (2006). Accurate modeling wave mechanisms and design considerations of a substrate integrated waveguide. *IEEE Transactions Microwave Theory Technology, 54*(6), 2516–2526.
23. Bozzi, M., Georgiadis, A., & Wu, K. (2011). Review of substrate integrated waveguide circuits and antennas. *IET Microwave Antennas Propagation, 5*(8), 909–920.
24. Huang, L., & Cha, H. (2015). Compact ridged half-mode substrate integrated waveguide bandpass filter. *IEEE Microwave Wireless Components Letter, 25*(4), 223–225.
25. Ji, L., Li, X.-C., He, X., et al. (2021). A slow wave ridged half-mode substrate integrated waveguide with spoof surface plasmon polaritons. *IEEE Transactions on Plasma Science, 49*(6), 1818–1825.
26. Luo, Q., Zhang, A. L., & Peng, H. P. (2021). Design and simulation of a surface wave-based cylindrical hollow plasma cavity for wakefield booster for future e+e− colliders. *IEEE Transactions on Plasma Science, 49*(6), 1848–1853.

27. Girka, I. O., Pavlenko, I. V., & Thumm, M. (2021). Zeroth radial modes of azimuthal surface waves in dense plasma loaded, coaxial helix traveling-wave-tube-like waveguides. *Physics of Plasmas, 28*(4), 043106.
28. Girka, I. O., & Thumm, M. (2021). Azimuthal surface waves in low-density plasma loaded, coaxial helix travelling-wave-tube-like waveguides. *Problems of Atomic Science and Technology. Series Plasma Electronics and New Methods of Acceleration, 4*(134), 24–29.
29. Annaka, Y., Ogura, K., Rachi, K., et al. (2021). Design and fabrication of 200-GHz oversized surface-wave oscillator. *IEEE Transactions on Plasma Science, 49*(1), 33–39.
30. Liu, Y., Liu, W., Yu, Z., et al. (2021). Efficient and tunable terahertz emission by transforming surface waves to coherent Smith-Purcell radiation. *IEEE Transactions on Plasma Science, 49*(1), 455–459.
31. Sun, S., He, Q., Xiao, S., et al. (2012). Gradient-index meta-surfaces as a bridge linking propagating waves and surface waves. *Nature Materials, 11*, 426–431.
32. Paknys, R., & Jackson, D. R. (2005). The relation between creeping waves, leaky waves, and surface waves. *IEEE Transactions on Antennas and Propagation, 53*(3), 898–907.
33. Zheng, L., Yang, X., Gong, W., et al. (2021). Scattering of surface traveling wave and its attenuation in absorbing coating. *Journal of Applied Physics, 130*(10), 105304.
34. Dai, S., Ma, Q., Andersen, T., et al. (2015). Subdiffractional focusing and guiding of polaritonic rays in a natural hyperbolic material. *Nature Communications, 6*, 6963.
35. Zhou, K., Lu, L., Li, B., et al. (2021). Hyperbolic plasmon–phonon dispersion and tunable spontaneous emission enhancement in Ge2Sb2Te5-based multilayer graphene and hBN system. *Journal of Applied Physics, 130*(9), 093102.
36. Thomaschewski, M., Yang, Y., Wolff, C., et al. (2019). On-chip detection of optical spin–orbit interactions in plasmonic nanocircuits. *Nano Letters, 19*(2), 1166–1171.
37. Rudé, M., Simpson, R. E., Quidant, R., et al. (2015). Active control of surface plasmon waveguides with a phase change material. *ACS Photonics, 2*(6), 669–674.
38. Wu, S., Hogan, N., & Sheldon, M. (2019). Hot electron emission in plasmonic thermionic converters. *ACS Energy Letters, 4*(10), 2508–2513.
39. Splawinski, M., Bostock, S., Chau, K. J., et al. (2021). Superlens coupling to object and image: A secondary resonance mechanism to improve single-negative imaging of electromagnetic waves. *Journal of Applied Physics, 129*(16), 163102.
40. Beg, F. N., Bell, A. R., Dangor, A. E., et al. (1997). A study of picosecond laser–solid interactions up to 1019 W cm^{-2}. *Physics of Plasmas, 4*(2), 447–457.
41. Héron, A., Adam, J. C., & Mora, P. (2020). Generation of ultra-energetic ions by interaction of petawatt lasers with micrometer-scale foils. *Physics of Plasmas, 27*(1), 013103.
42. Marini, S., Kleij, P. S., Amiranoff, F., et al. (2021). Key parameters for surface plasma wave excitation in the ultra-high intensity regime. *Physics of Plasmas, 28*(7), 073104.
43. Anderson, T. (2011). *Plasma antennas.* Artech House.
44. Jenn, D. C. (2003). *Plasma antennas: Survey of techniques and. the current state of the art.* San Diego, CA: SPAWAR PMW 189.
45. Borg, G. G., Harris, J. H., Miljak, D. G., et al. (1999). Application of plasma columns to radiofrequency antennas. *Applied Physics Letters, 74*, 3272–3274.
46. Moisan, M., Shivarova, A., & Trivelpiece, A. W. (1982). Experimental investigations of the propagation of surface waves along a plasma column. *Plasma Physics, 24*(11), 1331–1400.
47. Borg, G. G., Harris, J. H., Martin, N. M., et al. (2000). Plasmas as antennas: Theory, experiment and applications. *Physics of Plasmas, 7*(5), 2198–2202.
48. Alexeff, I., Anderson, T., Parameswaran, C., et al. (2006). Experimental and theoretical results with plasma antennas. *IEEE Transaction on Plasma Science, 34*(2), 166–172.
49. Alexeff, I., Anderson, T., Farshi, E., et al. (2008). Recent results for plasma antennas. *Physics of Plasmas, 15*, 057104.
50. Kumar, R., & Bora, D. (2010). A reconfigurable plasma antenna. *Journal of Applied Physics, 107*, 053303.
51. Fathy, A. E., Rosen, A., Owen, H. S., et al. (2003). Silicon-based reconfigurable antennas—concepts, analysis, implementation, and feasibility. *IEEE Transactions on Microwave Theory and Techniques, 51*(6), 1650–1661.

52. Rayner, J. P., Whichello, A. P., & Cheetham, A. D. (2004). Physical characteristics of plasma antennas. *IEEE Transactions on Plasma Science, 52*(1), 269–281.
53. Wu, M., Wen, B. Y., & Zhou, H. (2009). Ionospheric clutter suppression in HF surface wave radar. *Journal of Electromagnetic Waves and Applications, 25*(10), 1265–1272.
54. Russo, P., Cerri, G., & Vecchioni, E. (2011). Self-consistent analysis of cylindrical plasma antennas. *IEEE Transactions on Antennas and Propagation, 59*(5), 1503–1511.
55. Xiong, J., Li, H., Wang, B. Z., et al. (2011). Theoretical investigation of rectangular patch antenna miniaturization based on the DPS-ENG bi-layer super-slow TM wave. *Progress In Electromagnetics Research, 118*, 379–396.
56. Ye, H. Q., Gao, M., & Tang, C. J. (2011). Radiation theory of the plasma antenna. *IEEE Transactions on Antennas and Propagation, 59*(5), 1497–1502.
57. Dynin, E. E., Kostrov, A. V., Smisnov, A. I., et al. (2000). Influence of the magnetized plasma shell on efficiency of radiation of a short antenna. In N. Novgorod (Ed.), *Proceedings of the 4th International Workshop "Strong Microwaves in Plasmas"* (Vol. 2, pp. 481–484). Russian Federation.
58. De Carlo, P., Magarotto, M., Mansutti, G., et al. (2021). Experimental characterization of a plasma dipole in the UHF band. *IEEE Antennas and Wireless Propagation Letters, 20*(9), 1621–1625.
59. Milligan, T. (2005). *Modern antenna design* (2nd ed., p. 512). John Wiley & Sons, Inc.
60. Liao, Z., Che, Y. Z., Luo, G. Q., et al. (2021). Enhanced radiation characteristics for Vivaldi antenna using spoof surface plasmon polaritons. *IEEE Transactions on Plasma Science, 49*(9), 2730–2736.
61. Singletary, P. J., & Cohen, M. B. (2021). Using a high-speed plasma as a conducting channel to enable a novel antenna approach. *IEEE Transactions on Plasma Science, 49*(2), 794–804.
62. Xu, K.-D., Lu, S., Guo, Y.-J., et al. (2021). High-order mode of spoof surface plasmon polaritons and its application in bandpass filters. *IEEE Transactions on Plasma Science, 49*(1), 269–275.
63. Bashkuev, Y. B., Khaptanov, V. B., & Dembelov, M. G. (2010). Experimental proof of the existence of a surface electromagnetic wave. *Technical Physics Letters, 36*(2), 136–139.
64. Bashkuyev, Y. B., Angarkhaeva, L. K., Naguslayeva, I. B., et al. (2016). Frequency domain criterion of appearance of an electromagnetic surface wave above the laminar ice-salt water structure. *Radiophysics and Quantum Electronics*, 59(5), 361–368.
65. Bashkuev, Y., Khaptanov, V., Dembelov, M., et al. (2019). Surface electromagnetic waves over the "ice-sea" structure. In: *2019 Russian Open Conference on Radio Wave Propagation (RWP)* (pp. 232–235). Kazan, Russia.
66. Wait, J. R. (1962). *Electromagnetic waves in stratified media.* Pergamon Press.
67. Xu, H., Sun, H., Zhang, B., et al. (2015). Transformation cloaks for surface electromagnetic waves. In: *2015 IEEE 4th Asia-Pacific Conference on Antennas and Propagation (APCAP)* (pp. 41–43). Bali, Indonesia.
68. Smolyaninov, I. I., Balzano, Q., Davis, C. C., et al. (2018). Surface wave based underwater radio communication. *IEEE Antennas and Wireless Propagation Letters, 17*(12), 2503–2507.
69. Zhang, X., Xu, Y., Tian, Z., et al. (2015). Launching terahertz surface wave with desired directions. In: *2015 40th International Conference on Infrared, Millimeter, and Terahertz waves (IRMMW-THz)* (pp. 1–2). Hong Kong, China.
70. Munk, B. A. (2003). *Finite antenna arrays and FSS.* John Wiley & Sons.
71. Weynants, R. R. (2009). ICRF review: From ERASMUS to ITER. *Proceedings of AIP Conference on Radio Frequency Power in Plasmas, 1187*, 3–12.
72. Messiaen, A. M., & Weynants, R. R. (1982). In: Heating in Toroidal Plasmas. *Proceedings of 3rd Varenna-Grenoble Int. Symposium* (Vol. 3, p. 1106).
73. Ballico, R. C., & Cross, M. J. (1990). Probe measurements of ICRF Alfven surface waves in the TORTUS tokamak. *Fusion Engineering and Design, 12*(1), 197–201.
74. Murphy, A. B. (1990). Waves in the edge plasma during ion cyclotron resonance heating. *Fusion Engineering and Design, 12*(1, 2), 79–92.

75. Porkolab, M. (1990). Parametric instabilities in the tokamak edge plasma in the ion cyclotron heating regimes. *Fusion Engineering and Design, 12*(1,2), 93–103.
76. Noterdaeme, J.-M., ICRH-team, ASDEX-team, NI-team (1990). Experimental results on edge effects during ICRF heating of ASDEX plasmas. *Fusion Engineering and Design, 12*(1,2), 127–201.
77. Lin, Y., Rice, J. E., Wukitch, S. J., et al. (2009). ICRF mode conversion flow drive on Alcator CMod and projections to other tokamaks. In: *Proceedings of AIP Conference on Radio Frequency Power in Plasmas* (Vol. 1187, pp. 57–64).
78. Lyssoivan, A., Douai, D., Koch, R., et al. (2012). Simulation of ITER full-field ICWC scenario in JET: RF physics aspects. *Plasma Physics and Controlled Fusion, 54*, 074014.
79. Aliev, Y. M., Schlüter, H., & Shivarova, A. (2000). *Guided-wave-prvduced plasmas*. Springer.
80. Czarnecka, A., Durodie, F., Figueiredo, A. C. A., et al. (2012). Impurity production from the ion cyclotron resonance heating antennas in JET. *Plasma Physics and Controlled Fusion, 54*, 074013.
81. Katanuma, I., Yagi, K., Haraguchi, Y., et al. (2011). The stability of flute modes in the GAMMA 10 A-divertor. *Plasma and Fusion Research, 6*, 2403080.
82. Chen, F. F. (1974). *Introduction to plasma physics*. Plenum Press.
83. Katanuma, I., Masaki, S., Sato, S., et al. (2011). Flute instability in the tandem mirror with the divertor/dipole regions. *Physics of Plasmas, 18*, 112304.
84. Manz, P., Ribeiro, T. T., Scott, B. D., et al. (2015). Origin and turbulence spreading of plasma blobs. *Physics of Plasmas, 22*(2), 022308.
85. Myra, J. R., & D'Ippolito, D. A. (2010). Scattering of radio frequency waves by blob filaments. *Physics of Plasmas, 17*(10), 102510.
86. Tierens, W., Zhang, W., Myra, J. R., et al. (2020). Filament-assisted mode conversion in magnetized plasmas. *Physics of Plasmas, 27*(1), 010702.
87. Messiaen, A., & Weynants, R. (2011). ICRH antenna coupling physics and optimum plasma edge density profile. Application to ITER. *Plasma Physics Controlled Fusion, 53*(8), 085020.
88. Messiaen, A., & Maquet, V. (2020). Coaxial and surface mode excitation by an ICRF antenna in large machines like DEMO and ITER. *Nuclear Fusion, 60*(7), 07601.
89. Grishanov, N. I., & Azarenkov, N. A. (2021). On the fast waves in a cylindrical current-carrying plasma. *Physics of Plasmas, 28*(4), 042106.
90. Girka, V., Girka, I., Pavlenko, I., et al. (2012). Coupled azimuthal modes propagating in current-carrying plasma waveguides. *Journal of Plasma Physics, 78*(2), 105–123.
91. Girka, V. O., & Girka, I. O. (1992). Coupled azimuthal surface waves in a nonuniform current—carrying plasma cylinder. *Soviet Journal of Communications Technology and Electronics, 37*(4), 23–29.
92. Girka, I. A., & Stepanov, K. N. (1990). Influence of azimuthal magnetic field on the spectra of magnetosonic oscillations of plasma cylinder. *Reports of Ukrainian Academy of Sciences Series, A*(3). 66–71. (in Russian).
93. Girka, I. A., & Stepanov, K. N. (1991). Influence of plasma column's toroidicity and ellipticity on the spectra of MHD waves. *Ukrainian Physical Journal, 36*(7), 1051–1058. (in Russian).
94. Girka, I. A., & Stepanov, K. N. (1992). MHD plasma oscillations in a bumpy magnetic field. *Ukrainian Physical Journal, 37*(1), 69–75. (in Russian).
95. Girka, I. A., Lapshin, V. I., & Stepanov, K. N. (1998). Splitting of the spectra of MHD plasma oscillations in a rippled magnetic field. *Plasma Physics Reports, 24*(11), 948–955.
96. Girka, I. O., & Kovtun, P. K. (2000). Effect of the helical nonuniformity of the confining magnetic field on the MHD eigenmodes in straight stellarators. *Plasma Physics Reports, 26*(1), 33–40.
97. Qin, C. M., Zhao, Y. P., Wang, H. Q., et al. (2013). Experimental investigation of the potentials modified by radio frequency sheaths during ion cyclotron range of frequency on EAST. *Plasma Physics and Controlled Fusion, 55*(1), 015004.
98. Kohno, H., Myra, J. R., & D'Ippolito, D. A. (2013). Radio-frequency sheath-plasma interactions with magnetic field tangency points along the sheath surface. *Physics of Plasmas, 20*(8), 082514.

99. Myra, J. R., & Kohno, H. (2019). Calculation of RF sheath properties from surface wave-fields: a post-processing method. *Plasma Physics and Controlled Fusion, 61*(9), 095003.
100. Anders, A. (2007). Metal plasmas for the fabrication of nanostructures. *Journal of Physics D: Applied Physics, 40*, 2272–2284.
101. Maier, S. A. (2007). *Plasmonics: Fundamentals and applications*. Springer.
102. Beck, F. J., Polman, A., & Catchpole, K. R. (2009). Tunable light trapping for solar cells using localized surface plasmons. *Journal of Applied Physics, 105*, 114310.
103. Catchpole, K. R., & Polman, A. (2009). Photovoltaic cell with surface plasmon resonance generating nanostructures, *US Patent Application 20110197959*. http://WWW.patentstorm.us/applications/20110197959/claims.html.
104. Soler, M., & Lechuga, L. M. (2021). Principles, technologies, and applications of plasmonic biosensors. *Journal of Applied Physics, 129*(11), 111102.
105. Wood, R. W. (1901). On a remarkable case of uneven distribution of light in a diffraction grating spectrum. *Proceedings of the Physical Society of London, 18*(1), 269–275.
106. Wood, R. W. (1912). XXVII. Diffraction gratings with controlled groove form and abnormal distribution of intensity. London, Edinburgh, Dublin Philos. *Magazine Journal Science, 23*(134), 310–317.
107. Strutt, J. W. (1907). On the dynamical theory of gratings. *Proceeding of the Royal Society London. Series A Contain Paper Mathematical Physics Character, 79*(532), 399–416.
108. Wood, R. W. (1935). Anomalous diffraction gratings. *Physical Review, 48*(12), 928–936.
109. Palmer, C. H. (1952). Parallel diffraction grating anomalies*. *Journal of the Optical Society of America, 42*(4), 269–276.
110. Palmer, C. H. (1956). Diffraction grating anomalies. II. Coarse gratings*. *Journal of the Optical Social of America, 46*(1), 50–53.
111. Fano, U. (1941). The theory of anomalous diffraction gratings and of quasi-stationary waves on metallic surfaces (Sommerfeld's waves). *Journal of the Optical Society of America, 31*(3), 213–222.
112. Bohm, D., & Pines, D. A. (1951). Collective description of electron interactions. I. Magnetic interactions. *Physical Review, 82*(5), 625–634.
113. Pines, D., & Bohm, D. A. (1952). Collective description of electron interactions: II Collective vs individual particle aspects of the interactions. *Physical Review, 85*(2), 338–353.
114. Pines, D. A. (1953). Collective description of electron interactions: IV. Electron interaction in metals. *Physical Review, 92*(3), 626–636.
115. Kretschmann, E., & Raether, H. Z. (1968). Radiative decay of non radiative surface plasmons excited by light. *Zeitschrift fur Naturforschung a, 23*(12), 2135–2136.
116. Otto, A. (1968). Excitation of nonradiative surface plasma waves in silver by the method of frustated total reflection. *Zeitschrift für Physik, 216*(4), 398–410.
117. Liedberg, B., Nylander, C., & Lunström, I. (1983). Surface plasmon resonance for gas detection and biosensing. *Sensors Actuators, 4(C)*, 299–304.
118. Cullen, D. C., Brown, R. G. W., & Lowe, C. R. (1987). Detection of immuno-complex formation via surface plasmon resonance on gold-coated diffraction gratings. *Biosensors, 3*(4), 211–225.
119. Pockrand, I., Swalen, J. D., Gordon, J. G., et al. (1978). Surface plasmon spectroscopy of organic monolayer assemblies. *Surface Science, 74*(1), 237–244.
120. Malmqvist, M. (1999). BIACORE: An affinity biosensor system for characterization of biomolecular interactions. *Biochemical Society Transactions, 27*(2), 335–340.
121. Liedberg, B., Nylander, C., & Lundström, I. (1995). Biosensing with surface plasmon resonance—How it all started. *Biosensors & Bioelectronics, 10*(8), i–ix.
122. Green, R. J., Frazier, R. A., & Shakesheff, K. M. (2000). Surface plasmon resonance analysis of dynamic biological interactions with biomaterials. *Biomaterials, 21*, 1823–1835.
123. Morrow, R., McKenzie, D. R., & Bilek, M. M. M. (2007). Electric field effects on adsorption/desorption of proteins and colloidal particles on a gold film observed using surface plasmon resonance. *Physica B: Condensed Matter, 394*, 203–207.

124. Feltis, B. N., Sexton, B. A., & Glenn, F. L. (2008). A hand-held surface plasmon resonance biosensor for the detection of ricin and other biological agents. *Biosensors & Bioelectronics, 23*, 1131–1136.
125. Ederra, I., Iriarte, J. C., Gonzalo, R., et al. (2011). Surface waves of finite size electromagnetic band gap woodpile structures. *Progress in Electromagnetics Research B, 28*, 19–34.
126. Joannopoulos, J. D., Meade, R. D., & Finn, J. N. (1995). *Photonic crystals: Molding the flow of light*. Princeton: Princeton University Press.
127. Kawashima, S., Ishizaki, K., & Noda, S. (2010). Light propagation in three-dimensional photonic crystals. *Optics Express, 18*(1), 386–392.
128. Sozuer, H. S., & Dowling, J. P. (1994). Photonic band calculations for woodpile structure. *Journal of Modern Optics, 43*, 231–234.
129. Gonzalo, R., Ederra, I., Maim, C. M., et al. (2001). Radiation properties of terahertz dipole antenna mounted on photonic crystal. *Electronics Letters, 57*(10), 613–614.
130. Ederra, I., Gonzalo, R., & Alderman, B. E. J. (2008). Electromagnetic band-gap based planar imaging array for 500 GHz. *IEEE Transactions Microwaves Theory Technology, 56*(11), 2256–2265.
131. Weily, A. R., Esselle, K. P., & Sanders, B. C. (2003). Photonic crystal horn and array antennas. *Physical Review E, 68*(1), 016609.
132. Saxler, J., Rivas, J. G., & Janke, C. (2004). Time-domain measurements of surface plasmon polaritons in the terahertz frequency range. *Physical Review B, 69*, 155427.
133. Gerasimov, V. S., Cherkassky, V. S., Knyazev, B. A., et al. (2012). Study of surface plasmons travelling along straight and curved metal-dielectric interfaces: Experiment and theory. In *37th International Conference on Infrared, Millimeter and Terahertz Waves*, Wollongong, Australia, September 23–28, Tue-A-4-4.
134. Ilin, N. V., Kondratiev, I. G., & Smirnov, A. I. (2009). Planar wave guiding systems based on resonant meta materials. In N. Novgorod (Ed.), *Proceedings of the 7th International Workshop "Strong Microwaves in Plasmas"* (Vol. 2, pp. 550–554). Russian Federation.
135. Losurdo, M., Moreno, F., Cobet, C., et al. (2021). Plasmonics: Enabling functionalities with novel materials. *Journal of Applied Physics, 129*(22), 220401.
136. Vanraesa, P., & Bogaerts, A. (2021). The essential role of the plasma sheath in plasma–liquid interaction and its applications—A perspective. *Journal of Applied Physics, 129*(22), 220901.
137. Gabbani, A., Petrucci, G., & Pineidera, F. (2021). Magneto-optical methods for magnetoplasmonics in noble metal nanostructures. *Journal of Applied Physics, 129*(21), 211101.
138. Koenderink, A. F., Alu, A., & Polman, A. (2015). Nanophotonics: Shrinking light-based technology. *Science, 348*(6234), 516–521.
139. Gellé, A., Jin, T., de la Garza, L., et al. (2020). Applications of plasmon-enhanced nanocatalysis to organic transformations. *Chemical Reviews, 120*(2), 986–1041.
140. Larsson, E. M., Langhammer, C., Zorić, I., et al. (2009). Nanoplasmonic probes of catalytic reactions. *Science, 326*(5956), 1091–1094.
141. Ueno, K., Oshikiri, T., Sun, Q., et al. (2017). Solid-state plasmonic solar cells. *Chemical Reviews, 118*(6), 2955–2993.
142. Mayer, K. M., & Hafner, J. H. (2011). Localized surface plasmon resonance sensors. *Chemical Reviews, 111*(6), 3828–3857.
143. Mejía-Salazar, J. R., & Oliveira, O. N. (2018). Plasmonic biosensing: Focus review. *Chemical Reviews, 118*(20), 10617–10625.
144. Stewart, M. E., Anderton, C. R., Thompson, L. B., et al. (2008). Nanostructured plasmonic sensors. *Chemical Reviews, 108*(2), 494–521.
145. Langer, J., Jimenez de Aberasturi, D., Aizpurua, J., et al. (2020). Present and future of surface-enhanced Raman scattering. *ACS Nano, 14*(1), 28–117.
146. Ding, S.-Y., Yi, J., Li, J.-F., et al. (2016). Nanostructure-based plasmon-enhanced Raman spectroscopy for surface analysis of materials. *Nature Reviews Materials, 1*(6), 16021.
147. Itoh, T., Yamamoto, Y. S., & Ozaki, Y. (2017). Plasmon-enhanced spectroscopy of absorption and spontaneous emissions explained using cavity quantum optics. *Chemical Society Reviews, 46*(13), 3904–3921.

148. Nie, S., & Emory, S. R. (1997). Probing single molecules and single nanoparticles by surface-enhanced Raman scattering. *Science, 275*(5303), 1102–1106.
149. Manera, M. G., Ferreiro-Vila, E., Garcia-Martin, J. M., et al. (2014). Enhanced antibody recognition with a magneto-optic surface plasmon resonance (MO-SPR) sensor. *Biosensors and Bioelectronics, 58*, 114–120.
150. https://www.marketsandmarkets.com/Market-Reports/biosensors-market-798.html.
151. Lopez, G. A., Estevez, M.-C., Soler, M., et al. (2017). Recent advances in nanoplasmonic biosensors: Applications and lab-on-a-chip integration. *Nanophotonics, 6*(1), 123–136.
152. Zhou, X., Wong, T., Song, H., et al. (2014). Development of localized surface plasmon resonance-based point-of-care system. *Plasmonics, 9*, 835–844.
153. Lee, K.-L., You, M.-L., Tsai, C.-H., et al. (2016). Nanoplasmonic biochips for rapid label-free detection of imidacloprid pesticides with a smartphone. *Biosensors & Bioelectronics, 75*, 88–95.
154. Langer, J., Novikov, S. M., & Liz-Marzán, L. M. (2015). Sensing using plasmonic nanostructures and nanoparticles. *Nanotechnology, 26*(32), 322001.
155. Attridge, J. W., Daniels, P. B., Deacon, J. K., et al. (1991). Sensitivity enhancement of optical immunosensors by the use of a surface plasmon resonance fluoroimmunoassay. *Biosensors & Bioelectronics, 6*(3), 201–214.
156. Kinkhabwala, A., Yu, Z., Fan, S., et al. (2009). Large single-molecule fluorescence enhancements produced by a bowtie nanoantenna. *Nature Photonics, 3*, 654–657.
157. Chen, H., Zhang, H., Liu, M., et al. (2017). Realization of tunable plasmon-induced transparency by bright-bright mode coupling in Dirac semimetals. *Optical Materials Express, 7*, 3397–3407.
158. Han, L., Tan, Q., Li, H., et al. (2021). Applications of chip-scale semiconductor metamaterials based on plasmon-induced transparency in modulation and sensing. *Journal of Applied Physics, 129*(13), 133105.
159. Zhou, Y.-C., Zhang, X.-C., Hu, Y.-X., et al. (2021). High-Q perfect absorption induced by the coupling of LSP and SPP modes. *Journal of Applied Physics, 129*(8), 083103.
160. Agrawal, A., Cho, S. H., Zandi, O., et al. (2018). Localized surface plasmon resonance in semiconductor nanocrystals. *Chemical Reviews, 118*(6), 3121–3207.
161. Jamilan, S., Semouchkin, G., & Semouchkina, E. (2021). Analog of electromagnetically induced transparency in metasurfaces composed of identical dielectric disks. *Journal of Applied Physics, 129*(6), 063101.
162. Lin, L., Bi, X., Gu, Y., et al. (2021). Surface-enhanced Raman scattering nanotags for bioimaging. *Journal of Applied Physics., 129*(19), 191101.
163. Mahigir, A., & Veronis, G. (2021). Light coupling structures and switches for plasmonic coaxial waveguides. *Journal of Applied Physics., 129*(19), 193103.
164. Kauranen, M., & Zayats, A. (2012). Nonlinear plasmonics. *Nature Photonics, 6*, 737–748.
165. Wurtz, G. A., & Zayats, A. V. (2008). Nonlinear surface plasmon polaritonic crystals. *Laser Photonical Review, 2*, 125–135.
166. Ionin, A. A., Kudryashov, S. I., Seleznev, L. V., et al. (2013). Nonlinear regime of the excitation of a surface electromagnetic wave on the silicon surface by an intense femtosecond laser pulse. *JETP Letters, 97*(3), 121–125.
167. Uryupin, S. A., & Frolov, A. A. (2013). Excitation of surface waves by a short laser pulse in a conductor. *Quantum Electronics, 43*(12), 1132–1138.
168. Uryupin, S. A., & Frolov, A. A. (2014). Generation of surface waves and low-frequency radiation under exposure of a conductor to a laser pulse focused by a cylindrical lens. *Quantum Electronics, 44*(9), 866–872.
169. Frolov, A. A., & Uryupin, S. A. (2017). Excitation of THz surface waves in the conductor by a drag current generated by a focused femtosecond pulse. In *Proceedings of the 10th International Workshop "Strong Microwaves and Terahertz Waves: Sources and Applications"* (pp. 333–334). Nizhny Novgorod, Russian Federation, 17–22 July, 2017.
170. Sederberg, S., Firby, C. J., Greig, S. R., et al. (2017). Integrated nanoplasmonic waveguides for magnetic, nonlinear, and strong-field devices. *Nanophotonics, 6*(1), 235–257.

171. Bortchagovsky, E., Demydenko, Y., Bogoslovskaya, A., et al. (2021). Microellipsometry study of plasmonic properties of metal–insulator–metal structures with ordered lattices of nanoparticles. *Journal of Applied Physics, 129*(12), 123104.
172. Ray, P. C. (2010). Size and shape dependent second order nonlinear optical properties of nanomaterials and their application in biological and chemical sensing. *Chemical Reviews, 110*, 5332–5365.
173. Ray, P. C. (2006). Diagnostics of single base-mismatch DNA hybridization on gold nanoparticles by using the hyper-Rayleigh scattering technique. *Angewandte Chemie International Ed., 45*, 1151–1154.
174. Neely, A., Perry, C., Varisli, B., et al. (2009). Ultrasensitive and highly selective detection of Alzheimer's disease biomarker using two-photon Rayleigh scattering properties of gold nanoparticle. *ACS Nano, 3*, 2834–2840.
175. Kim, Y., Johnson, R. C., & Hupp, J. T. (2001). Gold nanoparticle-based sensing of "spectroscopically silent" heavy metal ions. *Nano Letters, 1*, 165–167.
176. Singh, A. K., Senapati, D., Wang, S., et al. (2009). Gold nanorod based selective identification of *Escherichia coli* bacteria using two-photon Rayleigh scattering spectroscopy. *ACS Nano, 3*, 1906–1912.
177. Lu, W., Arumugam, S. R., Senapati, D., et al. (2010). Multifunctional oval-shaped gold-nanoparticle-based selective detection of breast cancer cells using simple colorimetric and highly sensitive two-photon scattering assay. *ACS Nano, 4*, 1739–1749.
178. Emboras, A., Niegemann, J., Ma, P., et al. (2016). Atomic scale plasmonic switch. *Nano Letters, 16*(1), 709–714.
179. Moisan, M., & Zakrzewski, Z. (1991). Plasma sources based on the propagation of electromagnetic surface waves. *Journal of Physics D: Applied Physics, 24*, 1025–1048.
180. Margot-Chaker, J., Moisan, M., Chaker, M., et al. (1989). Tube diameter and wave frequency limitations when using the electro-magnetic surface wave in the m=l (dipolar) mode to sustain a plasma column. *Applied Physics, 66*(9), 4134–4148.
181. Zhelyazkov, I., & Atanassov, V. (1995). Axial structure of low-pressure high-frequency discharges sustained by travelling electromagnetic surface waves. *Physics Reports, 255*, 79–201.
182. Sugai, H., Ghanashev, L., & Nagatsu, M. (1998). High-density flat plasma production based on surface waves. *Plasma Sources Science and Technology, 7*, 192–205.
183. Margot-Chaker, J., Moisan, M., & Teichmann, J. (1991). A new approach to the development of ECR plasma sources. In *Proceedings of the 1st International Workshop "Strong Microwaves in Plasmas"* (Vol. 1, pp. 473–478). Suzdal, Russian Federation.
184. Russo, P., Primiani, V. M., Cerri, G., et al. (2011). Experimental characterization of a surfaguide fed plasma antenna. *IEEE Transactions on Antennas and Propagation, 59*(2), 425–433.
185. Durandet, A., Arnal, Y., Margot-Chaker, J., et al. (1989). Investigation of a plasma source sustained by an electromagnetic surface wave at 2.45 GHz under free-fall regime. *Journal of Physics D. Applied Physics, 22*(9), 1288–1299.
186. Delprat, S., Chaker, M., & Margot, J. (2001). Investigation of the gas pressure influence on patterned platinum etching characteristics using a high-density plasma. *Journal of Applied Physics, 89*(1), 29–33.
187. Shibkov, V. M., Chernikov, V. A., Ershov, A. P., et al. (2003). Surface microwave discharge in a supersonic airflow. In N. Novgorod (Ed.), *Proceedings of the 5th International Workshop "Strong Microwaves in Plasmas"* (Vol. 2, pp. 737–742). Russian Federation.
188. Shibkov V. M., & Shibkova L. V. (2011). Surface microwave discharge in air. In N. Novgorod (Ed.), *Proceedings of the 8th International Workshop "Strong Microwaves and Terahertz Waves"* (pp. 341–342) Sources and Applications, Russian Federation.
189. Kossyi, I. A. (2011). New microwave plasma sources based on the latest advances of fundamental gas—discharge physics. Ibid, 25.
190. Ivanov, O. A., & Koldanov, V. A. (2000). Self-consistent model of the pulsed SW-discharge in hydrogen. In N. Novgorod (Ed.), *Proceedings of the 4th International Workshop "Strong Microwaves in Plasmas"* (Vol. 1, pp. 319–324). Russian Federation.

191. Skovoroda, A. A., & Zhyl'tsov, V. A. (1996). The microwave discharge in a plasma neutralizer of ITER injector. In N. Novgorod (Ed.), *Proceedings of the 3rd International Workshop "Strong Microwaves in Plasmas"* (Vol. 1, pp. 368–373). Russian Federation.
192. Skovoroda, A. A., Kulygin, M. V., Smirnov, V. P., et al. (2006). Application of steady state microwave discharges in nuclear technology. In N. Novgorod (Ed.), *Proceedings of the 6th International Workshop "Strong Microwaves in Plasmas"* (Vol. 2, pp. 647–656), Russian Federation.
193. Vikharev, A. L., Gorbachev, A. M., Ivanov, O. A., et al. (1996). Long plasma columns with high electron densities produced by pulsed surface waves discharges. In N. Novgorod (Ed.), *Proceedings of the 3rd International Workshop "Strong Microwaves in Plasmas"* (Vol. 1, pp. 395–400), Russian Federation.
194. Ishijima, T., Toyoda, H., Takanishi, Y., et al. (2011). Design of large-area surface wave plasma excited by slotted waveguide antennas with novel power divider. *Japanese Journal of Applied Physics, 50*, 036002.
195. Boas, C., Himmel, G., & Hirsch, S. (1993). Diagnostics on the plasma created in a pulsed microwave resonator. In N. Novgorod (Ed.), *Proceedings of the 2nd International Workshop "Strong Microwaves in Plasmas"* (Vol. 1, pp. 304 308). Russian Federation.
196. Dine, S., Booth, J.-P., Curley, G. A., et al. (2005). A novel technique for plasma density measurement using surface-wave transmission spectra. *Plasma Sources Science and Technology, 14*, 117–786.
197. Moisan, M., Sauve, G., Zakrzewski, Z., et al. (1994). An atmospheric pressure waveguide-fed microwave plasma torch: The TIA design. *Journal of Plasma Sources Sciences Technology, 3*(4), 584–592.
198. Gamero, A. (1995). *Phenomena in ionized gases*. AIP Press.
199. Lao, C., Gamero, A., Sola, A., et al. (2000). Populations of excited atomic states along argon surface-wave plasma columns at low and intermediate pressures. *Journal of Applied Physics, 87*(11), 7652–7659.
200. Lao, C., Cotrino, J., Palmero, A., et al. (2001). Electron temperature measurement in a surface-wave-produced argon plasma at intermediate pressures. *European Physical Journal D, 14*(3), 361–366.
201. Kang, I.-J., Cho, C.-H., Chang, H., et al. (2021). Characteristics of plasma flow for microwave plasma assisted aerosol deposition. *Nanomaterials, 11*, 1705.
202. Kovačević, M. S., Kuzmanović, L., Milošević, M. M., et al. (2021). An estimation of the axial structure of surface-wave produced plasma column. *Physics of Plasmas, 28*(2), 023502.
203. Nagatsu, M., Naito, K., Ogino, A., et al. (2005). Characteristics of surface-wave and volume-wave plasmas produced with internally mounted large-area planar microwave launcher. *Applied Physics Letters, 87*(16), 161501.
204. Boni, F., Jarrige, J., Désangles, V., et al. (2021). The curling probe: A numerical and experimental study. Application to the electron density measurements in an ECR plasma thruster. *Review of Scientific Instruments, 92*(3), 033507.
205. Sugai, H., Ghanashev, I., Hosokawa, M., et al. (2001). Electron energy distribution functions and the influence on fluorocarbon plasma chemistry. *Plasma Sources Science and Technology, 10*(2), 378–385.
206. Girka, V., Girka, I., & Pavlenko, I. (2001). Electrodynamic model of the gas discharge sustained by azimuthal surface waves. *Contributions to Plasma Physics, 41*(A), 393–406.
207. Girka, V., Girka, I., Girka, A., et al. (2011). Theory of azimuthal surface waves propagating in non-uniform waveguides. *Journal of Plasma Physics, 77*(4), 493–519.

Index

© The Editor(s) (if applicable) and The Author(s), under exclusive license to Springer Nature Switzerland AG 2022

I. Girka and M. Thumm, *Surface Flute Waves in Plasmas*, Springer Series on Atomic, Optical, and Plasma Physics 120, https://doi.org/10.1007/978-3-030-98210-2

C

D

E

F

G

H

I

J

K

L

M

U

V

W

Z

The manufacturer's authorised representative in the EU is Springer Nature Customer Service Centre GmbH, Europaplatz 3, 69115 Heidelberg, Germany. If you have any concerns regarding our products, please contact ProductSafety@springernature.com

Printed and bound by CPI Group (UK) Ltd, Croydon, CR0 4YY
30/09/2026
02225492-0019